Remote Sensing and Geographical Information Systems for Resource Management in Developing Countries

EURO

COURSES

A series devoted to the publication of courses and educational seminars organized by the Joint Research Centre Ispra, as part of its education and training program.
Published for the Commission of the European Communities, Directorate-General Telecommunications, Information Industries and Innovation, Scientific and Technical Communications Service.

The EUROCOURSES consist of the following subseries:

- Advanced Scientific Techniques
- Chemical and Environmental Science
- Energy Systems and Technology
- Environmental Impact Assessment
- Health Physics and Radiation Protection
- Computer and Information Science
- Mechanical and Materials Science
- Nuclear Science and Technology
- Reliability and Risk Analysis
- Remote Sensing
- Technological Innovation

REMOTE SENSING

Volume 1

The publisher will accept continuation orders for this series which may be cancelled at any time and which provide for automatic billing and shipping of each title in the series upon publication. Please write for details.

Remote Sensing and Geographical Information Systems for Resource Management in Developing Countries

Edited by

Alan S. Belward

Commission of the European Communities,
Joint Research Centre, Institute for Remote Sensing Applications,
Ispra, Italy

and

Carlos R. Valenzuela

International Institute for Aerospace Survey and Earth Sciences (ITC),
Enschede, The Netherlands

KLUWER ACADEMIC PUBLISHERS
DORDRECHT / BOSTON / LONDON

Based on the lectures given during the Eurocourse on
'Remote Sensing and Geographical Information Systems for Management of Natural Resources in Developing Countries'
held at the Joint Research Centre Ispra, Italy, May 14–25, 1990

ISBN 0-7923-1268-6

Publication arrangements by
Commission of the European Communities
Directorate-General Telecommunications, Information Industries and Innovation,
Scientific and Technical Communication Unit, Luxembourg

EUR 13494

Published by Kluwer Academic Publishers,
P.O. Box 17, 3300 AA Dordrecht, The Netherlands.

Kluwer Academic Publishers incorporates the publishing programmes of
D. Reidel, Martinus Nijhoff, Dr W. Junk and MTP Press.

Sold and distributed in the U.S.A. and Canada
by Kluwer Academic Publishers,
101 Philip Drive, Norwell, MA 02061, U.S.A.

In all other countries, sold and distributed
by Kluwer Academic Publishers Group,
P.O. Box 322, 3300 AH Dordrecht, The Netherlands.

Printed on acid-free paper

Printed in the Netherlands

Table of Contents

PREFACE

There is a growing realization that our expanding population and thirst for industrial and technological development directly threaten our environment. This awareness has emerged as the nations of the world have developed a global outlook; the earth was seen and photographed as a whole for the first time during the Apollo missions of the late 1960s; also there is increasing recognition that the earth functions as a single system in which every living organism interacts with every other, and that ecological processes do not recognize geo-political boundaries. At the same time however, concern with our environment must be balanced against the need to provide food, housing, health care, education and employment for the growing populace.

This conflict can only be resolved through careful management of the earth's resources. Management of natural resources implies knowledge regarding their state, distribution and dynamics. This in turn implies the active collection and recording of environmental data. In fact such data have been collected throughout history.

From the earliest civilizations inventories, through *field surveys*, have been made concerning the spatial organization of the environment. These have then been translated, through the process of *cartography*, into maps; maps for navigation, for military use, for agriculturalists, maps of a general nature such as topographic maps, and thematic maps showing more specific themes such as land use.

In the past only qualitative use could be made of these data as appropriate mathematical and statistical methods for spatial analysis were lacking. However, this situation is changing. Suitable methods are being developed, and advances in computer technology mean these can be applied to extensive data bases. Thus quantitative use of map data is now a possibility.

Maps, either on paper or in digital form, are expensive to produce; expensive in terms of both time and money. Consequently they tend to be updated and revised infrequently. In addition, maps, other than the most general, can only cover a limited area of the earth's surface. However, sensors on satellite platforms can provide consistent data sets on a regular basis for virtually all the earth's surface.

Since the early 1970s hundreds of images of the earth have been taken every day. These can be used for describing non-renewable earth resources (minerals and fossil fuels for example) and renewable resources (such as the forests and food) alike. The satellite images can be processed to give accurate location with respect to a geodetic grid so that the position and spatial relationships of earth features can be described. The images also contain information concerning other attributes, such as temperature and colour, and when used in sequence can provide data from the temporal domain. The acquisition, processing and interpretation of satellite sensor data are all aspects of the technology of *remote sensing*.

The well established traditions of field survey and cartography (both conventional and digital) are often combined with remote sensing in *Geographical Information Systems* (GIS). GIS are computer based information storage, processing, retrieval and display systems designed specifically to handle spatial data sets such as maps and related non-graphic information. Special software packages allow concurrent analysis of spatial data sets from diverse sources. Thus they can for example, be used to study the interactions between environmental processes, they can be used for predicting trends and can be used for modelling the outcome of different management decisions.

Remote sensing and GIS technologies are of particular relevance to developing countries, where areas of interest are often large, communications sometimes difficult and existing data bases can be incomplete. Moreover, the requirement for management of the natural resources in these areas is particularly keenly felt as many of the world's most fragile and threatened ecosystems are found in these countries.

In parallel with the technological development there is a need for training and education in the application of these technologies if they are to be fully exploited.

Recognizing this need Directorate - General VIII (Development) and Directorate - General XII (Science, Research and Development) of the Commission of the European Communities (CEC) invited the Institute for Remote Sensing Applications at the CEC Joint Research Centre in Ispra, Italy in collaboration with GIS specialists from the International Institute for Aerospace Survey and Earth Sciences in Enschede, The Netherlands to organize a course specifically for personnel from developing countries directly involved in the management of natural resources. The material presented here is drawn from this course.

The principles, instrumentation and techniques behind remote sensing and GIS are presented, along with a review of specific applications relevant to natural resource management in developing countries. Given the breadth of GIS and remote sensing it is impossible for a single book to deal in depth with all aspects of these subjects. The course, and thus this book, was designed to provide an introduction to remote sensing and GIS, to highlight recent advances, and to examine future developments. The overall objective was not to provide remote sensing and GIS "recipes" but to make environmental scientists and managers aware of the uses of these technologies in their work.

In the first two chapters the physical principles underlying remote sensing are discussed, and the concept of using spectral data to obtain information on earth surface features is introduced. In chapter three the key mechanisms controlling the spectral properties of vegetation, soils and water are presented and the idea of remote sensing systems capturing spatial/temporal variations in spectral behaviour introduced. Chapter four then reviews the main remote sensing systems operating in the "optical" wavelengths, and chapter five goes on to discuss how computer aided "image processing" can be used to extract information from these images. Chapter six expands on the theme of quantitative analysis introduced in the previous chapter, and shows how by applying the principles covered in the first two chapters, we can arrive at consistent recording by satellite sensors of the spectral characteristics introduced in chapter three. Chapter seven introduces "radar" remote sensing, and reviews the role of these data in natural resources surveys.

The rest of the remote sensing section of the book looks at example applications of remote sensing relevant to developing countries; chapter eight elaborates on the mechanisms controlling plant / soil reflectance, with particular emphasis placed on agricultural crops, and then shows how remote sensing can be used to make "surrogate" measurements of agronomic variables such as plant development or crop yield. Chapter nine discusses some of the practical considerations to carrying out such work on regional and continental scales, and chapter 10 reviews the role of remote sensing in operational agricultural production forecasting in sahelian Africa. Chapter 11 presents an overview of the use of remote sensing for continental rainfall estimation. Chapter 12 reviews the use of remote sensing for regional hydrological studies, and Chapter 13 discusses the use of remote sensing in tropical deforestation assessment.

The second part of the book is devoted to a review of Geographical Information Systems and starts with an introduction to the principles and concepts of GIS. Subsequent chapters discuss data structures needed for GIS, data base management, data input and analysis, data error detection and verification, and finally how GIS and remote sensing can be used in expert systems.

The various contributions assume that readers have a general scientific background, though no prior knowledge concerning the specific subject matter is expected. However, recognizing that some readers may require more detailed information the authors, throughout the book, refer to other text books and key papers in specialist publications such as journals, conference proceedings and reports.

The authors of this book come from a range of European establishments actively engaged in remote sensing and GIS research and applications. The time and effort they have put into their contributions, and the work of the staff of the EUROCOURSES secretariat at Ispra and the staff of DG VIII and XII is gratefully acknowledged and due thanks offered.

ASB, Ispra, May 1990

List of contributors

F. Baret, INRA, Station de Bioclimatologie, Centre de Rechereches d' Avignon, BP 91, France. (Tel. 90 31 61 03/090, Telex 432870 INRAAVIF, Fax 90 31 62 98)

E. Bartholomé, Commission of the European Communities, Joint Research Centre, Institute for Remote Sensing Applications, 21020 Ispra, Varese, Italy (Tel 0332 789111, Fax 0332 789001)

A. S. Belward, Commission of the European Communities, Joint Research Centre, Institute for Remote Sensing Applications, 21020 Ispra, Varese, Italy (Tel 0332 789111, Fax 0332 789001)

A. C. Bird, Cranfield Institute of Technology, Silsoe College, Silsoe, Bedfordshire, MK45 4DT U.K. (Tel. 0525 60428, Fax. 0525 61527))

G. Bocco, International Institute for Aerospace Survey and Earth Sciences (ITC), 350 Boulevard 1945, PO box 6, 7500 AA Enschede, The Netherlands (Tel 053 320330, Fax 053 304396)

R. Bonifácio, Dept of Meteorology, University of Reading, Reading, RG6 2AU U.K. (Tel 734 875123, Fax 734 352604).

P. N. Churchill, Commission of the European Communities, Joint Research Centre, Institute for Remote Sensing Applications, 21020 Ispra, Varese, Italy (Tel 0332 789111, Fax 0332 789001)

H. de Brouwer, International Institute for Aerospace Survey and Earth Sciences (ITC), 350 Boulevard 1945, PO box 6, 7500 AA Enschede, The Netherlands (Tel 053 320330, Fax 053 304396)

A. Fabbri, International Institute for Aerospace Survey and Earth Sciences (ITC), 350 Boulevard 1945, PO box 6, 7500 AA Enschede, The Netherlands (Tel 053 320330, Fax 053 304396)

J-M. Grégoire, Commission of the European Communities, Joint Research Centre, Institute for Remote Sensing Applications, 21020 Ispra, Varese, Italy (Tel 0332 789111, Fax 0332 789001)

J. Hill, Commission of the European Communities, Joint Research Centre, Institute for Remote Sensing Applications, 21020 Ispra, Varese, Italy (Tel 0332 789111, Fax 0332 789001)
Laboratory for Image Processing, JRC.

J. P. Malingreau, Commission of the European Communities, Joint Research Centre, Institute for Remote Sensing Applications, 21020 Ispra, Varese, Italy (Tel 0332 789111, Fax 0332 789001)

A. Meijerink, International Institute for Aerospace Survey and Earth Sciences (ITC), 350 Boulevard 1945, PO box 6, 7500 AA Enschede, The Netherlands (Tel 053 320330, Fax 053 304396)

A. J. Sieber, Commission of the European Communities, Joint Research Centre, Institute for Remote Sensing Applications, 21020 Ispra, Varese, Italy (Tel 0332 789111, Fax 0332 789001)

C. R. Valenzuela, International Institute for Aerospace Survey and Earth Sciences (ITC), 350 Boulevard 1945, PO box 6, 7500 AA Enschede, The Netherlands (Tel 053 320330, Fax 053 304396)

M. J. C. Weir, International Institute for Aerospace Survey and Earth Sciences (ITC), 350 Boulevard 1945, PO box 6, 7500 AA Enschede, The Netherlands (Tel 053 320330, Fax 053 304396)

G. G. Wilkinson, Commission of the European Communities, Joint Research Centre, Institute for Remote Sensing Applications, 21020 Ispra, Varese, Italy (Tel 0332 789111, Fax 0332 789001)

J. A. Zinck, International Institute for Aerospace Survey and Earth Sciences (ITC), 350 Boulevard 1945, PO box 6, 7500 AA Enschede, The Netherlands (Tel 053 320330, Fax 053 304396)

PRINCIPLES OF REMOTE SENSING : ELECTROMAGNETIC RADIATION, REFLECTANCE AND EMISSIVITY

A.C. BIRD
Silsoe College,
Cranfield Institute of Technology,
Silsoe,
Bedfordshire, MK45 4DT
United Kingdom

1. Introduction

1.1. DEFINITION OF REMOTE SENSING

Remote sensing is generally defined as the technology of measuring the characteristics of an object or surface from a distance. In the case of earth resource monitoring the object or surface is on the land mass of the earth or on the sea and the observing sensor is in the air or in space. The observation of the object is then made through the atmosphere along a vertical or near-vertical path. The process of measuring an object's characteristics from a distance is governed by the laws of physics and it is necessary to have a basic understanding of some of these laws (Whelan and Hodgson 1972). This requirement arises because the emphasis in remote sensing today is on deriving quantitative rather than qualitative data from images obtained from sensors on satellites. The many factors contributing to the signal produced at the satellite have, therefore, to be acknowledged.

1.2. INFORMATION FLOW IN REMOTE SENSING

In order for an observing sensor to acquire knowledge about a remote object, there must be a flow of information between the object and the observer. There has to be a carrier of that information. That carrier is electromagnetic radiation (EMR). The main elements in the process of data collection in remote sensing are the object to be studied, the observer or sensor, the EMR that passes between the two and the source of the EMR. In fact, EMR appears in many stages of the overall data acquisition and analysis as is illustrated in figure 1 where the arrow indicates the flow of information.

At four of the five stages where information moves from one place to another the carrier is electromagnetic radiation. Visible light can be the form that it takes between the sun, the earth and the sensor and between the image processor monitor and the human eye. The form is man-made radio signals between the sensor and a receiving station.

A basic understanding of EMR is central to dealing with the physical principles of remote sensing, see for example Campbell (1987), Curran (1985), Elachi (1987), Lillesand and Kiefer (1987), Sabins (1986) and Swain and Davis (1978).

A. S. Belward and C. R. Valenzuela (eds.),
Remote Sensing and Geographical Information Systems for Resource Management in Developing Countries, 1–15.

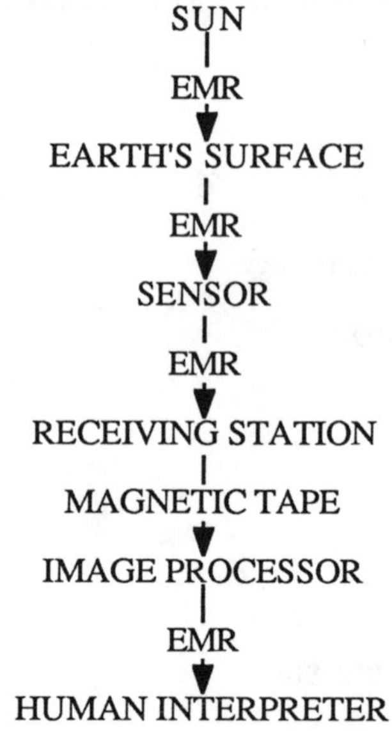

Figure 1. The flow of information in a remote sensing system

2. Models of Electromagnetic Radiation

2.1. WAVE MODEL

The concept of EMR describes the way in which energy is transferred from one object to another through space. All objects at temperatures above absolute zero (0 K, or -273°C) radiate energy as EMR and there are two main models which are used to describe the characteristics of that energy.

The first is the wave model where the oscillation of the electric and magnetic fields around an object lead to the concept of energy being carried as a wave of electromagnetic radiation. The waves can then be characterized using the usual notations.

2.1.1 Wavelength. The waves correspond to repetitive cycles of the electric field in space. The wavelength is the distance occupied by one complete cycle and is usually denoted by the Greek symbol 'λ'. λ is measured in metres 'm', centimetres 'cm', micrometres 'μm' (ie. 10^{-6} m) or nanometres 'nm' (ie. 10^{-9} m).

2.1.2 Frequency. The time occupied by one cycle at a point in space leads to the number of cycles per second which is the frequency. This is denoted by the Greek symbol 'υ'. υ is measured in s^{-1}.

NAME	WAVELENGTH RANGE
GAMMA RAYS	< 0.03 nm
X RAYS	0.03 nm - 3 nm
ULTRA-VIOLET	3 nm - 0.3 μm
VISIBLE	0.3 μm to 0.7 μm
INFRARED	0.7 μm to 300 μm
MICROWAVE	300 μm to 20 cm
RADIO	> 20 cm

Table 1: Regions of the electromagnetic spectrum

2.1.3. Speed. The speed of the wave through space is constant and is related to the wavelength and frequency by the equation

$$\lambda \upsilon = c \tag{1}$$

In space c is known as the 'speed of light' and has a value of 3×10^8 ms^{-1}.

2.2. ELECTROMAGNETIC SPECTRUM

Given that c in Equation (1) is a constant, EMR can be characterized either by wavelength or by frequency. Wavelength is commonly used to describe the electromagnetic spectrum. Table 1 lists the regions of that spectrum.

The main regions that are of use in remote sensing are the visible, the infrared and the microwave. Images of the same object generated by energy carried by each of these forms of EMR will appear very different because each wavelength carries different information. For example, the visible wavelengths might describe the colour of the object, the infrared might describe the temperature, while the microwave might describe the surface roughness. Nevertheless, all are a form of EMR and all are governed by the same basic physical principles.

The main wavelength ranges that are used in remote sensing are described in more detail in Table 2.

NAME	WAVELENGTH RANGE	ABBREVIATION
VISIBLE	0.4 μm - 0.7 μm	VNIR
NEAR INFRARED	0.7 μm - 1.3 μm	VNIR
SHORT WAVE INFRARED	1.5 μm - 2.5 μm	SWIR
MID INFRARED	3 μm - 5 μm	MIR
THERMAL INFRARED	8 μm - 14 μm	TIR
MICROWAVE	1 cm - 20 cm	--

Table 2: Regions of the Spectrum used in Remote Sensing

Most satellite sensors will acquire data in one or more of these wavebands but not in all of them. The sensor is designed to employ certain wavebands according to the information that is required from the final imagery.

In addition to the regions listed in Table 2 some use is made of gamma rays and ultraviolet radiation. Clearly, there are gaps in the coverage of the wavelengths and these will be discussed in a later section dealing with the effects of the atmosphere.

2.3. PARTICLE MODEL

The second model of electromagnetic radiation describes the movement of energy as the movement of a stream of particles. These particles are known as photons and have zero mass. The energy carried by each photon depends on the source of EMR that has produced the photon. The usefulness of this concept lies in the fact that a particular energy value can then be associated with the EMR. The two models are related by the equation

$$\text{Photon energy} \quad Q = h\upsilon \tag{2}$$

where h is known as Planck's Constant and has a value of 6.626×10^{-34}Js.
Equations (1) and (2) combine to give

$$Q = \frac{hc}{\lambda} \tag{3}$$

Clearly, each photon of thermal infrared radiation will have less energy than each photon of visible radiation because in equation (3) the greater value of λ for thermal infrared leads to a smaller value of Q. This is one reason why sensors working in the visible tend to be more advanced than those working in the thermal infrared. The technology of detecting the lower energy thermal infrared photons is more complex.

3. Radiation Laws

3.1. PLANCK'S RADIATION EQUATION

It was stated earlier that all objects with a temperature above absolute zero emit electromagnetic radiation. As 0 K is equivalent to -273°C this holds for all the objects observed in remote sensing.

Planck's radiation equation is an important way of describing the energy carried by EMR at different wavelengths. The equation describes the maximum amount of energy that can be emitted by an object

$$M_\lambda = \frac{c_1}{\lambda^5 (\exp[c_2/\lambda T] - 1)} \tag{4}$$

where M_λ is the spectral radiant emittance of the object measured in $Wm^{-2}\ \mu m^{-1}$, T is the absolute temperature, c_1 and c_2 are constants and are given by $3.74 \times 10^8\ Wm^{-2}\ \mu m^{-4}$ and $1.44 \times 10^4\ \mu mK$ respectively. Thus, from equation (4) we can see that M_λ represents the rate at which energy is emitted from a unit surface area of the object at a particular wavelength. The importance of this equation can best be appreciated by reference to Figure 2.

Clearly the temperature of the object affects the level of energy emitted at any particular wavelength.

3.2. STEFAN-BOLTZMANN EQUATION

Equation (4) can be integrated over all wavelengths to calculate the value of the radiant emittance M. This is measured in Wm^{-2} and is the overall rate at which energy is emitted from a unit surface of an object. The result of the integration gives the Stefan-Boltzmann radiation law

$$M = \sigma T^4 \tag{5}$$

where σ has a value of $5.67 \times 10^{-8}\ Wm^{-2}\ K^{-4}$. M increases rapidly with temperature and this is borne out by Figure 2 where M is the area under the curves describing M_λ.

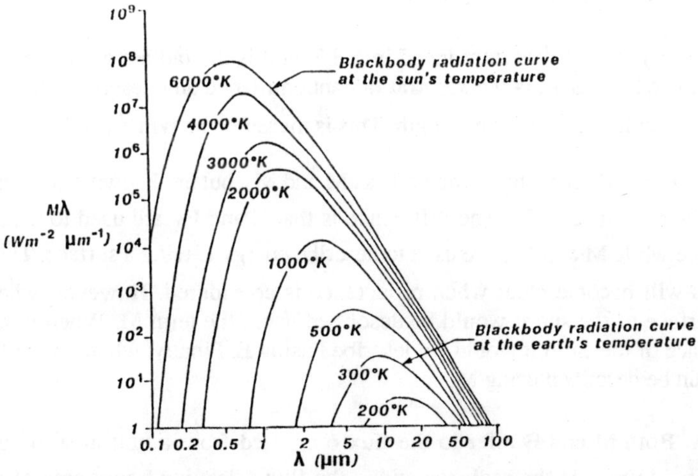

Figure 2: Curves showing relationship between M_λ and λ for objects at different temperatures

3.3. WIEN'S DISPLACEMENT LAW

Equation (4) can also be used to work out the wavelength λ_{max} at which the maximum value of M_λ occurs. This leads to the equation

$$\lambda_{max} = \frac{A}{T} \tag{6}$$

where A = 2898 μmK

λ_{max} decreases as T increases and this, again, can be seen in Figure 2. The maximum value of M_λ for the sun occurs at around 0.5 μm in the middle of the visible region of the spectrum. For the earth, it occurs at around 9.7 μm in the middle of the thermal-infrared region. Overall, as would be expected, the sun emits a great deal more energy than the earth.

These equations enable us to appreciate the characteristics of EMR that lead to the design and use of satellite sensors.

3.4. TERMINOLOGY

3.4.1. Radiant Energy. Terminology used in many areas of remote sensing requires explanation before any discussions of radiance or reflectance can begin. Q is the radiant energy measured in Joules (J) and has already been used.

3.4.2. Radiant Flux. ϕ is the radiant flux and is the rate at which energy is transferred from one surface to another. ϕ is measured in Watts (W) which are Js^{-1}.

3.4.3. Radiant Emittance. M was introduced in § 3.2. and is the radiant emittance measured in Wm^{-2}. M_λ, introduced in § 3.1., is the spectral radiant emittance and describes the way in which energy is distributed with respect to wavelength. This is measured in $Wm^{-2}\ \mu m^{-1}$.

3.4.4. Irradiance. E and E_λ have the same units as M and M_λ but are known as the irradiance and the spectral irradiance respectively. The difference is that E and E_λ are used to describe energy arriving at a surface while M and M_λ are used to describe energy leaving a surface. The importance of this distinction will become clear when reflectance is considered. However, when energy is emitted by the surface of the sun it would be described using the term M. When it subsequently arrives at the surface of the earth it would be described using E. Finally, when it is reflected by the earth it would again be described using M.

3.4.5. Radiance. Both M and E refer to the flux ϕ emitted from a unit area of surface in all directions. The final term L is the radiance and is the flux ϕ leaving a unit area of surface in a particular direction and into a particular solid angle. L is measured in $Wm^{-2}\ sr^{-1}$ where a steradian (sr) is a unit of solid angle. L_λ is the spectral radiance and, as usual, refers to L at a particular wavelength measured in $Wm^{-2}\ sr^{-1}\ \mu m^{-1}$. The significance of radiance is that this is what is actually measured by a satellite. The satellite looks in a particular direction at the surface and its detector defines a restricted solid angle. Therefore, the digital number recorded in the satellite image corresponds to L and not M or ϕ or Q. The physical characteristics of the object that are being measured by the sensor are defined in terms of M_λ and E_λ, so it is necessary to understand how these relate to the actual measurements of L.

4. Reflectance

4.1. DEFINITION OF REFLECTANCE

There are two processes that can lead to a satellite sensor detecting the presence of an object and measuring its characteristics. The first is the process by which a source of EMR, for example the sun, sends radiation to the earth's surface which is reflected by the earth. The reflected radiation is then detected by the satellite sensor.

The irradiance will be given by E, while the reflected radiant emittance will be given by M. However, a measurement of M would not, on its own, provide useful information about an object, as it would increase or decrease as E increased or decreased.

The reflectance of an object is a physical property of the object independent of the illumination conditions. It is a unitless quantity and is given by

$$\text{reflectance } \rho = \frac{M}{E} \tag{7}$$

similarly spectral reflectance is given by

$$\rho_\lambda = \frac{M_\lambda}{E_\lambda} \tag{8}$$

ρ cannot be greater than 1 as it is not possible for more energy to be reflected than was incident.

Reflectance is a useful property because it describes a characteristic of the object alone. Therefore, it is desirable to find a relationship between the reflectance ρ of the object and the radiance L as measured by the satellite.

4.2. SPECULAR AND DIFFUSE REFLECTION

4.2.1. Specular reflection. A smooth, shiny surface leads to the specular reflection of radiation. In this case the geometric laws of reflection are obeyed with the angle of incidence equaling the angle of reflection. This type of reflection would be of limited use in remote sensing as the satellite and the sun would have to be at exactly equal angles to the object. This usually only happens when the sun is reflected by water into the field of view of the sensor and this is usually known as glint.

4.2.2. Diffuse Reflection. A diffuse reflector scatters incoming radiation in all directions. A perfectly diffuse reflector is one where the radiance L from the surface is constant for any look angle. This is known as a Lambertian surface and is the ideal situation for remote sensing where it is preferable for the radiance from a target to be independent of look-angle and, so, independent of its location across the image swath.

In general, natural surfaces are closer to being Lambertian surfaces than smooth surfaces, and the Lambertian assumption is frequently made to simplify models of reflection.

4.3. RADIANCE AND REFLECTION

It is possible to derive a relationship between radiance and reflectance if a Lambertian surface is assumed. By integrating the outgoing flux ϕ from a point over a hemisphere and dividing it by the incoming flux it can be shown that

$$\rho = \frac{\pi L}{E} \tag{9}$$

and consequently that

$$\rho_\lambda = \frac{\pi L_\lambda}{E_\lambda} \tag{10}$$

If the sun is at an angle of θ to the surface then the equation modifies to

$$\rho_\lambda = \frac{\pi L_\lambda}{E_\lambda \cos\theta} \tag{11}$$

These equations raise the possibility of calculating the reflectance of a surface from a satellite image. This is provided that it is possible to calibrate the digital numbers of the satellite sensor to give spectral radiance values. Calibration data should be supplied by the distributor of the image (Price 1987); this subject is discussed in more detail by Hill, later in the book. E_λ could be measured at the time of the image acquisition but it is unusual for this to happen. However, the characteristics and movements of the sun are well understood and it is possible to calculate a value for E_λ for a particular day and time.

By assuming a Lambertian surface and ignoring atmospheric effects it is therefore possible to use Equation (11) to measure the reflectance of an object from a satellite image (Robinove 1982).

5. Emissivity

5.1. DEFINITION OF EMISSIVITY

The second process which can lead to a sensor detecting the presence of an object is where the object itself acts as the source of the EMR. It has already been stated that all objects radiate EMR and Equations (4) and (5) described the values of M_λ and M that result from this. In fact, these equations represent the maximum possible values that M_λ and M can have and relate to an ideal object known as a 'blackbody'. Most natural objects are not blackbody emitters and, so, the values of M_λ and M are lower at any given temperature. The degree of approximation to a blackbody is given by the emissivity ε where

$$\varepsilon = \frac{M}{M_{BB}} \tag{12}$$

M is the radiant emittance of the object with emissivity ε, and M_{BB} is the radiant emittance of a blackbody at the same temperature. ε is a unitless quantity between 0 and 1.

5.2. EMISSIVITY AND TEMPERATURE

A sensor which measures the energy emitted by an object rather than that reflected is able to measure either the emissivity of the object or the temperature of the object.
Given Equation (12) then Equation (4) for an object of emissivity ε becomes

$$M_\lambda = \frac{\varepsilon_\lambda c_1}{\lambda^5 (\exp [c_2/\lambda T] - 1)} \tag{13}$$

Assuming a Lambertian surface and ignoring atmospheric effects it can be shown that

$$M_\lambda = \pi L_\lambda \tag{14}$$

M_λ can therefore be derived from the satellite measure of L_λ in which case if ε_λ and λ are known, then Equation (13) rearranges to

$$T = \frac{c_2}{\lambda(\ln[(\varepsilon_\lambda c_1 \lambda^{-5}/M_\lambda) + 1])} \tag{15}$$

On the other hand, if λ and T are known, then Equation (13) rearranges to

$$\varepsilon_\lambda = \frac{M_\lambda \lambda^5(\exp[c_2/\lambda T] - 1)}{c_1} \tag{16}$$

In summary, then, an understanding of the physical principles behind EMR allows the user of satellite data to apply the appropriate processes to an image to derive certain physical properties of the objects in the image. These properties can be reflectance, emissivity or temperature. Information on reflectance or emissivity and, in particular, how they vary with wavelength, can assist in the identification of the object. Knowledge of an object's temperature will not necessarily help in identifying its constituent materials, but it can be important in such studies as those of the sea-surface for weather forecasting or for the identification of hazards such as forest fires.

6. Wavelength Dependence of Physical Properties

6.1. ENERGY LEVELS

Remote sensing for the identification of objects works on the basis that some of the physical properties of objects depend on the wavelength of the reflected or emitted radiation. In order to understand this, very brief consideration should be given to the quantum theory of radiation.

A very simple model of the atom, which is the building block of materials, is of a central positively charged nucleus around which negatively charged electrons are 'orbiting'. Orbits near the nucleus have a low energy level, while orbits further from the nucleus have a high energy level. If an electron moves from a high orbit to a low orbit it loses energy by emitting a photon with the required energy difference between the two orbits. In other words, the atom emits EMR with a wavelength obeying Equation (3) given the photon energy.

Quantum theory limits the number of possible energy states that the atom can occupy and therefore the number of energy values that the photon can have.

This also works in reverse. If an electron is to move from a low energy level to a high energy level, then it must gain energy from an incoming photon. Quantum theory, again, limits the energy levels so only photons of certain energies will be able to be absorbed by the atom. This effectively means that the atom may absorb EMR of some wavelengths but not others.

6.2. SPECTRAL PROPERTIES OF MATERIALS

The absorption or reflection of EMR by any material therefore varies with wavelength. When radiation is incident on an object then it can be absorbed or reflected according to the energy of its photons, which is directly related to the wavelength. The atoms and molecules that make up the material control the wavelengths at which absorption and emission occur. Absorptions in the visible and near infrared tend to be controlled by the changes in energy levels of orbital electrons, while in the thermal infrared they are controlled by the vibrational and rotational energy levels of the molecules (Hunt 1977, Kahle and Goetz 1983).

It is therefore found that the reflectance and the emissivity of an object can vary with wavelength and that the variation is related to the nature of the materials forming that object.

7. Remote Sensing of Reflected or Emitted Energy

7.1. REFLECTED ENERGY

The point has already been made that objects can be detected either by identifying the energy that they reflect or the energy that they emit. Clearly it would be confusing to receive both reflected and emitted energy at our sensor, so much of the time this situation is avoided.

Figure 2 indicates that the sun, which has a temperature of around 6000 K, has a maximum M_λ at a wavelength of around 0.5 µm. Therefore remote sensing in the visible and near-infrared is based on the measurement of reflected radiation from the earth's surface.

The physical properties of the object that are of interest in this case are the reflectance ρ and the spectral reflectance ρ_λ.

7.2. EMITTED ENERGY

The irradiance from the sun decreases as the wavelength of the radiation increases. In the thermal infrared the solar irradiance is relatively low (see § 9). However, Figure 2 shows that the earth, with a temperature of around 300 K, has a maximum M_λ at a wavelength of around 9.7 µm. Therefore remote sensing in the thermal infrared is based on the measurement of radiation emitted by the earth's surface.

The physical properties of the object that are of interest are, then, emissivity ε, spectral emissivity ε_λ, and temperature T.

7.3. REFLECTED VERSUS EMITTED

The techniques of remote sensing by reflection or emission are equally valid providing as they do complementary information. To date, the identification of spectral variations of properties have concentrated on reflectance, given that the simpler technology associated with shorter wavelengths has allowed the development of multi-spectral measurements in the visible and near-infrared. Remote sensing by emission has concentrated on the measurement of temperature, although it has been shown that the variation of emissivity with wavelength can be related to the constituent minerals of rocks. This has led to the development of multi-spectral thermal infrared sensors.

Detectors that collect reflected solar energy can clearly only operate during the daylight hours for the object, whereas emitted thermal-infrared radiation can be acquired day or night.

Satellite sensors that detect radiation in the middle-infrared (3 μm - 5 μm) receive both reflected and emitted components. The confusion between the two can only be resolved by night time operations where the reflected component disappears.

8. Summary

Remote sensing is the technology of measuring various characteristics of an object without physical contact, usually from air-borne or from space platforms. It is a source of data carrying information about the earth system which is of value to ecologists, hydrologists, meteorologists and managers of the earth's renewable and non-renewable resources. Effective use of these data depends on an understanding of the physical processes that control the movement of electromagnetic radiation which is the carrier of the information.

Electromagnetic radiation can be characterized by wavelength and the electromagnetic spectrum contains several regions which are useful in remote sensing. Objects are identified by the extent to which they reflect incident radiation or emit their own radiation. Sensors detect the radiance from an object and this can be related to either the reflectance or the emissivity of the object. Both ρ and ε vary with wavelength and the analysis of this variation can lead to the identification and characterization of the object.

9. A practical exercise

In this final section the theory introduced above is used to explain why remote sensing sensors operating in the visible wavelengths measure reflected energy whilst those operating in the thermal-infrared measure emitted energy.

9.1. THE PROBLEM

Calculate the spectral radiant emittance from a surface due to reflection of solar irradiance and the spectral radiant emittance due to emission from the surface. This calculation should be carried out for a wavelength of 0.5 μm (visible) and a wavelength of 10 μm (thermal-infrared). The surface has a reflectance of 0.65 at $\lambda = 0.5$ μm and an emissivity of 0.95 at $\lambda = 10$ μm.
Assume that the earth has a temperature of 300 K and that the sun is a Lambertian source with an emissivity of 1, a temperature of 6000 K and a diameter which subtends an angle of 10 milliradians at the earth. Assume a solar zenith angle of 0°.

9.2. THE SOLUTION

Let S be the earth-sun distance and α be the angle subtended by the sun at the earth.
The spectral radiant flux from the sun can be given by

$$\phi_\lambda = L_\lambda.\Omega.A \tag{17}$$

where L_λ is the spectral radiance, Ω is the solid angle subtended by a surface of area B on the earth and A is the area of the solar disk observed from the earth. Equation (17) arises from the definition of radiance given in § 3.4.5.
From the definition of a solid angle

$$\Omega = \frac{B}{S^2} \tag{18}$$

From the data provided

$$A = \pi\,(\alpha S/2)^2 \tag{19}$$

Therefore

$$\phi_\lambda = L_\lambda.\frac{B}{S^2}.\pi\frac{\alpha^2 S^2}{4} \tag{20}$$

Given that the sun is a Lambertian source and thus using Equation (14), we have

$$\phi_\lambda = M_{\lambda sun}.B.\frac{\alpha^2}{4} \tag{21}$$

From the definition given in § 3.4.4., the irradiance for the area B on the earth is

$$E_{\lambda earth} = \frac{M_{\lambda sun}.\alpha^2}{4} \tag{22}$$

M_λ can be calculated using Equation (13) with T = 6000 K

At 0.5 µm $M_{\lambda sun} = 9.94 \times 10^{7}$ Wm^{-2} μm^{-1}

At 10 µm $M_{\lambda sun} = 1.37 \times 10^{4}$ Wm^{-2} μm^{-1}

$\alpha^2/4$ is 2.5×10^{-5} and therefore at 0.5 µm:

$$E_{\lambda earth} = 2.48 \times 10^{3}\ Wm^{-2}\ \mu m^{-1}$$

and given that $\rho = 0.65$

reflected $M_{\lambda earth} = \rho E_{\lambda earth} = 1.61 \times 10^{3}$ Wm^{-2} μm^{-1}

At 10 µm

$$E_{\lambda earth} = 0.34\ Wm^{-2}\ \mu m^{-1}$$

and given that $\varepsilon = 0.95$, then $\rho = 0.05$ and
reflected $M_{\lambda earth} = 1.7 \times 10^{-2}$ Wm^{-2} μm^{-1}

The radiant emittance due to emission from the earth is given by Equation (13) with T = 300 K.

At 0.5 µm emitted $M_{\lambda earth}$ is effectively zero.

At 10 µm

emitted $M_{\lambda earth} = 29.5$ Wm^{-2} μm^{-1}

The exercise shows that in the visible bands only the reflected M_λ is significant while in the thermal infrared the reflected M_λ is significant, but much smaller than the emitted M_λ. This explains why reflectance is measured in the visible and emissivity in the thermal infrared. While the sun emits more thermal infrared radiation than the earth only a small portion of this is intercepted by the earth.

References

Campbell, J.B., 1987. *Introduction to Remote Sensing.* (U.S.A.: The Guildford Press).

Curran, P.C., 1985. *Principles of Remote Sensing.* (U.S.A.:Longman Group Limited).

Elachi, C., 1987. *Introduction to the Physics and Techniques of Remote Sensing.* (New York: John Wiley & Sons Inc.)

Hunt, G.R., 1977. Spectral signatures of particulate Minerals in the Visible and Near Infrared. *Geophysics,* **42**, 5O1-513.

Kahle, A.B., and Goetz, A.F.H., 1983. Mineralogic Information from a New Airborne Thermal Infrared Multispectral Scanner.*Science,* **222**, 24-27.

Lillesand, T.M., and Kiefer, R.W., 1987. *Remote Sensing and Image Interpretation.* 2nd ed. (New York: John Wiley and Sons Inc.).

Price, J.C., 1987. Calibration of Satellite Radiometers and the Comparison of Vegetation Indices. *Remote Sensing of Environment,* **21**, 15-27.

Robinove, C.J., 1982. Computation with Physical Values from Landsat Digital Data. *Photogrammetric Engineering and Remote Sensing.* **48**, 781-784.

Sabins, F.F.,1986. *Remote Sensing: Principles and Interpretation.* 2nd ed. (New York: W.H. Freeman & Co.).

Swain, P.H., and Davis, S.M., 1978. *Remote Sensing: the Quantitative Approach.* (New York: McGraw-Hill Inc.).

Whelan, P.M., and Hodgson, M.J., 1972. *Essential Pre-University Physics.* (UK: John Murray Ltd).

PRINCIPLES OF REMOTE SENSING : INTERACTION OF ELECTROMAGNETIC RADIATION WITH THE ATMOSPHERE AND THE EARTH

A.C. BIRD
Silsoe College,
Cranfield Institute of Technology,
Silsoe,
Bedfordshire, MK45 4DT
United Kingdom

1. Introduction

1.1. ACTIVE AND PASSIVE REMOTE SENSING

There are two main categories of remote sensing which employ quite different technologies. Active remote sensing is the term used to describe the situation where the satellite or airborne system provides the source of illumination. This technique is used when no suitable natural source of radiation exists.

Passive remote sensing is the term used to describe systems where the source of the object illumination is independent of the sensor and is a natural source (Campbell 1987).

1.2. VISIBLE AND NEAR INFRARED (VNIR)

Remote sensing in the visible and near-infrared wavelengths usually falls into the 'passive' category. Here the sun is the source of the irradiance on the object being observed. The sensor collects the solar radiation which is reflected by the object. Active remote sensing occurs at these wavelengths only in the rare case where an aircraft carries a laser as the source of illumination.

1.3. THERMAL INFRARED

Remote sensing in the thermal-infrared wavelengths also falls into the 'passive' category but in this case the object itself is the source of the radiation. There is no irradiance and the sensor detects radiation which has been emitted by the object.

Thermal-infrared images can be acquired during daytime or night time hours but the lower night time temperatures will lead to a lower overall signal.

1.4. MICROWAVE

Planck's radiation curves seen in the previous chapter indicate that neither the sun nor the earth are a strong source of radiation at microwave wavelengths. Passive remote sensing techniques are used

A. S. Belward and C. R. Valenzuela (eds.),
Remote Sensing and Geographical Information Systems for Resource Management in Developing Countries, 17–30.

with the earth as a source of naturally emitted radiation but the wavebands have to be wide and the spatial resolution fairly coarse in order to give sufficient signal to form an image.

Active remote sensing systems are common in microwave work and here the satellite or aircraft carries an antenna which emits a microwave signal. This signal is reflected by the ground and the return signal is detected again by the antenna. The most common imaging active systems are the Side Looking Airborne Radar (SLAR) and the Synthetic Aperture Radar (SAR), while non-imaging systems include scatterometers and altimeters. Microwave remote sensing is discussed in more detail in the chapter by Sieber and Churchill later in the book.

2. Solar Irradiance

2.1. THE SUN

The sun plays a major role in data acquisition by remote sensing. Solar energy is the source of electrical power for satellites but more importantly the sun supplies the irradiance for VNIR and Short Wave Infrared (SWIR) images.

The earth supplies its own radiation in the thermal-infrared but Planck's law tells us that this is dependent on the earth's temperature which is, in turn, controlled by solar illumination.

The sun can be thought of as a ball of burning gas, although it is believed that nuclear reactions generate the high temperatures at its centre. It has an effective diameter of 1.4×10^6 kilometres and, on average, is a distance of 150×10^6 kilometres from the earth. The earth-sun distance changes throughout the year and, therefore, the value of E_λ at the earth also changes.

A temperature of 6000 K at the surface explains the magnitude of the solar irradiance intercepted by the earth. It has already been seen in the previous chapter that this leads to a maximum M_λ at 0.5 μm of about 10^8 Wm^{-2} μm^{-1} falling with increasing wavelength to an M_λ at around 10 μm of about 1.5×10^4 Wm^{-2} μm^{-1}. The earth then intercepts an E_λ of 2.5×10^3 Wm^{-2} μm^{-1} at 0.5 μm falling to 0.5 Wm^{-2} μm^{-1} at 10 μm.

The Solar Constant is an expression of the irradiance E due to the sun. This is defined as the amount of irradiance on a unit flat surface normal to the direction of the suns rays in free space at the earth's mean distance from the sun. The value of the Solar Constant is 1375 Wm^{-2}.

2.2. RADIANT INTENSITY

The radiant intensity I is the flux ϕ proceeding from a source per unit solid angle and in a particular direction. I is expressed in Wsr^{-1}. This definition leads to the expression shown in equation (1), where Ω is the solid angle.

$$\phi = I.\Omega \tag{1}$$

If an area A is illuminated by the radiant flux then

$$E = \frac{\phi}{A} = \frac{I.\Omega}{A} \tag{2}$$

The solid angle Ω is defined as the area of the object illuminated divided by the square of the object-source distance. This then gives

$$\Omega = \frac{A \cos \theta}{S^2} \tag{3}$$

where S is the object-source distance and θ is the angle between the normal to the object surface and the object-source direction. Equations (2) and (3) combine to give

$$E = \frac{I \cos \theta}{S^2} \tag{4}$$

2.3. INVERSE SQUARE LAW

Equation (4) means that

$$E \,\alpha\, \frac{1}{S^2} \tag{5}$$

and therefore E falls as the earth-sun distance increases. If S doubles then E falls by a factor of 4. This explains the earlier statement that E and E_λ change throughout the year as the earth-sun distance changes. However, if the Solar Constant is known then it is possible to use Equation (5) to calculate E for values of S other than the average.

2.4. COSINE LAW

Equation (4) leads to the rule that

$$E \,\alpha\, \cos \theta \tag{6}$$

and therefore E decreases as θ increases. In other words, the solar irradiance decreases as the solar zenith angle increases leading to the sun becoming lower in the sky. Again, if the Solar Constant is known, then E can be recalculated for any solar zenith angle.

The previous chapter indicated that it is necessary to know E at the earth's surface if the reflectance ρ is to be derived from the radiance L. These equations together with Planck's radiation equation enable one to calculate a value for E for a satellite image provided the date and time of the image is known so that the solar zenith angle and the earth-sun distance can be calculated.

3. The Atmosphere

3.1. ABSORPTION

All discussions so far have ignored the quite significant effects of the medium through which the electromagnetic radiation is travelling - the atmosphere (Campbell 1987).

The atmosphere is an important consideration in remote sensing using any of the active or passive techniques. VNIR radiation from the sun passes through the atmosphere before reaching the earth and after reflection passes through again before reaching the satellite. The same is true of microwave signals produced by an active instrument on a satellite. Thermal-infrared and microwave radiation emitted by the earth pass through the atmosphere once before reaching a sensor.

Absorption is the mechanism by which electromagnetic radiation of particular wavelengths is attenuated by the atmosphere. The concept of absorption was introduced in the previous chapter where it was explained that photons of radiation corresponding to particular wavelengths can be absorbed by atoms or molecules if the photon energy corresponds to a particular gap between their energy levels. The fact that the photons are absorbed means that they can no longer contribute to the irradiance at the earth's surface.

The main gases responsible for such absorption in the atmosphere are water vapour (H_2O), carbon dioxide (CO_2) and ozone (O_3). Each molecule of these gases leads to the absorption of several wavelengths of radiation. The atmosphere is therefore opaque to these wavelengths and transparent to others.

3.2. ATMOSPHERIC WINDOWS

The wavelength regions for which the atmosphere is transparent are known as atmospheric windows. Table 1 summarizes the main windows that are used in remote sensing. Remote sensing of the earth's surface is not possible outside these windows.

A clear consequence of the existence of these atmospheric windows is that sensors are designed to work in wavebands corresponding to the transparent atmosphere. The exceptions to this are sensors which are designed to measure characteristics of the atmosphere rather than the earth. For example, Meteosat forms an image using the 6 μm - 7 μm waveband. This is known as the water vapour waveband because H_2O makes the atmosphere opaque in this region and the Meteosat images show the distribution of this water vapour.

WAVELENGTH RANGE	ABBREVIATION
0.4 µm -> 1.3 µm	VNIR
1.5µm -> 1.8 µm	SWIR
2.0 µm -> 2.6 µm	SWIR
3.0 µm -> 5.0 µm	MIR
8.0 µm -> 14 µm	TIR
>1 mm	MICROWAVE

Table 1: Atmospheric windows where effects of absorption are weak

The atmospheric windows and non-windows exist to protect the earth and maintain its energy balance. For example, ozone keeps out the ultra-violet radiation which would be harmful to life on earth. The VNIR window allows the sun to supply life-giving radiation at its waveband of maximum emittance while the TIR window allows the earth to lose the heat it gains from the sun by radiating at the earth's waveband of maximum emittance. One of the concerns associated with discussions of the 'greenhouse' effect is that increasing levels of CO_2 in the atmosphere are blocking the TIR window. The concerns over the so-called 'hole' in the ozone layer relate to the appearance of an unwelcome atmospheric window.

3.3. DISTRIBUTION OF GASES

Despite the fact that atmospheric windows are used in remote sensing the transmittance will never be 100% and therefore the radiance that reaches the satellite is not the same as that which leaves the earth. The radiance is attenuated by the atmosphere. When attempts are made to relate the spectral radiance L_λ at a satellite to the spectral reflectance ρ_λ of the earth's surface it is necessary to calculate the L_λ that would have left the earth's surface. In order to be able to do this it is necessary to be able to model the effects of absorption on L_λ and a knowledge of the distribution of the absorbing gases of the atmosphere is needed.

CO_2 only accounts for some 0.03% of the dry gas of the atmosphere and is distributed uniformly up to an altitude of about 100 kilometres.

O_3 is concentrated in a layer between 20 kilometres and 50 kilometres above the earth. O_3 causes absorption in a narrow band between 9 µm and 10 µm in the thermal infrared. Due to the altitude of the ozone layer this has no effect on airborne remote sensing given that the earth is the source of the radiation and the aircraft operate no higher than 10 kilometres. However, it is important for satellite operations.

H_2O is the most difficult gas to model as its distribution decreases exponentially as altitude increases and the climate controls the amount of water that can be held as a gas in the atmosphere. The hotter the atmosphere, the more water vapour it can contain. In cold, dry climates, the density of water vapour is of the order of 10^{-2} gm^{-3} while in hot humid climates it is of the order of 30 gm^{-3}.

A later section will look at how models of the atmosphere can be used to correct radiance values measured at a satellite.

3.4. SCATTERING

Scattering is the mechanism by which electromagnetic radiation is reflected or refracted by 'objects' in the atmosphere. These 'objects' can be gas molecules, dust particles or water droplets. The result is that while the radiation continues to move through the atmosphere it is no longer moving on its original straight line to the earth's surface. The three main types of scattering are Rayleigh, Mie and non-selective.

Rayleigh scattering occurs when the wavelength λ of the radiation is much larger than the size of the 'object' responsible for the scattering. The 'objects' in this case are gas molecules and it can be shown that

$$\sigma \propto \frac{1}{\lambda^4} \tag{7}$$

where σ is the scattering cross-section. This means that the shorter wavelengths are scattered more strongly as σ increases with decreasing λ. The greater scattering of short wavelength light is responsible for the blue colour of a clear sky.

Mie scattering occurs when the wavelength of the radiation is of the same order of magnitude as the size of the scattering 'object'. This type of scattering is caused by water vapour, dust and aerosols. The scattering cross-section is less easy to predict and ranges over

$$\sigma \propto \frac{1}{\lambda^2} \text{ and } \sigma \propto \frac{1}{\lambda^{0.5}} \tag{8}$$

The effects of both Rayleigh and Mie scattering decrease as the wavelength of the radiation increases. This accounts for the fact that the images in the visible region of the spectrum are frequently more noisy than those in the near-infrared. It is a general rule that haze has less effect on the longer wavebands and these tend to give the clearest images of a study area.

Non-selective scattering occurs when the wavelength of the radiation is much smaller than the size of the scattering 'object'. The scattering is now independent of wavelength and is caused by fog or water droplets in clouds. This type of scattering has the most drastic effect on satellite sensor images and, with the exception of those from microwave systems, generally renders them useless for applications other than meteorology.

3.5. SCATTERING PARTICLE DISTRIBUTION

The gas molecules distributed throughout the atmosphere are responsible for Rayleigh scattering. This has most effect on the visible wavebands. Areas in shadow on the earth's surface are illuminated by a predominantly blue light which is scattered into them. All areas are illuminated by a mixture of direct sunlight and scattered sunlight.

Mie scattering can affect wavebands from the visible to the thermal-infrared. The particle sizes vary from 0.1 μm to 10 μm and can be soil and rock debris, forest fire debris, sea salt, volcanic debris or industrial pollution. These particles are suspended in the lowest few kilometres of the atmosphere and the wind controls their movement and distribution.

Non-selective scattering is caused by fog and cloud particles of sizes from 1 μm to 100 μm. Obviously, the distribution of these is highly variable.

Microwave wavelengths are considerably longer than any of the atmospheric particle sizes and scattering is not an important effect for this part of the spectrum.

3.6. SUMMARY

3.6.1. Less than 0.3 μm. All radiation is absorbed with the ozone layer responsible for absorption of the ultra-violet below 0.3 μm.

3.6.2. 0.3 μm to 0.4 μm. Ultra violet radiation in this range is transmitted through the atmosphere but severely affected by Rayleigh scattering.

3.6.3. 0.4 μm to 1.3 μm. The atmosphere is transparent but Rayleigh and Mie scattering can affect the image as will clouds and fog. Many satellites use this window.

3.6.4. 1.3 μm to 1.5 μm. Water vapour causes absorption of radiation.

3.6.5. 1.5 μm to 1.8 μm. An atmospheric window used by the Landsat Thematic Mapper. Mie scattering can occur.

3.6.6. 1.8 μm to 2.0 μm. Water vapour causes absorption of radiation.

3.6.7. 2.0 μm to 2.5 μm. An atmospheric window used by the Landsat Thematic Mapper. Mie scattering can occur.

3.6.8. 2.5 μm to 3 μm. Water vapour and carbon dioxide cause absorption.

3.6.9. 3 μm to 5 μm. An atmospheric window used by the United States National Oceanic and Atmospheric Administration's meteorological satellite sensors. High temperature (~ 500 K) objects can be detected at these wavelengths and this window has military applications.

3.6.10. 5 μm to 8 μm. Water vapour causes absorption of radiation. Meteosat uses this waveband to image the water vapour distribution in the atmosphere.

3.6.11. 8 μm to 14 μm. An atmospheric window used by many sensors. Scattering by large particles can be a problem and absorption by water vapour and carbon dioxide has to be allowed for.

3.6.12. 14 μm to 1 mm Water vapour and carbon dioxide cause absorption.

3.6.13. Greater than 1 mm. An atmospheric window for microwave remote sensing. Clouds are also transparent in this window.

4. Accounting for Atmospheric Effects

4.1. RADIANCE AT THE SENSOR

The spectral radiance which reaches the satellite sensor can be derived from the spectral radiant emittance that leaves the earth's surface (Forster 1984)

$$M_{T\lambda} = M_{R\lambda} + M_{E\lambda} \tag{9}$$

where $M_{T\lambda}$ is the total spectral radiant emittance and $M_{R\lambda}$ and $M_{E\lambda}$ are the spectral radiant emittance due to reflection and emission respectively.

As seen in the previous chapter $\pi L_\lambda = M_\lambda$ and from the definitions of reflectance and emissivity

$$L_{S\lambda} = \frac{1}{\pi}(\rho_\lambda E_\lambda + \varepsilon_\lambda M_{BB\lambda}) \tag{10}$$

where $L_{S\lambda}$ is the spectral radiance at the sensor. Depending on the waveband being used one of the terms on the right of equation (10) may be negligible. By measuring $L_{S\lambda}$ at the sensor it is possible to calculate either ρ_λ or ε_λ.

However this ignores the atmospheric effects that have just been discussed. In order to account for these, Equation (10) needs to be expanded

$$L_{S\lambda} = L_{T\lambda} + L_{P\lambda} \tag{11}$$

where $L_{T\lambda}$ is the spectral radiance transmitted through the atmosphere and $L_{P\lambda}$ is the spectral radiance scattered or emitted into the field of view of the sensor by the atmosphere (Fraser and Kaufman 1985)

$$L_{T\lambda} = \frac{T_\lambda}{\pi}(\rho_\lambda E_\lambda + \varepsilon_\lambda M_{BB\lambda}) \tag{12}$$

where T_λ is the spectral transmittance of the atmosphere which will be less than 1 if absorption occurs. E_λ is the irradiance at the earth's surface and can be expressed as

$$E_\lambda = E_{0\lambda} T_\lambda \cos\theta + E_{D\lambda}$$

(13)

where $E_{0\lambda}$ is the solar irradiance for zero zenith angle and $E_{D\lambda}$ is the diffuse component of the solar irradiance caused by scattering in the atmosphere. Equation (11) now becomes

$$L_{S\lambda} = \frac{T_\lambda}{\pi} (\rho_\lambda [E_{0\lambda} T_\lambda \cos\theta + E_{D\lambda}] + \varepsilon_\lambda M_{BB\lambda}) + L_{P\lambda}$$

(14)

In order to calculate ρ_λ it is necessary to ignore $M_{BB\lambda}$ as being small, to calculate $E_{0\lambda}$ (see earlier chapter), to know θ and to model T_λ, $E_{D\lambda}$ and $L_{P\lambda}$. In order to calculate ε_λ it is necessary to ignore E_λ as being small, to calculate $M_{BB\lambda}$ (see earlier chapter) and to model T_λ and $L_{P\lambda}$. Satellite sensors actually measure L_S and therefore the above equations have to be integrated over the waveband of the sensor. This should only be done once all wavelength dependent variables have been accounted for. The result of this will be to find ρ or ε for particular ranges of wavelength. The narrower the satellite wavebands the nearer one gets to identifying the shape of ρ_λ.

4.2. MODELLING THE ATMOSPHERE

T_λ, $E_{D\lambda}$ and $L_{P\lambda}$ need to be modelled as there is rarely any direct measurement made of atmospheric conditions at the time of image acquisition. T_λ can usually be expressed in the form

$$T_\lambda = \exp(-\tau_\lambda \sec\theta)$$

(15)

where τ_λ is known as the optical path and has the components

$$\tau_\lambda = \tau_{Mie} + \tau_{Rayleigh} + \tau_{Absorption}$$

(16)

where

$$\tau_{Absorption} = \tau_{H_2O} + \tau_{CO_2} + \tau_{O_3}$$

(17)

Computer programs exist which allow the calculation of τ_λ. The Lowtran series is an example of such programs. This series is based on standard models of the atmosphere and uses any local meteorological data which may be available (Wilson and Anderson 1986).

$E_{D\lambda}$ and $L_{P\lambda}$ are complex to calculate as they require processing of the equation of radiative transfer and can also be dependent on the overall background reflectance of a study area. Various models and programs have been developed to calculate these factors.

4.3. THE NEED FOR CORRECTION

The analysis in the previous sections at first sight appears to be complex. However, if suitable computer programs are available, then it is possible to relate absolute physical properties of an object to the radiance measured at a sensor. Before such a process is attempted it is a good idea to ask the question "Is it really necessary?"

The answer to this question is going to be "Yes" if absolute radiometric calibration is required for an image. This is not always the case and it may be possible to avoid the above analysis.

Visual interpretation of an image is a common form of analysis of data. No atmospheric correction is required if the interpretation is limited to one image as the information derived is essentially qualitative.

Atmospheric correction and radiometric calibration may be needed if one image is to be compared with another. This is because the solar irradiance, the zenith angle and the atmospheric conditions are likely to change from one image to another. If absolute numbers are to be extracted from each image and compared or if classification of the two images together is to be carried out then correction is necessary (Fraser *et al.* 1977). However, if only relative differences in numbers are needed then it may be possible to use image processing techniques to suppress some of the atmospheric effects.

The correction packages discussed in the previous section rarely, if ever, allow for atmospheric variations within an image. Where these occur image processing techniques such as ratioing and moving-window contrast enhancement can be used. Such corrections are based on 'cosmetic' processing rather than physical principles and are only suitable for cases where visual interpretation is required or relative differences are to be analysed (Schowengerdt 1983, Richards 1986).

When one image is to be compared with another, but again on a relative basis, it is possible to apply image processing techniques such as dark-object subtraction and histogram matching. These approximately balance the ranges of grey levels in the images.

The need for atmospheric correction becomes greater as the spectral bands provided by sensors become narrower and attempts are made to identify subtle variations in ρ_λ. The computing and thinking time can be extensive and it is necessary to assess the benefit to be gained against the cost. Extra processing is undesirable if the extra accuracy it introduces is irrelevant in the overall context of a project. On occasion it may be possible to ignore the atmospheric effects but such a step should be based on sound argument rather than wishful thinking. The concepts developed in § 3 and § 4 provide a basis for making such decisions.

5. Interactions with the Earth's Surface

5.1. ENERGY CONSERVATION

The interaction of electromagnetic radiation with the earth's surface is an important aspect of remote sensing as to some extent this defines the information that will be carried to the sensor.

At the point at which the EMR hits the earth the energy conservation law dictates that all energy should be accounted for

$$\phi_{i\lambda} = \phi_{r\lambda} + \phi_{t\lambda} + \phi_{a\lambda} \tag{18}$$

where $\phi_{i\lambda}$ is the incident flux, $\phi_{r\lambda}$ is the reflected flux, $\phi_{t\lambda}$ the transmitted flux and $\phi_{a\lambda}$ the absorbed flux. By dividing through by $\phi_{i\lambda}$ we have

$$1 = \rho_{\lambda} + t_{\lambda} + \alpha_{\lambda} \tag{19}$$

where ρ, t and α are the reflectance, transmittance and absorptance respectively. Kirchoff's law states that $\alpha_{\lambda} = \varepsilon_{\lambda}$ and so equation (19) becomes

$$\rho_{\lambda} + t_{\lambda} + \varepsilon_{\lambda} = 1 \tag{20}$$

For many materials $t_{\lambda} = 0$ and this holds true in the thermal infrared where we have

$$\rho_{\lambda} + \varepsilon_{\lambda} = 1 \tag{21}$$

In the thermal region it can, therefore, be said that good emitters are poor reflectors and vice versa.

Note that water has a non-zero t_{λ} for the visible wavelengths and so Equation (20) cannot be simplified.

5.2. SPECTRAL CURVES

The subscript λ has been used throughout the discussions of physical principles to indicate that a particular characteristic varies with wavelength. The spectral reflectance ρ_{λ} of an object is important

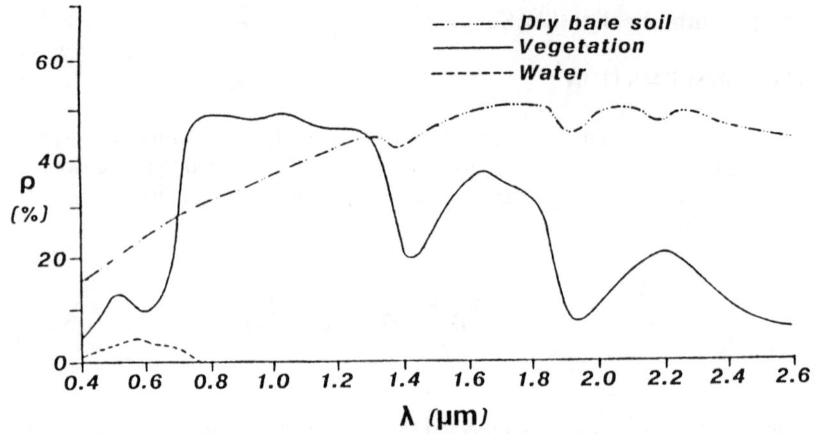

Figure 1. Spectral curves for vegetation, soil and water.

in this context. It is an attribute of an object which is controlled by the material forming that object and so if ρ_λ can be measured in sufficient detail then it may be possible to identify the material.

Graphs can be drawn which plot the spectral reflectance against wavelength. These graphs produce characteristic shapes for various materials and these are usually known as "spectral curves". The term "spectral signature" has been used in the past but some have claimed that it is misleading because it infers that all surfaces of a particular type, say wheat, would have a clearly defined identical signature. In fact, the term is not misleading if it is interpreted as meaning that each separate example of wheat will have a subtly unique signature. However, the term spectral curve is, perhaps, safer and equally expressive (Thomas *et al.* 1987).

5.3. VEGETATION, SOIL AND WATER

Figure 1 gives an example of three spectral curves where ρ_λ is plotted against λ for vegetation, soil and water in the visible to short wave infrared range.

Water has a characteristic curve which can be explained by reference to Equation (19). In the visible region of the spectrum, the transmittance of water is significant and so both the absorptance and the reflectance are low. The reflectance of water in the visible scarcely rises above 5% and sensors which are designed to measure water colour have to have much more sensitive detectors than those looking at land.

The absorptance of water rises rapidly in the near-infrared where both reflectance and transmittance are low. The signal received at a sensor in the near-infrared over water can usually be attributed to atmospheric path radiance rather than reflection from the surface.

The transmission of visible radiation through water means that if the water is sufficiently shallow the radiation can be reflected at the lake or sea bottom, transmitted back through the water and detected by the sensor. It is then not clear from the visible image whether or not the detected surface is above or below water. However, inspection of a near-infrared image will resolve the issue as a submerged surface will not be detected in the near-infrared because transmission does not occur.

Soil has very different characteristics to water with ρ_λ increasing with wavelength in the VNIR. The increase in reflectance with wavelength in the visible is consistent with the human eye's observation that soils can have a red or brown colour to them. There is usually a strong correlation between the value of ρ_λ in the visible and the value of ρ_λ in the near-infrared for soils whether the soils are inherently light or dark. The overall level of reflectance is of the order of 20% in the visible and 30% in the near-infrared. For this reason the land-water boundary is simple to identify on a satellite image.

Green vegetation has a more complex curve than either soil or water. In the blue and red part of the spectrum the absorptance is high due to the presence of chlorophyll in the plant and the reflectance is low. As expected, the spectral reflectance rises in the green part of the spectrum and can reach a level of around 10%. This is clear from the fact that the human eye sees vegetation as green.

The reflectance of vegetation is actually at its highest in the near-infrared where it can rise to more than 50%. The level of this reflectance is characteristic of particular species of vegetation. Measurements of ρ_λ for the red and the near-infrared are effective in identifying the presence of vegetation because they highlight both the absorptance feature and the high reflectance feature of the spectral curve.

The reflectance falls with increasing wavelength in the SWIR and there are two major absorption features at 1.4 μm and 1.9 μm which are caused by water in the plant. However, these do not correspond to atmospheric windows and so cannot be observed by satellite sensor remote sensing.

5.4. IDENTIFICATION FROM SPECTRAL CURVES

Measurements of ρ_λ can lead to the identification of surface materials only if they are sufficiently detailed. In the past satellites could provide images in the VNIR based on four fairly wide wavebands (~ 100 nm). Currently images can be acquired in the VNIR and SWIR but still with wide wavebands (~ 50 nm).

In the future there will be imaging spectroradiometers providing hundreds of images in the VNIR with narrow wavebands (~ 2 nm). These will allow the formation of spectral radiance curves. However they will only be related to spectral reflectance provided that the appropriate atmospheric corrections and radiometric calibrations are carried out on the data. This leads us back to the importance of the concepts discussed in this chapter.

6. Summary

Active remote sensing is where the sensor provides the source of irradiance while passive remote sensing is where the sensor collects radiation originating from a natural source. In the VNIR and SWIR the sun is the source of irradiance and the reflectance characteristics of an object are measured. In the TIR the earth is the source and its emittance characteristics are measured.

The atmosphere is important in remote sensing because it is the medium through which the EMR passes. H_2O, CO_2, and O_3 in the atmosphere absorb EMR and make it opaque to certain wavelengths. The regions of transparency are known as atmospheric windows and these are generally used by sensors. EMR observed in these windows is still liable to be affected by scattering from particles within the atmosphere.

The measurement of absolute values of ρ or ε requires that the values of L recorded by the sensor must be corrected for atmospheric effects.

Measurement of ρ_λ in sufficient detail can lead to the construction of spectral curves which are characteristic of materials. Soil, vegetation and water have clearly differing spectral curves which are discussed in more detail in subsequent chapters of the book.

References

Campbell, J.B., 1987. *Introduction to Remote Sensing.* (U.S.A.: The Guildford Press).

Forster, B.C., 1984. Derivation of Atmospheric Correction Procedures for LANDSAT MSS with Particular Reference to Urban Data. *International Journal of Remote Sensing,* **5**, 799-817.

Fraser, R.S. and Kaufman, Y.J., 1985. The Relative Importance of Scattering and Absorption in Remote Sensing. *I.E.E.E. Transactions on Geoscience and Remote Sensing*, **23**, 625-633.

Fraser, R.S., Bahethi, O.P. and Al-Abbas, A.H. 1977. The Effect of the Atmosphere on Classification of Satellite Observations to Identify Surface Features. *Remote Sensing of Environment*, **6**, 229-249.

Richards, J.A., 1986. *Remote Sensing Digital Image Analysis: An Introduction.* (Berlin: Springer-Verlag).

Schowengerdt, R.A., 1983. *Techniques for Image Processing and Classification in Remote Sensing* (London: Academic Press Inc.).

Thomas, I.L., Benning, V.M. and Ching, N.P., 1987. *Classification of Remotely Sensed Images.* (U.K.: Adam Hilger).

Wilson, S.B. and Anderson, J.M., 1986. The Applicability of LOWTRAN 5 computer Code to Aerial Thermographic Data Correction. *International Journal of Remote Sensing*, **7**, 379-388.

SPECTRAL CHARACTERISTICS OF VEGETATION, SOIL AND WATER IN THE VISIBLE, NEAR-INFRARED AND MIDDLE-INFRARED WAVELENGTHS

A. S. BELWARD
Commission of the European Communities,
Joint Research Centre,
Institute for Remote Sensing Applications,
21020 Ispra (Varese) Italy.

1. Introduction

In the previous chapter the concept of identifying and characterizing earth surface features on the basis of spectral behaviour was introduced. It was shown that electromagnetic radiation (EMR) incident with an object is either reflected, absorbed or transmitted. Furthermore, the relative proportion of each of these was shown to vary with the wavelength of the incident EMR and with the nature of the material making up the object. Spectral reflectance curves were presented as a means of describing the spectral properties of objects, and example curves of vegetation, soil and water were given. This chapter describes in more detail the basic mechanisms which influence the spectral properties of these major earth surface features in the visible, near and middle infrared wavelengths; data recorded through these parts of the electromagnetic spectrum are most frequently available to the user of remote sensing, and if the controlling mechanisms are properly understood then we can begin to make inferences regarding the physical and physiological nature of earth objects from measured spectral response.

2. Spectral properties of vegetation

The spectral properties of vegetation are quite distinctive, as Gates (1970) observed "A plant leaf reflects and transmits incident radiation in manner that is uniquely characteristic of pigmented cells containing water solutions". Figure 1 shows a generalized spectral reflectance curve for a green plant over the wavelength range 0.4 µm - 2.6 µm. Spectral reflectance is low in the visible and middle-infrared wavelengths, which are dominated by absorption, whilst the near-infrared is characterized by high levels of reflectance and transmission, with minimal absorption. These features can be explained in terms of the physical and physiological characteristics of vegetation.

2.1 VISIBLE LIGHT

Plants use light energy from the visible part of the spectrum to synthesize the organic compounds necessary for maintenance and growth. This is brought about by the process of photosynthesis.

A. S. Belward and C. R. Valenzuela (eds.),
Remote Sensing and Geographical Information Systems for Resource Management in Developing Countries, 31–53.

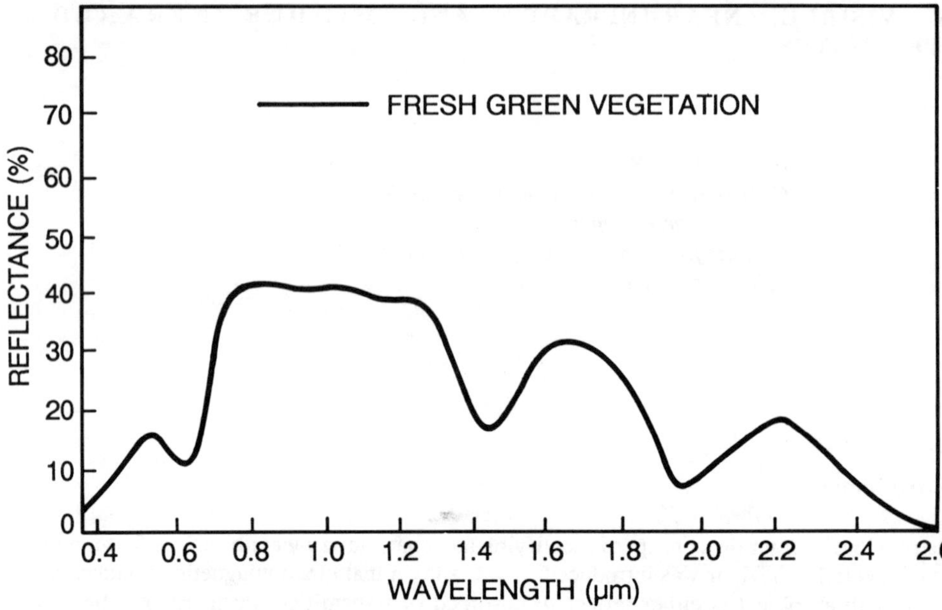

Figure 1. Spectral reflectance curve of fresh green vegetation

Cell organelles, called chloroplasts, contain pigments capable of absorbing light energy and using this to convert energy-poor carbon dioxide (CO_2) and water (H_2O) into energy-rich carbohydrates (CH_2O) and oxygen (O_2). This is shown in equation (1)

$$CO_2 + H_2O \xrightarrow[\text{Plants}]{\text{Sunlight}} CH_2O + O_2 \tag{1}$$

Leaf structure is such that the raw materials needed for this process are brought into contact with the chloroplasts; carbon dioxide, present in air, enters the leaf through the leaf openings (stomata) and moves through the leaf in inter-cellular air spaces; water is present in all hydrated leaf cells; and leaf optical properties are such that most light penetrates the leaf surface rather than being reflected (Knipling 1970, Woolley 1971).

As light enters the leaf it is scattered by refraction and Rayleigh / Mie scattering (see Bird, chapter 2). Many cell organelles, lysosomes for example, are smaller than 1 μm in size and are therefore capable of causing Rayleigh scatter. However since early this century refraction has been identified as the main cause of scattering of light within leaves (Willstätter and Stoll 1918, Allen *et al.* 1973). This occurs because of refractive index differences between hydrated plant cells - refractive index of 1.47, and inter cellular air - refractive index of 1.0, (Gausman 1985).

The light, being scattered and refracted throughout the leaf, comes into contact with the chloroplasts. The pigments in the chloroplasts, such as the chlorophylls and carotenes absorb light

Type of pigment	Characteristic absorption maxima (μm) - in organic solvents
Chlorophyll a	4.20, 6.60
Chlorophyll b	4.35, 6.43
β-Carotene	4.25, 4.50, 4.80
α-Carotene	4.20, 4.40, 4.70
Violaxanthol	4.25, 4.50, 4.75

Table 1. Absorption maxima of the main photosynthetic pigments found in higher plants.

of specific energy, causing electron transitions within the pigment molecular structure, and the resulting transfer of energy is used to drive the photochemical reactions of the photosynthetic cycle.

As a result of the quantum requirements for energy absorption (discussed in chapter 1) the plant pigments absorb light strongly at some wavelengths and not at all at others. The main pigments found in higher plants and their absorption maxima are listed in table 1.

All the pigments show strong absorption of blue light, with the chlorophylls also showing strong absorption of red light. Absorption, by individual leaves, at these wavelengths is between 70 and 90% (Gausman 1985). As there is usually 5 to 10 times as much chlorophyll as carotenoid pigments in higher plants, chlorophyll tends to dominate spectral response.

At the wavelengths shown in table 1 each photon has the precise energy needed to cause the electron transitions in the photosynthetic pigments. The energy of green light does not cause these electron transitions and so tends to be reflected and transmitted, rather than absorbed. Hence maximum reflectance from plants in the visible wavelengths occurs in the green part of the spectrum, and so we see plants as green.

Towards the end of the functional life of leaves and other plant organs there is a stage where the tissues deteriorate. This is called senescence. During senescence the chlorophyll breaks down, which leads to a marked increase in reflectance at red wavelengths as absorption by this pigment is reduced. Other pigments such as the orange carotenes and yellow xanthophylls (violxanthol is a xanthophyll), plus non-photosynthetic pigments such as red anthocyanins and yellow flavones are often still present. No longer masked by the presence of the chlorophylls, these pigments are clearly seen in the brown and yellow colours of senescent vegetation, for example ripe cereal crops such as Sorghum (*Sorghum vulgare*), Millet (*Pennisetum typhoides*) and Wheat (*Triticum sativum*).

2.2 NEAR-INFRARED LIGHT

As we saw in the first chapter, the longer the wavelength of EMR, the lower the energy value of each photon. In contrast to visible wavelength EMR, the energy levels of near-infrared light (0.8-1.1 μm) are not great enough to drive the photochemical reactions within the photosynthetic cycle, thus the pigment containing chloroplasts are transparent to near-infrared light.

With absorption typically around 5%, the spectral response of plants in the near-infrared is dominated by transmission, about 40% and reflection, 55% (Figures for a single citrus leaf, *Citrus sinensis*, Gausman 1985). The proportion of each of these varies with plant species; Gates *et al.* (1965) and Sinclair *et al.* (1971), among other early workers, observed that internal leaf structure is dominant in controlling the spectral response of plants in the near-infrared.

Variations in the internal structure of leaves causes differences in the amount of internal scattering, and thus differences in transmission and reflection. This can be seen in figure 2a which is an image of an area of woodland recorded through the near-infrared waveband (0.91 - 1.05 μm). Figure 2b is the corresponding forest stand map identifying tree species. The coniferous trees (with needle leaves), mainly Scots pine (*Pinus sylvestris*) and Corsican pine (*Pinus nigra*) appear as dark areas on the image, (ie they have low reflectance), the broad leafed trees, mainly Beech (*Fagus sylvatica*) appear much brighter, (ie they have much higher levels of reflectance), note, in particular, the striking contrast between the high reflectance from the Beech trees along the curving road which runs through the wood, and the low reflectance from the surrounding conifers.

During senescence, near-infrared reflectance also changes. Sinclair *et al.* (1971) note that reflectance increases marginally in the early stages. This may occur because the leaf internal cells become dehydrated, and the cell walls pull apart and change in orientation. This leads to increase in cell / air interfaces (Daughtry and Biel 1985) and, as stated above, liquid (cell) / air discontinuities largely control the internal scattering of light, and thus reflectance.

In the advanced stages of senescence near-infrared reflectance falls; the internal tissues completely collapse, forming a series of horizontal layers of cell walls with the resulting reduction in air spaces being the probable cause of decreased reflectance (Knipling 1967).

Because much of the incident radiation at near-infrared wavelengths passes through a single leaf the multiple leaf layers of a developing vegetation canopy have an additive effect on near-infrared reflectance. If a figure of 50% near-infrared transmission for a single leaf is taken then a simple diagram, shown in figure 3 can be used to illustrate this (Hoffer 1978). Of the total incident near-infrared EMR at the first leaf layer 50% is reflected and 50% transmitted. The transmitted component is then incident upon the second leaf layer where again 50% is transmitted and 50% reflected. The reflected component from the second layer is passed back to the first where again 50% is reflected and 50% transmitted. Thus, in this simple model if one leaf layer returns 50% of the incident near-infrared EMR, two leaf layers will return 62.5%. Myers (1970) working with cotton plant leaves recorded increases in near-infrared reflectance from 50% with one leaf up to 84% with six leaves. There comes a point when further increases in the number of leaf layers no longer affects reflectance. This point is referred to as the near-infrared infinite reflectance ρ^{∞}; in the example cited from Myers (1970) ρ^{∞} is 84%.

2.3 MIDDLE-INFRARED LIGHT

In the middle-infrared the internally scattered light is absorbed by water in the plant. Liquid water, like pigments, absorbs EMR with specific energy levels. Unlike pigments the water absorptions are caused by transitions in the vibrational and rotational states of the water molecules, not electron transitions which require much higher energy levels. It is primarily the longer wavelength, lower energy, middle-infrared light which is absorbed by water, and the shorter wavelength, high energy, visible light absorbed by the plant pigments.

Fundamental moisture absorption bands occur at 2.66, 2.73 and 6.27 μm. Objects containing appreciable amounts of water will have extremely low reflectance levels at these wavelengths. Progressively weaker "overtone" bands are found at 1.9, 1.4, 1.1 and 0.96 μm.

Water absorption is, of course, most pronounced at these specific wavelengths, and can reach more than 90%. However, the absorption bands are strong enough to exert a carry over effect, such that reflectance at wavelengths between the bands is also influenced. In other words, when reflectance is low (or high) in the moisture absorption bands it is also low (or high) between them.

Figure 2a Near-infrared (0.91 - 1.05 μm) data from an airborne multi spectral scanner. The image is not a photograph but a digital image which has been displayed using computer image processing methods (see Wilkinson, chapter 5). The image shows the sharply defined contrast between the high NIR reflectance from the beech trees and the much lower reflectance from the conifers. The area covered by the image is about 1 km^2. The "resolution" of the scanner (ie the smallest ground area viewed) is around 2 m. This provides images with sufficiently high resolution for individual tree crowns to be seen.

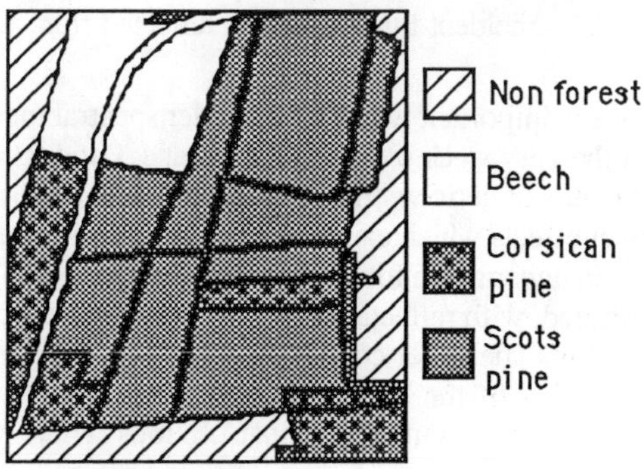

Figure 2b Forest stand map

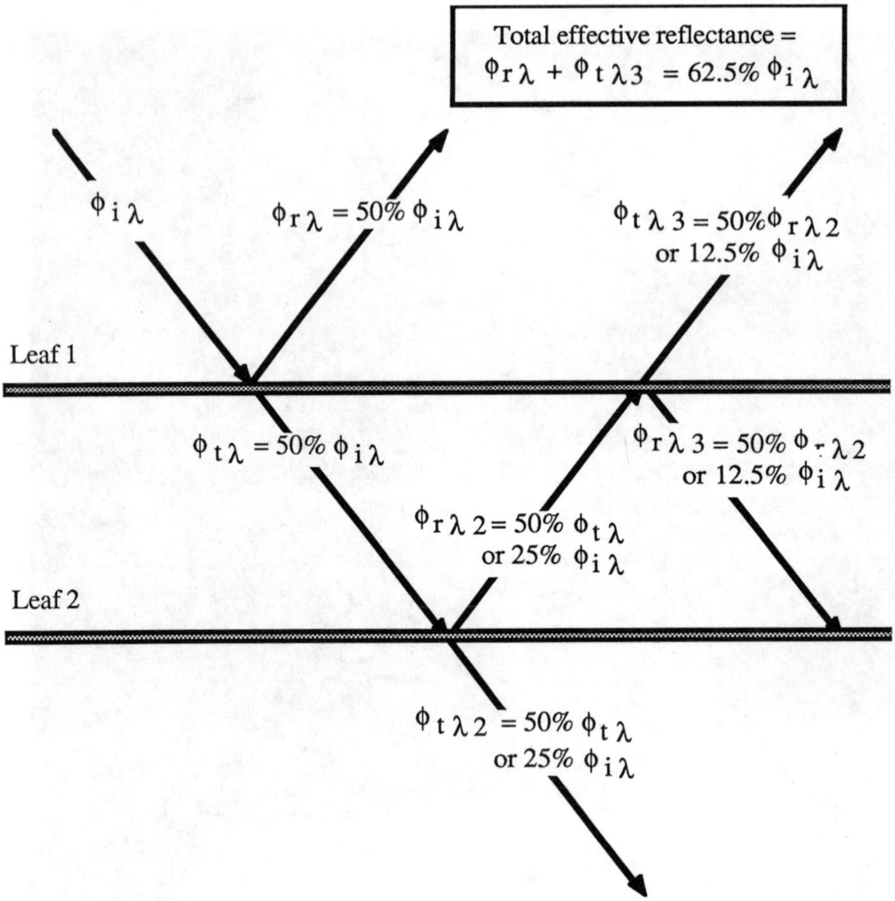

Figure 3. The effect of multiple leaf layers on the effective reflectance from vegetation. $\phi_{i\lambda}$ = the incident flux, $\phi_{r\lambda}$= the reflected flux, $\phi_{t\lambda}$= the transmitted flux, (after Hoffer 1978)

This is important when we consider spectral measurement from space platforms; with the exception of the very weak overtone bands, at 1.1 and 0.96 μm, all the moisture absorption bands fall outside the atmospheric windows used in remote sensing (see Bird, chapter 2, §3.) thus it is only the carry over effects which can be detected from satellite sensors.

In general, the lower the water content of a plant the higher the reflectance will be in the middle-infrared, with reflectance peaks occurring between the moisture absorption bands at around 1.6 and 2.2 μm. The water content of a plant will depend upon a) the percent moisture content, and b) the thickness of the leaves. The effects of water absorption are so strong that changes in middle-infrared reflectance are most obvious where water content is very low.

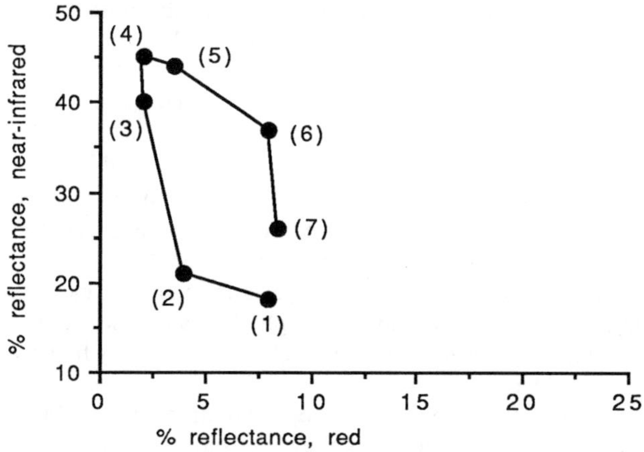

Figure 4. Changes in red near-infrared reflectance for an annual cereal crop through a single growing season. Figures in brackets identify specific stages, see text for explanation

Hoffer and Johannsen (1969) working with Maize (*Zea mays*) found that pronounced increases in middle-infrared reflectance were most evident where moisture content fell below around 54%.

2.4 TEMPORAL AND SPATIAL CONSIDERATIONS

All plants undergo two processes of change, those of growth and differentiation. Growth is applied to quantitative changes (ie changes in volume, not form or function), differentiation is applied to qualitative changes such as the the changes from vegetative to reproductive phase in a plant growth cycle.

The term phenology is often used to describe the seasonal changes in vegetation. Different plant species in the same geographical location can have contrasting phenologies; in temperate regions, for example, Barley (*Hordeum vulgare*) reaches maturity two to three weeks before Wheat. The yellow / brown barley fields contrast sharply with the still green wheat fields. Earlier in the season the phenological contrast between the two is less well defined, with both crops appearing lush green. Similarly, the phenology of the same plants in different locations can contrast; for example Maize (*Zea mays*) can reach maturity 4 months after sowing at low altitudes in Kenya, yet can take over a year to mature at higher altitudes.

We have seen from the preceding subsections that the spectral reflectance properties of vegetation change during growth and development. Multi spectral measurements taken at different times can, therefore, be used to track the phenology of different vegetation types.

Figure 4 shows spectral changes in red and near-infrared wavelengths for an annual crop such as wheat. Reflectance is shown in the feature space defined by axis representing these two spectral regions. The numbers in brackets relate to points on the figure. Soon after emergence crop cover is very low and the resulting reflectance is strongly dominated by soil (1). The precise position in the red near-infrared space will depend on the reflectance characteristics of the soil (see §3.4, below). As the vegetative cover develops, the increase in chlorophyll content causes spectral response in the red wavelengths to drop because of the increased effects of pigment absorption, and near-infrared reflectance to increase because more highly reflective vegetation, and less soil is seen (2). As cover continues to develop red reflectance usually reaches a minimum value; so much red light is absorbed by the upper canopy (§ 2.1) that additional leaf layers, appearing with canopy development, have little influence, ie red infinite reflectance is soon reached (3 & 4). Usually, near-infrared reflectance continues to increase with increasing number of leaf layers, even after red ρ^{∞} has been reached (3 & 4), this continues until near-infrared ρ^{∞} too is reached (4). During senescence the breakdown of chlorophyll leads to a marked increase in the levels of red reflectance, and the re-orientation of the leaf cells also causes a drop in near-infrared (5). Note that in the early stages of senescence a slight increase may be apparent in near-infrared reflectance. Red reflectance continues to increase as senescence progresses and chlorophyll breaks down, further changes in leaf structure cause further reduction in near-infrared reflectance (6). Finally harvest returns us to bare soil and post-harvest crop residue (7).

Graphs, such as this, are thus a temporal spectral representation of the plant's phenology. Because different plants have different phenologies they will have different temporal spectral profiles. Characterization of temporal spectral profile can be used for example, as an aid to crop mapping by identifying periods of maximum spectral separation between crop types.

Figure 5 shows false colour infrared satellite images from an arable area of the United Kingdom near the beginning (February), in the middle (April and May) and towards the end (August) of a growing season. The intensity of the red tones is related to the level of near-infrared reflectance. Field (W) was planted with winter wheat. In February the wheat has emerged, cover is not complete and the pale pink tones indicate the low levels of near-infrared reflectance. During April and May the crop cover develops, and we can see how the red tones deepen as NIR reflectance increases. In August the crop has been harvested, we are left with bare soil and stubble, NIR reflectance is correspondingly low. Field (R) was planted with winter oilseed rape (*Brassica napus*) This follows a similar growth pattern to the wheat, with the obvious exception of the May image. Winter oilseed rape flowers in May; fields of brilliant yellow rape flowers stand out dramatically against the more uniform green of other crops in the United Kingdom at this time. The flowers have very high reflectance in visible and near-infrared wavelengths, and the fields appear almost white on satellite images. By August the rape too has been harvested and the fields have the same appearance as the wheat. Field (B) was planted with sugar beet (*Beta vulgaris var. altissima*). This crop is sown far later than the other two crops and throughout February, April and May the fields of beet appear as bare soil on the images. In August however, the beet is well developed and the fields now take on the spectral response of vegetation.

Although spectral temporal profiles are variable, all green plants follow the general pattern of decreasing red reflectance, and increasing near-infrared reflectance, with an increase in plant cover and phytomass. Various ratios and combinations of red and near-infrared reflectance (collectively known as vegetation indices) have been used to quantify the relative changes in reflectance that occur at these wavelengths during plant development. Vegetation indices are widely used as surrogate measurements of vegetation production and as a means of studying vegetation phenology; these subjects are discussed in later chapters.

The highly variable nature of plant reflectance is not just due to the processes of growth and differentiation but also due to spatial variations such as background soil type (Huete *et al.* 1984) and, in the case of crops, managerial difference such as nutrient inputs and planting density (Christ 1984). In addition, because leaves are not diffuse, Lambertian reflectors (Grant 1987), variations in sun-target-sensor geometry are known to influence the reflectance characteristics of vegetation. This point is discussed in more depth by Baret in a later chapter.

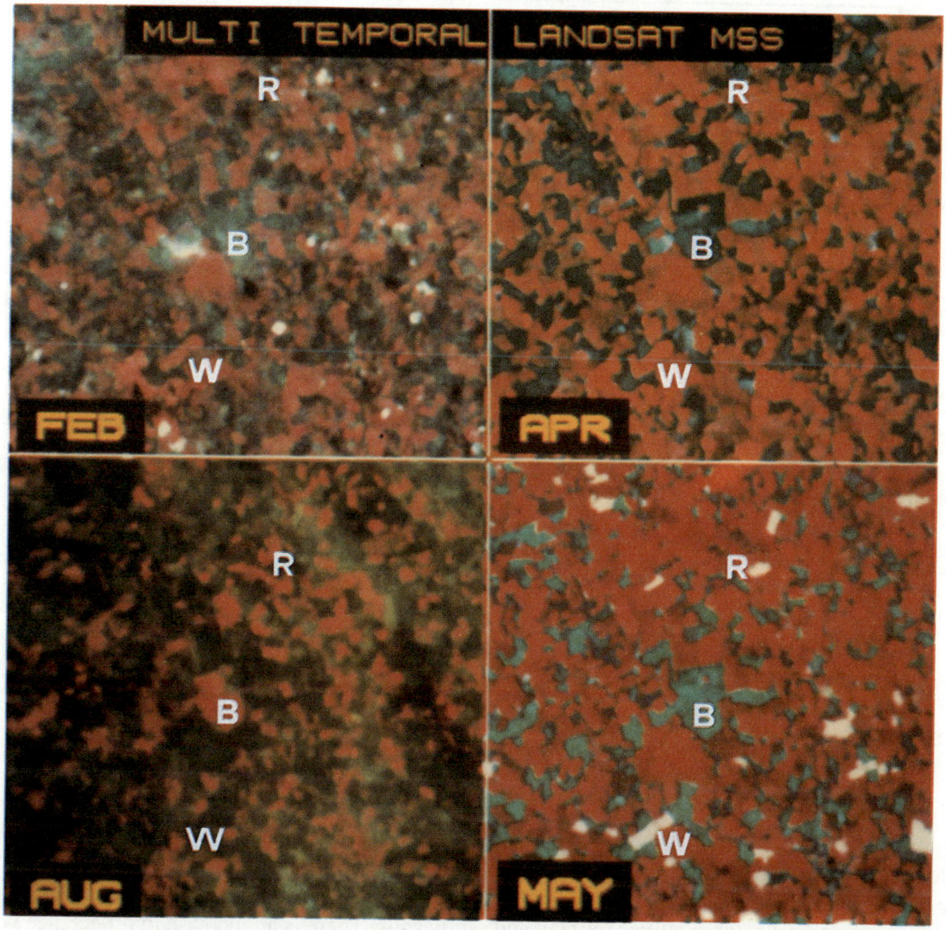

Figure 5. Four Landsat Multi Spectral Scanner (MSS) false colour composite (FCC), 12 km^2 images of an agricultural area of UK, from 1980. Hill, chapter 4, describes the MSS. FEB = 18th February, APR = 12th April, MAY = 18th May, AUG = 16 August. FCCs are created by displaying data from the green (0.5 - 0.6 μm), red (0.6 - 0.7 μm) and NIR (0.8 - 1.1 μm) wavelengths through the blue, green and red channels of a colour monitor. Green vegetation always appears bright red on FCC images because the high reflectance from vegetation in the NIR is displayed in red. Soils never appear bright red on FCCs. Here the soils appear blue / grey. The white patches in the February image are clouds; note the dark cloud shadows on the ground (above, and slightly to the left of each cloud). The images show how reflectance changes of an agricultural area can be captured by satellite sensors. The increasing NIR reflectance of the developing winter wheat (W) can be seen in the increasing intensities of red, until August when the wheat is harvested and we are left with bare soil plus stubble. The rape (R) follows the same pattern, except in May the rape comes into flower. The flowering rape has quite distinct reflectance characteristics and appears white on the May image (confusion with clouds is avoided; a) there are no shadows, b) the fields have a more regular shape). The August image is the "inverse" of the earlier scenes, the winter sown crops have been harvested and appear as bare soil, all the later sown crops, such as beet (B) have emerged, and no longer appear as bare soil. Multi temporal analysis requires careful image preparation: images must have the same geometry, sensors must be calibrated, and if possible, atmospheric effects (Bird, chapter 2), must be accounted for - these operations are discussed in chapters 5 and 6.

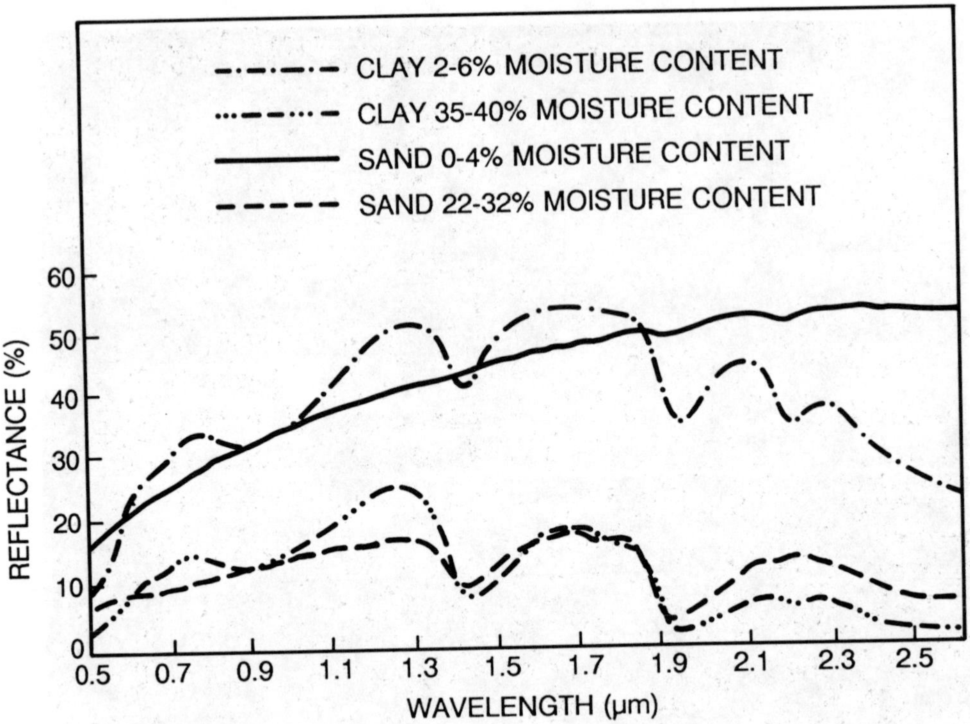

Figure 6. Spectral reflectance curves for bare soil (after Hoffer and Johannsen 1969)

3. Spectral properties of soil.

The spectral properties of soils are much simpler than those of vegetation; transmission of EMR in the optical wavelengths is non existent and all incident energy is either absorbed or reflected. Figure 6 shows spectral reflectance curves for clay and sandy soils. There is an increase in reflectance with increasing wavelength for both soils, though the effect of variations in water holding capacity upset this trend in the moisture sensitive middle-infrared. The basic form of the curves holds for many soils (Condit 1970), though the magnitude varies. This variation is due to the physical properties of any particular soil. The main soil variables found to influence spectral properties are texture, moisture content, organic matter content and iron oxide content.

3.1 SOIL TEXTURE AND MOISTURE CONTENT

The main criterion used in classifying soils is texture. The term soil texture is used to describe the size and distribution of soil particles. There are a number of different classifications in current use, all of which identify three major groupings; sand, silt and clay. Table 2 shows name and size range of constituent particles for two widely used systems.

Name	Size range (mm)	
	International system	MIT * system
Clay	< 0.002	<0.002
Silt	0.002 - 0.02	0.002 - 0.006 0.006 - 0.02 0.02 - 0.06
Fine sand	0.02 - 0.2	0.06 -0.2
Medium sand		0.2 - 0.6
Coarse sand	0.2 - 2.0	0.6 - 2.0
Gravel	>2.0	2.0 - 60.0

* Massachusets Institute of Technology

Table 2. Particle size range and name according to two widely used soil classification schemes.

Early laboratory analysis of soil reflectance properties showed that increasing particle diameter was related to decreasing reflectivity (Bowers and Hanks 1965). However, laboratory analysis is concerned with dispersed soils, not the aggregates found in natural conditions; it has been observed that decreasing particle size increases reflectance from sandy soils, however the opposite is true for fine textured soils. Outside the laboratory, factors such as moisture content cannot be easily controlled and there are very strong relationships between soil texture and water holding properties.

The relative proportions of sand, silt and clay affect the movement of air and water through a soil. If moisture is present in a soil then each particle is covered by a thin film of water. Thus soils with a small particle size, and therefore a large total surface area per unit volume, will hold a large amount of water. In addition some of the air spaces in the soil will be occupied by water. The ease with which water drains or evaporates from a soil is very much influenced by the size of the pores. Sands for example, with large particles and large pores dry out easily whilst clays do not. Marshal and Holmes (1979) provide a good review of this process.

The effect of this on soil reflectance is particularly marked when the spectral reflectance curves of air dry and moist soils are compared. The spectral reflectance curves in Figure 6 are for air dry and moist sand and clay. The large particle size of the sand allows more complete drying than the clay, consequently an air dry sand contains less water than an air dry clay (in this case 0 - 4% for the sand, compared with 2 - 6% for the clay); the sand shows little in the way of moisture absorption in the water absorption bands at 1.4 and 1.9 μm. In comparison the air dry clay still retains sufficient moisture for pronounced dips in reflectance to occur at these points. On wetting the sand too displays these dips.

The changes in soil reflectance on wetting are most pronounced in the moisture absorption wavebands, but note that reflectance at all wavelengths is lower for the wet soils; we are all familiar with the way in which soil goes darker after rain.

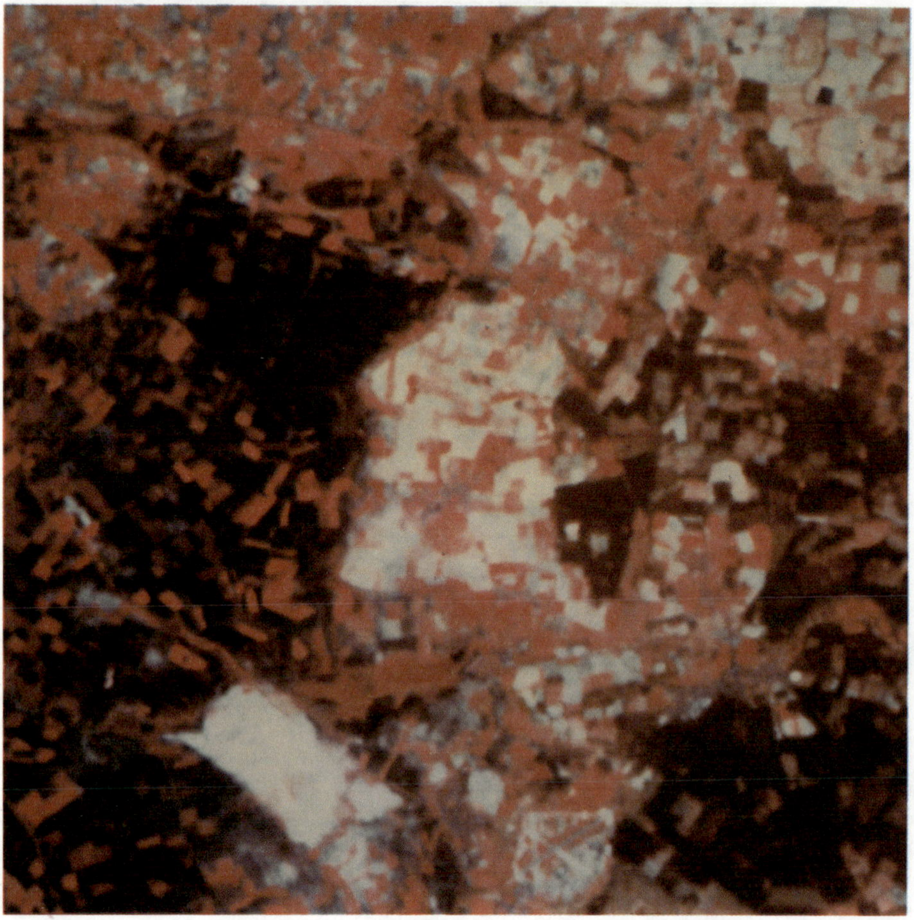

Figure 7. Landsat Multi Spectral Scanner (MSS) false colour composite (FCC), 50 km^2 image of part of the United Kingdom, April 1982. As with the FCCs shown in figure 5, vegetation, with the exception of the extensive coniferous woodland, appears red. Here the red areas are grassland and winter cereal crops, *eg* wheat. Coniferous woodland, as shown in figure 2, has relatively low levels of NIR reflectance and appears as dark red / brown on the FCC shown here. The bare soils, again largely supporting late sown crops like beet, appear almost black where the organic content is high (humified peat), and almost white where organic content is very low (brown earth). Note how the concrete runways of the airfield can easily be confused with the brown earth soils. Spectral confusion between bare soil and man made features such as roads, or urban areas in general, is common with these images.

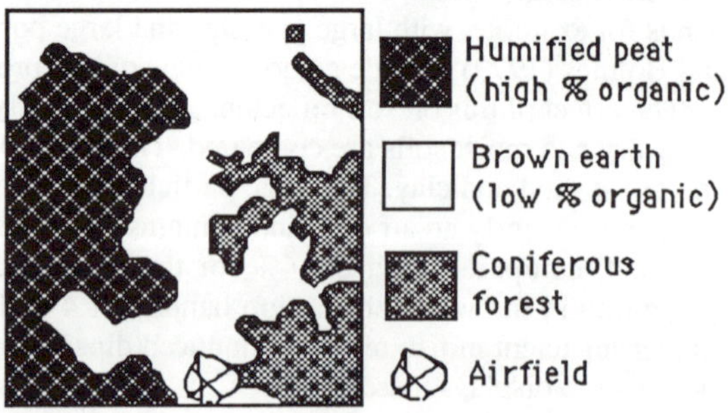

Figure 7. Soil distribution

In addition to the moisture absorption bands, the spectral reflectance curves for the clay dip at 2.2 µm. This wavelength corresponds to a hydroxyll absorption band (Hunt and Salisbury 1970) The other hydroxyll absorption band at 1.4 µm is confused with moisture absorption. The hydroxyll absorption bands tend to give soils with a high clay content quite distinctive spectral curves.

Apart from it's effect on air and water movement through a soil, texture also has an effect on surface roughness. In theory small particle size leads to a smooth soil surface, which in turn leads to higher levels of reflectance. However, in practice, clay soils form aggregates more easily than sandy soils. Thus natural processes and cultivation operations tend to mask the simple effect of texture by introducing soil surface roughness on a macro scale. A rough surface, with a high incidence of interaggregate cracks and pores, will trap more of the incoming light and reflectance will be correspondingly reduced.

3.2 ORGANIC MATTER CONTENT

Soils containing a high percentage of organic matter generally have low levels of reflectance, and appear quite dark in colour, usually dark brown or black. Soils with little or no organic material tend to reflect more light and are often light brown or grey in colour. In general, the higher the percentage of organic material, the darker the soil will be. The colour will also be influenced by the degree of decomposition of the organic material; well decomposed material gives a black soil colour, whilst less well decomposed material generally imparts a brown colour to the soil. Figure 7 shows a false colour infrared image from a satellite sensor of an area of contrasting soils. The very dark grey / black areas on the left hand side of the image are soils from the Adventurers series of humified peat soils group (these are soils with a high organic matter content). The almost white areas on the right hand side of the image are soils from the Freckenham and Worlington series of the brown earth group (these are soils containing very little organic matter). The red areas are cereal crops and grassland, the brown areas on the extreme right are extensive conifer plantations.

The above observations on the relationship between colour and organic matter content are mainly based on work done on soils which have developed in temperate climates (Page 1974 for example). As such they may not hold true for all climatic conditions in which soils may develop. Care should be taken in applying these observations to the interpretation of spectral reflectance data for soils from other climatic regions.

3.3 IRON OXIDE CONTENT

Many of the red soils of the world owe their colour to a high iron content. In general, the higher the iron oxide content the lower the reflectance, particularly in the visible wavelengths. Figure 8 shows a false colour infrared image from a satellite sensor of an area of contrasting soils in the West African Sahel. The dark grey / black areas are Laterite crust, the grey areas are clay outwash and the pale yellow tones are sands. Vegetation again appears bright red. The laterite has much lower reflectance than the sands or clays in all the wavelengths used to make this image, hence the almost black appearance. The low reflectance of the laterite areas may be due to the specific absorption characteristics of the metallic ions in the soil.

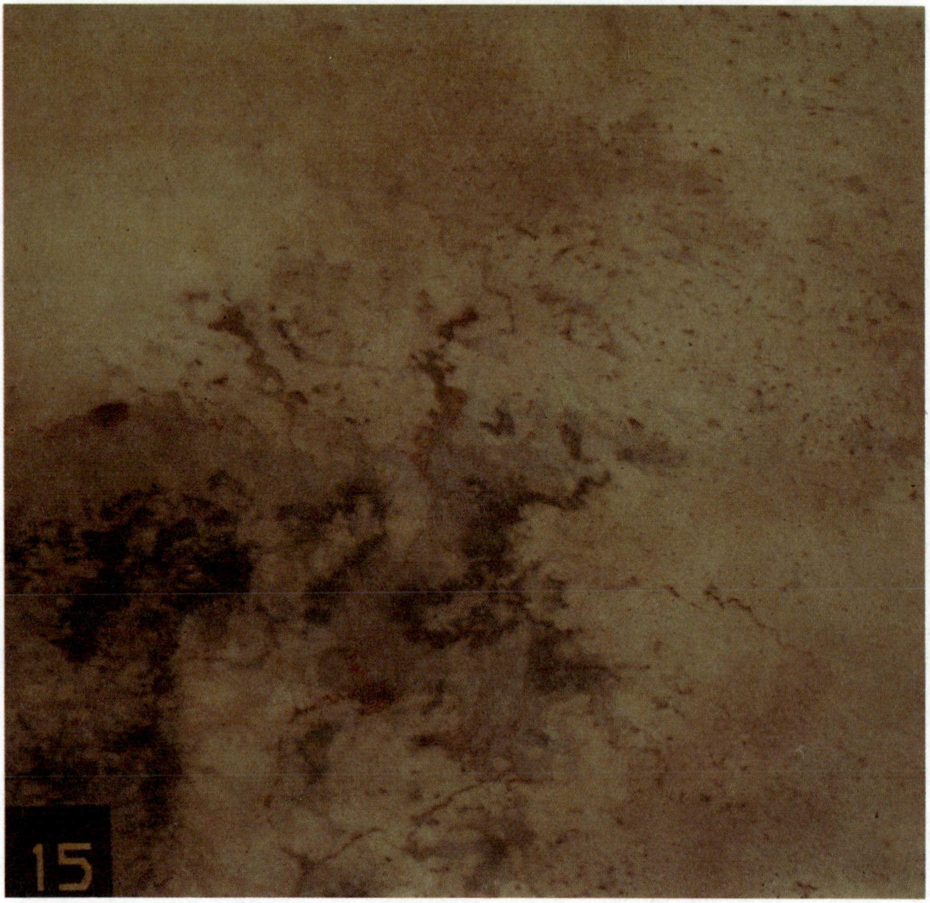

Figure 8. Landsat Multi Spectral Scanner (MSS) false colour composite (FCC), 50 km^2 image of the Sahel rangelands of Niger, October 1984. The faint pink blush over much of the image is from a sparse grass cover after the poor rains of '84. Denser vegetation occurs where soil moisture is greater; along the rivers and in the inter - dunal depressions (the latter are marked by the scatter of red dots across the image). The sands show through the poor vegetation cover as yellow tones. The almost black laterite crust has a high % iron oxide (Fe_2O_3) and correspondingly low reflectance. These soils largely follow escarpments of low lying hills. The clays, appearing as grey, are an "outwash" from the edges of these hills. The limited crop production in this area (appearing as the brightest reds) is mainly confined to growing Sorghum at the edges of the clay outwash fans.

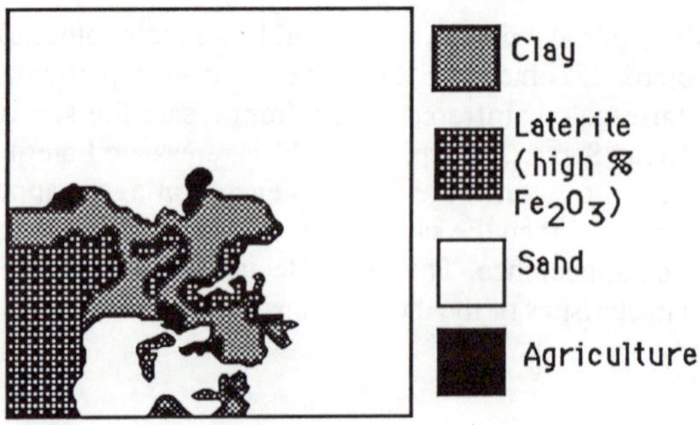

Figure 8. Soil distribution

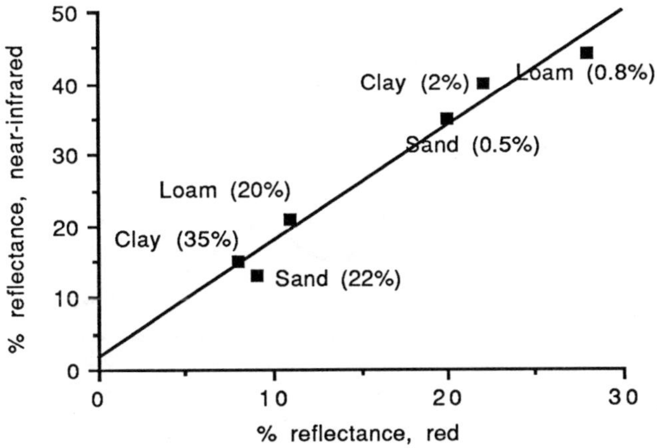

Figure 9. Red near-infrared reflectance for Loam, Sand and Clay soils at different moisture contents. The numbers in brackets are % moisture content

3.4 TEMPORAL AND SPATIAL CONSIDERATIONS

Soils develop over many years, Myers (1984) estimates that the formation of 2.5 cm of top soil can take anything from 100 to 2500 years depending on the soil type and climate. Thus within the framework of most natural resource surveys, factors such as organic matter content and iron oxide content may be considered as stable. However, there are considerable variations in soil moisture content, and, as Baret discusses in a later chapter, sun-target-sensor geometry has to be considered for soils as well as vegetation.

Because soils are a mixture of complex and varied materials, reflectance from an area of unknown soil could be attributed to a range of physical factors. Is an area of dark soil seen on a satellite image high in organic matter content, high in iron oxide content, or just wetter than surrounding areas ?

The use of imagery from different times can help to resolve such confusion; tonal characteristics which are constant with time are likely to be due to soil physical properties such as texture or organic matter content. In contrast, variations in tone for a given area are likely to indicate changes in soil moisture status. An alternative is the use of thermal measurements, as wet soils usually have a lower surface temperature than dry soils due to the cooling effect of evapotranspiration.

Even though the spectral curves of soil change with texture, moisture content, organic matter content and iron oxide content it has been shown that there is a linear relationship between reflectance at any two wavelengths for soil, and that by and large, this feature of soil reflectance is spatially and temporally constant. Thus, if a soil has relatively high reflectance in the visible wavelengths, then reflectance in the near-infrared will also be high, and vice versa, if visible reflectance is low, then near-infrared reflectance is low.

This can be seen in figure 9 which plots the reflectance of a number of soils in red near-infrared feature space. Note how the position varies not just as a function of soil type but also the moisture

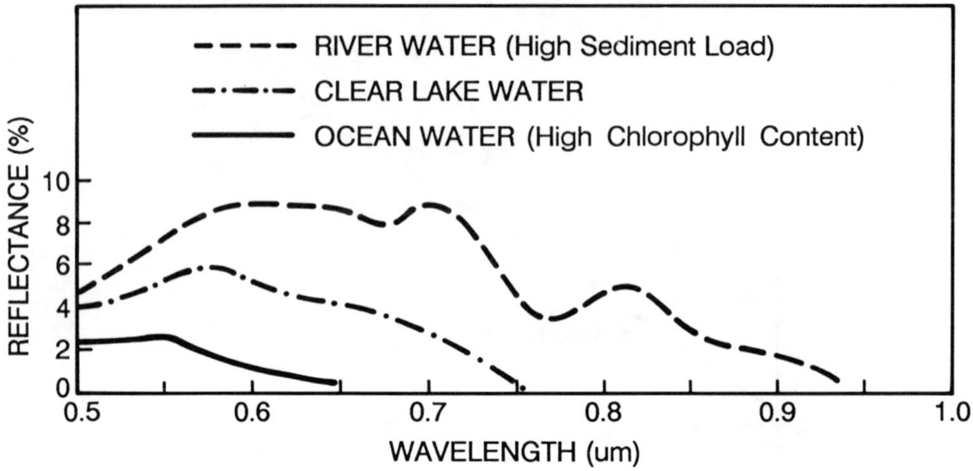

Figure 10. Spectral reflectance curves for water. River and lake water after Bartolucci *et al.* 1977, ocean water after Clarke *et al.* 1970.

content of any given soil. As moisture content increases, regardless of soil type, then the soils move down closer to the origin.

The nature of the soil background will, of course, influence the reflectance characteristic of any vegetated area at low percentage vegetation cover. Thus position (1) in figure 4 (§2.4) could be anywhere along the linear distribution for soils seen in figure 9. Some authors have even suggested that for crops growing on very bright soils reflectance from the developing canopy can fall in the early stages due to shadowing effects; there is considerable reduction in the reflectance from a shaded bright soil, compared with non-shaded, see Kauth and Thomas (1976), for example. The linear properties of soil reflectance have important implications concerning the use of remotely sensed data for the quantitative estimation of plant biophysical parameters such as % canopy cover and leaf area index, and are discussed by Baret later in the book.

4. Spectral properties of water

As with vegetation and soil the spectral characteristics of water change with wavelength. Changes in spectral response also occur with changes in physical, chemical and biological state; just as dry soil and wet soil, green vegetation and senescent vegetation have contrasting spectral properties, so too do clear rivers and turbid rivers, shallow lakes and deep oceans.

The spectral properties of pure water, as measured in the laboratory, have been known for many years. Clear water is known to have excellent transmission properties and weak absorption characteristics in the visible wavelengths, and weak transmission, strong absorption characteristics in the infrared. This can be seen from the spectral reflectance curve for clear lake water shown in figure 10. Reflectance peaks in the green wavelengths, and tails off to almost zero in the infrared.

Water bodies are three dimensional, therefore any measure of reflectance may include a component from the bottom, from the surface and from the water column itself. Water can behave as a specular reflector. Specular reflection, from the surface, is usually perceived as sun glint from air or space platforms, and has the effect of obscuring all sub-surface features. Variations in the bottom material and the depth and quality of the water column are associated with changes in the spectral properties of water. These changes are only apparent to any great extent in the visible wavelengths. In the infrared, relationships between reflectance and water quality parameters tend to be random because of the effect of the fundamental and overtone water absorption bands noted in § 2.3 (Lathrop and Lillesand 1986).

4.1 CLEAR WATER

The depth to which light will penetrate water varies with wavelength. Lepley *et al.* (1975) estimate the depth of penetration for clear water to be 10 m in the 0.5 - 0.6 μm waveband, falling to only 10 cm in the 0.8 - 1.1 μm range. Values approach zero at longer wavelengths.

As depth of water increases then reflectance for any wavelength decreases, because there is a greater path length through the water column over which absorption and scatter can occur. As a result it is possible to use remote sensing measurements to update the bathymetric information on navigation charts. Because of the greater depth of penetration of blue and green wavelength EMR it is data recorded through these parts of the spectrum which are used for such work. For example, Hammack (1977) working with satellite sensor data recorded through the 0.5 - 0.6 μm waveband reports the accurate measurement of depths to 22 m, and even the effective mapping of shoals as deep as 40 m in the clear waters off Berry Island in the Bahamas, in less clear water, off the Florida coast, shoals only as deep as 10 m could be detected.

From clear water the signal will not only depend on the depth of water but also on the nature of the bottom material; a light sandy bottom will give stronger reflectance than a bottom composed of decaying organic matter. The reflectance detected from clear water can therefore be difficult to separate from the reflectance from bare soil. However, water absorbs almost all EMR at infrared wavelengths, and, in contrast to bare soil, reflectance is very low. Because of this, coastal and lacustrine boundaries and even small rivers and water bodies can be reliably identified from infrared remotely sensed data, (Work and Gilmer 1976 for example).

4.2 TURBID WATER

Water bodies under natural conditions usually contain a range of organic and inorganic materials, some dissolved, others in suspension. Turbid waters can be divided into those containing phytoplankton and their derivative products, and those containing organic and inorganic sediments (Aranuvachapun and Robertson 1985).

4.2.1 Phytoplankton and their derivative products have a profound influence on the spectral properties of water, particularly in the blue and green wavelengths. The ability to identify and map plankton rich waters is important for commercial reasons because phytoplankton support large zooplankton and fish populations, and for environmental reasons, as evidenced by the concern with algal pollution in the Adriatic in summer 1989. Algal blooms are often associated with upwelling events. These occur where nutrient rich cold bottom water is brought to the surface. At the surface the water receives more sunlight, which leads to the development of the blooms. These sites are commercially important for the fishing industry, and can be mapped using satellite images, such work is almost impossible from *in situ* measurements as the blooms soon decay and are dispersed.

As with higher plants, chlorophyll, and other photosynthetic pigments, in phytoplankton absorb blue light for photosynthesis, and as with higher plants increase in chlorophyll concentration also leads to increased levels of green reflectance. These effects can be directly seen on the reflectance from water with high phytoplankton, and therefore high chlorophyll concentrations. Reflectance in blue wavelengths is reduced and green enhanced. Note how the reflectance in the red wavelengths drops sharply for the ocean water with high chlorophyll content shown in figure 10. Many empirical equations have been developed which use changes in reflectance at these wavelengths for the quantitative estimation of the pigment concentration of water (Verdin 1985, Lathrop and Lillesand 1986 for example).

4.2.2 organic and inorganic sediments in water tend to affect reflectance in the green and red wavelengths. Monitoring and mapping of sediment loads in water is also of commercial and environmental importance. For example, sediment transport studies provide information on rates of upstream soil erosion, on silting rates in irrigation storage ponds, on harbour silting patterns, on estuary sediment distributions, they provide information for maintenance of shipping lanes and data needed in planning causeway or bridge construction.

Water with high inorganic sediment loads usually has higher reflectance than clear water, and peak reflectance usually occurs at the longer red wavelengths, rather than green. This shift can be seen in the spectral reflectance curve for river water with a high sediment load shown in figure 10. Many investigators have found that reliable quantitative estimates of suspended sediment concentration can be made from measurements of reflected solar radiation in the visible wavelengths, particularly in the red wavelengths (Scarpace *et al.* 1979 for example). These are usually derived from linear multiple regression techniques between some measure of radiance and water quality measurements such as turbidity (Whitlock *et al.* 1982).

4.2.3 Other factors influencing water quality such as chemical variations are, in themselves, nearly impossible to detect with current optical remote sensing systems. For example, Scherez (1971) showed that no spectral reflectance differences in the optical wavelengths were apparent when gases such as oxygen and carbon dioxide, or inorganic salts such as sodium chloride were dissolved in distilled water. Detection of chemicals such as nitrogen or phosphorous may, however, be possible, if variations in their concentration correlate well with some measurable factor such as turbidity.

Salinity has been identified by Johnson and Munday (1983) as being a chemical variable in water which can be measured by remote sensing, though from passive microwave radiometry, not from EMR in the optical part of the spectrum. Microwave remote sensing, both active and passive, has also been successfully used for the detection of oil films on water (Gruner *et al.* 1986). Though, unlike salinity, oil films on water can also be detected using satellite sensor measurements of EMR from the visible and near-infrared wavelengths (Deutsch and Estes 1980).

4.3. ICE AND SNOW

The detection and mapping of ice and snow is of great importance to hydrologists interested in estimating runoff from snow melt. Monitoring iceberg production and movement is important to oil companies concerned with location and protection of expensive rigs, and to the protection of shipping in general.

The spectral properties of water in a frozen state are quite distinct from those of liquid water, with reflectance levels being generally very high. Indeed in the visible and near-infrared wavelengths the reflectance from snow can be so high that the detectors on some satellite sensors reach saturation (Bartolucci *et al.* 1975).

The high levels of reflectance can lead to spectral confusion with clouds which also reflect strongly across the wavelength range 0.4 - 3.0 µm. This spectral overlap also exists in the thermal-infrared as both cloud tops and snow are cold.

However, O'Brien and Munis (1975) showed that snow exhibits a unique transition in spectral response from around 100% at wavelengths less than 0.8µm to nearly 0% at wavelengths above 1.4 µm; at these wavelengths clouds still reflect strongly, thus middle-infrared wavelengths can be used to separate the two.

The condition of snowpacks varies in terms of wetness and density / grain size. In general older wetter snow is characterized by a slight decrease in reflectance at wavelengths less than 0.8µm and a more pronounced drop in the near-infrared (O'Brien and Munis 1975). The difference between visible and near-infrared reflectance also increases as the density of snow changes from fresh powdery snow through to more compact snowpacks.

Birnie (1986) suggests that the changes in reflectance may be due to discontinuities in cover as the snow ages and melts. Thus, when viewed from space platforms, a sensor would see a mixture of snow free and snow covered ground, rather than continuous snow cover.

4.4. TEMPORAL AND SPATIAL CONSIDERATIONS

The monitoring of many water quality variables, such as sediment load, chlorophyll content, bathymetry and snowcover and ice mapping is relevant on local scales; the references cited throughout the section are nearly all concerned with local situations. However, more than seventy percent of the planet surface is covered with water; oceans, snow, ice, lakes and rivers, which along with ground water, aquifers, clouds and atmospheric water vapour make up the the global hydrologic cycle. The water component of our planet effectively forms a continuous ecosystem, in a state of flux, which has a profound influence on the functioning of the earth system. The need for coherent monitoring of the "blue planet" ecosystems is now keenly felt (NASA undated). The spatial scale is global, which introduces problems of data acquisition and processing, the temporal scales to be considered present other difficulties. Snowfields melt, form rivers which flow into lakes and oceans, evaporation returns the water to the atmosphere where it forms clouds and falls as rain or snow; on average water can be resident for around 10 days in the atmosphere and a similar period in rivers, tens of years in lakes, centuries in deep aquifers or glaciers and anything from 10 to more than 1,000 years in the oceans. The monitoring of a global ecosystem on a temporal scale ranging from days to years represents an immense challenge.

The oceanography and hydrology communities have, over the years, amassed an impressive range of remote sensing capabilities to meet this challenge, a full discussion of which is beyond the scope of this book. Johnson and Munday (1983), and Maul (1985) provide good introductions to this subject.

5. Summary

Reflectance from a vegetation canopy varies both with wavelength and with the growth and development of the plants. Absorption by pigments controls plant reflectance in the visible wavelengths; leaf structure, controlling the scattering of light in the leaf, is the dominant factor in the near-infrared; and in the middle-infrared absorption by water in the leaf is most important.

Soil reflectance is less variable. All soils generally show an increase in reflectance with increasing wavelength, particularly at visible and near-infrared wavelengths. Increasing moisture content, iron oxide content and organic matter content reduce soil reflectance in all wavelengths considered in this chapter, with water absorption having a particularly pronounced effect in the middle-infrared. There is a positive correlation between soil reflectance in one wavelength and another, and this is, by and large, spatially and temporally constant.

Most spectral information concerning water characteristics is gleaned from the visible wavelengths, though the land / water boundary is most obvious in the infrared. The blue / green wavelengths are best for bathymetric studies in clear water, and for the mapping of phytoplankton rich waters. The green / red wavelengths are best for detecting inorganic sediments.

Snow and cloud both have very high visible and near-infrared reflectance, though they can be separated in the middle-infrared where reflectance from snow is minimal. The near-infrared wavelengths contain information relating to snow aging and condition.

6. Conclusions

This chapter has introduced the basic concepts needed for the interpretation and analysis of remotely sensed data recorded through visible, near and middle-infrared wavelengths. Subsequent chapters develop these ideas and show how such data can be applied to aspects of natural resource management such as agricultural production monitoring, estimation of tropical deforestation rates, and for hydrological studies.

Such use of spectral information is only possible if it is captured is some permanent form such as a spectral curve. The spectral curves, used above to describe the spectral characteristics of earth surface features, are from non imaging electronic sensors. Much of remote sensing relies on the use of multi spectral images. Currently the sensors used to produce such images only record data through a few wavelengths. However, if the waveband is known, then the spectral characteristics described above can be used in the interpretation process. The following chapters describe the sensors and platforms from which these images are derived, and introduce some of the data processing techniques in current use.

As Bird points out in the first chapter the technology for recording longer wavelength, lower energy EMR is complex, and so has a shorter history than that of the wavelengths discussed here. After all, the photographic process has been known since 1839. This situation is changing, thermal data are becoming far more widely used and sensors operating in the microwave regions of the electromagnetic spectrum will provide much of the data for remote sensing by the mid 1990's. The use of thermal data in rainfall estimation and in crop yield forecasting are presented in later chapters, whilst elsewhere, energy matter interactions in the microwave region are combined with a review of present and future Radar remote sensing systems.

References

Allen, W. A., Gausman, H. W., and Richardson, A. J., 1973, Willstatter - Stoll theory of leaf reflectance evaluated by ray tracing, *Applied Optics*, **12**, 2448 - 2453.

Aranuvachapun, S., and Robertson, Y. C., 1985, Satellite ocean colour scanners, their performance, application and potential for commercial exploitation, volume 2, part 1 case studies : Ocean colour imaging. Report published by Oxford Computer Services Ltd, Fordybrook, East Hendred U.K.

Bartolucci, L. A., Robinson, B. F., and Silva, L. F., 1977, Field measurements of the spectral response of natural waters, *Photogrammetric Engineering and Remote Sensing*, **43**, 595 - 598.

Bartolucci, L. A., Hoffer, R. M., and Luther, S. G., 1975, Snowcover mapping by machine processing of SKYLAB and Landsat MSS data, in *Operational Applications of Satellite Snowcover Observations* edited by A. Rango (NASA SP - 391), (Washington D.C.: Scientific and technical information office, NASA) pp 295 - 312.

Birnie, R. V., 1986, Pixel-mixing effects and their significance to identifying snow condition from Landsat MSS data, *International Journal of Remote Sensing*, **7**, 845 - 853.

Bowers, S. S., and Hanks, R. J., 1965, Reflection of radiant energy from soils, *Soil Science*, **100**, 130 - 138.

Christ, R. E., 1984, Cultural and environmental influences on temporal - spectral development patterns of corn and soybeans, *Remote Sensing of Environment*, **14**, 3 - 12.

Clarke, G. L., Ewing, G. C., Lorenzen, C. J., 1970, Spectra of back scattered light from the sea obtained from aircraftas a measure of chlorophyll concentrations, *Science*, **167**, 1119 - 1121.

Condit, H. R., 1970, The spectral reflectance of American soils, *Photogrammetric Engineering*, **36**, 955 - 966.

Daughtry, C. S. T., and Biehl, L. L. 1985, Changes in spectral properties of detached birch leaves, *Remote Sensing of Environment*, **17**, 281 - 289.

Deutsch, M., and Estes, J. E., 1980, Landsat detection of oil from natural seeps, *Photogrammetric Engineering and Remote Sensing*, **46**, 1313 - 1322.

Gates, D. M., 1970, Physical and physiological properties of plants, in *Remote Sensing with special reference to agriculture and forestry* (Washington D.C. : National Academy of Sciences) pp 224 - 252.

Gates, D. M., Keegan, H. J., Schleter, J. C., and Weidner, V. R., 1965, Spectral properties of plants, *Applied Optics*, **4**, 11 - 20.

Gausman, H. W., 1985, *Plant Leaf Optical Properties*, (Lubbock: Texas Tech Press).

Grant, L., 1987, Diffuse and specular characteristics of leaf reflectance, *Remote Sensing of Environment*, **22**, 309 - 322.

Gruner, K., Bartsch, N., Witte, F., Schreiber, H. and Keydel, W., 1986, Contributions to oil spill detection and analysis with RADAR and microwave radiometry, results of the Archimedes II campaign, in *proceedings of international geoscience and remote sensing symposium held in Zurich 8th 11th september 1986* (Paris: ESA publications division) pp 1375 - 1380.

Hammack, J. C., 1977, Landsat goes to sea, *Photogrammetric Engineering and Remote Sensing*, **43**, 683 - 691.

Hoffer, R. M., 1978, Biological and physical considerations in applying computer aided analysis techniques to remote sensor data, *in Remote sensing: the quantitative approach* edited by P. H. Swain and S. M. Davis (New York: McGraw Hill) pp 227-289.

Hoffer, R. M., and Johannsen, C. J., 1969, Ecological potentials in spectral signature analysis, in *Remote Sensing in Ecology* edited by P. L. Johnson (Athens, U.S.A.: University of Georgia Press).

Huete, A. R., Post, D. F., and Jackson, R. D., 1984, Soil Spectral Effects on 4-Space Vegetation Discrimination, *Remote Sensing of Environment*, **15**, 155-165.

Hunt, G. R., and Salisbury, J. W., 1970, Visible and near-infrared spectra of minerals and rocks, *Modern Geology*, **1**, 283 - 300.

Johnsen, R. W., and Munday, J. C., Jr., 1983, The marine environment, in *Manual of Remote Sensing*, editor in chief R. N. Colwell (Falls Church: American Society of Photogrammetry) pp 1371 - 1496.

Kauth, R. J. and Thomas, G.S., 1976, The tasselled cap - a graphic description of the spectral - temporal development of agricultural crops as seen by Landsat, *Proceedings of the 2nd International Symposium on Machine Processing of Remotely Sensed Data* (West Lafayette: Purdue University), pp 4B41 - 4B51.

Knipling, E. B., 1967, Physical and physiological basis for differences in reflectance of healthy and diseased plants. Clearinghouse for Federal Science and Technology information, U. S., Department of Commerce. Ad 652 679. 24pp

Knipling, E. B., 1970, Physical and physiological basis for the reflectance of visible and near-infrared radiation from vegetation, *Remote Sensing of Environment,* **1,** 155-159.

Lathrop, R. G., and Lillesand, T. M., 1986, Use of Thematic Mapper data to assess water quality in Green Bay and Central Lake Michigan, *Photogrammetric Engineering and Remote Sensing*, **52**, 671 - 680.

Lepley, L. K., Foster, K. E., and Everett, L. G., 1975, Water quality monitoring of reservoirs on the Colorado river utilizing ERTS - 1 imagery, in *Remote Sensing and Water Resources Management* edited by K. P. B. Thompson, R. L. Lane, and S. .C. Csallany (Urbana: American Water Resources Association) pp 105 - 110.

Marshall, T. J., and Holmes, J. W., 1979, *Soil Physics* (Cambridge: Cambridge University Press).

Maul, G. A., 1985, *Introduction to Satellite Oceanography* (Dordrecht: Martinus Nijhoff).

Myers, V. I., 1970, Soil, water and plant relations, in *Remote Sensing with special reference to agriculture and forestry* (Washington D.C. : National Academy of Sciences) pp 253 - 297.

Myers, N., 1984, *GAIA, an atlas of planet management* (New York: Doubleday)

NASA, undated, From pattern to process: the strategy of the Earth Observing System. Eos science steering committee report, volume II, National Aeronautics and Space Administration, Washington D.C.

O'Brien, H. W., and Munis, R. H., 1975, Red and near-infrared spectral reflectance of snow, in *Operational Applications of Satellite Snowcover Observations* edited by A. Rango (NASA SP - 391), (Washington D.C.: Scientific and technical information office, NASA) pp 345 - 360.

Page, W. R., 1974, Estimation of organic matter in atlantic coastal plain soils with a colour difference meter, *Agronomy Journal,* **66**, 652 - 653.

Scarpace, F. L., Holmquist, K. W., and Fisher, L. T., 1979 Landsat analysis of lake quality, *Photogrammetric Engineering and Remote Sensing*, **45**, 623 - 633.

Scherez, J. P., 1971, Remote sensing considerations for water quality monitoring, in *proceedings of the seventh international symposium on remote sensing of environment* (AnnArbor: Environmental Research Institute of Michigan) pp 1071 - 1087.

Sinclair, T. R., Hoffer, R. M., and Schreiber, M. M., 1971, Reflectance and internal structure of leaves from several crops during a growing season, *Agronomy Journal*, **63**, 864-868.

Verdin, J. P., 1985, Monitoring water quality conditions in a large western reservoir with Landsat imagery, *Photogrammetric Engineering and Remote Sensing*, **51**, 343 - 353.

Whitlock, C. H., Kuo, C. Y., and LeCroy, S.R., 1982, Criteria for the use of regression analysis for remote sensing of sediment and pollutants, *Remote Sensing of Environment,* **12**, 151 - 168.

Willstätter, R., and Stoll, A., 1918, in *Untersuchungen Über die Assimilation der Kohlensäure.* (Berlin: Springer-Verlag) pp 122 - 127.

Work, E. A. Jr, and Gilmer, D. S. 1976, Utilization of satellite data for inventorying prairie ponds and lakes, *Photogrammetric Engineering and Remote Sensing*, **42**, 685 - 694.

Woolley, J. T., 1971, Reflecance and transmittance of light by leaves, *Plant Physiology*, **47**, 656-662.

REMOTE SENSING SYSTEMS: SENSORS AND PLATFORMS

J. HILL
Commission of the European Communities,
Joint Research Centre,
Institute for Remote Sensing Applications,
21020 Ispra (Varese) Italy.

1. Introduction

The previous chapters have introduced the concept of studying earth surface phenomena through the analysis of their spectral behaviour. This requires the capture and recording of spectral data in a form suitable for analysis. Permanent, or near-permanent records of spectral information can be made using both imaging and non-imaging instruments. Both measure, either continuously or within discrete wavelength bands of a specific width, the reflected or emitted electromagnetic radiation across certain wavelength regions. Non-imaging spectro-radiometers permit the acquisition of detailed target measurements in the field or the laboratory, and thereby provide us with the basic knowledge about the spectral properties of targets. Imaging instruments usually operate from airborne and space-borne platforms and, as the name implies, produce images of the earth which can then be used in inventorying, mapping and monitoring the earth's resources.

The imaging systems can be divided into three general categories: framing cameras, scanning systems, and pushbroom imagers (Elachi 1987). In the past framing camera systems have been used on a number of space platforms and indeed have formed the basis of remote sensing in Eastern Europe for the last 15 years (Legg 1990). Scanning systems, used in a number of currently operational sensors, use a scanning mirror which projects the image of the surface resolution onto electronic detectors. The side to side scanning of the mirror provides an image swath across the ground track of the satellite, and the platform motion provides the along track imaging. In some cases a number of detectors are used to cover a set of across-track lines with each scan, instead of only a single one. The scanning process is limited by the accuracy of the rotating mirror and the relatively short integration time for each pixel which may affect the within-scan geometry and cause signal-to-noise problems.

Pushbroom imagers, also in current use, need no scanning mechanism, but use a linear array of detectors to cover all pixels in the across-track dimension. Compared to the scanning devices the pushbroom systems have no moving parts, and they record each scan line at once by means of a whole line array of detectors for each spectral bandpass: each pixel in a scan corresponds to an individual charge-coupled device (CCD) detector. The major advantage of these instruments is that no rotating mirrors are involved in the data collection process, which potentially increases the geometric quality of the imagery and also reduces the costs of the instrument. However, pushbroom scanners require the precise calibration of many more detectors in order to minimize along-track striping effects, and suitable detectors for the short-wave infrared region tend to be very expensive and are not yet readily available in large numbers.

A. S. Belward and C. R. Valenzuela (eds.),
Remote Sensing and Geographical Information Systems for Resource Management in Developing Countries, 55–70.

2. The Ideal Remote Sensing System

The ideal remote sensing satellite system provides global coverage and daily repeat cycles. Many high quality sensors on-board the same platform produce stereoscopic and multi-spectral images at a high spatial resolution in many narrow spectral bands, from visible to microwave wavelengths (Curran 1985). The system calibration is known, reliably monitored during the operation period, and no atmospheric effects modify the radiative transfer of energy either on its way to the earth's surface or coming from it. And last, but not least, the transmitted signal can be related to unique matter/energy interactions at the earth's surface. This system is of course reliable and economical to operate, the acquired data are processed through a real-time data handling system, and are received by a user community with skills both in their respective disciplines and in remote sensing data acquisition and analysis techniques (Lillesand & Kiefer 1979).

For many technical reasons this ideal system will never be launched. Real remote sensing satellite systems have been designed according to their prime applications in either Earth's resources analysis, meteorology or planetary research, and their design and construction usually involved many compromises between technically feasible solutions and costs. It must be recognized that this imposes constraints on the use of data from specific systems, for example when data from meteorological satellites are used in earth's resources applications.

3. Earth Resources Satellites

The first photographs of the earth's surface from space were taken from the United States' Mercury, Gemini and Apollo satellites. These were acquired initially for reasons of curiosity and publicity and only later for the study of the earth's resources. Following the success of these manned space missions, the U.S. National Atmospheric and Space Administration (NASA) developed an experimental earth resources satellite series to evaluate the utility of images collected from an unmanned satellite. The first satellite in this series, a modified Nimbus weather satellite, was launched in 1972 and carried two types of sensors: a four waveband multispectral scanning system (MSS) and three Return Beam Vidicon (RBV) television cameras. It was first called the Earth Resources Technology Satellite-1 (ERTS-1) and, later in 1975, renamed Landsat-1. This satellite operated for almost six years until January 1978; during this period two almost identical successors, Landsat-2 and Landsat-3, were launched in 1975 and 1978 respectively. The early Landsat MSS programme gave us more than 32 *billion* km^2 of images of the earth's surface; unique in time, though many images covering the same locations on earth. Each MSS image covers an area of 185 by 185 km at a nominal ground resolution of 79 m, and they provide data in four spectral bands (initially identified as 4, 5, 6, and 7, the three RBVs were bands 1, 2 and 3) covering the green, red and twice the near-infrared parts of the reflective optical spectrum. Landsat-3 also had a band operating in the thermal-infrared (known as band 8). A slightly modified MSS system is still in operation today thus giving us consistent data sets extending back over some 18 years.

The 1970s established hopes that modern space images could be used efficiently for mapping the earth's surface at small and medium scales. In particular, the Landsat system has provided spectral data in digital and photographic formats for use in many applications, but a real breakthrough in thematic mapping was not achieved, essentially because of poor geometric performance and inadequate spatial and spectral resolution for medium scale work in Europe. Other highly promising experimental missions involving space-borne thermal sensors, the Heat Capacity Mapping Mission (HCMM) and microwave sensors, for example SEASAT, suffered from poor continuity in data delivery (see Churchill and Sieber, later in this book).

Based on these years of basic and applied research, improved electro-optical sensors have been developed and placed in orbit (Hempenius *et al.* 1983). In July 1982, Landsat-4 was launched as the first of the so-called "Second Generation Earth Observation Satellites". This carried the slightly modified MSS[1], and an additional new instrument, the Thematic Mapper (TM). The TM also provides 185 by 185 km images, though of improved spatial and spectral resolution. Due to technical failures regular TM transmission from Landsat-4 ended in February 1983 (two of the spacecraft's four solar panels failed, severely limiting the available power, and failure of both X-band transmitters precluded the direct transmission of TM data). MSS transmission was however maintained. An identical satellite, Landsat-5, was launched on 1st March 1984. This satellite is still operating in 1990 (well beyond the end of its three-year design life) so that today continuous TM data for a six-year-period are already available.

In 1986 the french SPOT-1 (Système Probatoire de l'Observation de la Terre) was successfully launched from ESA's Kourou establishment, and an identical SPOT-2 satellite followed in 1990. The satellites carry two identical pushbroom scanners called the 'High Resolution Visible' instruments (HRV-1 & HRV-2), these can be used to record either panchromatic or multi-spectral data.

4. Meteorological Satellites

Meteorological satellites have been in use longer than Earth resources satellites. The prime use of the data collected by these sensors has always been for the study and prediction of short-term changes in weather, usually on a time scale of a few hours or days. Environmental scientists, however, rarely need these images for meteorological applications. They usually wish to use them for studies of climate and terrain, and have concentrated on the use of data from polar orbiting satellites.

The first generation of these satellites operated from 1960 to July 1966, and were followed by a series of second and third generation instruments which are currently in use. The third generation of this series include the satellites called TIROS-N and NOAA-6 to NOAA-13. They have similar orbital characteristics to Landsat, that is a near-polar, sun-synchronous orbit at an altitude of 830 km. Among other instruments they carry the Advanced Very High Resolution Radiometer (AVHRR),the data from which have been used for weather forecasting, and also for a range of environmental applications such as the mapping of sea surface temperatures and locating forest fires. The data became especially well known to the non-meteorological community when environmental scientists started to exploit the red and near-infrared wavebands for vegetation monitoring (Townshend *et al.* 1984, Tucker *et al.* 1984, Justice *et al.* 1985, Malingreau 1986).

---

[1] Spectral bands remain the same though resolution is slightly different at 82 m (c.f. 79 m). Note also bands are now identified as 1, 2, 3 and 4. The RBV was dropped from the second generation Landsats, hence the change in band identification.

5. Characterization of Space-borne Remote Sensing Systems

Space-borne remote sensing systems can be characterized on the basis of orbital characteristics and spectral and spatial resolution capabilities.

5.1 ORBITAL CHARACTERISTICS

Orbit selection and sensor characteristics are closely related to the strategy required to achieve observation patterns such as continuous monitoring, repetitive coverage of different periodicities, global mapping or selective imaging (Elachi 1987). The vast majority of earth-orbiting remote sensing platforms have been placed into circular orbits, the most important of which are the geostationary and polar orbits.

5.1.1 *Geostationary orbits.* An important special case of geosynchronous orbits is the geostationary orbit. This is achieved when a satellite is placed into a circular orbit with an altitude of 35,800 km, located above the equator. In this case the satellite velocity compensates for the rotational speed of the earth and appears to be fixed in the sky. In other words the satellites have a fixed view of the earth. Of course, because they occupy a 'fixed' position, no one geostationary satellite can view the entire earth's surface. As an absolute minimum, two satellites perfectly positioned on opposite sides of the planet would provide a view of the whole earth. However, the resolution becomes very coarse towards the edges of each view of the earth disk, so for acceptable images we really need more than two satellites. It is important to remember that no matter how many satellites we place in geostationary orbit we will never get acceptable resolution towards, and over, the polar regions.

Geostationary orbits are much demanded for telecommunication satellites, but also used for meteorological satellites that are intended to provide coverage of the earth between 60° N to 60° S. At present four of these satellites are operational, the Geostationary Operational Environmental Satellites (GEOS) west and east, Meteosat and Himawari (Curran 1985).

5.1.2 Sun-synchronous polar orbits. Earth observation satellites usually operate in repetitive, circular, near-polar orbits which, with the exception of polar areas above about 81° provide almost global coverage. If the orbit precession exactly compensates for the earth's rotation around the sun, the orbit is sun-synchronous thus providing constant illumination geometry during the whole year. The orbit plane rotates at the same rate as the mean rate of the earth about the sun, and the angle between the orbit plane and the line that connects the earth's centre with the sun's centre remains constant. This is shown in figure 1. The Landsat, SPOT and NOAA satellites all occupy near-polar, sun-synchronous orbits.

5.2 RADIOMETRIC CHARACTERISTICS

Sensors can be characterized on the basis of four factors. Firstly the number of spectral bands. Secondly the position of the spectral band within the electromagnetic spectrum, bands are usually selected within the atmospheric windows (see Bird, chapter 2). Thirdly the width of each band, i.e. the wavelength range covered by each band. Fourthly the quantization levels of each band, this is related to the range of digital values potentially available within a band (see Wilkinson, chapter 5).

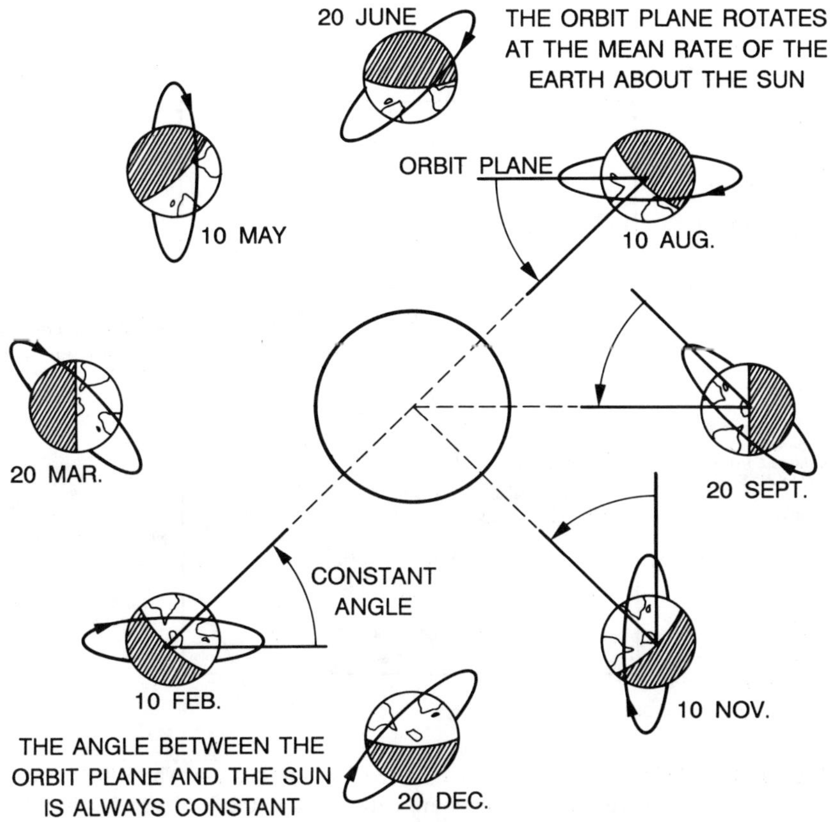

Figure 1. Characteristics of a sun-synchronous polar orbit

5.3 SPATIAL CHARACTERISTICS

The spatial characteristics of a sensor are manifest in the area covered by a single scene and the finest detail that can be seen within the scene. The area covered by a scene depends upon the scanner geometry and the orbital characteristics, thus the geostationary satellite sees half of the earth as a single scene and the polar orbiters image a swath. The swath width is variable and dictates the maximum across track dimensions of any image. The along track dimensions are not fixed, being determined by the forward motion of the satellite, though conventionally the imaged swath is chopped up into square images such that the along track and across track dimensions are equal. Thus one can imagine the earth's surface to be covered by a grid made up of the orbit ground tracks and the arbitrary along track image divisions. This leads to the "framed world concept" and is the basis on which most satellite images are located. Figure 2 shows an example for the early Landsat satellites, and gives details concerning data distribution for the main satellite systems introduced in the previous sections.

Figure 2. Satellite image identification and data ordering.

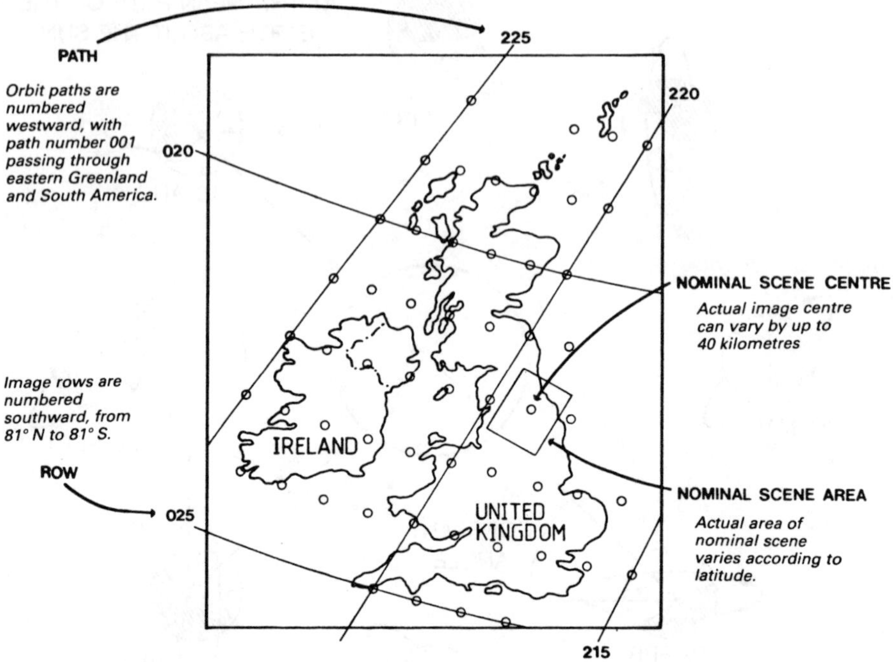

Figure 2A The Worldwide Reference System (WRS) for Landsats 1-3. The WRS is made up of a global grid in which the vertical elements are called paths, the horizontal, rows. The paths correspond to the ground track of the satellite orbit. The rows correspond to the nominal scene centres. Scenes are identified on the basis of two three digit numbers. Thus in the example shown P219/R022 identifies the nominal scene centre for an image covering part of north east England.

The different orbit pattern followed by Landsats 4-5 means that these satellites have their own WRS. It is important therefore to specify which WRS any given path / row coordinate pair refers to. For Landsats 4-5 both the TM and MSS scenes can be identified from the one WRS. Note that TM images may be ordered as quarter scenes. Recent changes have lead to more flexibility in scene ordering. Although the paths are fixed by the orbits, the rows are no longer fixed; scene centre can be located anywhere along a path, not just at the path row intercepts (n.b. this is only valid for new orders, archived data are still only available according to the fixed WRS coordinates.

SPOT too has its own location system though this is complicated by the fact that there are two sensors in operation on one satellite, and the fact that the sensors can be pointed. Figure 2B shows an example of the grid. The SPOT reference system attributes one row (J) and two path (K) coordinates in each case, accounting for both HRV-1 and HRV-2 frame locations in twin vertical viewing configuration. Off-nadir view scenes always correspond to a given row coordinate, but may not correspond to any particular path.

Figure 2B.

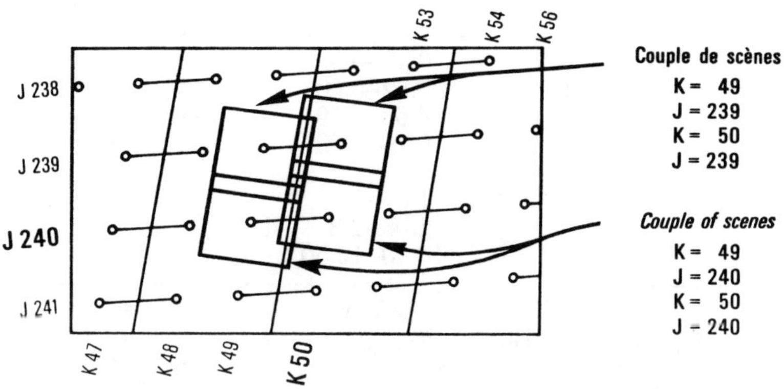

There is no standard for ordering AVHRR images, though these are usually specified on the basis of so many minutes of data along an orbit. More details are available from the contact addresses shown in figure 2C.

Figure 2C.

Landsat	SPOT	NOAA
from European Receiving Stations (Fucino, Kiruna)	worldwide	from Receiving Stations in Tromso, Maspalomas, Fucino
EURIMAGE Via Galileo Galilei, 5 I–00044 Frascati (Roma) Italy Phone: +6 941801 Fax: +6 9426285	SPOTIMAGE Ave. Eduard Belin F–31055 Toulouse Cedex France Phone: +61 539976 Fax: +61 281354	ESRIN Via Galileo Galilei, 5 I–00044 Frascat (Roma) Italy Phone: +6 941801 Fax: +6 94180361
from Non–European Receiving Stations EOSAT 4300 Forbes Boulevard Lanham, MD 20706 U.S.A. Phone: +301 5520537 Fax: +301 5520507		from Non–European Receiving Stations NOAA/NESDIS/NCDC Satellite Data Services Division Rm. 100 Princeton Executive Center Washington, D.C. 20233 Phone: +301 7638400

Table 1. Comparison of the Landsat TM and MSS sensors

	Landsat–TM	Landsat/MSS
Spectral bands (μm)	0.45–0.52	
	0.52–0.60	0.50–0.60
	0.63–0.69	0.60–0.70
	0.76–0.90	0.70–0.80
		0.80–1.10
	1.57–1.78	
	2.08–2.35	
	10.40–12.50	
Swath width	185 km	185 km
Ground resolution	30 (120 m)	80 m
Data rate	85 Mbits/sec	15 Mbits/sec
Quantization	8 bits	6 bits

A digital satellite image is made up of individual picture elements, called pixels. The dimensions of the pixel to a large extent determine the spatial resolution of the sensor. Fine resolution sensors can image small objects on the ground, coarse resolution sensors can only image larger features. Whilst the resolution of the sensor is determined by system factors, mainly the IFOV of the individual detectors, the ability to resolve is dependent upon the interaction between system factors, viewing conditions and terrain characteristics (Forshaw *et. al.* 1983).

6. Operational Polar-Orbiting Satellite Sensor Systems

6.1 THE LANDSAT THEMATIC MAPPER

The Landsat TM and MSS are multi-spectral imaging systems of the scanning type. Both systems use similar configurations except for the larger number of bands and the higher spatial resolution on the TM (table 1), both operate simultaneously from the Landsat-4 and -5 spacecraft (see §3).

The Thematic Mapper instrument collects radiometric data (256 radiance levels) in 7 spectral bands, and it offers a ground resolution of 30 by 30 m in the six reflective bands, and of 120 by 120 m in the emissive thermal infrared channel (Blanchard & Weinstein 1980).

A scan mirror moves the view of the sensor's telescope back and forth across the ground track during which the TM continuously records data. The ground track for the first 6 bands consists of 16 lines, which correspond to an array of 16 along-track detectors per spectral channel. For comparison, the Landsat MSS has only 6 detectors per band. There is one such array of detectors per channel, each with an optical filter to define the corresponding spectral band for the visible, near and short-wave infrared. The seventh channel, recording in the thermal-infrared, has a four-element array. Thus, for each scan, 16 across-track lines (and 4 lines for the thermal-infrared) are mapped simultaneously. The sensor has a 40-cm aperture, the focal length is 2.4 m, and each detector is 0.1 mm across, this gives an instantaneous field of view (IFOV) of 42.5 µrad, which corresponds to a 30 m ground resolution element from 705 km altitude (Elachi 1987). These system characteristics are summarized in figure 3.

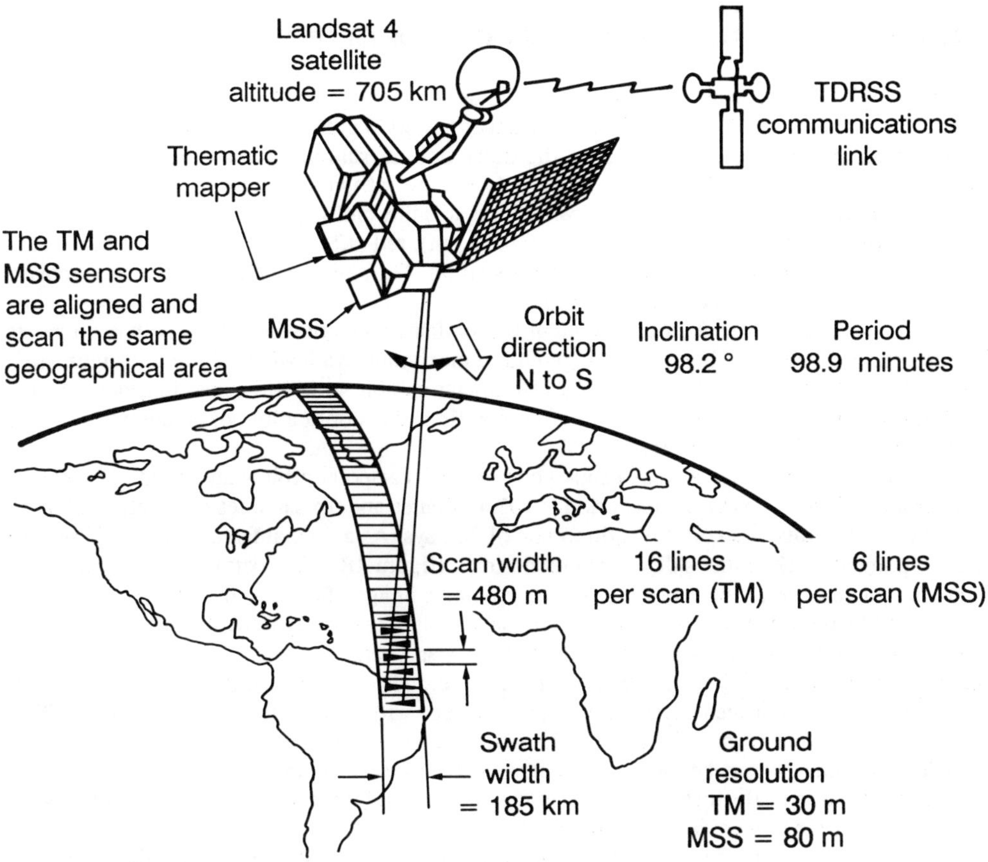

Figure 3. Landsat 4 and 5 system characteristics

The satellite operates in a repetitive, circular, near-polar orbit (inclination 98.22) at a nominal altitude of 705.3 km. It is sun-synchronous at approximately 9:30 local time in central Europe. The orbital period is 98.8 minutes, the satellite completes 14.6 orbits per day and views the entire earth within 16 days. This means that all ground locations are viewed at least once every 16 days. In fact although the TM instrument images a swath of approximately 185 km, at the equator adjacent orbits are only 172 km apart. Thus there is an overlap (called side-lap) of 7 %. Towards higher latitudes, the side-lap increases to a maximum of 84 % (table 2) which, by reducing the repetition period for the overlapping areas to 8 days, in many cases compensates for the more frequent cloud coverage in these regions (Kontoes & Stakenborg 1990). Note that adjacent orbits to the west are imaged with a delay of seven days, those to the east 9 days later.

Table 2. Sidelap of adjacent Landsat 4 & 5 ground swaths

Latitude	0°	10°	20°	30°	40°	50°	60°	70°	80°
Sidelap %	7.3	8.7	12.9	19.7	29.0	40.4	53.6	68.3	83.9

6.2 SYSTEME PROBATOIRE DE L' OBSERVATION DE LA TERRE

The two SPOT HRV pushbroom scanners can be used to acquire imagery either in panchromatic or multi-spectral mode (Begni 1982). Each scanner has 6000 individual detectors (c.f. 16 for the TM). In the panchromatic mode all of the detectors are sampled which provides a spatial resolution of 10 m. In multi-spectral mode only one out of two detectors is sampled giving a spatial resolution of 20 m. The HRV instruments acquire multi-spectral data in three spectral bands, covering the visible green (XS1, 0.50-0.59 μm), red (XS2, 0.61-0.68 μm) and the near-infrared (XS3, 0.79-0.89 μm) portions of the spectrum. The single panchromatic channel covers the spectral range between 0.51 and 0.73 μm.

Each instrument covers a ground swath of approximately 60 km which, once both instruments are operating in nadir-looking mode, can provide a maximum swath width of 117 km (figure 4A). Due to the fact that the HRV-1 and HRV-2 instruments exhibit some differences in bandwidth and calibration, the use of such data set requires first the application of some radiometric normalization methods (Schott *et al.* 1988).

The SPOT satellites are also in a sun-synchronous, near-polar orbit at an altitude of 832 km. The orbital inclination is 98.7°, its period is 101.4 minutes and the satellites complete 14.2 orbits per day. This results in an orbit repetition rate of 26 days. As a special feature, the HRV sensors can be pointed to 27° either side of the ground track (figure 4B). This permits the acquisition of stereoscopic imagery; if the same area on the ground is imaged from two different positions, i.e. orbits, and then the resulting two images are viewed simultaneously through a stereoscope (see Wolf 1983) a three dimensional view of the ground is apparent. The sensor can also be pointed to cloud-free areas, providing one knows where the clouds are, and the pointability offers the capacity to acquire repeated coverage of a specified area with greater frequency than the satellites 26-day repeat cycle (figure 4C and D). It is theoretically possible to view the same ground location nine times within the same repetition cycle, but there exist practical restrictions: due to the commercialisation of the instruments a user/customer would need to have very high priority for successfully placing orders for maximum multi-temporal coverages.

The panchromatic mode, together with the possibility of acquiring stereoscopic data, is mainly used for cartographic updating, at scales up to 1:50,000, while the multi-spectral mode is generally prefered for environmental studies.

6.3 THE TIROS-N/NOAA AVHRR

The U.S. National Oceanic and Atmospheric Administration's (NOAA) TIROS-N/NOAA satellites have a near-polar and sun-synchronous orbit at an altitude of 830 km. The orbital inclination is 99.09°. The nominal duration is about 102 minutes which produces 14.1 orbits per day. The satellite has a 9 day repeat cycle. Since the number of orbits per day is not integer, the nadir tracks do not repeat on a daily basis; however, the local solar time of each satellite's passage is approximately constant for any latitude (Malila & Anderson 1986).

These satellites carry the AVHRR with an IFOV of about 1.4 milliradians which, at nadir, translates to a spatial resolution of 1.1 x 1.1 km. With a scan angle of 55.4° to either side of nadir it covers a swath width of around 3,000 km, where ground resolution degrades to 2.4 x 6.9 km at maximum scan angles (see the later chapter by Belward). Because of the large swath width the effective repeat time is one daylight pass per day, but there are always two of these satellites in orbit providing the collection of data twice daily (table 3).

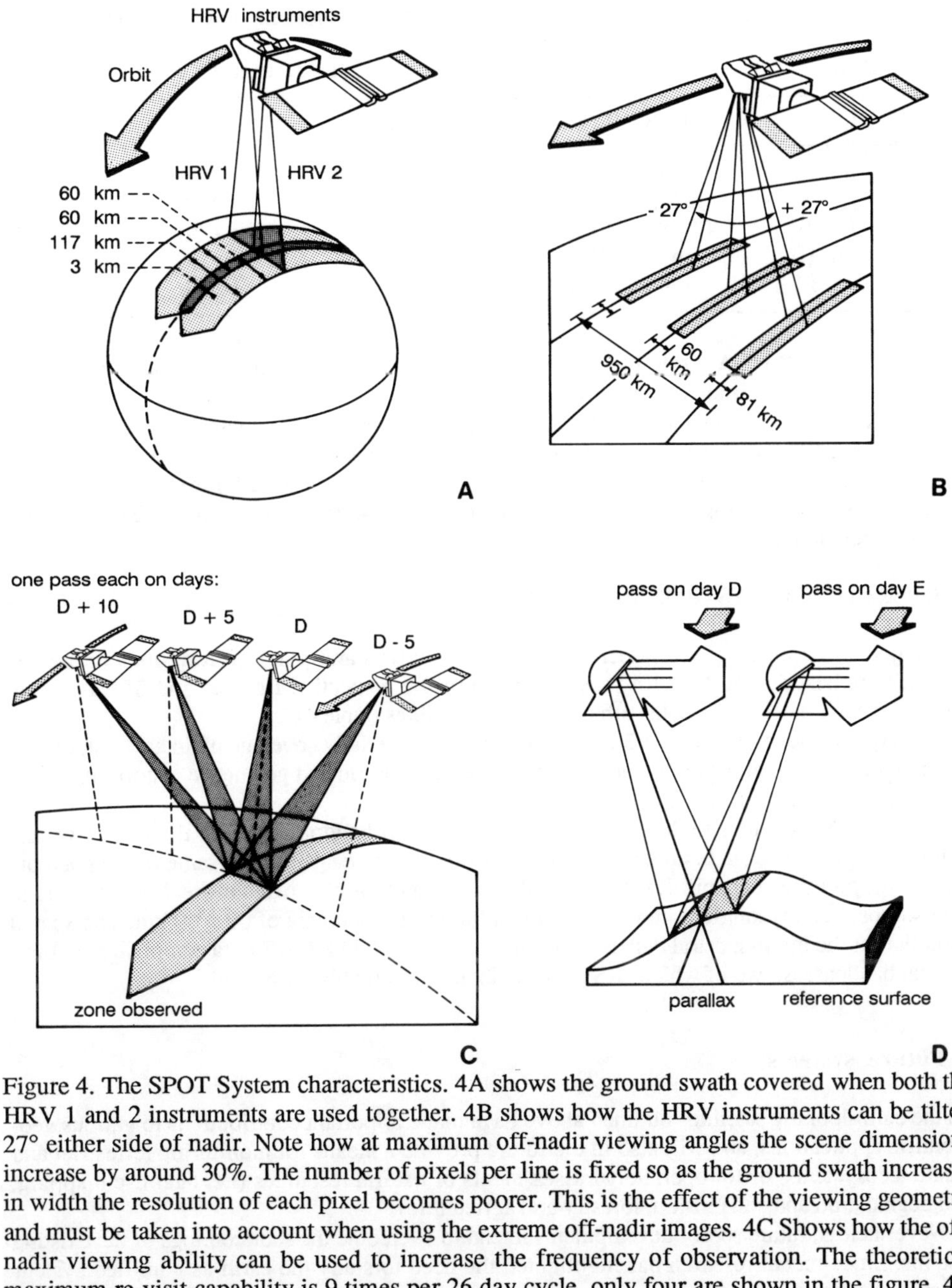

Figure 4. The SPOT System characteristics. 4A shows the ground swath covered when both the HRV 1 and 2 instruments are used together. 4B shows how the HRV instruments can be tilted 27° either side of nadir. Note how at maximum off-nadir viewing angles the scene dimensions increase by around 30%. The number of pixels per line is fixed so as the ground swath increases in width the resolution of each pixel becomes poorer. This is the effect of the viewing geometry and must be taken into account when using the extreme off-nadir images. 4C Shows how the off-nadir viewing ability can be used to increase the frequency of observation. The theoretical maximum re-visit capability is 9 times per 26 day cycle, only four are shown in the figure. 4D shows how these different views of the same area from different orbits create parallax which gives a stereo image, thus giving topographic information.

Table 3. Approximate times of NOAA satellites equator crossing (local solar time)

Satellite	Oper. Period	Ascending Node	Descending Node
TIROS N	10/78–01/80	15.00	03.00
NOAA 6	06/79–06/86	19.30	07.30
NOAA 7	08/81–01/65	14.30	02.30
NOAA 8	05/83–05/86	19.30	07.30
NOAA 9	02/85–11/88	14.30	02.30
NOAA 10	since 11/86	19.30	07.30
NOAA 11	since 11/88	14.30	02.30

The odd-numbered NOAA satellites provide afternoon coverage (ascending node), and even-numbered satellites morning coverage (descending node).

The AVHRR instrument has been produced in several versions, all of which sense radiation in the visible red (0.58-0.68 μm), near-infrared (0.72-1.1 μm) and short-wave infrared (3.55-3.93 μm). From NOAA-7 on, two thermal-infrared (10.3-11.3 μm, 11.5-12.5 μm) bands provide a direct means of correcting for radiant contributions from water vapour when measuring sea-surface temperatures.

6.4 OPERATIONAL SYSTEMS SUMMARIZED

The major differences between operational earth resources and polar-orbiting meteorological satellites are best demonstrated by a comparison of the respective Landsat and SPOT system parameters with those of the polar-orbiting NOAA satellites (table 4).

It becomes evident that the AVHRR instruments provide daily coverage of an area, but only at the price of low spatial resolution and considerable radiometric and geometric distortions which are due to the large off-nadir scanning geometry.

Landsat-TM provides high spatial resolution in seven bandpasses from the reflective and emmissive optical region, but it needs a 16-day-period to arrive again at the same frame position. Due to its pointing capabilities, the SPOT systems potentially offer more frequent image acquisitions, but any single frame covers only a limited ground area of 60 x 60 km. The spatial resolution of the multi-spectral data is superior to that of TM, but SPOT data comprise only three spectral bandpasses, two of which (XS1 and XS2) are even highly correlated.

7. Future systems

All the earth-looking satellites outlined above have made important contributions to various geo-scientific applications, and will also in the future provide a means for monitoring terrestrial and aquatic ecosystems, or for operational assessments of natural resources (i.e. land use mapping, land degradation and soil erosion, forestry and agriculture).

Nevertheless, future sensor developments intend to overcome the shortcomings of the existing systems. Since recent research has highlighted the need to study the earth as a unified system future instrument platforms will carry extensive sets of instruments rather than one unique sensor system. Both NASA and the European Space Agency (ESA) have committed themselves to

Table 4. Comparison of the Landsat TM, SPOT HRV and NOAA AVHRR sensors

Technical Specifications	Landsat–TM	SPOT–HRV	NOAA–AVHRR
Spectral bands (μm)	0.45–0.52		
	0.53–0.61	0.50–0.59	
	0.63–0.69	0.61–0.68	0.58–0.68
	0.78–0.90	0.79–0.89	0.72–1.10
	1.57–1.78		
	2.08–2.35		3.55–3.93
	10.40–12.50		10.30–11.30
			11.50–12.50[1]
Orbit type	sun–synchronous	sun–synchronous	sun–synchronous
Equator crossing	$9.^{40}$ (D)	$9.^{30}$ (D)	$19.^{30}$ (A)/$7.^{30}$ (D)[2] $14.^{30}$ (A)/$2.^{30}$ (D)[1]
Inclination	98.25°	98.7°	98.8°
Period	98.9 min	101.4 min	101–102 min
Nominal altitude	705 km	832 km	853.7 km
Total FOV	15°	4.1°	110.8°
Swath Width	185 km	60 km	3000 km
Ground resolution	30 x 30 m	20 x 20 m	1.1 x 1.1 km [3] 2.5 x 6.9 km [4]
Radiometric Resolution	8–bit (0–255)	8–bit (0–255)	10–bit (0–1023)
Revisit capability	16 (8) days	26 (3) days	1 day

1) odd–numbered 2) even–numbered NOAA satellites 3) nadir 4) off–nadir

launch, within this decade, so-called "Earth Observation Polar Platforms" (EOPP). On these platforms we will find multispectral imaging devices which, in some way, may be considered successors of second generation earth observation satellites (High Resolution Imaging Spectrometers). A number of airborne systems are already today available for research and application purposes, but the next major move will be to put these instruments into space.

NASA is supporting the development of "Moderate Resolution Imaging Spectrometers" (MODIS-N, MODIS-T) and the "High Resolution Imaging Spectrometer" (HIRIS), the ESA equivalent is a "Medium Resolution Imaging Spectrometer" (MERIS) which is designed to have 15 spectral bands in the 0.4-1.1 µm range. The majority of these bands will have a width of 10 nm, the ground resolution at nadir will be selectable to 250 or 1000 m, and the instrument will have off-nadir pointing capabilities. The MODIS-T design is similar (1100 m ground resolution, 32 bands between 0.41 and 0.89 µm, but MODIS-N should also include bands in the short-wave and thermal-infrared region (Goetz 1989). The low spatial resolution, the restricted spectral range and the large swath (2,330 km with MODIS) make these instruments appear as successors of AVHRR-type sensor, basically suited for developing long-term global data bases for terrestrial (vegetation monitoring, desertification) and oceanographic (phytoplankton distribution, sea-surface temperatures) applications (table 5).

Table 5. Planned EOS Imaging Spectrometer Systems

	MODIS–N	MODIS–T	MERIS	HIRIS
No. Spectral Bands	36	32 (contiguous)	15	192 (contiguous)
Spectral Range (microns)	0.41–0.95 3.75–14.24	0.41–0.89	0.41–1.05	0.40–2.45
Spectral Sampling Interval/Bandwidths (nm)	10–50 50–500	15	1.25–20	9.4 (VNIR) 11.7 (SWIR)
Radiometric Range	12 bits	12 bits	12 bits	12 bits
Nominal Pixel Size (GIFOV) (m)	214/428/856	1100	250 (land) 1000 (ocean)	30
Swath Width (FOV) (km)	2780	2200	1500	24
Off–Nadir Pointing	–	± 50 along ± 45 across track	± 20 along track	+ 52/–30 along/± 26 across track

source: C. Banninger, Inst. Image Processing & Computer Graphics, Graz, Austria, (pers. commun., 1990)

Given its spatial resolution (30 m at nadir), the HIRIS instrument stands in the continuation of the existing TM instrument. It will be pointable, and cover the continuous spectral region between 0.4 and 2.5 μm with 196 spectral bands of 10 nm bandwidth (table 5). Due to the horrendous data transmission rate, the swath width must be restricted to 24 km. HIRIS is mainly meant to permit more detailed examinations and monitoring of ecosystems, once anomalies are identified from the MODIS/MERIS instruments.

There are also a number of future remote sensing systems operating in the microwave parts of the electromagnetic spectrum. These are discussed in a later chapter by Churchill and Sieber.

References

Anuta, P. L. Bartolucci, E. Dean, F. Lonzano, E. Malaret, C. McGillem, J. Valdes & C. Valenzuela, 1984, Landsat-MSS and Thematic Mapper data quality and information content analysis, *Proceedings of IGARSS '84 Symposium, Strasbourg (France), 27-30 August 1984, ESA SP-215*, pp 85-92.

Begni, G., 1982, Selection of the optimum spectral bands for the SPOT satellite, *Photogrammetric Engineering and Remote Sensing*, **48**, 1613-1620.

Blanchard, L.E. & O. Weinstein, 1980, Design challenges of the Thematic Mapper, *IEEE Transactions Geoscience & Remote Sensing*, **GE-18**, 146-160.

Curran, P.J., 1985, *Principles of Remote Sensing* (London & New York: Longman).

Elachi, C., 1987, *Introduction to the physics and techniques of remote sensing* (New York: Wiley).

Forshaw, M.R.B., Haskell, A., Miller, P.F., Stanley, D.J., and Townshend, J.R.G., 1983, Spatial resolution of remotely sensed imagery, A review paper. *International Journal of Remote Sensing*, **4**, 497-520.

Goetz, A.F.H., 1989, Future sensor development for airborne and spaceborne imaging spectrometers, *Proceedings of the Ispra-Course on "Imaging Spectroscopy: Fundamentals and Prospective Applications", held at JRC Ispra, Italy, 23-27 October 1989, Commission of the European Communities, Joint Research Centre (JRC), IFA/89/ 5*, 16.

Hempenius, S.A., B.S. Marwaha, A. Murialdo & Wang Ren-Xiang, 1983, Characteristics of the second generation earth-observation satellites, *ITC Journal*, **1983-1**, 21-33.

Justice, C.O., J.R.G. Townshend, B.N. Holben & C.J. Tucker, 1985, Analysis of the phenology of global vegetation using meteorological satellite data, *International Journal of Remote Sensing*, **6**, 1271

Kontoes, C. & J. Stakenborg, 1990, Availability of cloud free Landsat images for operational projects. The analysis of cloud cover figures over the countries of the European Communities, *International Journal of Remote Sensing*, **11**, 1599-1608.

Legg, C. A., 1990, Digital satellite imagery from the Soviet Union; an important new tool for environmental monitoring, in *Remote Sensing and Global Change the proceedings of the Remote Sensing Society 16th Annual Conference, Swansea 19th 21st September 1990* (Nottingham: Remote Sensing Society) pp 65 - 74.

Lillesand, T.M. & R.W. Kiefer, 1979, *Remote Sensing and Image Interpretation* (New York: John Wiley & Sons).

Malila, W.A. & D.M. Anderson, 1986, Satellite data availability and calibration documentation for land surface climatology studies, International Satellite Land Surface Climatology Project (ISLSCP), ISLSCP Report No. 5, NAS5-28715, Environmental Research Institute of Michigan, Ann Arbor, Michigan.

Malingreau, J.-P., 1986, Global vegetation dynamics: satellite observations over Asia, *International Journal of Remote Sensing*, **7**, 1121-1146.

Schott, J.R., C. Salvaggio & W.J. Volchok, 1988, Radiometric scene normalization using pseudo-invariant features, *Remote Sensing of Environment*, **26**, 1-16.

SPOT IMAGE, 1986, Guide des utilisateurs de données SPOT, révision no. 01 (décembre 1988), CNES Toulouse.

Townshend, J.R.G., C.O. Justice, B.N. Holben & C.J. Tucker, 1984, Monitoring and mapping global vegetation cover using data from meteorological satellites, *Proceedings of the ESA/EARSeL Symposium on Integrated Approaches in Remote Sensing, Guildford, UK, 8-11 April, ESA SP-214*, pp 75-79.

Townshend, J.R.G., J.R. Gayler, J.R. Hardy, M.J. Jackson & J.R. Barker, 1983, Preliminary analysis of Landsat-4 Thematic Mapper products, *International Journal of Remote Sensing*, **4**, 817-828.

Tucker, C.T., A. Gatlin & S.R. Schneider, 1984, Monitoring vegetation in the Nile delta with NOAA-6 and NOAA-7 AVHRR, *Photogrammetric Engineering and Remote Sensing*, **50**, 53.

Wolf, P. R., 1983, *Elements of photogrammetry* (Tokyo: McGraw Hill Book Company).

THE PROCESSING AND INTERPRETATION OF REMOTELY-SENSED SATELLITE IMAGERY - A CURRENT VIEW

G. G. WILKINSON
Commission of the European Communities,
Joint Research Centre,
Institute for Remote Sensing Applications,
21020 Ispra (Varese), Italy.

1. The Basic Structure of a Digital Image

The techniques for processing imagery acquired by modern remote sensing satellites are now fairly advanced, although new developments are occurring more or less continuously. The purpose of this chapter is to review the main principles behind the most commonly-used analysis techniques and to give an indication of where new advances are currently being made.

1.1 IMAGE CAPTURE AND QUANTISATION

Satellite sensor systems are complex optical and electronic devices. The internal structure of such devices does not concern us directly here, however it is important to understand the main functionality of these devices, which is to convert the incoming radiance from a spot view of the Earth's surface into an electrical signal and then to quantize this signal to generate a digital number representing one picture element (pixel) -see figure 1.

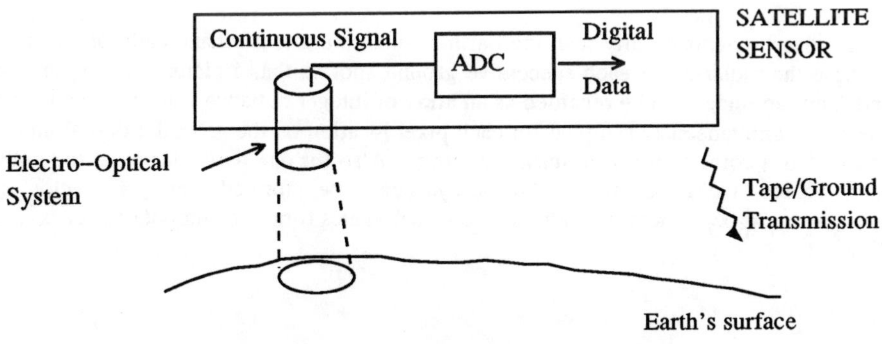

Figure 1. Typical satellite sensor system.

A. S. Belward and C. R. Valenzuela (eds.),
Remote Sensing and Geographical Information Systems for Resource Management in Developing Countries, 71–96.

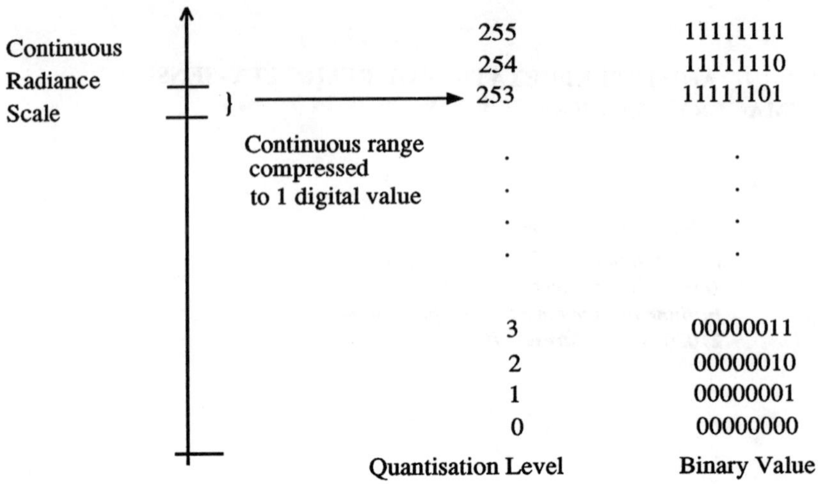

Figure 2. Digitisation of channel radiance values

Normally the sensor system includes "beam splitters" to separate components of the incoming radiance from different parts of the electromagnetic spectrum (usually in the visible, near infra-red or thermal infra-red regions). These separate components (7 in the case of modern generation systems like the Landsat Thematic Mapper) each give rise to an electronic signal which after quantisation or digitisation become the "channel values" for the particular pixel. For each channel a binary number is produced by the digitisation process (figure 2) usually in the range 0-255 so that 8 bits can be used to store one channel value.

Some information is lost in the digitisation process but there are considerable benefits from handling the pixel channel values as single byte integer numbers rather than decimal real numbers.

1.2 IMAGE ARRAYS

Imaging sensor systems usually scan the Earth's surface either mechannically or electronically and sample the radiances at each successive ground spot so that a picture mosaic or image is created. Thus an image can be regarded as an array of integer radiance values. Since the various channels are simultaneously sampled for each pixel location on the ground a digital image may be regarded as a coincident set of numerical arrays. Also for one particular pixel the individual channel values extracted from the arrays can be formed into a vector $\underline{x}_p = (x_1,x_2,x_3,x_4,x_5,x_6,x_7)$ where the individual channel values form the components x_1, x_2, ...x_7. See figure 3.

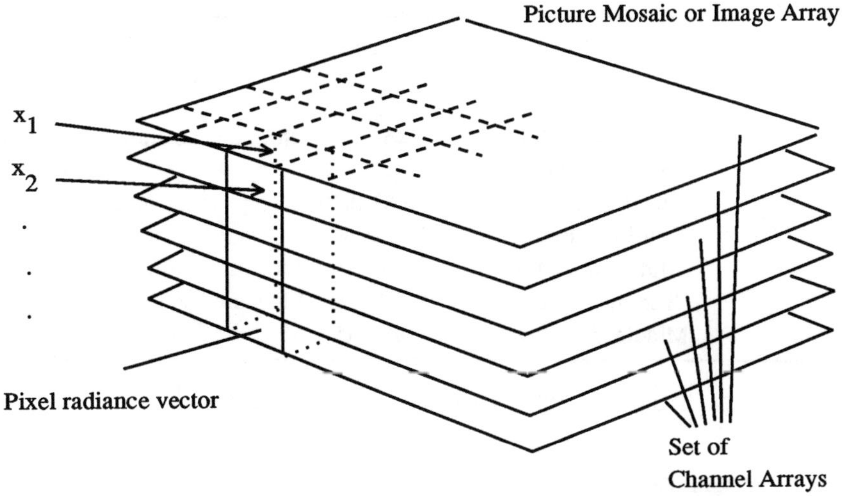

Figure 3. Structure of a digital image.

1.3 SPECTRAL SAMPLING

Besides the reduction in precision which occurs when continuous radiance values are converted to quantised digital values, there is also an information reduction when a continuous reflectance spectrum from the ground target is converted to a set of channel values. A continuous spectrum is therefore sampled in only a narrow region and each channel provides one digital value which can not be regarded as a true representation of a complex spectrum. Nevertheless satellite sensor channels are selected so that they fall at appropriate parts of the continuous spectrum -i.e. some where vegetation reflectance is low and others where it is high. Channel differences and ratios then give a good opportunity to map vegetation. See figure 4 and table 1.

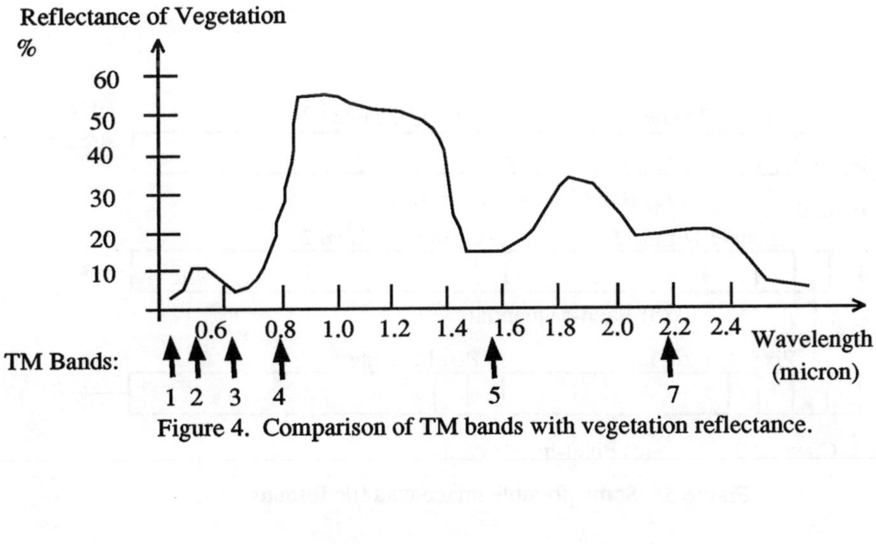

Figure 4. Comparison of TM bands with vegetation reflectance.

TM Band No.	Spectral Region	Channel Limits (microns)	
1	Blue/Green Visible	0.45-0.52	Strong vegetation absorption
2	Green Visible	0.52-0.60	Strong vegetation reflection
3	Red Visible	0.63-0.69	Very strong vegetation reflection
4	Near Infra-Red	0.76-0.90	"
5	Middle Infra-Red	1.55-1.75	Moisture sensitive
6	Thermal Infra-Red	10.4-12.50	Moisture and temperature sensitive
7	Middle Infra-Red	2.08-2.35	

Table 1. Landsat Thematic Mapper Channels.

1.4 IMAGE DATA FILES

A typical digital image in principle consists of a set of large coincident arrays of integer pixel values (one array per channel). However when such an image is stored as a computer file (e.g. on magnetic disc or tape) it is basically held as a continuous stream of bytes where 1 byte holds one channel value for one pixel location. Normally image files also contain *housekeeping information* (e.g. about spacecraft status, position, and calibration details). Clearly the arrangement of a set of two-dimensional arrays (plus housekeeping data) can be done in many ways when the images are written as byte streams. Figure 5 illustrates several possibilities for this. Normally the byte stream is broken into sections (records) and special input/output software is needed to retrieve and/or store images in these different formats. A user of the imagery will usually be interested in retrieving an array of pixel values from some selected part of the overall image in one or more channels for display, enhancement and analysis. Input/output software therefore has to be defined to access the correct memory locations in the continuous byte stream for this purpose.

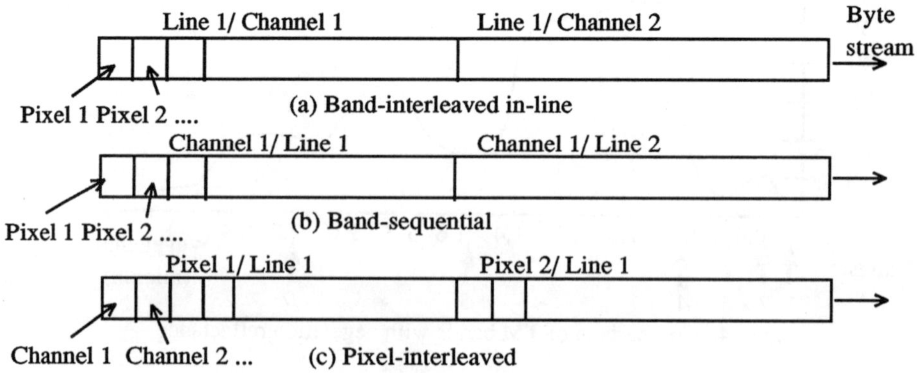

Figure 5. Some possible image data file formats.

2. Pre-Processing and Image Rectification

The relationship between the radiance of the electromagnetic radiation reflected or emitted by ground targets and that measured by the satellite instrument is a complex one. Firstly the radiance emitted at ground level has to pass through the intervening atmosphere on its way to the satellite. Clouds are normally only semi-transparent or even opaque at visible and infra-red wavelengths thus obscuring ground features. Remote sensing in cloudy conditions is limited to use of radar/microwave sensor systems. But even in cloud-free conditions the atmosphere contains many components (moisture, particulate matter, pollutants, trace gases) which can absorb or scatter the radiance from the ground (and from the sun) besides generating their own additional radiance through emission or absorption. In addition in the visible and near infra-red parts of the spectrum the pixel radiances arise primarily from reflection of solar radiation -this too has to pass through the atmosphere and will arrive at different elevation angles through the day. The strength of the solar beam also varies. Bird, elsewhere in this book discusses these issues in more detail.

In addition the atmospheric radiance measured by the satellite system has to pass through various optical components and electrical amplification circuits -the former may slowly degrade in flight and the gain characteristics of the latter may alter. The overall sensitivity of the satellite sensor is in a constant state of fluctuation, so that even the relationship between incoming radiance *at the satellite* and the recorded pixel value is non-trivial. Pixel images may also contain noise arising from the electronic systems of the spacecraft or sensor apparatus. They may also be geometrically distorted by imperfections in the sensor optical components and by the effects of varying ground terrain. Overall there is a major requirement for calibration, pre-processing, correction and rectification of imagery before it can be used in any quantitative remote sensing application.

2.1 PROBLEMS OF IN-FLIGHT CALIBRATION

In practical terms, remote sensing with environmental satellites is concerned with measuring the absolute relectivity (or albedo) of surface features (rocks, soils, crops, natural vegetation, man-made structures etc.) in different spectral bands with a view to identification and hence mapping. This requires an accurate calibration so that the measured radiance at the satellite can be related faithfully and simply to the radiance and reflectivity of the ground target. There are two initial stages in this process -calibration for sensitivity changes and atmospheric correction for the unavoidable scattering and absorption processes that occur. Satellite sensor systems are typically provided with on-board calibration apparatus to enable some checks to be made on the gross changes in sensitivity of the main detector components. This apparatus usually consists of either calibration lamps (for visible/near infra-red channels) or calibration black-bodies (for thermal infra-red channels) which emit known radiances. These lamps or black bodies can be viewed by the sensor during calibration scans -usually by means of a rotating mirror. Calibration scans can be made regularly during the life-time of a satellite to check on degradation. Results from the Nimbus-7 Coastal Zone Colour Scanner (CZCS) for example indicate sensitivity changes of up to 25% over a 4-year period -see figure 6. In principle calibration must be performed to give an absolute accuracy in radiance measurement of <0.5% (1 digital count) and can be carried out by adjusting the actual pixel values with a linear function e.g.:

$$x' = \alpha x + c$$

(where x = original pixel value, α = calibration gain factor, c = zero offset).

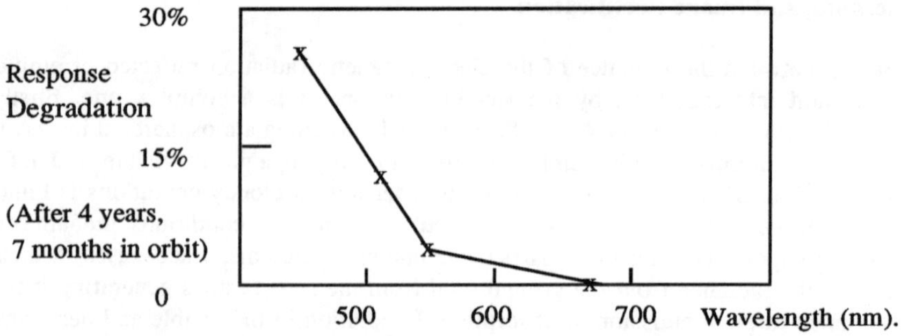

Figure 6. CZCS channel sensitivity degradation (After Hovis *et al.* 1985)

The calibration offset can be obtained by viewing deep-space (or alternatively water bodies, lakes etc. for near infra-red channels) which have minimal radiance. Unfortunately the change in gain factor of the entire satellite sensor system can not be reliably obtained from an internal calibrator because some of the main optical components are external to the calibration viewing mirror in many systems. This has led to the development of reflectance-based methods for calibration (Slater *et al.* 1987) in which special ground test targets with high reflectivity are selected (e.g. flat sands) and coincident measurements of the surface radiance are made from helicopter-borne radiometers. To do this effectively a high altitude site must be used to minimise the atmospheric attenuation of the radiance. Normally calibration parameters are provided by data distribution organisations (e.g. as part of housekeeping data). These also indicate how to convert pixel count values to radiance values.

2.2 ATMOSPHERIC CORRECTIONS

Many users of satellite imagery require the analysis of pixels from different geographic locations and on different dates. Hence corrections must be applied for atmospheric variability -i.e. each satellite scene requires independent correction to give accurate surface reflectivity values. Several methods have been proposed for this such as Hill and Sturm (1988). The apparent at-satellite reflectivity of solar radiation is obtained as a ratio between the measured upwards radiance and the downwards solar irradiance:

$$r_{sat}(\lambda) = \frac{\pi L(\lambda).d^2}{E_0(\lambda).\mu_0}$$

(where L(λ) = total radiance measured by the satellite; d = Earth-Sun distance in A.U.;
E_0 = mean solar irradiance incident above the atmosphere; μ_0 = cosine of solar zenith angle).
The true ground target reflectance is instead given by:

$$r_{GT}(\lambda) = \frac{[L(\lambda)-L_P(\lambda)].\pi}{t(\lambda).E_G(\lambda)} - \rho_{env}(\lambda)$$

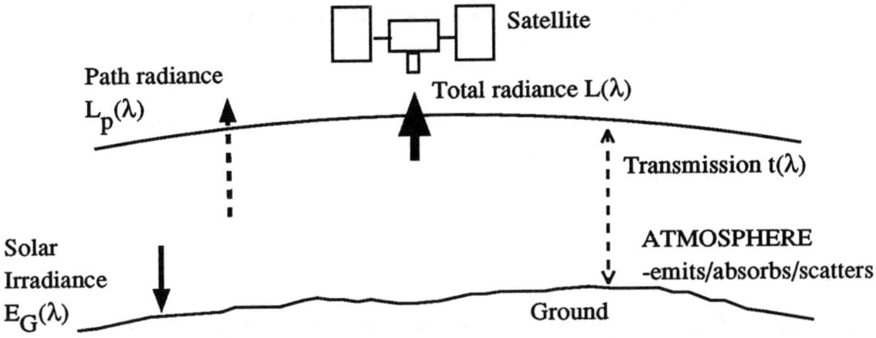

$$\pi L(\lambda)=\pi L_p(\lambda)+t(\lambda)E_G(\lambda)$$

Figure 7. Radiative balance for solar reflection from ground target (in visible or near infra-red channels).

where $L_p(\lambda)$ is radiance arising from the atmospheric path between ground and satellite, $t(\lambda)$ is the transmission from ground to satellite, $E_G(\lambda)$ is solar irradiance at ground level -see figure 7. The factor $\rho_{env}(\lambda)$ is a correction for the reflectance effects of pixels in the vicinity of the ground target -these may scatter some radiance which affects the radiance detected.

In practice models can be used to estimate the atmospheric effects taking into account the backscattering of the solar irradiation (by molecular processes and by particulate matter), and the absorption effects of atmospheric gases (all of which are wavelength dependent). Such models can generate transmission figures $t(\lambda)$ and path radiances $L_p(\lambda)$. For practical applications models can be verified by simultaneous ground measurements and can then be used on different images on different dates. For multi-temporal crop studies the model of Hill and Sturm (1988) developed at the JRC has yielded accuracies averaging 8.5% in the seven bands of the Landsat Thematic Mapper instrument.

For thermal infra-red channels (e.g. Landsat-TM band 6 or Meteosat IR window channel for example) the atmospheric correction procedure is rather different to that used for visible and near infra-red channels. At thermal infra-red wavelegths (~10 microns) there is minimal scattering in the atmosphere and minimal reflection of solar radiation. Images at these wavelengths represent thermal black body emission from the ground features and indicate their temperatures. However, even though the atmosphere is relatively transparent at these wavelengths, water vapour in particular (and especially in warm moist tropical conditions) can have an important effect on modifying detected surface temperatures. The radiance in a thermal infra-red channel leaving the Earth's surface is given by:

$$L_G(\lambda) = B(T_S,\lambda)d\lambda$$

-i.e. an effective black body function tor an emitting surtace at temperature T_S. However this is attenuated in the atmosphere, and also the atmosphere itself emits infra-red radiation (at its own temperature at different levels). The radiance detected by the satellite is thus:

$$L_S(\lambda) = B(T_S,\lambda)t_0(\lambda) + \int_z \epsilon(z)B(T(z),\lambda)t_z(\lambda)dz.$$

(where t_0 = transmission from ground to satellite; t_z = transmission from level z in the atmosphere to the satellite; T(z) = temperature at level z; ε(z) = emissivity at level z).
See figure 8. If this equation is inverted so that a measured radiance L_S is assumed to come from a black body at temperature T i.e.:

$$L_S(\lambda) \equiv B(T,\lambda)$$

then the temperature T derived will be less than the true surface temperature T_S (because the atmosphere is cooler than the ground). The difference between T and T_S is known as the temperature deficit and is typically a few degrees K in tropical zones. True atmospherically corrected surface temperatures can be obtained by adding the estimated deficit value to the measured brightness temperatures (from calibrated pixel values). This estimation can be done with climatological models and checked by ground surface measurements such as ocean surface temperature measurements by ships (Williamson and Wilkinson 1980). Some instruments provide dual channels in the thermal infra-red at slightly different wavelengths where there is differential absorption. The comparison of the channel radiances may be used to form a regression to determine the average atmospheric absorption.

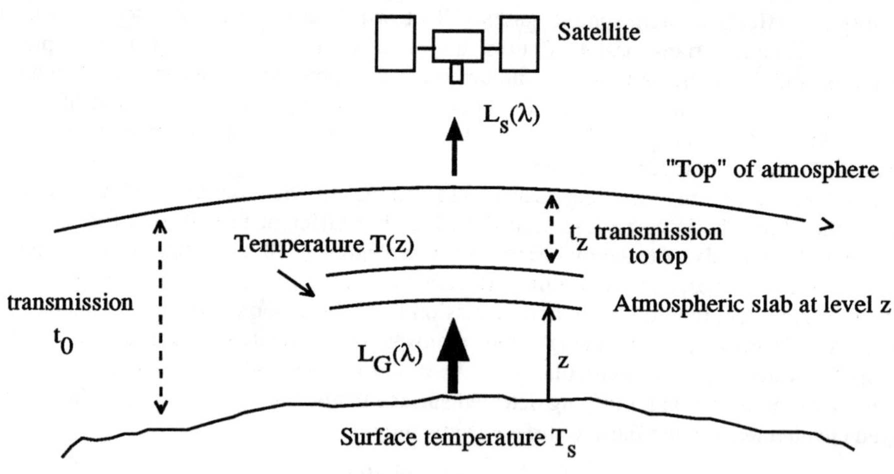

Figure 8. Radiative balance in thermal infra-red channels.

2.3 GEOMETRICAL OPERATIONS ON IMAGERY

Since satellite images suffer some unavoidable distortions owing to sensor optical aberrations, satellite instabilities and terrain variations etc. it is often important to geometrically correct them. This is especially necessary when images are to be matched to digital maps or cartographic grids (e.g. in geographical information systems) and when images of the same area but from different dates have to be used in multitemporal analysis. Also there can be some distortion variations between the different spectral bands. Geometrical corrections are therefore an important part of image pre-processing.

In principle an input image with pixel array coordinates x,y has to be "deformed" into an output image with a new pixel coordinate system (say p,q). This is achieved by a deformation mapping (M) which may include a combination of scaling, translation and rotation of the original image -see figure 9.

If a mapping is applied directly to input image coordinates (x,y), the generated output image coordinates (p,q) will in general take non-integer values. Since this output array has to be produced at regular integer grid points the generated values of p and q have to be rounded to the *nearest* integer values. It is then possible in this process that some coordinates in the output image will not have pixels assigned to them from the input image and "holes" will appear in the output image. For this reason it is preferable to base the mapping on a transformation *from output coordinates to input coordinates* -i.e. to identify which input image pixels are needed to create every single output image pixel. The output-input mapping procedure is the most commonly used with satellite images.

There are many ways of defining mappings between the coordinate systems, but a particularly common approach is to use a polynomial model e.g. mapping M is defined as:

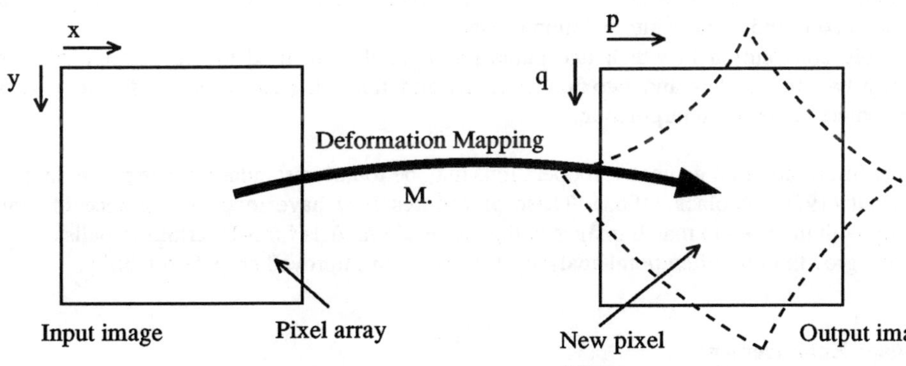

Figure 9. Geometrical deformation.

$$x = a_0 + a_1p + a_2q + a_3p^2 + a_4pq + a_5q^2 + \ldots$$

$$y = b_0 + b_1p + b_2q + b_3p^2 + b_4pq + b_5q^2 + \ldots$$

This explicitly states how to find a pixel coordinate (x,y) that must be used to create an output image pixel (p,q). The main problem of this method is that we have to find the polynomial coefficients (a_0, a_1, ...b_0, b_1,.....) in order to define the mapping. This can be done in several ways. Firstly if we know exactly what geometric warping is necesary because we have knowledge of the image distorting function the coefficients can be computed explicitly. However this situation is rare. A more common approach, especially with satellite images in the context of multitemporal analysis or map-image matching is to use *ground control points* (GCP's). These are points which are readily identified on the input image and and on the reference image or map which we wish to match. They could typically be points on coastlines or road junctions. The idea is then that we find a suitable polynomial which effectively transforms the coordinates of the GCP's in the input image to the required coordinates for these same GCP's in the output image with the minimum error. This is effectively a polynomial fitting procedure.

Once the polynomials are accurately defined the appropriate mapping can be carried out pixel-by-pixel. Sometimes however this procedure is slow and it is more suitable to map every other pixel (for example) and to interpolate the new grid coordinates for the intermediate ones.

It is also often necessary to resample a geometrically-transformed image so that the number of pixels created exactly matches some pre-defined array size (e.g. so that it can be displayed using all points of a display screen). This requires a process of resampling (or general interpolation) to create intermediate pixels. This interpolation of the image can be carried out in several ways, all of which involve forming a mathematical approximation to the original image. There are three common technique for this:

(i) nearest neighbour sampling in which intermediate values are obtained directly from the nearest neighbour pixel,
(ii) bilinear sampling in which the image grey-level values are assumed to change linearly between pixels and a linear interpolation is used,
(iii) cubic convolution in which two cubic polynomials are fitted to the surrounding pixels (usually two to the left and two to the right) and the fitted function is used to give any intermediate values on interpolation.

Further information on geometrical operations may be found in standard image processing texts e.g. Pratt (1978), Niblack (1985). These procedures now have much importance in remote sensing with the need to match images to digital terrain models for 3-D terrain visualisation and with the growth in multitemporal analysis of datasets for improved crop detectability.

3. Image Enhancement

Once images have undergone preprocessing for calibration, atmospheric correction and geometric deformation, they can then be used directly for application studies. However it is usually necessary to apply various enhancement and/or data reduction procedures to them to assist in visualisation and interpretability.

3.1. CONTRAST MODIFICATION

Although as we have seen pixel data is usually digitised to integer values in the range 0-255 and computer-based image display systems are capable of displaying all 256 pixel intensity levels on a monitor, it is often found that in particular channels the image data do not occupy the full dynamic range from 0-255. In this situation a displayed image will not appear to the viewer to have much *contrast.* Ideally this contrast should be improved by modifying the pixel values to exploit the full dynamic range of the display system:

$$x' = f(x)$$

Each new pixel value x' is obtained from the original pixel value x by applying a contrast modifying function f. In principle f may be very simple -e.g. a linear function of x. The behaviour of this can be readily appreciated from an examination of the image histogram (which shows how many pixels have particular integer values). Figure 10 shows the effect on the histogram of a simple linear contrast modification function.

In principle any monotonic function may be used to apply contrast modifications. For example it would be possible to apply a non-linear function which 'stretched' the contrast over certain pixel value ranges and 'narrowed' it over others so that features of interest stood-out in the display for the user's attention.

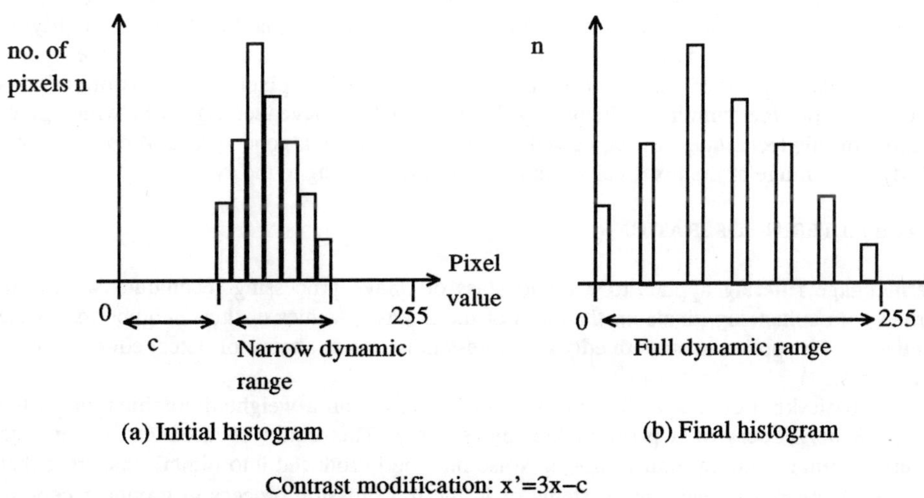

Figure 10. Effect of linear contrast stretch on image histogram

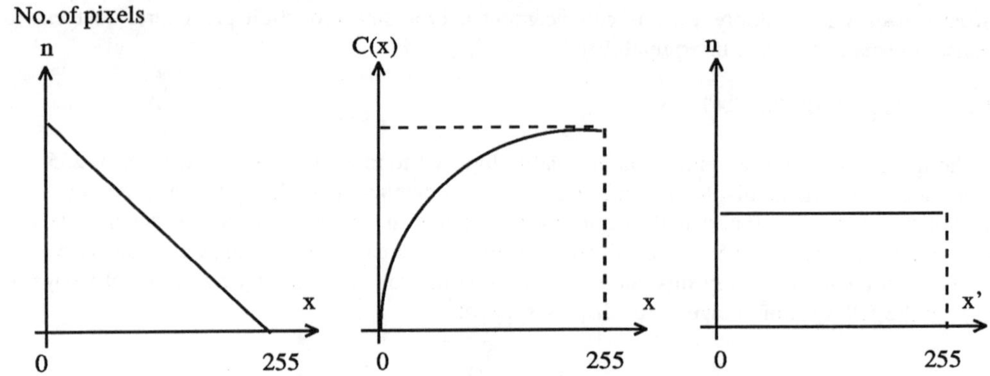

(a) Initial image histogram (b) Cumulative frequency diagram (c) Equalised histogram

$$\text{Here} \quad x' = C(x) = \int_0^x n(x)dx$$

Figure 11. Histogram equalisation technique.

3.2 HISTOGRAM METHODS

A particularly useful non-linear contrast modification technique is the method of histogram equalisation. This uses the image histogram directly in formulating the contrast modifying function. The purpose of this technique is to level the image histogram as far as possible -that is to assign an equal number of pixels to each integer pixel value. The pixel value transformation is taken directly from the cumulative frequency diagram for the image -see figure 11 which gives an example of this technique. An "equalised" image should have a good spread of contrast over the full dynamic range of the pixel values and is therefore pleasing to the eye.

3.3 IMAGE FILTERING OPERATIONS

The term image filtering applies to a broad class of image processing techniques which are concerned with enhancing pixels on the basis of the intensity values of their neighbours. These techniques can be applied to smooth edges, suppress noise, and enhance or detect edges or other linear features.

The most basic kind of image filtering operation is based on a weighted combination of the immediate 8 neighbours of a pixel in the image array. This could be used for example to "clean-up" an image -i.e. to remove noise. Noise may be introduced into digital imagery either by the satellite on-board electronic systems or in the transmission process to ground receiving

p_1	p_2	p_3
p_4	x	p_5
p_6	p_7	p_8

$$\textit{If}\ \left(x - \frac{1}{8}\sum_{i=1}^{8} p_i\right) > \epsilon \qquad \textit{then}\ x = \frac{1}{8}\sum_{i=1}^{8} p_i$$

(ε = threshold)

Figure 12. Simple noise cleaning filtering operation

stations. Noise usually manifests itself in images as pixels whose values have been corrupted in some way. Isolated noisy pixels therefore appear to have intensity values which do not match the values of their immediate neighbours. A typical noise filtering operation would be to compare a pixel value with the average of its neighbours. If it differs by more than a reasonable amount (the *threshold*) then the pixel is replaced by the neighbour average -see figure 12. There is of course a possibility that isolated pixels with unusual intensity do arise from some genuine ground feature and in noise filtering it is possible to erroneously remove them. Therefore before any filtering is carried out it is important to assess carefully what kind of noise cleaning and threshold level is necessary.

Noise filtering is essentially an operation which smoothes imagery -i.e. large deviations from the local pixel average are smoothed out. Often it is useful to do the opposite -that is to sharpen imagery to produce enhanced local pixel intensity differences. This can be useful to pick-out

p_1	p_2	p_3
p_4	p_5	p_6
p_7	p_8	p_9

Image neighbourhood

-1	0	1
-2	0	2
-1	0	1

1	2	1
0	0	0
-1	-2	-1

3x3 Sobel operators

p_5 becomes $|p_3-p_1| + 2\,|p_6-p_4| + |p_9-p_7| + |p_1-p_7| + 2\,|p_2-p_8| + |p_3-p_9|$

Figure 13. Sobel edge enhancement.

lines and edges for example. Lines are important because they are often of much interest in images -indicating communications infrastructure, geological features etc. In modern high resolution imagery (SPOT or Landsat TM) communications features are readily visible although they are usually only 1-2 pixels wide. They are often rather indistinct in images because the pixels containing them usually overlap neighbouring regions (fields, houses etc.). Line or edge sharpening is thus an important and useful technique. This is usually done by applying a filter locally to the image which accentuates in some way the local image gradient (i.e. gradient of the pixel values). A typical filter is the *Sobel operator* which is applied to a 3x3 pixel neighbourhood in both horizontal and vertical directions -see figure 13.

The Sobel method replaces each pixel by a value indicating the strength of the image gradient at its location using neighbour differences in both horizontal and vertical directions. In fact by taking a 3x3 "window" the Sobel operator achieves the combined effect of enhancing local gradient and also reducing noise (because an averaging process is implicitly applied in the direction orthogonal to the gradient computation). Over the years many different local filtering operators have been proposed to enhance edge or line features. Some of these use larger neighbourhood windows e.g. 5x5 pixels or more. Many of these are reviewed in standard image processing texts such as Pratt (1978) or Ballard and Brown (1982).

After a typical edge-enhancing filter has been applied to an image it is sometimes useful to apply a thresholding procedure to further enhance edges where they have been detected and to suppress pixel values where they have not been detected. Such a procedure can lead to an edge-map which is potentially useful for extracting ground control points or for map revision applications for example. The basis of the thresholding procedure is illustrated in figure 14.

Unfortunately edge-detection remains a difficult problem in remotely-sensed image processing, primarily because any filtering and thresholding operation can either miss useful edges (because the threshold is set too high) or alternatively pick-out too many (because the threshold is set too low) -i.e. noise and small local intensity variations are detected as edges. These problems of omission and commission still remain to be resolved although various methods have been proposed for this (for example by erring on the side of omission and then adopting a line following and linking procedure e.g. Nevatia and Babu, 1980).

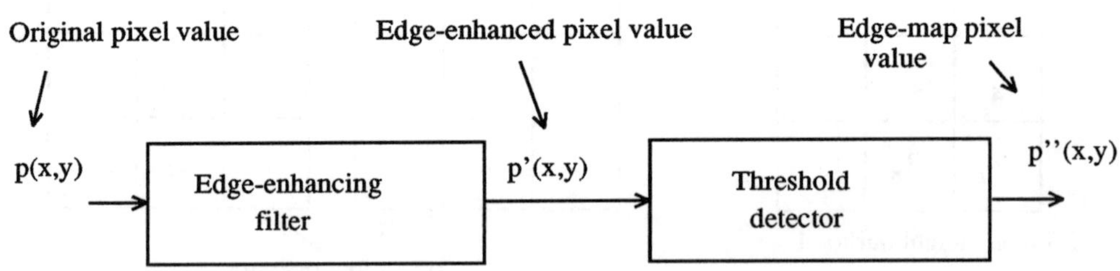

If p'(x,y) > T then p''(x,y) = 255
If p'(x,,y) <= T then p''(x,y) = 0

Figure 14. Typical edge-thresholding procedure.

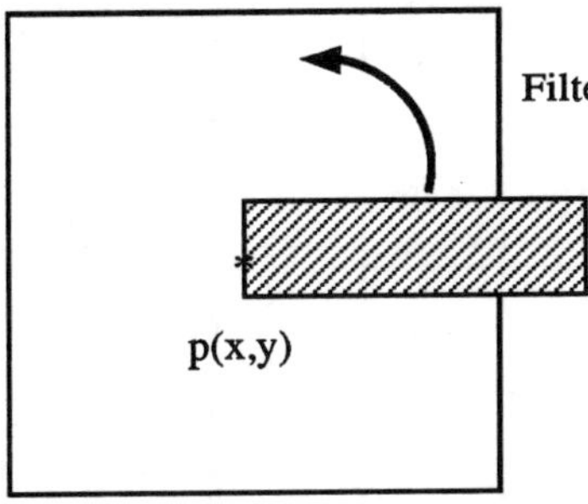

Figure 15. Rotating bar filter for edge-preserving smoothing. (After Nagao and Matsuyama , 1979).

It is interesting to note that in principle the twin aims of (i) noise suppression and (ii) edge enhancement are somewhat incompatible. The former requires a smoothing of the image, the latter a sharpening. Interestingly Nagao and Matsuyama (1979) proposed a method to overcome this incompatibility by applying edge-preserving smoothing. Instead of using a filter centred on the pixel of interest, they proposed the use of an elongated bar-shaped filter which is rotated around the pixel -see figure 15. The variance of the pixel values is calculated at each different position of the filter and the direction of the filter is identifed for which the variance is a minimum. The pixel value is then replaced by the average of those along the direction of the filter. In this way the smoothing procedure avoids averaging pixels in directions where there could be significant edge structure present.

3.4 IMAGE SEGMENTATION

Another important technique in the analysis and enhancement of satellite images is segmentation. The purpose of segmentation is to group pixels into regions where each pixel has approximately the same characteristics -i.e. similar spectral intensity values in each channel if it is a multispectral image. By dividing an image into homogeneous zones it is often easier to interpret. For example individual fields will then appear completely uniform. Also once divided into homogeneous segments it is then possible to digitise the boundaries of the segments to create an areal topographic feature which may be inserted into a digital cartographic database or a GIS. In addition segmentation helps in classification of areal land cover features and in performing inventories (e.g. crop area estimation). The main benefit of segmentation is that local pixel variation is removed and whole regions are uniformly assigned an average pixel value -the image becomes simpler and yet more meaningful in analysis. See figure 16.

Many different methods have been proposed to carry out segmentation. They include edge detection methods, region growing methods and clustering methods. Edge detection methods attempt to find edges and then to identify closed segment boundaries, but this is difficult because of the problems inherent to edge detection mentioned earlier. Region growing is the most common approach in which a combination of *splitting* and *merging* is used -i.e. pixels are

110	108	101	116	119
113	105	110	120	118
112	119	111	201	200
113	107	116	204	206
104	110	109	203	205

(a) Original pixel data

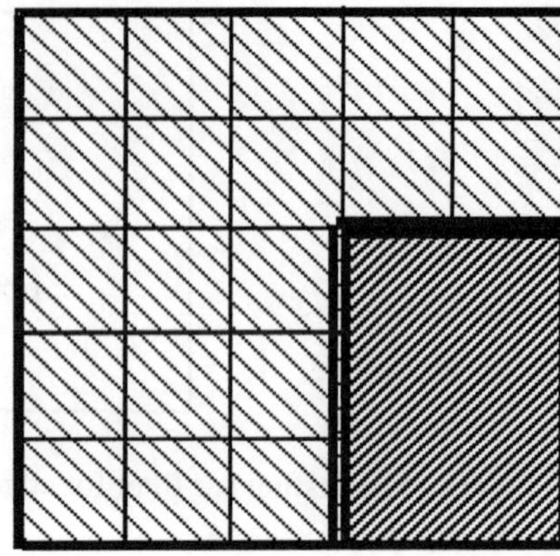

(b) Derived segments

Figure 16. Segmentation.

merged with their neighbours if they obey a similarity constraint and also regions which are formed may be split up if they fail to match the uniformity measure. Clustering is a procedure in which all pixels in the image are placed into groups of spectral similarity. However before they can be put into segments, pixels have to be identified in the groups which are spatially-contiguous -only these can form valid segments.

All these methods of segmentation have their drawbacks and segmentation is still being vigorously researched, not least because of its value in simplifying the structure of digital images for computer vision applications (e.g. in robotics). In remote sensing it has recently been found useful to use a knowledge-based approach in which background map information (e.g. in a GIS) provides a guide to the image segmentation procedure (Tailor *et al.* 1988).

3.5 DATA REDUCTION TRANSFORM METHODS

One of the most important aspects of remotely-sensed satellite imagery is its multispectral quality. Ground features have different reflectivities in different spectral channels and some channels are better at discriminating land cover types than others. The full set of multispectral image arrays conveys a great deal of information about the land surface. However it is actually rather difficult to view say six different image channels at the same time (as might be useful in analysing Thematic Mapper imagery). It is useful therefore to be able to form some kind of composite image which reduces the original 6 or 7 spectral bands to one or two but at the same time preserving most of the useful visual information from all of the bands. There are many methods to do this. In this section we shall concentrate on just two techniques which are commonly used in remote sensing and which give good results.

The first of these is the Principal Components Transform (PCT) method. This approach mathematically transforms the original spectral channels into a new set of *synthetic channels* which are effectively formed by a linear combination of the original channels. However this

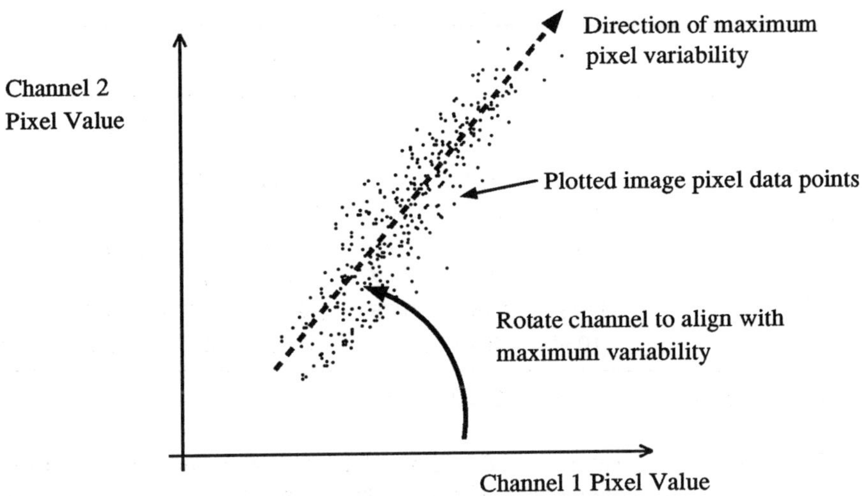

Figure 17. Basis of the PCT method.

linear combination is chosen in such a way that the first synthetic channel holds the maximum variability of the pixel intensities taken from the original channels -it must therefore be computed after analysis of the original image band statistics. The PCT is represented graphically in figure 17 where we consider only a restricted case of 2-band imagery.

In this example it can be seen that the multispectral pixel values (represented as dots in the image "feature space" diagram) have greatest spread along the direction of the dashed arrow. Therefore if we mathematically rotate the channels so that channel 1 aligns with this direction then that one channel will have the greatest variation in pixel values and will give an enhanced image with good contrast that takes advantage of the maximum spectral variability in all the original channels. This mathematical "rotation" is achieved as a linear combination of the original bands and basically gives a synthetic channel -i.e. it has not been explicitly sensed by the satellite. In fact other synthetic channels can be formed by taking subsequent directions of maximum variability after the first rotation has been carried out. This technique can be regarded as a *data reduction* procedure because visual interpretation can be readily performed using the first PCT synthetic channel rather than using all the original channels. The PCT is sometimes called the Karhunen-Loeve transform (KLT). Further details of its operation can be found in standard texts such as Pratt (1978).

Another useful data reduction transformation which uses a linear combination technique is the Tasseled Cap transform developed for TM data (Crist and Cicone, 1984). This has some similarity to the PCT but is designed to create 3 useful synthetic channels which are directly related to specific land cover characteristics namely: brightness, greenness and wetness. The coefficients used to form the linear combinations of Thematic Mapper channels are given in Table 2.

Synthetic Channel	TM Band Coefficients					
	1	2	3	4	5	7
Brightness	.3037	.2793	.4743	.5585	.5082	.1863
Greenness	-.2848	-.2435	-.5436	.7243	.0840	-.1800
Wetness	.1509	.1973	.3279	.3406	-.7112	-.4572
Fourth	-.8242	.0849	.4392	-.0580	.2012	-.2768
Fifth	-.3280	.0549	.1075	.1855	-.4357	.8085
Sixth	.1084	-.9022	.4120	.0573	-.0251	.0238

Table 2. Tasseled Cap Transform Coefficients (After Crist and Cicone, 1984).

The synthetic "brightness" channel is derived from a weighted linear combination of all six TM reflection bands (i.e. excluding the thermal infra-red). The "greenness" channel weights channel 4 most strongly in which vegetation reflectance is known to be strongest (see table 1). The "wetness" channel weights are designed to contrast the middle infra-red channels (5 and 7) with the shorter wavelength channels (1-4), primarily because the longer wavelength channels are found to be more sensitive to soil moisture. Overall the principle of the Tasseled Cap transform is to achieve data reduction as with the PCT and to do it in such a way that the synthetic channels so produced have some physical meaning -in this respect it has enjoyed much success. The name of this transform comes from the fact that a 2-dimensional feature space plot of pixels on brightness and greenness axes usually shows a cap-shaped distribution for vegetated surface images.

4. Image Classification and Information Extraction.

So far we have considered ways of improving imagery and enhancing features for better visual interpretation. However the most important aspect of satellite remote sensing is actually identifying surface features by quantitative means. We thus turn to the important topics of image classification and information extraction.

4.1 A MODEL OF PATTERN CLASSIFICATION

Pattern classification is the technique by which we analyse images to put pixels (or perhaps segments) into meaningful environmental classes (e.g. "forest", "wheat-field", "bare-rock", etc.). This is a key application of remote sensing for it enables maps of land cover features to be created. Pattern recognition is normally envisaged to involve two phases -one is learning, the other classification (see figure 18). The first phase is the procedure in which the user defines which land cover classes are meaningful for the application study and then identifies features of the image data which characterise those classes. This can usually be done by identifying pixels in the image which correspond to known classes at ground level and which have been categorised in field surveys for example. The learning procedure then has to produce a

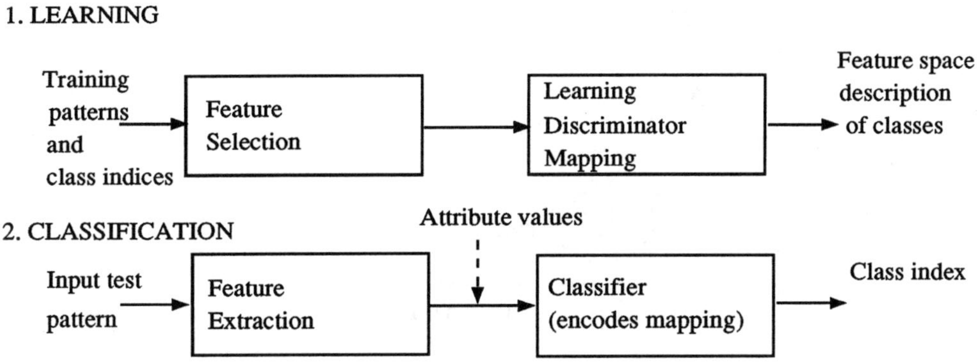

Figure 18. Standard model of pattern recognition. (After Fu, 1982).

meaningful quantitative analysis method which will reliably identify those same classes in other un-mapped areas solely from imagery. The second phase of pattern classification thus involves extracting the relevant aspects of features from new imagery and applying the decision-making or classifier procedure which was produced in the learning phase. This model applies to most application analysis of remotely-sensed imagery. In fact there are many ways of performing classification; some of the common ones will be reviewed in the next sections.

4.2 FEATURE SPACE CLUSTERS AND CLASSIFICATION

It is possible to regard the different spectral channels of a satellite sensor as providing 'axes' for image data values -i.e. in each channel we may have an axis of a graph representing the full range of pixel values that are possible. It is then possible to produce multi-dimensional graphs that represent image data -this is the feature space and pixels of the image can be plotted as points in this space. For a sensor such as the Landsat TM we would need 7-dimensional graphs to represent the full multi-spectral data for each pixel. These are impossible to draw in 2 dimensions although they are perfectly acceptable as a mathematical model. For ease of representation we shall examine feature space graphs with only two dimensions -i.e. with axes drawn from two spectral channels (these may be either real channels or synthetic ones).

The process of image classification is concerned essentially with firstly identifying the spectral characteristics of pre-defined land cover classes in the image feature space and then subsequently devising a method to "divide" the feature space so that any new pixel can be assigned to one of those pre-defined classes on the basis of its channel values. Ideally a feature space plot of some *training* pixels will show that the different pixels of each land cover class are grouped together in clusters -see figure 19. Often however pixels are not so tightly-grouped in feature space and various mathematical procedures must be used for classification in order to assign pixels to the optimum class.

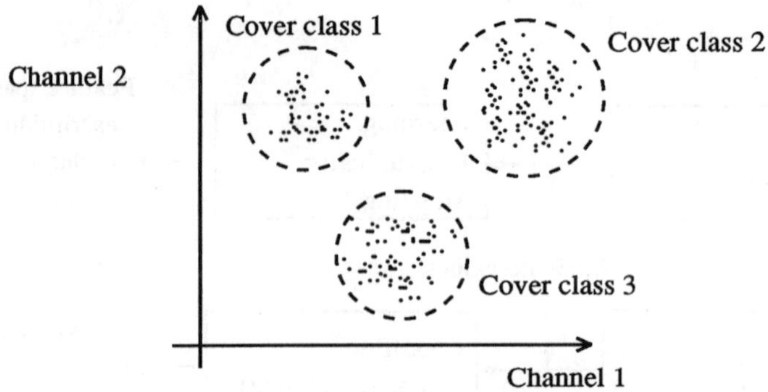

Figure 19. Pattern clusters in feature space.

4.3 UNSUPERVISED CLASSIFICATION - K NEAREST NEIGHBOURS

Sometimes it can be useful to classify image data without forcibly using any preconceived idea of how the pixels for the different cover classes are grouped in feature space -this is the approach of unsupervised classification. A typical algorithm for unsupervised classification is the k-nearest neighbours (k-means) algorithm. This is basically an iterative procedure for clustering the image data into meaningful groups. Initially the user has to specify a set of cluster centres $\underline{c}_1$, $\underline{c}_2$, ... $\underline{c}_k$ which are scattered in the feature space. It is possible to commence the algorithm with randomly-selected centres although choosing positions which can reasonably be expected to represent real cover classes is better. The basic algorithm then proceeds in the following way:

(i) choose a new pixel;

(ii) allocate it to the cluster centre to which it is closest in N-dimensional feature space;

(iii) recalculate the means of each cluster;

(iv) return to (i).

The k-means algorithm is thus designed simply to cluster the pixels. Since there is no guarantee that the initial cluster centres accurately represent the best *group* centres for an image, practical implementations of the algorithm include extra processing for:

(a) merging two clusters if their means "drift" together within a minimum distance;
(b) scrapping a cluster if it has too few pixels and then reallocating its members to other clusters;
(c) breaking up a cluster into two smaller ones if it has a large variance of pixel locations.

After an unsupervised algorithm has been executed it is necessary to attach some physical meaning to the clusters so produced. This is not always so easy. In practice unsupervised algorithms may be used as a mechanism to help identify representative pixels for *supervised* algorithms. They are not commonly used as the sole means of classifying images for a specific application.

4.4 SUPERVISED CLASSIFICATION - NON-PARAMETRIC METHODS

Supervised classification follows the general model of pattern recognition outlined in section 4.1 -i.e. there is an initial learning or training phase followed by the actual classification. In the training phase it is customary to identify pixels (or better still segments) in test images for which ground data is available in order to localise specific ground cover classes in feature space. Once this phase has been undertaken a *discriminator function* must be devised to divide the feature space into regions for classification purposes. A very simple method for doing this is to draw linear boundaries in feature space to create regions for different classes -this is the approach of the *parallelepiped* or *box classifier* (see figure 20).

A more useful classifier uses the Euclidean minimum distance criterion between a pixel and the means of the training clusters -i.e. a pixel is assigned to the class with nearest mean position in feature space. These methods for classification are regarded as *non-parametric* or *distribution-free* methods since there is no consideration of how the training data pixels are scattered or distributed within each class. Although these methods are simple to carry out, they suffer from some serious drawbacks. For example the box classifier takes no account of the shape of training clusters, and the minimum distance classifier will always choose the nearest cluster centre even if, on statistical grounds, it is more likely that the pixel belongs to a further more widely-spread cluster.

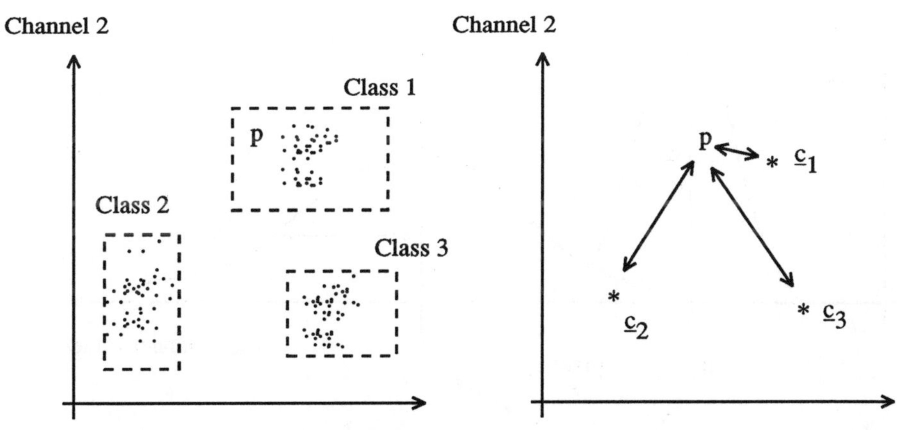

(a) Parallelepiped classifier

(b) Minimum distance classifier ($|\underline{x}_p - \underline{c}_i|$ is minimised)

Figure 20. Simple classifier methods.

4.5 SUPERVISED CLASSIFICATION - PARAMETRIC METHODS

Supervised parametric methods are designed to overcome some of the difficulties experienced with non-parametric classifiers. They rely on the use of a statistical model of the distribution of pixels in the classes.

A typical approach is the *maximum-likelihood classifier*. This is based on the *normal probability distribution function* which is used in a multivariate context -i.e. within the multiple dimensions of a multispectral image feature space. For an image with N channels an N-dimensional normal probability function is used to model a training class -i.e. the mean and standard deviation (or spread) of the training data points in the class are calculated in each dimension. Once the multivariate probability density functions have been obtained for each land cover class it is then possible to classify a pixel by assigning it to the distribution to which it has the highest probability of belonging. This is the basis of the maximum-likelihood classifier. Its behaviour is illustrated in one dimension in figure 21.

The probability of pixel x being in class i is given by:

$$p(x \mid i) = \frac{1}{(2\pi)^{\frac{1}{2}}\,\sigma_i} \cdot exp[\frac{-1}{2\sigma_i^2}(x - \mu_i)^2]$$

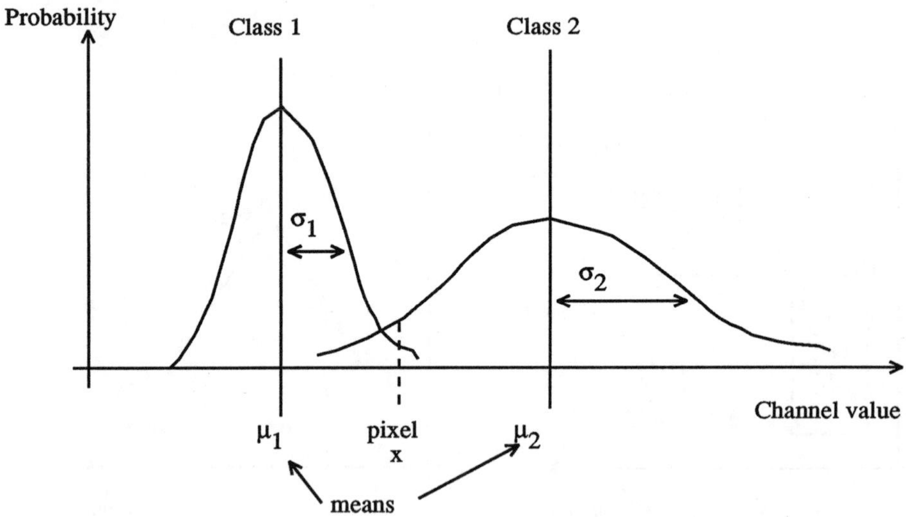

Figure 21. Maximum-likelihood classification. (Pixel x has greater probability of being in class 2, even though it is nearer the mean of class 1).

where μ_i and σ_i are the means and standard deviations for the distribution of class i. In practice this formula has to be applied in its more complex multivariate form.

The advantage of parametric methods over non-parametric classifiers can clearly be seen in figure 21. Pixel x is assigned to class 2 even though it is actually closer in distance to the mean of class 1. This is because class 2 is a more widely-spread class, whereas class 1 is very narrow. The disadvantage of parametric methods is that they attach a fixed model to the data clusters. This model may not truly represent reality. Further details may be found in texts such as James (1987) or Thomas, Benning and Ching (1987).

4.6 ACCURACY ASSESSMENT

The assignment of pixels to classes by any of the methods described above can not be done without error because all methods involve some form of approximation in the form of a model of class characteristics in feature space. Performance should therefore be estimated by using ground truth data. A suitable method for showing performance is to draw-up a class confusion matrix showing how pixels of each class (known from ground data) were actually classified. A typical matrix is shown in figure 22. There are several ways of interpreting the data in such a matrix. The simplest is to calculate the average of the leading diagonal figures. Ideally a classifier should have a score of 100% for each entry along the leading diagonal. In practice this is never achieved. Average performance exceeding 80% is rare.

A better estimate of accuracy is obtained by dividing the total of the leading diagonal entries (correct classifications) by the total of the off-diagonal entries (incorrect classifications). Story and Congalton (1986) also recommend calculating the 'map users' and 'map producers' accuracies. The former indicates how successfully a classified image cover type matches the cover type on the ground; the latter indicates how successfully a reference ground cover type is identified in the image. The map users accuracy is obtained by dividing the diagonal figure by the sum of the figures in the corresponding column and by contrast the map producers accuracy divides the diagonal figure by the sum of the corresponding row.

		CLASSIFICATION					
		Forest	Urban	Water	Agriculture	Uncultivated	None
TRUE CLASS	Forest	54	9	5	26	6	0
	Urban	11	62	6	8	4	9
	Water	8	1	75	8	3	5
	Agriculture	12	9	10	40	17	12
	Uncultivated	7	8	3	26	47	9

Average Performance = 55.6%

Figure 22. Typical class confusion matrix (numbers indicate percentage of pixels allocated to each class).

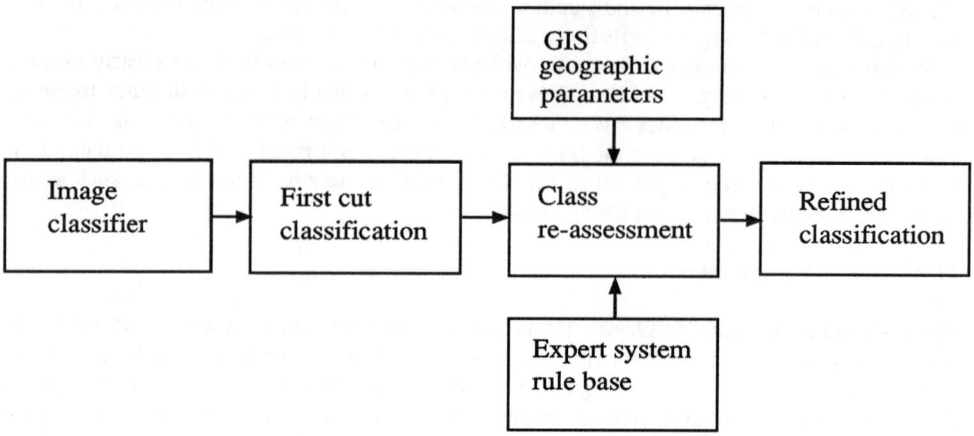

Figure 23. Concept of knowledge-based pixel class refinement.

4.7 PROBLEMS IN CLASSIFICATION AND NEW APPROACHES

Although methods for classification are now well-advanced, many real images lead to problems. Firstly, because pixels have some spatial extent, there will always be some which overlap several land cover classes. These "mixed" pixels can not be classified into any single class. Some attempts have been made recently to model mixed spectral signatures with the aim of assigning proportionate class membership.

Another difficulty is that it is often difficult to separate ground cover classes in feature space because they overlap significantly. This leads to class confusion i.e. it may be impossible to uniquely assign a pixel to one class. One new approach to solving this problem is to use expert systems and geographic information systems for contextual analysis (e.g. Wilkinson and Mégier, 1990). In this approach expert system rules are used to refine pixel classification on the basis of spatial or physical context (e.g. *"Vineyards do not exist above 1000m. altitude"*). In combination with geographic parameters drawn from a GIS expert system rules can help significantly in reducing class confusions -see figure 23.

5. Hard Copy Analysis

It is often useful in satellite image processing to produce hard copy products which can be used for application studies. These products may simply be photographic representations of the original image data or be classified "thematic maps". It is important to realise that products derived from imagery will normally preserve the raster array structure of the original satellite data. They are therefore very different from analogue form aerial photographs. Also the satellite hard copies may have been enhanced to highlight specific areas or cover classes.

There are several benefits to using hard copy products rather than digitally-stored image data, particularly for applications in developing countries. For example the images are recorded in a near-permanent form on a familiar medium like a map. They also do not need special computer equipment for examination and are therefore useful in field work. They are particularly useful in giving a synoptic overview of a region without the need for special processing.

If satellite data are used to create a thematic map or "satellite map" which directly shows land cover classes, it is important that those classes should be meaningful and exhaustive -i.e. that they describe a realistic land cover classification taxonomy. Also for ease of interpretability the classes should be represented in a cartographically pleasing way -i.e. with sensible selection of colours, symbols, textures etc.

The combined use of remote sensing and geographic information systems now enables thematic maps to be produced which have satellite imagery (or derived classifications) as a background with map information (e.g. roads, soils, height contours etc.) as overlays. These maps have enormous potential in resource planning activities and the satellite imagery can often lead to useful revisions of the map information by identifying the locations of new features which should be added to the maps. The merging of the twin technologies of remote sensing and geographic information systems is now yielding considerable benefits in many disciplines. This approach is likely to expand considerably in the next few years. Other chapters in this book discuss the technology and techniques of geographic information systems in more detail.

References

Ballard, D. H. and Brown, C. M., 1982, *Computer Vision* (Englewood Cliffs, NJ: Prentice-Hall).

Crist, E. P. and Cicone, R. C., 1984, A physically-based transformation of thematic mapper data -the TM tasseled cap, *IEEE Trans. on Geoscience and Remote Sensing*, **GE-22**, 3, 256-263.

Fu, K. S., 1982, Hybrid approaches to pattern recognition, in *Pattern Recognition Theory and Applications*, edited by J. Kittler, K. S. Fu and L. F. Pau, (Dordrecht: D. Reidel), 139-155.

Hill, J. and Sturm, B., 1988, Image-based atmospheric correction of multitemporal thematic mapper data for agricultural land cover classification, *Proc. IGARSS-88, held at Edinburgh, Scotland, 13-16 September 1988* (Noordwijk: ESA Publications Division) 895-899.

Hovis, W. A., Knoll, J. S. and Smith, G. R., 1985, Aircraft measurement for calibration of an orbiting spacecraft sensor, *Applied Optics*, **24**, 407-410.

James, M., 1987, *Pattern Recognition* (Oxford: Blackwell Scientific Publications).

Nagao, M. and Matsuyama, T., 1979, Edge preserving smoothing, *Computer Graphics and Image Processing*, **9**, 394-407.

Nevatia, R. and Babu, K. R., 1980, Linear feature extraction and description, *Computer Graphics and Image Processing*, **13**, 257-269.

Niblack, W., 1985, *An Introduction to Digital Image Processing* (Birkeroed, Denmark: Strandberg Publishing).

Pratt, W. K., 1978, *Digital Image Processing* (New York: John Wiley).

Slater, P. N., Biggar, S. F., Holm, R. G., Jackson, R. D., Mao, Y., Moran, M. S., Palmer, J. M. and Yuan, B., 1987, Reflectance and radiance-based methods for the in-flight absolute calibration of multispectral sensors, *Remote Sensing of Environment*, **22**, 11-37.

Story, M. and Congalton, R. G., 1986, Accuracy assessment; a user's perspective, *Photogrammetric Engineering and Remote Sensing*, **52**, 397-399.

Tailor, A. M., Corr, D. G., Cross, A., Hogg, D. C., Lawrence, D. H., Mason, D. C., Petrou, M. and Vango, R. D., 1988, Development of a knowledge-based segmentor for remotely sensed imagery, *Phil. Trans. Roy. Soc. Lond.*, **A 324**, 437-446.

Thomas, I. L., Benning, V. M. and Ching, N. P., 1987, *Classification of Remotely Sensed Images* (Bristol: Adam Hilger).

Wilkinson, G. G. and Mégier, J., 1990, Evidential reasoning in a pixel classification hierarchy -a potential method for integrating image classifiers and expert system rules based on geographic context, *International Journal of Remote Sensing* (in press).

Williamson, E. J. and Wilkinson, G. G., 1980, Sea surface temperature measurements from Meteosat, *Proc. Second Meteosat Scientific User Meeting, held at London, 26-27 March 1980*, (Darmstadt: European Space Agency), Annex J.

A QUANTITATIVE APPROACH TO REMOTE SENSING: SENSOR CALIBRATION AND COMPARISON

J. HILL
Commission of the European Communities,
Joint Research Centre,
Institute for Remote Sensing Applications,
21020 Ispra (Varese) Italy.

1. Introduction

The previous chapters have shown how information concerning the spectral properties of earth surface features can be captured by sensors on space platforms and recorded as a digital image. Furthermore these chapters have shown how atmospheric effects and sensor characteristics can influence the signal recorded in these digital images. Though this is known, in the past 20 years of remote sensing application research, most workers have used the integer values and ignored the physical magnitude and scientific units of the data. Unfortunately, comparisons between radiometric data obtained from different sources are only possible if the at satellite digital values are corrected to give some consistent physical measurement (Price 1987). This chapter examines ways in which this is possible. Quantitative approaches in remote sensing imply the assessment of a physical quantity and error limits on each step in the processing chain. Both radiometric and geometric fidelity are required in order to establish the ground reflectance at a given point on the earth and its location both to some known degree of accuracy.

Ideally, a comparison of several satellite scenes from the same area would show changes only in the scene's intrinsic properties and be invariant to other effects. But sensor measurements result from the interaction of several factors. Therefore the intrinsic spectral properties of surface materials must be separated from the effects of variations in sensor and viewing conditions (Robinove 1982, Woodham and Gray 1987). In addition, the different scenes of the same area may come from different sensors (see chapter four). Although more sensors mean increased opportunities for earth observation this necessarily complicates operational procedures.

For the use of multi-temporal or multi-sensor data sets one can choose to normalize radiance values to a standard set of conditions, or, instead, to retrieve the target reflectance factors by applying suitable radiometric corrections. Data normalization can be performed, to some extent by statistical methods, but these do not yield a quantitative measure of ground reflectance (Schott *et al.* 1988, Hill and Sturm 1990). Radiometric corrections to ground reflectance have the advantage of being directly comparable to radiometric measurements of spectral target signatures. These corrections must account for sensor characteristics (spectral bandpasses, calibration), scene-related effects (illumination conditions, observation geometry, atmospheric phenomena and topographic variations) and target reflectance characteristics, all of which are inter-related to some extent and have the same prerequisites, i.e. radiometric calibration, an atmospheric model and a target reflectance model (Teillet 1986).

A. S. Belward and C. R. Valenzuela (eds.),
Remote Sensing and Geographical Information Systems for Resource Management in Developing Countries, 97–110.

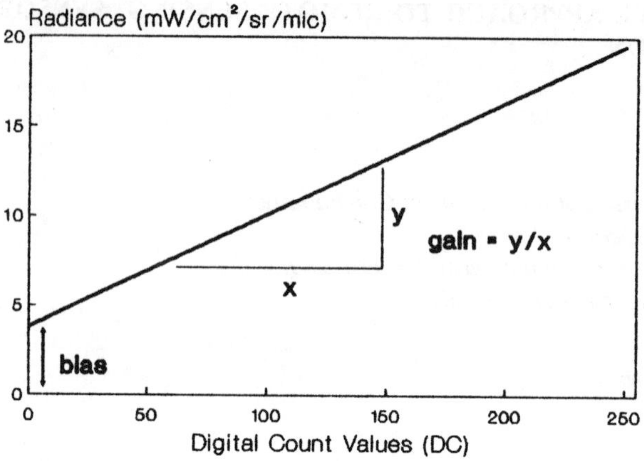

Figure 1. Calibration function relating radiance to digital count values.

2. Radiometric calibration of satellite sensors

The radiant flux leaving a unit area of a surface into a particular solid angle in a particular direction is called radiance (see Bird, chapter one). In optical remote sensing, it is the ground-reflected solar irradiance which carries information concerning the target spectral properties. After upward transmission through the atmosphere, the attenuated signal is registered by the satellite's detectors. Each of these detectors transforms the incident energy into an electric signal which is amplified, digitized, encoded in 8- (TM, SPOT), 10- (AVHRR) or 16-bit words and transmitted to various ground stations. From this digital count value, the at-satellite measured radiance can be retrieved by using a set of gain and bias pairs for each spectral band which represent the calibration of the remote sensing system (figure 1). Pre-launch laboratory testing of each instrument determines these calibration values. However, pre-launch values may change during launch, and later in space, due to heavy vibrations of the whole instrument, outgassing of the filters, response to pronounced temperature changes, slow deterioration of the electronic system, deterioration of the optical system or decreasing sensitivity of detector elements.

Calibration technology for satellite radiometers has been reviewed by Williamson (1977), the physical theory has been discussed by Wyatt (1978), and available calibration data for important operational sensor systems have been compiled by Robinove (1982), Malila and Anderson (1986), Markham and Barker (1986, 1987a) and Price (1987, 1988a, 1988b).

Some operational optical satellite systems have in-flight calibration facilities, others do not, or it is difficult to use this ancillary information. The quantitative use of satellite radiometry therefore suggests the need for ground verification of satellite measured radiances referring to ground areas of known reflectance. The large uniform area of gypsum sand at White Sands, New Mexico, has been thoroughly studied as a calibration site for the Landsat-4/-5 Thematic Mapper (Slater *et al.* 1986, 1987), the SPOT HRV (Begni *et al.* 1986, Dinguirard *et al.* 1988, Begni 1988) and the NOAA AVHRR sensors (Frouin and Gautier 1987, Teillet *et al.* 1990). White Sands has been selected because it is a mainly flat, extended area and, in the visible and near-infrared, exhibits high uniform reflectance. In addition, it is close to being a Lambertian reflector and it is at an elevation of about 1200 m in a region where the aerosol loading is low and the expectancy of clear weather is high (Slater *et al.* 1987).

2.1. IN-ORBIT CALIBRATION METHODS

Reflectance-based methods for the in-orbit calibration of optical satellite data basically comprise three approaches (Teillet *et al.* 1990). Firstly, ground and atmospheric measurements and reference to another calibrated satellite sensor. Secondly, ground and atmospheric measurements with no reference to another sensor. Thirdly, no ground and atmospheric measurements but reference to another sensor. Such methods have been applied to several data sets involving TM and SPOT HRV (Slater *et al.* 1986, 1987, Begni *et al.* 1986, Holm *et al.* 1989, Hill and Aifadopoulou 1990), NOAA-7 AVHRR(Frouin and Gautier 1987) and NOAA-9/-10 AVHRR imagery (Teillet *et al.* 1990).

In the case of coincident data acquisitions from two satellite sensors (the third method outlined above), the analysis can also be based upon the apparent at-satellite reflectance $\rho*$. This parameter compensates for differences in illumination geometry. As shown in earlier chapters, if we assume lambertian ground reflectance, $\rho*$ can be computed for a given spectral band with

$$\rho* = \frac{L\,\pi\,d^2}{E_0 \cos\theta} \tag{1}$$

where θ is the solar zenith angle and d is a correction coefficient for the actual sun-earth distance. The exoatmospheric solar irradiance E_0 for the Thematic Mapper and SPOT HRV bandpasses can be obtained from Markham and Barker (1987b) and Price (1988a); the corresponding values for the AVHRR channels are found in Planet (1988). L denotes the at-satellite measured spectral radiances.

The two other methods for in-orbit calibration require the use of radiative transfer computations to predict the radiance at the aperture of the satellite sensor which is then compared to the corresponding digital counts. A model based on the formulation of radiative transfer as developed from Tanré *et al.* (1979, 1985) is often used. This code provides corrections for atmospheric scattering, absorption and pixel adjacency effects, and it makes extensive use of analytical expressions and pre-selected atmospheric models, resulting in a high precision and a short execution time. It directly relates the apparent at-sensor reflectance $\rho*$ to the target reflectance factor ρ_t according to

$$\rho* = t_g(\mu) \left\{ \rho_{at} + \frac{T(\mu_0)\,[t_d(\mu)\,\rho_t + t_s(\mu)\,<\rho>]}{1 - <\rho>\,s} \right\} \tag{2}$$

where ρ_t is the target reflectance, $<\rho>$ the background contribution to the apparent reflectance and s gives the spherical albedo; ρ_{at} denotes the intrinsic atmospheric signal component, μ and

μ_0 are the cosines of the zenith angles of the direction vectors from the pixel towards satellite and sun. $T(\mu_0)$ gives the total downward, $t_d(\mu)$ the diffuse and $t_s(\mu)$ the scattered upward transmittance; $t_g(\mu)$ denotes the upward gaseous transmittance of the atmosphere (Tanré *et al.* 1987).

2.2 LANDSAT THEMATIC MAPPER SYSTEM CALIBRATION.

TM at-satellite measured radiance L [mW/cm^2/sr/μm] for each spectral bandpass is retrieved by using a set of gain and bias pairs according to

$$L_{tm} = a_0 + a_1 \, DC \tag{3}$$

Given that pre-launch calibration gains (a_1) and offsets (a_0) have changed, the on-board TM calibration device can provide effective adjustments, but only if decreasing system sensitivity is not due to deterioration of the transmission properties of the optical system. Because of this limitation, the TM in-flight calibration has been analyzed through a series of contemporary ground measurements at White Sands (Slater *et al.* 1986, 1987). The results from these ground verification experiments have been used to determine post-launch calibration constants for the visible and near-infrared TM bands (TM1-TM4). Respective updates for the short-wave infrared bands appear to be more ambiguous due to stronger variations of the measurement, and have not been applied (Hill and Sturm 1990).

These post-launch calibration constants have been further analyzed during two radiometric ground measurements campaigns in 1984 and 1988 (Maracci *et al.* 1986, Maracci *et al.* 1990, Hill and Aifadopoulou 1989). Both campaigns provided ground and atmospheric measurements close to the time of Landsat-5 overflights, which could be compared to the results of radiative transfer calculations. Based upon the atmospheric correction method mentioned above (equation 2), the validity of the post-launch TM calibration constants was first confirmed during a Landsat-5 TM overflight in 1984 (Maracci *et al.* 1986, Hill and Sturm 1990). Additional measurements, in 1988, during three different TM overpasses (Maracci *et al.* 1990), re-confirmed the long term stability of these post-launch values. The comparison of the top-of-atmosphere (ToA) and ground measured target reflectance factors exhibited typical distortions which are due to atmospheric effects, but the atmospheric corrections produced a remarkably good coincidence between satellite- and ground-measured reflectance factors (figure 2). In the case of the ToA relationships, the bias values (a0) decrease as we move from the visible (TM1) to near-infrared (TM4) wavelengths, illustrating the decreasing importance of additive atmospheric scattering with increasing wavelength (see Bird, chapter 2). In the case of the atmospherically corrected data the bias values tend to approach zero; in other words the atmospheric correction has successfully removed these effects. Note also how the slope (a1) tends more closely to the ideal 1 to 1 situation. These results indicate that no significant degradation of the TM system response has occurred during the 1984 - 1988 period.

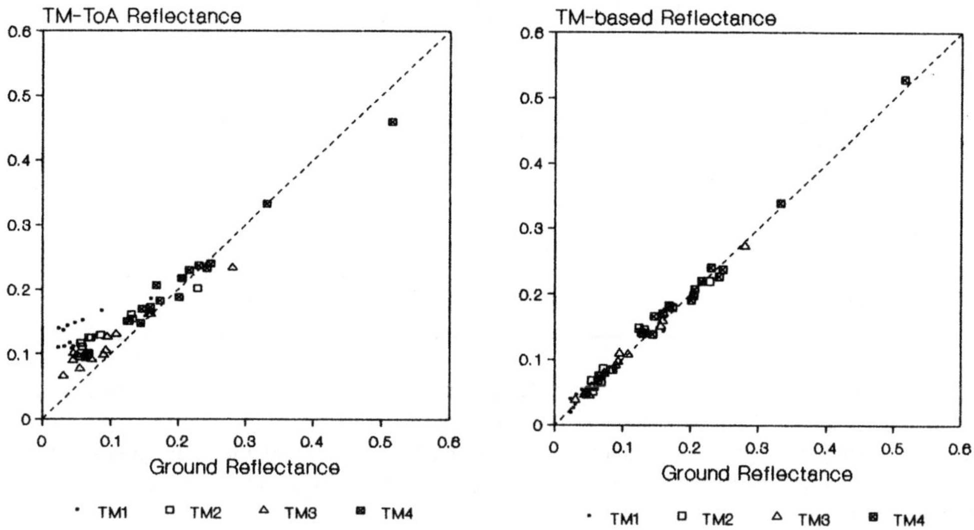

Band	ToA–reflectance a0	a1	r^2	RMSE	Ground reflectance a0	a1	r^2	RMSE
TM1	.108	.515	.778	.017	.009	.829	.966	.009
TM2	.070	.617	.941	.011	.007	.973	.983	.009
TM3	.054	.659	.974	.011	.005	.967	.983	.009
TM4	.046	.813	.989	.012	.002	1.002	.994	.009

Figure 2. Comparison of Thematic Mapper top-of-atmosphere (ToA) and modeled ground reflectances with associated ground measurements from the ground radiometry campaign of 1988

2.3 CALIBRATION OF THE SPOT-1 HRV-1 SENSOR

Because the TM instrument appears to be well calibrated, this can itself be used to help in the calibration of other instruments. For example, the radiometric accuracy of data from the SPOT HRV 1 has been examined through comparison with a contemporaneous Thematic Mapper scene. Both data sets were acquired on the 29th September 1986 with a time difference of only one hour between the satellite overpasses (Hill and Aifadopoulou 1990). This introduced minor changes in illumination geometry (table 1). Nadir viewing was assumed for the radiometric analysis although there was some small deviation from nadir viewing with the view angles for the overlapping parts of the two data sets being 0° to 5° for TM, and -4.1° to -2.0° for SPOT.

The at-satellite radiance L_{xs} [W/m^2/sr/µm] for each SPOT HRV band is calculated with

$$L_{xs} = a_1^{-1}\,DC \tag{4}$$

Table 1 Thematic Mapper and SPOT scene characteristics

	TM	SPOT
WRS frames	197 / 29	47 / 260&261
Acquisition date	29–Sep–1986	
Acquisition time	9.49 GMT	10.47 GMT
Scene location	44.6° N	44.2° N
	3.72° E	3.85° E
Sun elevation	37.42°	42.20°
Sun azimuth	145.92°	163.95°
Instr. tilt	–	R2.4°

The SPOT calibration constants a_1 are updated monthly by using an on-board sun calibrator and are provided on each CCT. In March 1986 this internal calibration device was tested at White Sands and, after some initial adjustments, a precision of 5.4 % was specified (Begni *et al.* 1986). Later on the calibration gain of the near-infrared bandpass (XS3) had to be increased by 8 % (Begni 1988).

Since the atmospheric conditions could be assumed not to have changed significantly between the TM and SPOT overpasses the radiometric analysis could be based on top-of-atmosphere (ToA) reflectances ($\rho*$), as shown in equation 1. The digital counts of the TM were thus converted to ToA values, and were then used to simulate equivalent ToA values for the SPOT HRV bands (Hill and Aifadopoulou 1990). By inverting equation 1 one can compute the corresponding radiance values (L), which, as shown earlier by Bird (chapter one) is what is recorded at the satellite. If the calibration of the TM instrument is correct then the simulated values should correspond to the actual values recorded by the SPOT HRV. Such a comparison, seen in figure 3, shows that in fact the agreement between the simulated and actual values is around 5%. In other words we can say that the in-flight calibration for the SPOT HRV is reliable, at least for the period covered by these images (Hill and Aifadopoulou 1990). Equally, these results provide evidence to confirm the precision in the reflectance-based calibration assessments and updates for both the TM and HRV instruments (Slater *et al.* 1986, 1987, Begni *et al.* 1986, Begni 1988), and they emphasize the importance of such approaches for the quantitative use of satellite data.

2.4 THE CALIBRATION OF THE NOAA-AVHRR INSTRUMENTS

The AVHRR only produces fully (in-flight) calibrated data for its thermal infrared channels (Planet 1988). At the time these sensors were conceived and designed a strong need for absolutely calibrated data in the visible and near-infrared channels was not foreseen. Therefore no in-flight calibration device for channels 1 and 2 was included in the instrument design. However, at least pre-launch calibrations were performed on each AVHRR instrument (Malila and Anderson 1986, Abel 1990).

The calibration functions for the visible channels directly give the so-called target albedo (A)

$$A = i + g\,DC \tag{5}$$

where gain (g) and intercept (i) are deduced from pre-launch calibration (Planet 1988).

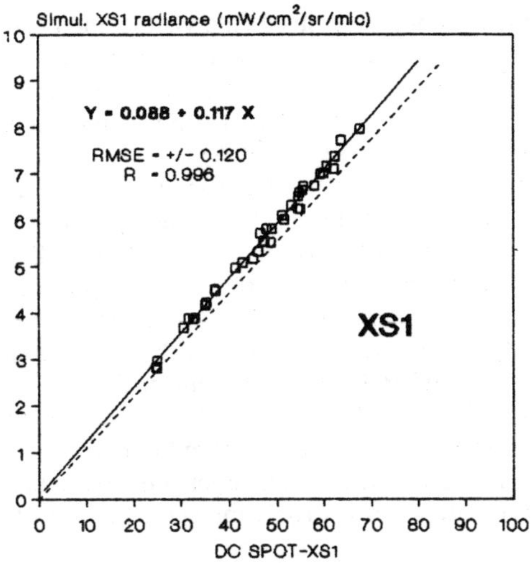

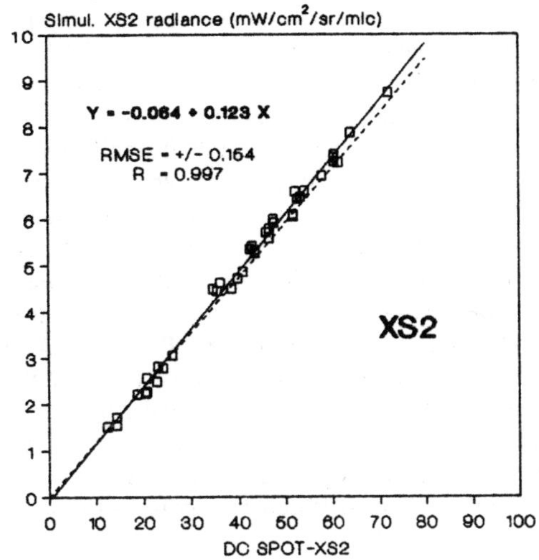

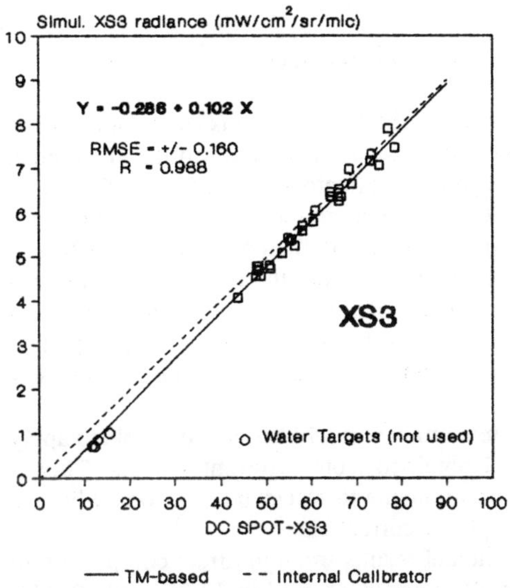

Figure 3. Comparison of TM-based and in-flight SPOT calibration coefficients as derived from simultaneously recorded satellite data

To convert from percent albedo A to radiance L [W/m^2/sr/μm] use the equation

$$L = \frac{E_0 A}{\Delta\lambda \, \pi \, 100} \tag{6}$$

where E_0 is the integrated exoatmospheric solar irradiance weighted by the spectral response function of the channel, and $\Delta\lambda$ is the equivalent width of the spectral response function of the channel (Planet 1988).

There is in fact strong evidence that the pre-launch calibration gains of the visible and near-infrared channels of the AVHRR on NOAAs 7 to 10 have decreased by between 10 to more than 30% after 2 years in orbit (Frouin and Gautier 1987, Smith *et al.* 1988, Teillet *et al.* 1990). There is also some evidence that the sensor degradation is usually in the range of 0 to 15 % after 1 year in orbit, and that the gains tend to stabilize after two years in orbit (Planet 1988).

Several studies have attempted to determine in-flight calibration of the visible and near-infrared NOAA-9, 10 and 11 AVHRR channels over low (ocean surfaces) and highly reflective (deserts) ground areas (Abel *et al.* 1988, Kaufman and Holben 1989, Holben *et al.* 1990, Teillet *et al.* 1990). The different results are consistent and seem to support the above-mentioned characteristics, but it is evident that reflectance-based methods are more difficult to apply than to high spatial resolution data of TM or SPOT-HRV type. This is mainly due to the low spatial resolution of the AVHRR sensor (see chapters by Hill and Belward elsewhere in this book): uniform reference targets must have at least the size of an AVHRR-pixel, it should not exhibit strong contrast to the surrounding areas, and the complete reference site should approach as closely as possible lambertian reflectance characteristics. Given that such a target could be found, it might still be difficult to precisely locate the target site in an AVHRR image(Teillet *et al.* 1990).

3. Comparison of Spectral Features for Multiple Sensor Approaches

If the raw digital counts are converted to calibrated physical measurements, using any of the methods introduced above, it becomes possible to directly compare spectral information recorded by different sensors. For example, greenness and brightness indicators have proven of value in numerous vegetation studies. The Normalized Difference Vegetation Index (NDVI) is one of the most widely used of these indices (see the chapters by Baret and Bartholomé later in this book). The NDVI is a linear combination of reflectance measurements from the red and near-infrared spectral bands. As seen in chapter 4 the position and width of these two bands varies slightly from system to system. Nevertheless, ground based measurements simulating the most important operational satellite sensors have shown that NDVIs from any of these systems are linearly related (Gallo and Daughtry 1987).

If we can verify such theoretical results by using real imagery it will become possible to apply these experimental relationships to combine NDVI obtained from different sensors. This of course requires correction of the satellite sensor measurements to reflectance factors, which in turn requires precise instrument calibration and atmospheric corrections.

Attempts have been made to validate the experimental results through direct comparison of the NDVI computed from contemporary SPOT and TM images, as used in §2.3. - table 2 (Hill and Aifadopoulou 1990). After having applied suitable atmospheric corrections, the SPOT and TM vegetation indices were calculated from the retrieved reflectance factors ρ_t in the red and near-infrared spectral bands. These NDVIs were then compared using a regression analysis based on more than 40 reference areas. Results (table 2) compare well with the experimental results of Gallo and Daughtry (1987).

The analysis described above was carried out using two sets of calibration coefficients for the SPOT data; one from its own in-flight calibrator, the other obtained from comparison to TM (§2.3.). Table 2 shows how these calibration coefficients influence the relationship between NDVI from the two sensors. Figure 4 shows this in terms of the distribution of NDVI values. This demonstrates that even in cases where excellent inter-calibration of satellite sensors exists,

Table 2. Coefficients for the linear relationships between the NDVI of the Landsat-5 TM and the SPOT HRV-1 systems

	Dep. Variable: $NDVI_{xs}$			Dep. Variable: $NDVI_{tm}$			
	a_0	a_1	r^2	a_0	a_1	r^2	SPOT Calibration
	.073	.907	.998	–.079	1.10	.997	in–flight
	.032	.932	.998	–.033	1.071	.999	TM–based
*	.040	.953	.999	–.042	1.049	0.999	

*) from spectral ground measurements (Gallo & Daughtry, 1987)

Figure 4. Histogram showing the distribution of NDVI values from the simultaneously acquired TM and SPOT scenes, where the SPOT NDVI has been transformed to TM-equivalent values using the function given by Gallo and Daughtry (1987)

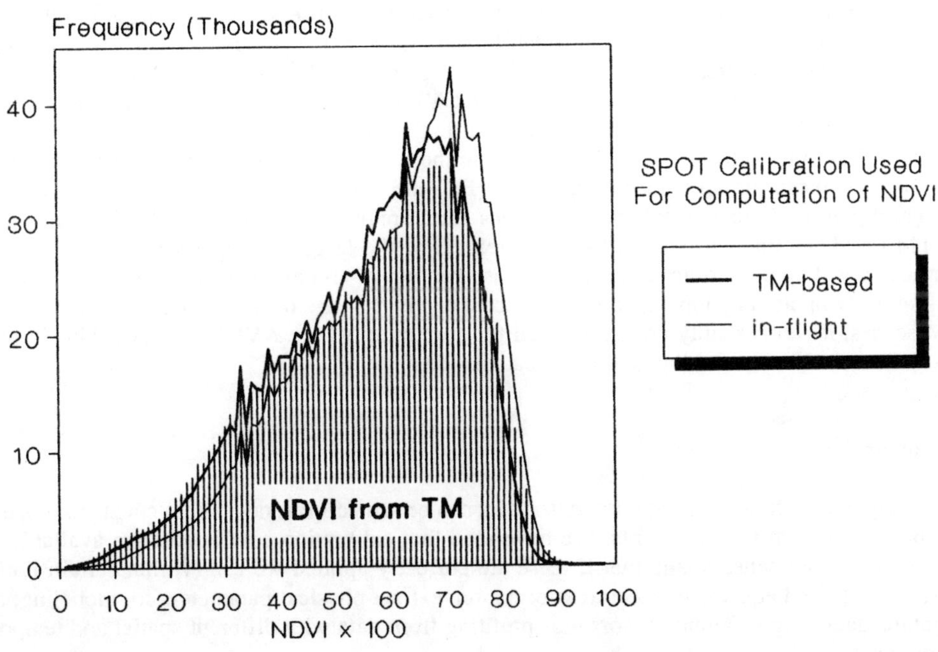

small calibration differences can lead to discrepancies in NDVI values. So long as these limits are borne in mind the study confirms the theoretical results and shows that NDVI from SPOT and TM can be used together, provided data are properly calibrated. These observations also hold true for other spectral features such as the Tasselled Cap equivalent greenness and brightness indicators (Hill and Aifadopoulou 1990).

This unfortunately is not true for the AVHRR. Although the theoretical study suggests that AVHRR vegetation indices are linearly related to those of the TM and SPOT wavebands, the use of real AVHRR data in multiple sensor configurations involves numerous complications.

Firstly, the calibration of the various AVHRR instruments can be monitored far less precisely than for the SPOT and TM sensors. This causes some problems for each individual AVHRR sensor, and may seriously affect approaches which have to use long-term time series of AVHRR-based vegetation indices from different NOAA satellites (see Belward, elsewhere in this book).

Secondly, the AVHRR scans across a swath of 2700 km. This changes sun-target-sensor geometry across an image; for example the solar zenith angle changes, at 30° latitude, amount to approximately 20° (Holben and Fraser 1984). The related variations in viewing and illumination characteristics across an AVHRR scan line result in variable sensor responses to the same target. These radiometric effects create even more complications once data from both odd- and even-numbered NOAA AVHRRs are used. Due to their orbital characteristics (morning and afternoon overflights), a specific site on the ground would, on the same day, be viewed under different conditions. The observations would include different satellite scan, satellite azimuth, solar elevation and solar azimuth angles which introduce modifications of the sun-target-satellite geometry, and the observations would most likely be made under different atmospheric conditions (Gallo and Eidenshink 1988). The influence of these effects on the red and near-infrared wavebands, and NDVI, of the AVHRR have been observed or simulated (Holben and Fraser 1984, Holben 1986, Gallo and Eidenshink 1988) and it appears that the visible and near-infrared responses increase with increased satellite scan angles. The vegetation index is less influenced by increased scan angles: The NDVI is generally minimized at the extreme backscatter angles, and maximized near the nadir in the forward scattering direction (Holben and Fraser 1984). For multi-temporal AVHRR data, the generation of maximum-value composites (Holben 1986) can further reduce these effects (Belward, chapter 9).

Thirdly, modelistic atmospheric corrections are more difficult to apply to AVHRR data than to TM or SPOT images. This is basically related to the large area coverage: over terrestrial surfaces a widely used simplification is the assumption of an homogeneous atmosphere. While this simplifying assumption appears to be acceptable in the case of SPOT and, to a lesser extent, TM scenes, it can certainly not be adopted for correcting whole AVHRR scenes. This implies restrictions to the use of radiative transfer calculations.

4. Conclusions

This chapter has shown that satellite sensor data can be turned from raw digital counts to properly calibrated physical measurements. We have seen that calibration coefficients are available for some, but not all sensors, and that if these are correctly applied we can (within certain limits) determine ground equivalent spectral information, extract physical parameters for modeling, and combine data from different sensors thus profiting from data with different spatial and temporal resolutions.

There is a need for better documentation of pre-flight instrument calibration. Improved on-board calibration devices are indispensable for future systems. Additionally the systematic assessment of the sensor's long term stability will still need verification through ground based methods. This is particularly so as the emphasis in remote sensing studies moves from the empirical approaches of the past to more quantitative techniques. For example, the vegetation canopy models used for the quantitative estimation of agronomic variables (such as biomass) require physical parameters such as radiance and reflectance (see Baret, chapter 8).

In reviewing sensor calibration this chapter has referred to much of the published work on this subject. Users of multi-sensor and multi-temporal remote sensing data should now consult the relevant papers for full descriptions of these techniques.

References

Abel, P., G.R. Smith, R.H. Levin and H. Jacobowitz, 1988, Results from aircraft measurements over White Sands, New Mexico, to calibrate the visible channels of spacecraft instruments in *Recent Advantages in Sensors, Radiometry, and Data Processing for Remote Sensing*, edited by P.N. Slater Proc. SPIE, 924, 208-214.

Abel, P., 1990, Prelaunch calibration of the NOAA-11 AVHRR visible and near IR channels, *Remote Sensing of Environment,* **31**, 227-229.

Begni, G., 1988, Absolute calibration of Spot data, Spot Newsletter, no. 10, 2-3.

Begni, G., M.C. Dinguirard, R.D. Jackson and P.N. Slater, 1986, Absolute calibration of the SPOT-1 HRV cameras, SPIE, 660, 66-76.

Dinguirard, M., G. Begni and M. Leroy, 1988, SPOT-1 calibration results after two years of flight, "Recent Advantages in Sensors, Radiometry, and Data Processing for Remote Sensing", P.N. Slater (ed.), Proc. SPIE, 924, 89-95.

Frouin, R. and C. Gautier, 1987, Calibration of NOAA-7 AVHRR, GOES-5, and GOES-6 VISSR/VAS solar channels, *Remote Sensing of Environment,* **22**, 73-101.

Gallo, K.P and C.S.T. Daughtry, 1987, Differences in vegetation indices for simulated Landsat-5 MSS and TM, NOAA-9 AVHRR, and SPOT-1 sensor systems, *Remote Sensing of Environment,* **23,** 439-452.

Gallo, K.P. and J.C. Eidenshink, 1988, Differences in visible and near-IR responses, and derived vegetation indices, for the NOAA-9 and NOAA-10 AVHRRs: a case study, *Photogrammetric Engineering and Remote Sensing,* **54,** 485-490.

Hill, J. and D. Aifadopoulou, 1989, Scene-based atmospheric correction of Thematic Mapper imagery acquired during 1988 over the Ispra/Novara region, Italy, Joint Research Centre Ispra, Institute for Remote Sensing Applications, Technical note, 69 pp.

Hill, J. and B. Sturm, 1990, Radiometric corrections of multi-temporal Thematic Mapper data for the use in land-cover classification and vegetation monitoring, *International Journal of Remote Sensing*, (in press).

Hill, J. and D. Aifadopoulou, 1990, Comparative analysis of Landsat-5 TM and SPOT HRV-1 data for use in multiple sensor approaches, *Remote Sensing of Environment*, (in press).

Holben, B.N. and R.S. Fraser, 1984, Red and near-infrared sensor response to off-nadir viewing, *International Journal of Remote Sensing*, **5**, 145-160.

Holben, B.N., 1986, Characteristics of maximum-value composite images from temporal AVHRR data, *International Journal of Remote Sensing*, **7**, 1417-1434.

Holben, B.N., Y.J. Kaufman and J.D. Kendall, 1990, NOAA-11 AVHRR visible and near-IR inflight calibration, *International Journal of Remote Sensing*, **11**, 1511-1519.

Holm, R.G., R.D. Jackson, B. Yuan, M.S. Moran, P.N. Slater and S.F. Biggar, 1989, Surface reflectance factor retrieval from Thematic Mapper data, *Remote Sensing of Environment*, **27**, 47-57.

Kaufman, Y.J. and B.N. Holben, 1990, Calibration of the AVHRR visible and near-IR bands by atmospheric scattering, ocean glint and desert reflection, *Journal of Climatology and Applied Meteorology*, in press.

Malila, W.A. and D.M. Anderson, 1986, Satellite data availability and calibration documentation for land surface climatology studies, International Satellite Land Surface Climatology Project (ISLSCP), ISLSCP Report No. 5, NAS5-28715, Environmental Research Institute of Michigan, Ann Arbor, Michigan.

Maracci, G., G. Andreoli and B. Hosgood, 1986, Measurements of spectral signatures in less favoured areas (LFA): a contribution to the definition of a remote sensing multi-temporal experiment, *ESA/EARSeL Symposium on Europe from Space, Lyngby, DK, 25-28 June 1986, ESA SP-258*, pp215-218.

Maracci, G., B. Hosgood, G. Andreoli and P. Grassi, 1990, Varese/Novara 88: Field Radiometry, Commission of the European Communities, Joint Research Centre Ispra, Institute for Remote Sensing Applications, Technical note T.N.I. 90.05.

Markham, B.L. and J.L. Barker, 1986, Landsat MSS and TM post-calibration dynamic ranges, exoatmospheric reflectances and at-satellite temperatures, EOSAT Landsat Technical Notes, no. 1, 3-8, (Lanham, MD: EOSAT).

Markham, B.L. and J.L. Barker, 1987a, Radiometric Properties of U.S. processed Landsat MSS data, *Remote Sensing of Environment*, **22**, 39-71.

Markham, B.L. and J.L. Barker, 1987b, Thematic Mapper bandpass solar exoatmospheric irradiances, *International Journal of Remote Sensing*, **8**, 517-523.

Planet, W.G., 1988, Data extraction and calibration of TIROS-N-NOAA radiometers, NOAA Technical Memorandum NESS 101 - Rev. 1, U.S. Department of Commerce, National Oceanic and Atmospheric Administration, Washington, D.C.

Price, J.C., 1987, Calibration of satellite radiometers and the comparison of vegetation indices, *Remote Sensing of Environment,* **21**, 15-27.

Price, J.C., 1988a, An Update on visible and near infrared calibration of satellite instruments, *Remote Sensing of Environment,* **24**, 419-422.

Price, J.C., 1988b, Errata: calibration of the NOAA-10 AVHRR, *Remote Sensing of Environment,* **26**, 303.

Robinove, C. J., Computation with physical values from Landsat digital data, *Photogrammetric Engineering and Remote Sensing,* **48**, 781-784.

Schott, J.R., C. Salvaggio and W.J. Volchok, 1988, Radiometric scene normalization using pseudo-invariant features, *Remote Sensing of Environment,* **26**, 1-16.

Slater, P.N., S.F. Biggar, R.G. Holm, R.D. Jackson, Y. Mao, M.S. Moran, M. Palmer and B. Yuan, 1986, Absolute radiometric calibration of the Thematic Mapper, *SPIE,* **660**, 2-8.

Slater, P.N., S.F. Biggar, R.G. Holm, R.D. Jackson, Y. Mao, M.S. Moran, M. Palmer and B. Yuan, 1987, Reflectance- and radiance-based methods for the in-flight absolute calibration of multispectral sensors, *Remote Sensing of Environment,* **22**, 11-37.

Smith, G.R., R.H. Levin, P. Abel and H. Jacobowitz, 1988: Calibration of the solar bands of the NOAA-9 AVHRR using high altitude aircraft measurements, *Journal of Atmospheric and Oceanic Tech.,* **5**, 631-639.

Tanré, D., M. Herman, P.Y. Deschamps and A. De Leffe, 1979, Atmospheric modelling of the background contribution upon space measurements of ground reflectance, including bi-directional properties, *Applied Optics,* **18**, 3587-3594.

Tanré, D., C. Deroo, P. Duhaut, M. Herman, J.J. Morcrette, J. Perbos and P.Y. Deschamps, 1985, Effets atmosphériques en télédétection, logiciel de simulation du signal satéllitaire dans le spectre solaire, *Proceedings 3rd International Colloquium on Spectral Signatures of Objects in Remote Sensing, Les Arcs, France, 16-20 Dec.*, ESA SP-247, pp 315-319.

Tanré, D., P.Y. Deschamps, P. Duhaut and M. Herman, 1987, Adjacency effect produced by the atmospheric scattering in Thematic Mapper data, *Journal of Geophysical Research,* **92**, 12000-12006.

Teillet, P.M., 1986, Image correction for radiometric effects in remote sensing, *International Journal of Remote Sensing,* **7**, 1637-1651.

Teillet, P., Slater, P. N., Y. Ding, R.P. Santer, R.D. Jackson and M.S. Moran, 1990, Three methods for the absolute calibration of the NOAA AVHRR sensors in-flight, *Remote Sensing of Environment*, **31**, 105-120.

Woodham, R.J. and M.H. Gray, 1987, An analytic method for radiometric correction of satellite multispectral scanner data, *IEEE Transactions on Geoscience and Remote Sensing*, **GE-25**, 258-271.

Williamson, L.E., 1977, Calibration technology for meteorological satellites, Atmospheric Sciences Laboratory Monograph No. 3, White Sands Missile Range, NM.

Wyatt, C.L., 1978, *Radiometric calibration: theory and methods*, (New York: Academic Press).

THE CURRENT STATUS OF ERS–1 AND THE ROLE OF RADAR REMOTE SENSING FOR THE MANAGEMENT OF NATURAL RESOURCES IN DEVELOPING COUNTRIES

P.N. CHURCHILL AND A.J. SIEBER

Commission Of The European Communities,
Joint Research Centre,
Institute Of Remote Sensing Applications,
21020 Ispra (Varese), Italy.

1. Introduction

Radar remote sensing is often portrayed as a new technique with enormous potential for providing data required by application specialists in a wide range of natural resources fields, but as yet technology of the future with no current proven applications and many unknown quantities. To an extent this is true; the interaction of microwave energy with a target is complex and not yet fully understood; there are currently no operational space platforms with radar sensors, though over the next decade a number of radar based spacecraft, starting with Europe's ERS–1, will be launched ensuring a continuous stream of radar data of the earth until beyond the year 2000.

However "new technique" is not completely correct. Radar remote sensing has been used in an operational sense since the late 1960's for both geological and vegetation mapping (Trevett 1986). In addition, the fact that the interaction of radars with targets is not fully understood is not unique within the remote sensing world; the understanding of the interaction of visible and infrared electromagnetic energy with targets is also still to be completed.

Radar remote sensing is also often seen as a technique apart from other remote sensing methodologies, to be used in place of visible and infrared data. This concept is also not valid; radar remote sensing is but one of a number of techniques for gathering data over the earth's surface, and should be seen as complementary to other remote sensing techniques, such as those described elsewhere in this book.

This Chapter provides a basic introduction to radar remote sensing and its application to the management of natural resources. As such the Chapter begins with an introduction to the fundamentals of radar remote sensing in order to explain its mode of operation and to describe the way in which radar energy interacts with a target. This is followed by a description of the history and future of radar remote sensing sensors, including a detailed description of the first radar based satellite since 1978, ERS–1. The final part of the Chapter describes the potential and the application of radar remote sensing in a range of natural resource applications.

2. Fundamentals of radar remote sensing

In order to fully appreciate the applicability of radar remote sensing in the management of natural resources it is important that the fundamentals of the sensor operation, the data acquisition geometry and the basic characteristics of radar scattering are understood. This section provides this background, and starts with a review, via the radar equation, of the basic relationships governing the operation of a radar and the interaction of the radar waves with a target. The section continues with a review of the operation, and in particular the geometry, frequency and polarisation, of different radar systems,

A. S. Belward and C. R. Valenzuela (eds.),
Remote Sensing and Geographical Information Systems for Resource Management in Developing Countries, 111–143.

an understanding of which is fundamental to the successful interpretation of the data. The Section concludes with a description of the statistical nature of the data, and in particular considers speckle, and the scattering of radar waves by targets.

2.1 THE RADAR EQUATION

The radar equation is fundamental to the description of the physical relationship between the radar sensor, the scattering properties of the target and the measured signal; as such it will be briefly described here. The familiar form of the radar equation is:

$$P_r = \frac{P_t G^2 \lambda^2 \sigma}{(4\pi)^3 R^4}. \qquad (1)$$

where P_r is the power received and P_t the power transmitted from a radar via the antenna; this is concentrated by the antenna in one direction. The ratio of this power to that from an isotropic antenna, i.e. one that radiates the same power over all solid angles, is called the antenna gain, G. The object is at range R from the transmitting antenna; λ describes the operating frequency.

In the general case the target consists of several elements with different scattering coefficients σ. The backscattered energy measured by the radar's receiver is therefore to be calculated with the following equation instead of (1):

$$P_r = \frac{\lambda^2}{(4\pi)^3} \sum_{i=1}^{N} \frac{P_{ti} G_i^2 \sigma_i}{R_i^4}. \qquad (2)$$

where the subscript i denotes the N different scattering elements. If the assumption is made that σ_i is constant within an element of surface area $\Delta A_i, \sigma_i$ can be written as:

$$\sigma_i = \frac{\sigma_i \Delta A_i}{\Delta A_i}. \qquad (3)$$

Differentiation of this gives a differential scattering cross section σ_0:

$$\sigma_0 = < \frac{\sigma_i}{\Delta A_i} >. \qquad (4)$$

The brackets < > denote averaging over the ensemble. Equation (2) can then be written in differential form as:

$$P_r = \frac{\lambda^2}{(4\pi)^3} \int_{\text{illuminated surface}} \frac{P_t G^2 \sigma^0 dA}{R^4}. \qquad (5)$$

This is the radar equation for a surface target over which the radar backscattering coefficient σ_0 can vary.

2.2 RADAR SENSOR OPERATION

A radar sensor is a system that measures the time it takes for electromagnetic waves to travel from the transmitting antenna to the target and back again to the receiving antenna. The travelling speed is well known, it being the speed of light. Therefore by measuring the time delay between transmit and receive the radar measures the actual distance between the radar and the target of interest. A radar looking sideways therefore maps an illuminated area in terms of a slant range map.

Active radar sensors can be divided into two categories; imaging and non–imaging. For the purposes of the management of resources the non–imaging type of sensor, such as the altimeter or scatterometer, has so far been found to be of less importance than the imaging type of sensor; only imaging sensors will be discussed here.

Imaging radar sensors can also be divided into two categories; real aperture radars (RAR) and synthetic aperture radars (SAR). These sensors and their operation are discussed in more detail below.
2.2.1 *Real aperture radars* The configuration of a real aperture radar system is shown in figure 1. The radar antenna illuminates a surface strip to one side of the nadir track. This side looking configuration is necessary in order to eliminate right – left ambiguities from two symmetrical equidistant points in that radar returns from two different points would be received at the same time. It is important to note that a down looking radar, a so called altimeter, is not imaging an area, only a side looking radar does this.
As the platform moves a continuous strip of swath width S is mapped along the flight line. The swath width is given by:

$$S \simeq \frac{h\beta}{\cos^2\theta} = \frac{\lambda h}{W\cos^2\theta} \tag{6}$$

where β is the antenna beam width in elevation , W is the width of the antenna, h is the height of the platform and θ is the look angle.

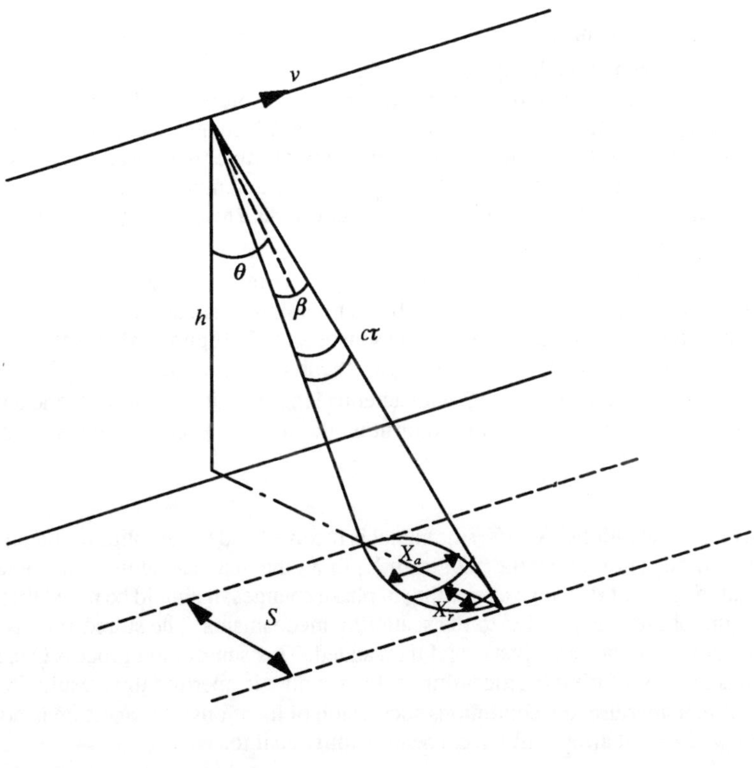

Figure 1. Geometry Of A Real Aperture Radar

The range resolution of this system, i.e. the resolution of the system across the swath, can be defined by examining the case of two points on a surface separated by a distance X_r and with their respective echoes separated by a time difference Δt of the form:

$$\Delta t = \frac{2X_r}{c}\sin\theta \tag{7}$$

The smallest discriminable time difference is equal to τ or $1/\beta$. On this basis the range resolution is given by:

$$2\frac{\Delta X_r}{c}\sin\theta = \tau$$

$$\Rightarrow \Delta X_r = \frac{c\tau}{2\sin\theta} = \frac{c}{2\beta\sin\theta} \tag{8}$$

The azimuth resolution, or the resolution of the system in the direction of the flight line, is equal to the width of the antenna footprint because the echoes from all the points along a line spanning that width are returned at the same time. On this basis the azimuth resolution is equal to:

$$\Delta X_a = \frac{h\beta'}{\cos\theta} = \frac{h\lambda}{L\cos\theta} \tag{9}$$

where β' is the antenna beam width in azimuth, h is the sensor height, λ is the wavelength, L is the length of the antenna and θ is the incidence angle.

From this equation it is clear that, firstly, the azimuth resolution will degrade across the swath; i.e. the scale of the data will not be consistent across the RAR image. Secondly it is also clear that if this type of system were to be applied on a spaceborne platform either the azimuth resolution would be in the order of hundreds of metres to kilometres, or the antenna would have to be considerably longer than the satellite in order to obtain a high resolution. As a result if high resolutions are required from space RAR's are not considered.

2.2.2 *Synthetic aperture radars* With the synthetic aperture technique the azimuth resolution is improved by the fact that the target remains in the beam for a significant amount of time, and is observed by the radar from numerous locations along the sensor's track. Figure 2 shows the geometry of the formation of a synthetic array by moving a single antenna along a track.

For the system described in figure 2 with antenna length L at a distance h and a wavlength λ the extension of the antenna main beam footprint on the surface F can be described as:

$$\mathrm{F} = \frac{2\lambda h}{L} \tag{10}$$

At each of these locations ($X_1 - X_i$) the signal is received and the amplitude and phase are stored as a function of time. Throughout the path from X_1 to X_n the distance between the antenna location X_i and the scattering point P changes, resulting in phase changes; it should be noted that this does not take into account phase changes due to the scattering mechanisms. The stored signals are then corrected so that they have the same phase and then added. This summation process is called coherent addition, and is the basis for SAR processing. A long synthetic aperture thus results from the movement of a short real aperture to a continuous succession of locations; this aperture is equal to F. On this basis the synthesised array will have a beam width equal to:

$$\theta_s = \frac{\lambda}{\mathrm{F}} = \frac{L}{2h} \tag{11}$$

with an along track extension of the footprint on the ground of:

$$\Delta X_a = h\theta_s = \frac{L}{2} \tag{12}$$

Two points are apparent from (12). Firstly, the further away the sensor is from the target, the larger the footprint is on the ground, and thus the longer the synthetic array. This results in a finer synthesised beam, which exactly counter balances the increase in distance; this overcomes the problem noted for the real aperture radar systems, whose azimuth resolution degrades with increasing range. The second point is that the smaller the antenna, the larger the footprint and the synthetic array. This again results in a finer synthesised beam and therefore a finer resolution. However, one should keep in mind that a smaller antenna results in a wider beam; this means that the impinging energy per unit ground area is lower. In order to reach the same level of energy per unit ground area with a smaller antenna requires more energy to be transmitted.

2.3 MEASUREMENT PARAMETERS

Besides the illumination geometry two principle parameters govern the transmission characteristics of the radar signal; the frequency and the polarisation. These are both described in more detail below.

2.3.1 *Frequency* Radars operate in a wide frequency range extending from approximately 220 MHz. to 40 GHz and higher. At an early stage in the development of radar a letter code was employed to denote frequency bands within this range. This was primarily undertaken so as to preserve military secrecy, but has been preserved to the present day and is frequently applied in radar studies. Table 1 lists the radar frequency band nomenclature adopted by the IEEE (IEEE 1976).

2.3.2 *Polarisation* A pulse of radar energy can be transmitted and / or received in different polarisation modes. The signal can be filtered in such a way that its electrical wave vibrations are restricted to a single plane perpendicular to the direction of wave propagation.

Band Designation	Nominal Frequency Range	Nominal Wavelength Range	Specific Radiolocation
HF	3 – 30 MHz.	100 – 10 m	
VHF	30 – 300 MHz.	10 – 1m	138 – 144 MHz. 216 – 225 MHz.
P	300 – 1000 MHz.	1000 – 300 mm	420 – 450 MHz. 890 – 942 MHz.
L	1000 – 2000 MHz.	300 – 150 mm	1215 – 1400 MHz.
S	2000 – 4000 MHz.	150 – 75 mm	2300 – 2500 MHz. 2700 – 3700 MHz.
C	4000 – 8000 MHz.	75 – 37.5 mm	5250 – 5925 MHz.
X	8000 – 12000 MHz.	37.5 – 25 mm	8500 – 10680 MHz.
K_u	12 – 18 GHz.	25 – 16.67 mm	13.4 – 14.0 GHz. 15.7 – 17.7 GHz.
K	18 – 27 GHz.	16.66 – 11.11 mm	24.05 – 24.25 GHz.
K_a	27 – 40 GHz.	11.11 – 7.5 mm	33.4 – 36.0 GHz.

Table 1. Standard Radar Frequency Band Nomenclature

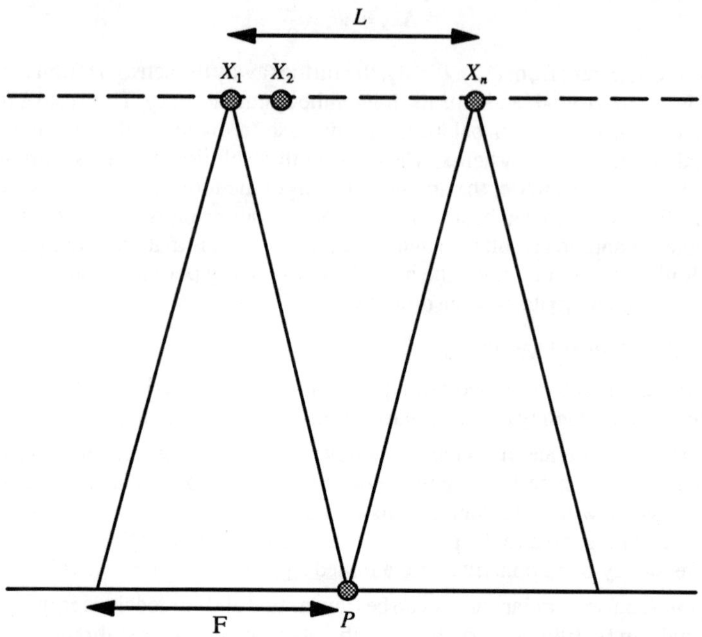

Figure 2. Geometry Showing The Formation Of A Synthetic Array By Moving A Single Antenna Along A Track .

Thus a radar signal can be transmitted in one polarisation while the backscattered signal can be received in a different polarisation. The most common polarisation combinations are those that utilise linearly polarised waves, usually characterised as horizontally (H) polarised or vertically (V) polarised waves. Using these linear orthogonal polarisations it is possible to have four different combinations of signal transmission and reception:

Transmit	**Receive**	**Notation**	**Type**
Horizontal	Horizontal	HH	Like
Horizontal	Vertical	HV	Cross
Vertical	Vertical	VV	Like
Vertical	Horizontal	VH	Cross

It should be noted that a general eliptical range of polarisations that can be applied are not restricted to linear states, circular polarisations can also be applied. Indeed a number of current radar systems are making full use of this capability by employing imaging radar polarimetry techniques, which measure the magnitude and the phase at four polarisations (HH, HV, VH, and VV) and from this construct a complete scattering matrix from which it is possible to synthesize any given polarisation (van Zyl *et al* 1987)

2.4 GEOMETRIC CHARACTERISTICS OF RADAR IMAGES

As noted in § 2.2 the geometry of imaging radar systems is fundamentally different from the majority of optical and infrared systems because of its side looking operation and its principle characteristic of measuring the travelling timeof signals. This has a number of effects on the data that require consideration, the first being relief displacement and the second being radar shadow.

2.4.1 *Relief displacement* Relief displacement in radar images is one dimensional and perpendicular to the line of flight. When a vertical feature on the ground is encountered by a radar pulse the top (point b in figure 3) of the feature A may be reached before the base (point a in figure 3), thus the backscatter from the top of the feature (point b_1) returns to the radar before that of the bottom (point a_1). As a result the feature will appear to "lean" towards the sensor in the image, a phenomena that is generally termed "layover".

For feature B, although no layover occurs, the radar pulse reaches the base of the feature (point c in figure 3) before the top (point d in figure 3). As such the feature will not be represented in its true size on the image (see points c_1 and d_1 in figure 3), but will in fact be compressed; this is termed "foreshortening".

2.4.2 *Radar shadow* The effects of the side looking operation of radars can have further effects, particularly in areas of exaggerated topographic relief. Figure 3 illustrates an example . For feature A the right hand side of the slope faces away from the incident radar wave, but is less steep than the depression angle and will therefore be illuminated by the radar, albeit weakly. Feature B shows a slope on the right hand side that is steeper than the depression angle. As a result the non–illumination will extend beyond the sloped area to incorporate features down range of the slope. This non–illumination is termed "radar shadow".

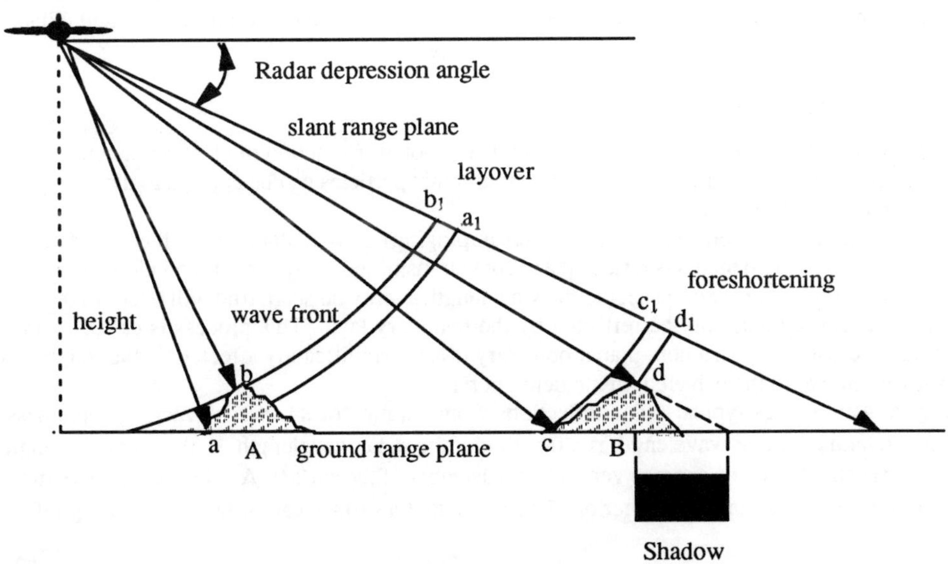

Figure 3. Effects Of Terrain Relief On Radar Images

It should also be noted that the depression angle changes across the swath, and that this will have an effect on the degree of radar shadow that will occur. Thus a feature that casts an extensive radar shadow at far range can be completely illuminated at near range.

2.5 SIGNAL FADING

The incident radar signal continually illuminates a given surface area which can contain many scattering points. All signals that come back to the receiver within a given time window Δt (recipricol of the system bandwidth $\frac{1}{\Delta F}$) are received coherently. The resulting backscattered signal is consequently a coherent addition of the echoes from a large number of scatterers. These returns add vectorally and result in a single vector which approximates to the amplitude V and phase ϕ of the total echo. The phase ϕ of each elementary vector is mainly related to the distance between the sensor and the corresponding scattering point, ignoring the phase changes due to the scattering mechanisms.The resulting signal is one sample out of a statistical ensemble. Even small changes in the position of the scatterers and / or movements of the receiver will result in a new and different coherent value, which means another sample out of a statistical ensemble. If the sensor or the scatterer moves by a small amount all the phases ϕ will change in the composite amplitude V. Thus successive observations of the same type of scatterers on the surface area as the sensor moves by will result in a different value of V. This variation is called fading.

As a result, in order to characterise the backscatter properties of a surface area many observations will be required and averaged. In the same vein for two neighbouring areas of similar overall scattering characteristics which are different at the level of fine detail the backscatter return will be different. Thus an image of an homogeneous area with a constant backscatter cross section will demonstrate a brightness variation from one pixel to the next. This effect is termed "speckle"; in order to measure the backscatter cross section of this area the return from many pixels should be averaged (Elachi 1987).

2.6 SCATTERING OF RADAR WAVES

The basic principles of reflection, scattering and diffraction or refraction of electromagnetic waves from plain boundary layers, rough surfaces and individual particles can be applied to the scattering of radar waves by natural or artificial targets.

Some of the energy of electromagnetic waves impinging on a boundary layer which is not necessarily planar between two media is reflected and some transmitted into the medium below. If the medium below is homogeneous in respect to the wavelength energy no scattering will occur in the medium, and the energy will only be reflected by the boundary layer. This process is called surface scattering. The scattering on a non–planar boundary layer is significantly affected by the roughness of this layer relative to the wavelength incident upon it.

Figure 4 illustrates typical radar reflections from a range of surface of varying roughness. "Rough" in terms of radar wavelength is a function of the radar wavelength λ, the incidence angle of the radar beam θ_i and the average vertical displacement of the surface Δh. One qualitative measure of roughness is the Rayleigh criterion. This states that a surface can be considered rough if:

$$\Delta h \cos\theta_i > \frac{\lambda}{8} \tag{13}$$

The "rougher" the target, the more diffuse the scatter and the more energy can be scattered back to the radar resulting in a brighter signature.

If the medium below the separating layer is inhomogeneous in terms of being comprised of materials of variable dielectric constant (see § 2.6.2) some of the energy entering it can be scattered back into the medium above the boundary layer. This is called volume scattering.

As a result it is clear that incident radars are modified by two quantities; the geometry and the electrical properties (or dielectric constant) of the target. Each of the quantities are dealt with in more detail below.

2.6.1 *Sensitivity to target structure* The wavelengths of electromagnetic radiation recorded by optical sensors, as described elsewhere in this book, are most sensitive to structure at the micrometer scale, whereas in contrast radar responds to structural variations of the centimeter and decimeter scales. Thus, a radar operating at a given wavelength is most sensitive to scatterers whose sizes are of the order of that wavelength (see figure 5).

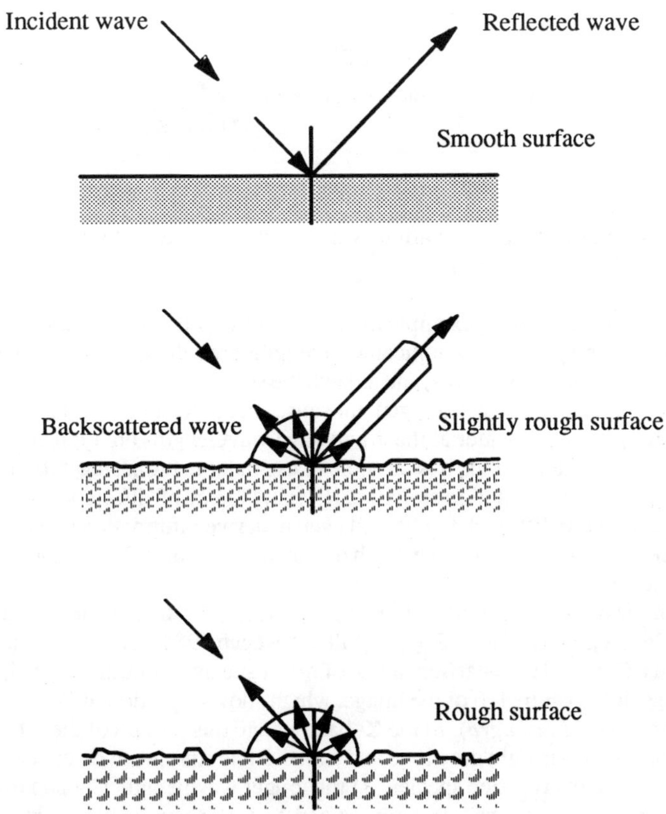

Figure 4. Radar Reflection From A Variety Of Surfaces

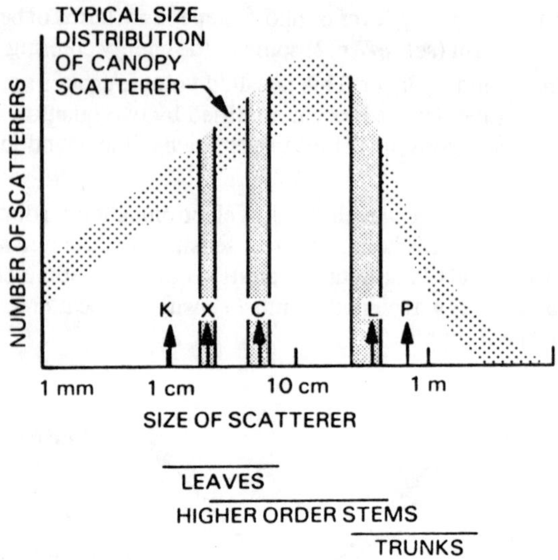

Figure 5. Distribution of scatterers at various wavelengths according to the size distribution of the canopy scatterers (after NASA 1986)

This has important bearing for example in the study of vegetation. The micrometer scale scatterers controlling spectral response in the optical wavelengths are cell organelles, cell walls and intercelular air space arrangements for example, and as such the spectral response provides little information about the vegetation structure in a more general sense. In contrast the centimetre / decimetre scale scatterers controlling the backscatter at the microwave wavelengths are typically the leaves, stems, fruits and branches of the target, thus it is possible to use radar to probe vegetation canopies and thus to derive information about vegetation structure. By using radars of different wavelengths it is possible to probe the canopies to different depths, and even to derive information about the type and nature of the ground beneath the vegetation. The depth of penetration will depend upon the frequency, polarisation and incidence angle used.

A good example of the use of multi–frequency SAR, and the different information each frequency is able to provide, is given in figure 6 (Churchill and Keech 1984). This example shows an X–band HH polarised and a C–band HH polarised image of forest and agricultural land in the east of England. The area of interest is the centre left of the image, which shows a portion of Scots pine (*Pinus sylvestris*) and Corsican pine (*Pinus nigra*). In the X–band image this portion of the scene is characterised by a relatively homogeneous light grey tone and a medium texture, with no difference in backscatter response apparent from the two tree species. At the longer wavelength C–band this is no longer the case with the Corsican pine, to the left, giving a darker tone than the Scots pine. In other words at X–band the structure of the Scots and Corsican pine appear similar, whereas at C–band they do not. This difference is possibly due to the fact that the C–band is sensitive to scatterers of different length and is thus being modified by different elements of the canopy.

(a) X–band HH

(b) C–band HH

Figure 6. X–Band HH Polarised And C–Band HH Polarised Airborne SAR Images Of Thetford Forest, UK (after Churchill and Keech 1984).

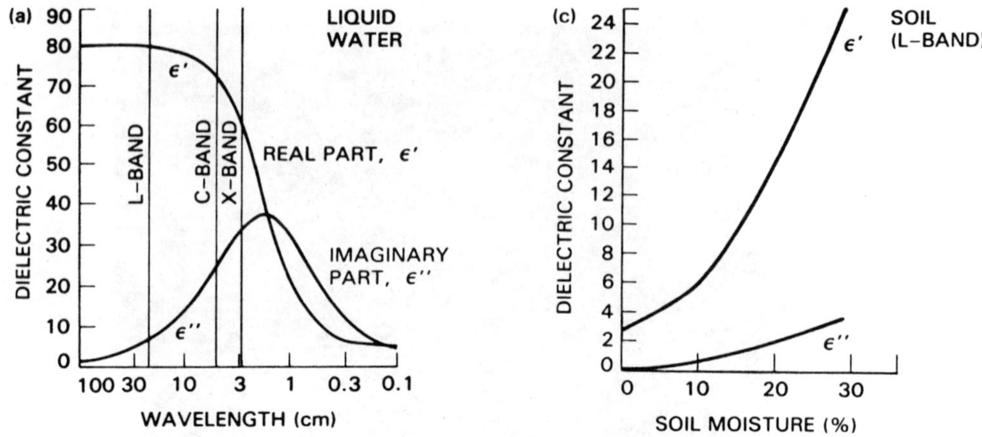

(a) (b)

Figure 7. Dielectric properties of (a) liquid water as a function of wavelength and (b) soil as a function of soil moisture (after Ulaby *et al* 1986)

2.6.2 *Sensitivity to target electrical properties* The electrical properties of the target also modify the incident radar wave. In general terms the electrical properties of the target are defined by the complex dielectric constant which is controlled to a major extent by the moisture content of the target. These relationships are demonstrated in figure 7.

In general this means that the presence of moisture in a target can significantly increase radar backscatter. This, of course, ignores any effects due to target structure.

2.7 RADAR MEASUREMENT CHARACTERISTICS

Radars have a number of unique measurement characteristics, including their ability to acquire temporal data on a regular basis, their independence from solar illumination and their side looking geometry. These provide a number of potential advantages to the management of natural resources, and are thus described in more detail below.

2.7.1 *Multi–temporal coverage* Sensors operating in the visible and infrared portions of the electromagnetic spectrum are dependent upon solar and thermal energy being scattered or emmitted from targets; adverse weather conditions can therefore severely hamper data collection for these sensors. In contrast radar sensors, by providing their own source of energy and by operating at relatively long wavelengths, can acquire data almost irrespectively of the prevailing weather conditions. This ability of radar sensors to provide year round coverage with a temporal resolution of a few days provides the possibility of monitoring natural resource change, in terms of both extent and condition.

2.7.2 *Independence from solar illumination angle* Further to the ability of radar sensors to collect data irrespective of the weather, radars can provide data in areas that provide problems to optical sensors in terms of low sun angles and extended periods of time without sun; this is especially true in areas near the polar regions. In addition, the ability to collect data irrespective of solar illumination means that data can be collected at night. Such night time data acquisitions may be advantageous in the assessment of moisture stress as day and night time data can be compared and the true moisture status of the vegetation may be ascertained.

2.7.3 *Side looking geometry* The fact that imaging sensors in radar remote sensing look to the side when collecting data (see § 2.2) can provide considerable advantages. The side looking mode of operation enhances vertical differences in the target area, for example in the case of topographic relief. Such an enhancement can provide significant advantages to applications that require data on changes in vertical structure, this is particularly true for applications such as geological mapping or erosion monitoring.

3. Radar based sensors for natural resources evaluation

3.1 EARLY DEVELOPMENT

The use of data from radars for the evaluation of the earth's surface has a relatively long history. Initial investigations were undertaken using real aperture Side Looking Airborne Radars (SLAR's) in the early 1950's. Much of this early work using imaging radar systems took place in regions of the world where other data collection techniques proved to be difficult, if not impossible to use, for example where excessive cloud cover limits the use of optical techniques, including the tropical forest regions of South America, West Africa and the Far East.

Radar imaging, long the province of the military, began to be seriously considered for earth resources applications following the limited declassification of radar systems in the late 1950's. Considerable impetus was provided by the flying of the Westinghouse AN/AQP–97 system over about 500,000 km^2 of the United States (1965–66). These data were only available on a limited basis, but generated a number of publications discussing geoscience applications (especially Dellwig *et al* 1966). Other milestone reports from this period dealing with the potential of radar data for resources applications include Beatty *et al* (1965), who evaluated the promise of radar data for preliminary resources inventories and economic evaluations in developing countries.

A further key development in the resources applications of radar was the mapping of Panama in 1967 and 1969, project "RAMP" (Viksne and Liston 1970). This was carried out using the Westinghouse system, and was accomplished in two stages. The study mapped 40,000 km^2 of eastern Panama and northwestern Columbia, and was the first real demonstration of radar's ability to overcome the problems of collecting data in cloud ridden tropical countries.

3.2 COMMERCIAL RADAR MAPPING PROJECTS

Project RAMP precipitated the release of radar systems for commercial operations at the beginning of the 1970's. As a result during this period approximately 13.5 – 14 million square kilometers were covered by airborne radar surveys. The most impressive of these was Project RADAM in Brazil (1970–76), which was initially designed to cover the Amazon basin, but was later expanded to cover the entire country (approximately 8.5 million square kilometers).

Project RADAM was followed by similar radar surveys in Nicaragua in 1971 (Martin–Kaye and Williams 1973), Peru (1974–75), Nigeria (1976–77) (Parry and Trevett 1979), Venezuela (1975) and South East Asia (various dates) (Froideaux, 1980). A summary of these, plus other smaller projects is given in table 2.

Country	Coverage (km^2)
Africa (5 countries)	1,014,000
Australia	52,000
Alaska	700,000
Brazil	8,500,000
British Solomon Islands	6,500
Canada	176,500
Central America	71,000
Columbia	476,000
Ecuador	130,500
Indonesia	97,500
New Guinea	182,000
Nicaragua	6,500
Panama	20,800
Peru	600,000
South East Asia	40,000
United Kingdom	3,000
United States	2.630,000
Venezuela	900,000

Table 2. Summary of SLAR system activities (up until 1980) (after Trevett 1986)

All these studies showed that the basic mapping of vegetation types could be undertaken using radar data. For example in the Nicaraguan study, 12 vegetation classes, ranging from swamp savannah with scattered woodland to hardwood mountain and hill forest, were successfully delineated (Martin–Kaye and Williams 1973).

These studies have been followed by the continued operational use of airborne synthetic aperture radars (SAR), in particular those used by Goodyear and Intera, both of which have flown substantial areas around the world, notably in South East Asia, Central America and Africa where the data have been used for operational vegetation surveys.

3.3 SPACEBORNE RADAR SYSTEMS

On the basis of the success of the airborne systems a number of spaceborne radar systems have been flown, or will be flown within the next decade. These systems are summarised in tables 3 and 4, and discussed in further detail below.

3.3.1 *Seasat* The first spaceborne imaging radar to be used for Earth observation was the L–band SAR on Seasat, an instrument package launched into an 800km altitude near polar orbit in June 1978. While the SAR was included in the Seasat payload primarily for the purpose of ocean wave imaging, radar images were also acquired over large areas of the Arctic ice pack and over land areas in the northern hemisphere. During its three month lifetime, Seasat acquired imagery of over 125 million km^2 of Central and North America, Western Europe, the North Atlantic, the North Pacific and the northern polar regions (Ford *et al* 1980).

As is apparent from table 4 these systems will provide SAR data from spaceborne platforms for the next decade, and beyond. It is also noticeable that, with the exception of Japan's JERS–1, C–band is common to all future spaceborne systems. This will provide continuity of data through the years, which has proved of value in optical based systems such as Landsat and AVHRR (see Belward Chapter 9).

3.3.2 *Shuttle imaging radars* The next spaceborne SAR to follow Seasat was the Shuttle Imaging Radar–A (SIR–A), ferried into a 38^0 inclination, 240 km altitude orbit in November 1981 by the Space Shuttle Columbia. The SIR–A technology was derived from Seasat, thus again using L–band and HH polarisation. However, since the SIR–A mission was to be used principally for geological research, the look angle was changed to a fixed angle of 47^0, thus ensuring an enhancement of topography. The resultant data from SIR–A led to the discovery of buried and previously uncharted dry river beds beneath the Sahara Desert in the Sudan and Egypt. This demonstrated the ability of an L–band radar to penetrate hyper–arid sand sheets to a depth of several metres. During its 2 day mission SIR–A acquired data over 10 million km^2 of the earth's surface, mostly over land (Cimino and Elachi 1982, Ford *et al* 1983).

The next NASA SAR mission was the Shuttle Imaging Radar–B (SIR–B), launched in October 1984 on the Space Shuttle Challenger. SIR–B into a 57^0 inclination orbit also used the Seasat / SIR–A technology with the same L–band frequency band and HH polarisation. However SIR–B was equipped with an pointing antenna so that selectable incidence angles could be obtained over the 15^0 to 60^0 range. This provided the first multi–incidence angle data set, useful for extracting geophysical information where there is a strong illumination angle signature. Despite problems with the Shuttle's K_u–band antenna and intermittent problems with the radar antenna, SIR–B provided good quality data over a number of sites and demonstrated the potential of radar for mapping surface features (particularly forests) using multiple incidence angle backscatter signatures and for topographic mapping. SIR–B data were also used to demonstrate the sensitivity of L–band radar data to soil moisture, to geological structural and lithologic features and to oceanic directional wave spectra (Ford *et al* 1986).

The successes of the SIR–A and SIR–B missions have led to a second generation SAR design that will be used for later shuttle flights and on the Earth Observing System (EOS). The Shuttle Imaging Radar–C (SIR–C) X–SAR mission, scheduled for two flights in the mid–1990's, will incorporate a multi–frequency, multi–polarisation, variable angle SAR. The SIR–C instrument will provide the power levels necessary for multi–polarisation imaging at L–band (23cm) and C–band (6cm) (NASA and JPL 1986). An X–band VV polarisation system will be provided by German and Italian institutes (Ottl 1986). For the L– and C–bands SIR–C will record the amplitude of the horizontally, vertically and both cross–polarised radar echoes, as well as the phase angle between the HH and VV echoes. These amplitude and phase data are expected to dramatically increase the information content of

Table 3. Summary Of Past Spaceborne Radars

	Seasat	SIR–A	SIR–B
Sponsor	NASA	NASA	NASA
Launch date	1978	1981	1984
Wave band	L	L	L
Polarisation	HH	HH	HH
Incidence angle	20^0	47^0	15^0–60^0
Swath width	100km	50–55km	20–55km
Azimuth resolution	25m	40m	25m
Range resolution	25m	40m	17–58m
No. of looks	4	4–7	4

Table 4. Summary Of Future Spaceborne Radars

	SIR–C/ X–SAR	ERS–1	ERS–2	JERS–1	Radarsat
Sponsor	NASA/ BMFT/ASI	ESA	ESA	Japan	Canada
Launch date	1994	1990	1994	1992	1994
Life time (yrs)		2–3	2–3	2	5
Repeat (days)		3–176	3–176	41	13
Inclination	$>57^0$	97.5^0	97.5	98.5^0	98.5^0
Wave band	X,C,L	C	C	L	C
Polarisation	Pol*	VV	VV	HH	HH
Incidence angle	15^0–55^0	23^0	23^0	35^0	20^0–$50+^0$
Swath width	15–90km	100km	100km	75km	50–500km
Azimuth res.	30m (4 looks)	30m (4 looks)	30m (4 looks)	18m (3 looks)	4–100m (1–8 looks)
Range res.	60–10m	30m (4 looks)	30m (4 looks)	18m (3 looks)	4–100m (1–8 looks)

Pol* = Polarimetric in C– and L–band only

radar images and to lead to expanded geoscientific applications. SIR–C / X–SAR will also provide electronic beam steering to allow acquisitions of swath widths in excess of 200km.

3.3.3 *ERS–1* The European Remote Sensing Satellite–1 (ERS–1) is described in full in § 4 of this Chapter. However it is worth noting that ERS–1, designed and built by European manufacturers on behalf of the European Space Agency (ESA), is the first radar based satellite to be flown since Seasat, and begins the series of radar satellites to be flown in the 1990's and beyond, and also begins the trend of using C–band SAR in space (ESA 1989).

3.3.4 *ERS–2* ERS–2 is basically a replica of ERS–1, with the opportunity to possibly improve the ATSR instrument by the addition of extra channels, particularly for application over land. ERS–2 has a planned launch for 1994, to coincide with the demise of ERS–1.

3.3.5 *JERS–1* The major objective of the Japanese Earth Resources Satellite–1 (JERS–1) is to explore non–renewable resources; in addition it is intended that this system will monitor other aspects such as land use, agriculture, forestry, fisheries, environmental protection, prevention of natural disasters and the surveillance of coastal regions. In order to undertake these tasks JERS–1 will have an L–band HH polarised SAR operating at 35^0 incidence angle (centre swath) with a 75km swath width and a 44 day repeat cycle. In addition JERS–1 will carry an optical sensor system comprising a visible and near infrared radiometer and a shortwave infrared radiometer; the optical sensors will have a capability for stereoscopic imaging (RRSS 1986, Ishiwada and Nemoto 1986).

3.3.6 *Radarsat* The Radarsat programme has gone through several significant modifications since its inception in 1980. The programme is now primarily Canadian with the USA as a junior partner supplying the launch and participating in the science programme. The Radarsat spacecraft is now scheduled for launch in 1994. The satellite will be placed in a sun–synchronous dawn–dusk orbit at an altitude of about 790km. The orbit selected has a 16 day repeat cycle containing a three day near repeat sub–cycle (Ahmed 1990).

The only instrument the satellite carries is a C–band SAR which is capable of an extensive range of operating modes including the steerable basic mode with incidence angles from 20^0 to 49^0, experimental modes at higher angles and ScanSAR modes with a maximum swath width of 500km at about 100m resolution. In order to map the Antarctic continent at fine resolution for the first time the satellite will also be capable of rotating itself to orient the SAR antenna left for a few days, and then returning to its normal configuration.

SAR data collected by Radarsat will be received by at least three receiving stations, at Gatineau and Prince Albert in Canada and Fairbanks in Alaska. In addition, Radarsat will carry an on–board tape recorder in order that world wide data acquisition can take place.

4. ERS–1 Status

The ERS–1 mission is the forerunner of a new generation of space missions for the 1990's employing radar sensors for the study of the environment (tables 2 and 3). As ERS–1 will be the first SAR satellite since 1978, and as it will also provide an indication of what can be expected from the future planned satellites it is worth noting its modes of operation in some detail . A schematic view of the ERS–1 satellite is given in figure 8.

The ERS–1 mission is designed to provide essential data for addressing a wide range of primary environmental problems, contributing to improved representation of ocean / atmosphere interactions in climate models; major advances in the knowledge of the ocean circulation and transfers of energy; more reliable estimates of the mass balance of the arctic and antarctic ice sheets; better monitoring of dynamic coastal processes and pollution; improved detection and management of land use change (ESA 1989).

4.1 INSTRUMENT PAYLOAD

The ERS–1 instrument payload is primarily devised for the measurement and monitoring of the oceans; therefore the main part of the sensor payload will only be dealt with briefly here. The SAR is the major sensor of interest for the management of terrestrial natural resources and will therefore be dealt with in more detail. The ERS–1 payload is made up of the following suite of sensors.

4.1.1 *Active microwave instrument (AMI)* The AMI operates in three different modes. Firstly in AMI Synthetic Aperture Radar Image mode. This will be utilised for the acquisition of wide swath images over the oceans, polar ices caps and land areas. In image mode the SAR will obtain strips of high resolution imagery 100km in width to the right of the satellite track. The 10m long antenna is aligned parallel with the satellite orbit in order to direct a narrow radar beam sideways and downwards onto the Earth's surface over the 100km swath. The operation of the mode is exclusive of other operating modes, and power considerations limit the operating time of the SAR to a maximum of 10 minutes per orbit. In addition the data rate is too high to allow on–board storage of the data, so images can only be acquired for areas within the reception zone of a suitably equipped ground station (see also § 4.3).

The mid–swath incidence angle of the radar in normal operation is 23^0. However the satellite has a 'roll tilt' capability which makes it possible to operate at 35^0 incidence angle on an experimental basis. This incidence angle is of special interest for some land applications. The geometry of the SAR image mode is shown in figure 9; the overall technical specifications are given in table 5.

The second mode of operation is the AMI Wave Mode. This mode yields 5 km by 5 km images at regular intervals along track for the derivation of the length and direction of ocean waves.

The third mode is the AMI Wind Mode. This uses three separate antennae for the measurement of sea surface wind speed and direction.

4.1.2 *Radar altimeter (RA)* The RA provides accurate measurements of sea surface elevation, significant wave heights, sea surface wind speeds and various ice parameters.

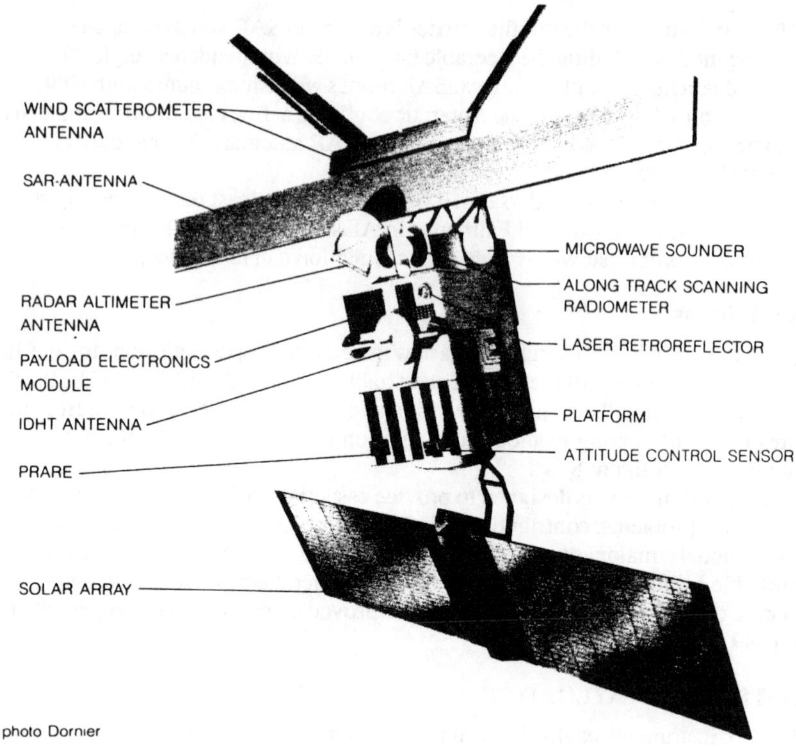

Figure 8. Schematic View Of The ERS–1 Satellite (after ESA 1989)

4.1.3 *Along track scanning radiometer (ATSR)* The ATSR combines infrared and radar sensors for the measurement of sea surface temperature, cloud top temperature, cloud cover and atmospheric water vapour.

4.1.4 *Precise range and range–rate equipment (PRARE)* The PRARE provides accurate determination of the satellite position and orbit characteristics, and the geodetic 'fixing' of the ground stations.

4.1.5 *Laser retro–reflector (LRR)* The LRR measures the satellite position and orbit using laser ranging stations on the ground.

4.2 ORBITS AND COVERAGE PATTERNS

ERS–1 has a sun synchronous circular orbit (quasi–polar) at a mean altitude of 785km and an inclination of 98.5^0. The local mean solar time at the crossing of the equator from north to south (i.e. descending node) is 10.30am.

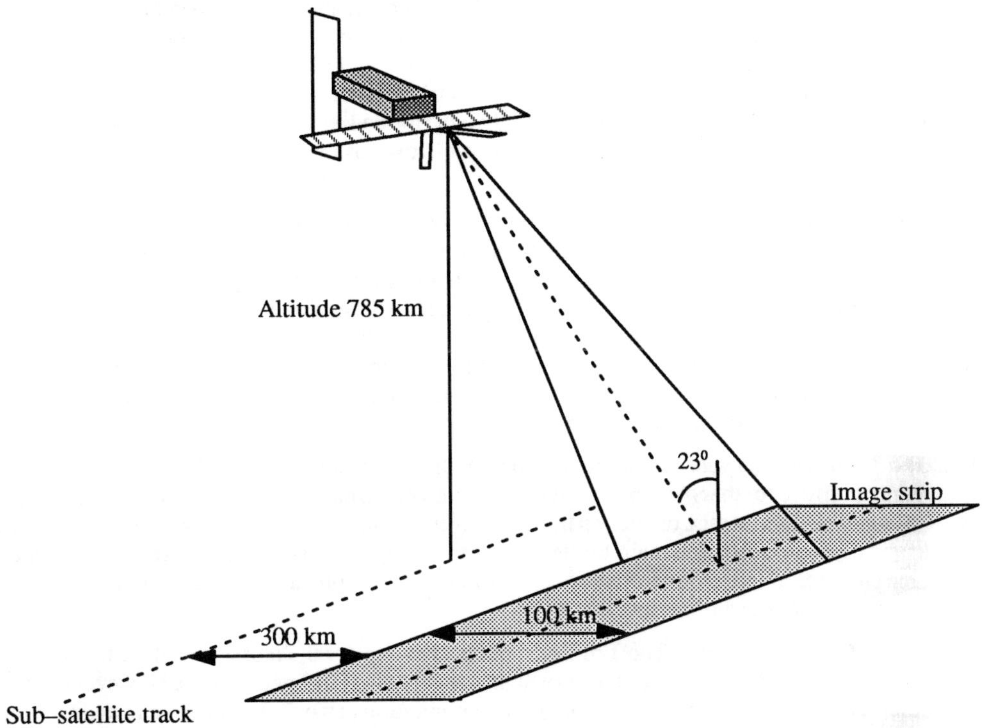

Figure 9. SAR Image Mode Geometry

Parameter	Value
Frequency	5.3 GHz. (C–band)
Polarisation	VV
Incidence Angle	23^0 (mid–swath)
Spatial Resolution	30m
Radiometric Resolution	2.5 dB at –18 dB
Swath Width	100 km
Data Rate	<105 Mbps

Table 5. ERS–1 SAR Image Mode Technical Specifications (after ESA 1989)

The ERS–1 mission will include a number of adjustment manoeuvres to synchronise the orbital period with various requirements for ground coverage. During the initial three month commissioning phase the satellite will have a 3 day repeat cycle at an altitude of 785km (called the reference orbit). Subsequent satellite height adjustments can provide various repeat cycles of between 3 and 176 days.

The majority of the rest of the mission will be performed in a 35 day repeat cycle, with some transition back to a 3 day repeat and a final 176 day orbit towards the end of the mission. Each of these orbits is discussed in more detail below.

4.2.1 *The 3 day repeat cycle* At a satellite altitude of 785km, the earth's rotation causes a spacing of 909km at the equator between successive tracks, and after completing 43 passes in 3 days the satellite orbit is such that it returns to exactly the same pattern of coverage. This cycle has been selected for the commissioning phase because it provides frequent revisiting of dedicated calibration sites under constant geometrical and illumination conditions. Similar 3 day cycles will operate for limited periods at other times during the mission with slightly different longitudinal phases to ensure highly repetitive coverage of ice zones during arctic winter (hence the name "ice phase"), and other areas (including land) that require high repetition frequencies.

4.2.2 *The 35 day repeat cycle* The main limitation of the 3 day repeat cycle is the restricted coverage of the earth's surface for the SAR and the wide separation of the radar altimeter tracks. A 35 day repeat cycle provides SAR coverage of every part of the earth's surface, with at least twice this coverage at middle and high latitudes. The main limitation of the 35 day cycle is the limited repeat coverage of any given land area, thus the possibilities of using these data in applications that required high number of observations, such as agriculture, are limited.

4.2.3 *The 176 day repeat cycle* The 176 day cycle is favoured for the measurement of the mean sea surface and ocean geoid, because of the very high intensity of the altimeter tracks. However conflicts with other requirements mean that such an orbit configuration can only be employed late in the mission.

4.3 DATA ACQUISITION

Ground stations at Kiruna (Sweden), Fucino (Italy), Gatineau (Canada) and Maspalomas (Canaries, Spain) will be the main facilities for the acquisition and rapid processing of ERS–1 data in Europe (figure 10).

The Kiruna station in northern Sweden will play a primary role because its high latitude position provides visibility to 10 out of every 14 orbits. Both instrument real time and (in the case of non–SAR data) on–board recorded data will be acquired by the station, with the raw data being stored on High Density Digital Tapes (HDDT). Fast delivery processing facilities comprise two processing chains devoted to SAR processing (i.e. image and wave mode) and one devoted to the processing of the data from the other instruments.

The Fucino station is to be used primarily for the acquisition of SAR image data over the Mediterranean Sea. Stations at Gatineau and Maspalomas will be used for the acquisition of low bit rate data (i.e. from all sensors apart from the SAR), although it is now hoped to up–grade the Maspalomas station to receive SAR data as well. All three station will have data recording and fast delivery processing facilities similar to Kiruna.

A large number of additional ground stations will increase the coverage for SAR data acquisition. Of particular interest are the proposed stations in Brazil, Ecuador, Antarctica, Australia, Japan, India, Indonesia and Pakistan. These will increase the SAR coverage of areas where problems of cloud cover have made the acquisition of other data sets difficult, if not impossible in the past.

Figure 10 shows the location of all the receiving stations.

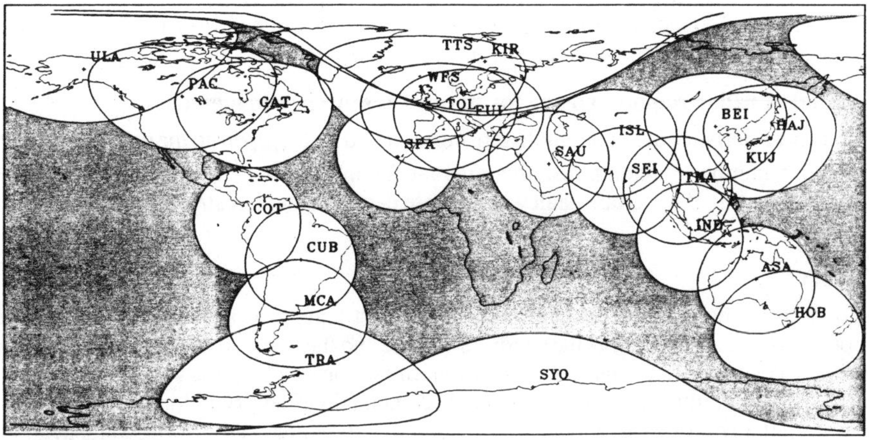

Key:

ASA: Alice Springs, Australia
BEI: Beijing, China
COT: Cotopaxi, Ecuador
CUB: Cuiaba, Brazil
FUI: Fucino, Italy
GAT: Gatineau, Canada
HAJ: Hatoyama, Japan
HOB: Hobart, Australia
IND: Jakarta, Indonesia
ISL: Islamabad, Pakistan
KIR: Kiruna, Sweden
KUJ: Kumamoto, Japan
MCA: Mar Chiquita, Argentina
PAC: Prince Albert, Canada
SAU: Riyahd, Saudi Arabia
SEI: Secunderabad, India
SPA: Maspalomas, Spain
SYO: Showa, Japan / Antarctica
THA: Bangkok, Thailand
TGS: Greenbelt, USA
TOL: Assaguel, France
TRA: Esperanza,Argentina / Antarctica
TTS: Tromso, Norway
ULA: Fairbanks, Alaska
WFS: West Freugh, United Kingdom

Figure 10. Coverage Zones For Proposed ERS–1 Ground Stations (after ESA 1989)

There will also be additional stations in some of the ESA member states, including Tromso in Norway, Aussaguel in France, West Freugh in the United Kingdom and Prince Albert in Canada . These will operate as national facilities for the reception of SAR data.

5. The potential of radar remote sensing for natural resource management

The particular characteristics of radar remote sensing, as noted in § 2, open up the possibility of utilising these data in a number of natural resource orientated applications. These include the following broadly defined areas, all of which are dealt with in more detail below: forestry; agricultural; hydrology; and geology

5.1 FORESTRY

The ability of SAR to observe target structure and target electrical properties, and thus to observe important biophysical properties such as the canopy structure and the tree trunks, are potentially significant inputs into forestry applications; such as the determination of a baseline survey of forest extent on a global basis; the monitoring of seasonal and inter–annual variability in forests; the monitoring of the nature and extent of forest responses to such stress agents as diseases, insects, water availability (drought and excess), airborne pollutants, wind damage, frost, fire and anthrpogenic practices (Sieber 1986, NASA 1986)

These potential inputs have not been defined purely on the basis of economic requirements, but also in the wider framework of the global monitoring of the earth's biota and their interaction with the hydrological, biogeochemical and climatic cycles (NRC 1985, NRC 1986a, NRC 1986b, Bretherton *et al* 1986). These applications require the accurate measurement of the biophysical properties of the earth's surface, including forest parameters such as biomass, canopy moisture content, water state and understorey characteristic; all of which can potentially be provided by radar remote sensing.

Detailed studies by a number of investigators have clearly demonstrated the potential of radar data to provide such basic information on the location of forested areas, and in some cases to give information on differences within the forest due to species, age and stress (Le Toan *et al* 1980, Mehta 1984, Sieber 1985, Cimino *et al* 1986, Hallikainen *et al* 1989, Hoffer and Lee 1989).

Other studies have gone further and have shown the potential use of radar data for detailed analysis of forest stands, with particular emphasis on the delineation of different parts of the tree's physiology (Sieber 1985, Paris 1986, Hoekman 1987, Richards *et al* 1987, Pitts *et al* 1988, Zoughi *et al* 1989).

Figure 11 gives an example of an airborne SAR image of a mixed agricultural\forestry area in the valley of the River Rhine near Freiburg in the Federal Republic of Germany. This image has been constructed by combining two single frequency (X–band HH polarisation, and L–band HH polarisation) images and artificially colouring them using digital image processing techniques. The image clearly demonstrates how the forest, typified by the light orange or grey image tone and relatively coarse image texture, can be delineated from the surrounding agricultural land which has a red or bright orange image tone and relatively smooth image texture. Further, within the forest to the west of the image, and on the banks of the river Rhine, gaps in the canopy in the form of clearcut areas, which demonstrate the bright orange tone of the agricultural land, can be delineated. In addition different stands of different species can be recognised within the forest by a difference of image tones (Sieber and Noack 1986, ESA/JRC 1988)

Figure 11. Airborne SAR image composite incorporating X–band HH polarisation (red) and L–band HH polarisation (blue) of forest in the Rhine valley (after Sieber and Noack 1986, ESA/JRC 1988).

NOTE: This image provides a good example of different facets of a SAR image. Note the River Rhine running top to bottom through the image shows a dark image tone due to the high dielectric constant and smooth surface of the water resulting in specular reflectance, Note also the light orange and gray toned forestry areas to the right of the Rhine. These demonstrate significant image texture due to the inhomogeneity of forest canopies to X–band frequencies. The urban area to the top of the forest also demonstrates image texture but of a more structured kind due to the fact that the radar response is mainly coming from the sides of houses, causing a "corner reflector" effect. Finally note the agricultural area to the right of the Rhine. Different tones are related to different crop types and growth stages; note also that the agricultural areas do not demonstrate significant image texture.

5.2 AGRICULTURE

Information concerning cropping and crop production is important for agricultural policy at all levels, both nationally and internationally, throughout the world (see for example the later Chapter by Bartholome). Within the European context, for example, this requires timely and accurate data in order to distinguish, identify and measure the area of crops of significant importance. On this basis it is then necessary to estimate production in a timely manner and to undertake production forecasting on an annual basis (Meyer–Roux 1987).

The all weather observation capabilities of radar are of particular significance for agricultural applications because this facility enables the monitoring of crops during critical time periods within the growing season. A number of studies have shown that radar remote sensing can play such a role. Radar has been shown to be sensitive to different crop types and conditions(Bush and Ulaby 1978, Hoogeboom 1982, Shanmugan *et al* 1983, Mehta 1983, Sieber and Trevett 1983, Guindon *et al* 1984, Cihlar *et al* 1986). In addition because of radar remote sensing's sensitivity to target structure and dielectric constant it has been shown that radar is sensitive to crop stage (De Loor *et al* 1982, Hoogeboom 1982, Ulaby *et al* 1986, Soares *et al* 1987), crop state including crop and soil moisture status (Ulaby and Bush 1976, Ulaby *et al* 1983, Brisco *et al* 1989) and physical crop parameters such as overall geometry, crop cover and plant density (Ulaby *et al* 1984, Sieber 1985).

Figure 12 shows an airborne SAR scene of an agricultural area in the east of England, incorporating X–band HH polarisation and L–band HH polarisation data that have been artificially coloured. Of major interest in this area are the fields of winter wheat, winter and spring barley. On the image these can generally be separated, with the winter wheat fields having a speckled yellow / green image tone, the winter barley is mainly dark orange / brown and the spring barley speckled green.

The fact that the differences between winter wheat and winter barley show up on this image, taken in early June, is particularly interesting because this is not a time when there is a striking visual difference between the two crops. Figure 13 shows the two crops on the day of the SAR data collection. Note both crops are green and in ear, with the main observed difference being the stronger stalks and broader flag leaves of the winter wheat. Thus differences in the radar image can be assumed to be due to these small differences in crop morphology and also likely differences in crop moisture content (Churchill *et al* 1984, ESA/JRC 1988).

5.3 HYDROLOGY

At point locations hydrological processes, such as infiltration and evapotranspiration, are well monitored and understood by hydrologists. However, the ability to give these measurements and processes spatial and temporal variability within large catchments is not currently possible on a reliable basis using extrapolation and modelling techniques (see Gregoire, Chapter 12). The unknown variable factors which confuse the models include soil properties and moisture, snowpack properties, topographic features, drainage patterns, land use, areas of standing water and subsurface water storage. Therefore in order to move beyond a detailed knowledge of the hydrological processes at individual locations to micro and macro scale analyses requires an ability to collect accurate data over a wider scale on a regular basis. Remote sensing, and radar remote sensing in particular, offers this possibility.

Radar remote sensing offers the opportunity to directly observe a number of important hydrological model parameters, such as land use, basin characteristics, channel geometry, slope information and the areal extent of standing water. Indeed procedures for using this type of information have been developed and modified for use in current hydrologic models (Engman 1981).

In addition to these parameters there is strong evidence to suggest that radar remote sensing can measure additional parameters that affect the terrestrial water balance.; these include soil moisture, snowpack conditions, frozen soil boundaries and vegetation moisture content. The measurement of

Figure 12 Airborne SAR image composite incorporating X–band HH polarisation (green) and L–band HH polarisation (red) of agricultural land in Norfolk, England (after Churchill *et al* 1984, ESA / JRC 1988)

Winter Wheat　　　　Winter Barley

Figure 13. Winter wheat and winter barley crops at the time of the SAR overflight in Norfolk, England (after Churchill et al 1984, ESA / JRC, 1988).

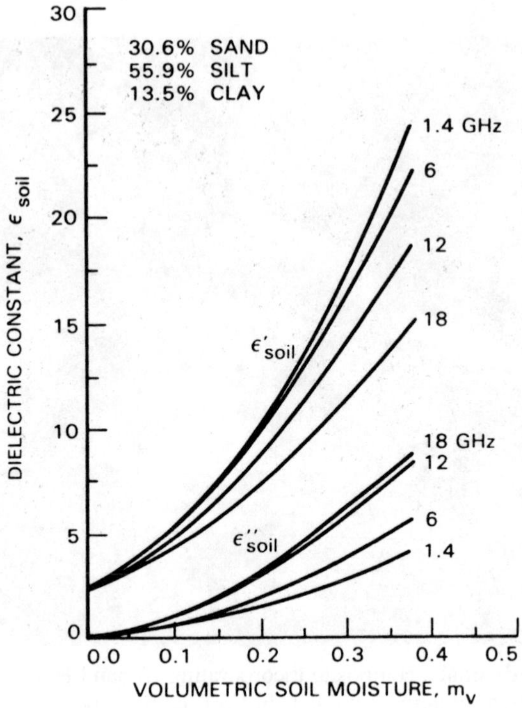

Figure 14. Measured dielectric constant as a function of volumetric moisture content for loamy soil at four radar frequencies (after Ulaby *et al* 1986).

these parameters may be possible as they all affect the dielectric constant of the target, and thus the backscattered signal.

In particular extensive work has been undertaken in the analysis of the relationship between soil moisture content and the backscattered radar signal, via the dielectric constant (Schmugge 1984, Ulaby *et al* 1986, Dobson and Ulaby 1986). An example of this relationship is given in figure 14.

A number of studies have demonstrated the potential of deriving soil moisture content from radar measurements. These have included the use of truck mounted scatterometers working in the frequency range of 1 – 18GHz. which identified radar sensitivity to near surface soil moisture as a function of surface roughness and soil texture for various combinations of radar sensor parameters (Ulaby 1974, Ulaby *et al* 1974, Ulaby and Batlivala 1976, Ulaby *et al* 1978 Dobson and Ulaby 1981).

In addition aircraft estimates of soil moisture from radar scatterometers have been reported by Ulaby *et al* (1979), Theis *et al* (1982), Jackson *et al* (1981) and Bernard *et al* (1984). Particularly good results have been obtained with low altitude airborne scatterometer experiments exemplified by Bradley and Ulaby (1981) and Jackson and O'Neill (1985); in particular the results demonstrated by Moe *et al* (1984) showed an excellent correlation between radar backscatter and soil moisture at both 1.6 GHz. and 4.75 GHz.

Orbiting radar sensors have also been tested for their ability to measure soil moisture content. Skylab with its 13.9 GHz. scatterometer (Ulaby *et al* 1974) and Seasat with the 1.275 GHz. SAR (Blanchard *et al* 1981) were the first systems to be tried. Seasat SAR images did demonstrate a sensitivity to soil

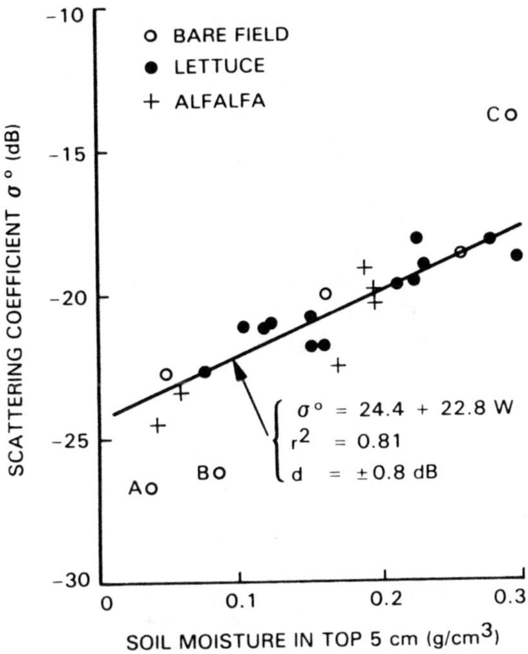

Figure 15. Shuttle Imaging Radar–B backscatter coefficient plotted as a function of volumetric soil moisture (after Wang *et al* 1986)

moisture (Ulaby *et al* 1983), whilst the measurement of soil moisture using the L–band SAR on SIR–B were also relatively successful (Wang *et al* 1986). The dependance of the SIR–B backscatter coefficient on volumetric soil moisture is shown in figure 15.

5.4 GEOLOGY

The use of radar remote sensing in geology can be considered in the context of a number of applications, including mineral exploration, arid land management and soil erosion monitoring. radar remote sensing offers such applications a number of potential advantages, among them the fact that the side looking geometry of radar data collection provides a sensitivity to surface morphology, and that the longer wavelengths used penetrate eolian sand veneers that often obscure near surface geological features in arid lands.

Imaging radar has proven to be useful for the delineation of surface indicators of tectonic and crustal activity, as well as for the discrimination of certain kinds of surface lithology. This is mainly due to radar's sensitivity to various scales of surface roughness (Dellwig *et al* 1968, MacDonald and Waite 1970, MacDonald and Waite 1973, Ford 1980, Elachi *et al* 1982, Ford *et al* 1983, Koopmans 1983, Sabins 1983). Stereo viewing (Leberl *et al* 1986) and the direct comparison of radar data with other data sets, such as Landsat (Chavez *et al* 1983, Ford *et al* 1983) have also been shown to significantly enhance the interpretability of structural and other geologic aspects of a terrain.

This ability has resulted in the considerable use of radar data in economic geologic exploration, and in particular the identification of potential petroleum or mineral resources. Indeed the private sector has made regular use of radar remote sensing in economic geologic exploration since the 1970's. As a result, much of this work has gone unreported in the scientific press, although a number of good examples can be found in the open literature, see for example Martin–Kaye and Williams (1973) in relation to Nicaragua, Froidevaux (1980) in relation to South East Asia and Viksne and Liston (1970), Wing (1970) and Wing and MacDonald (1973) in relation to Panama and Columbia.

For arid land management remote sensing investigations of deserts are made difficult by the presence of pervasive mantles of eolian sand, particularly in hyperarid to arid regions. These sand mantles commonly obscure the observation of the underlying materials and structures, and thus hamper the search for potable, near surface ground water and other mineral resources. The potential of long wavelength SAR sensors for the geologic mapping of the surface and shallow subsurface of these hyperarid desert terrains was shown with the discovery of the paleo–river channels in the Eastern Sahara following the SIR–A and SIR–B missions (McCauley *et al* 1982 1986, Elachi *et al* 1984, Schaber *et al* 1986).

In addition the ability to map the extent of the desert, and more importantly to map and monitor desertification via the degradation of arid ecosystems, is a further potential application of radar remote sensing.

6. Conclusions

Far from being a new technique, radar remote sensing is a methodology for collecting natural resources data that is to a large extent proven, in that it has been applied operationally for a large number of years. Yet this technology has great potential, best exemplified by the large number of spaceborne radar systems that will be flown over the next decade, and by the innovations in radar sensor technology such as the imaging polarimeters.

In addition, radar remote sensing offers unique information about targets, ranging from information about their structure to information about their moisture status. This combined with the ability of radar sensors to collect data irrespective of the prevailing weather conditions, available solar energy or sun illumination angle makes radar remote sensing one of the most important sources of information for natural resource management for the future.

7. References

Ahmed, S., 1990, Radarsat: a status report, *Proceedings of the Canadian Radar Data Development Program Workshop*, Gananoque, January.

Beatty, F.D., 1965, Geoscience potentials of side looking radar, *Raytheon Autometric Coorporation, (prepared for NASA by the US Army Coorps of Engineering under contract number DA–44–009–AMC–1040(X)).*

Bernard, R., Taconet, O., Vidal–Madjar, D., Thony, J.L., Vauclin, M, Chapoton, A., Wattrelot, F, and Lebrun, A., 1984, Comparison of three in situ surface soil moisture measurements and application to C–band scatterometer calibration, *IEEE Transactions on Geoscience and Remote Sensing*, **GE–22.**

Blanchard, B.J., Blanchard, A.J., Theis, S., Rosenthal, C.L., and Jones, C.L., 1981, Seasat SAR response for water resources parameters, *Final Report 3891*, Remote Sensing Center, Texas A&M University, Texas.

Bradley, G.A., and Ulaby, F.T., 1981, Aircraft radar response to soil moisture, *Remote Sensing of Environment*, **11**, 419.

Bretherton, F et al, 1986, Earth system science: a program for global change, Report to the Earth System Sciences Committee, NASA Advisory Council, Washington D.C.

Briscoe, B., Brown, R.J., and Pultz, T., 1989, The effects of free canopy water on SAR crop separability, *Proceedings of IGARSS'89 / 12th Canadian Symposium on Remote Sensing*, Vancouver, Canada, 10 –14 July, pp.424–428.

Bush, T.F., and Ulaby, F.T., 1978, An evaluation of radar as a crop classifier, *Remote Sensing of Environment*, **7**, 15–36.

Churchill, P.N., Deane, G.C., and Trevett, J.W., 1984, An evaluation of SAR 580 multi–frequency radar data over the Norfolk Test Site (GB4), SAR 580 Campaign Final Report, (edited by J.W.Trevett), JRC Ispra, Report No. S.A./04.E2.85.12/2.

Churchill, P.N., and Keech, M.A., 1984, Multifrequency analysis of SAR 580 imagery for woodland determinations in Thetford Forest, England, SAR 580 Campaign Final Report, (edited by J.W.Trevett), JRC Ispra, Report No. S.A./04.E2.85.12/2.

Chavez, P.S., Berlin, G.L., and Tarabzouni, M.A., 1983, Discriminating lithology and surficial deposits in the Al Hisma Plateau region of Saudi Arabia with digitally combined Landsat MSS and SIR–A images, In. *Remotely Sensed / Geographical Information Systems in Geologic Applications*, 1983 National Conference on Resource Management Applications, pp.22.

Cihlar, J., Brown, R.J., and Guindon, B., 1986, radar remote sensing of agricultural crops in Canada, *International Journal of Remote Sensing*, **7**, 195–212.

Cimino, J.B., Brandani, D., Casey, D., Rabassa, J.,and Wall, S.D., 1986, Multiple incidence angle SIR–B experiment over Argentina: mapping of forest units, *IEEE Transactions on Geoscience and Remote Sensing*, **GE–24**, 498–509.

Cimino, J.B., and Elachi, C., 1982, Shuttle imaging radar–A (SIR–A) experiment, JPL Publication 82–77, December.

De Loor, G.P., and Hoogeboom, P., 1982, The Dutch ROVE program, *IEEE Transactions on Geoscience and Remote Sensing*, **GE–20**, 3–10.

Dellwig, L.F., Kirk, J.N. and Walters, R.L., 1966, The potential of low resolution radar imagery in regional geologic studies, *Journal of Geophysical Research*, **71**, 4495–4498.

Dellwig, L.F., MacDonald, H.C., and Kirk, J.N., 1968, The potential of radar in geological exploration, *Proceedings of the Fifth Symposium on Remote Sensing of Environment*, University of Michigan, Ann Arbor, pp.747.

Dobson, M.C., and Ulaby, F.T., 1981, radar backscatter dependance on surface roughness, soil moisture and soil texture: Part III – Soil tension, *IEEE Transactions on Geoscience and Remote Sensing*, **GE–19**, 51–61.

Dobson, M.C., and Ulaby, F.T., 1986, Active radar soil moisture research, *IEEE Transactions on Geoscience and Remote Sensing*, **GE–24**.

Elachi, C., 1987, *Introduction to the physics and techniques of remote sensing*, (New York: John Wiley and Sons).

Elachi, C., Brown, W.E., Cimino, J.B., Dixon, T., Evans, D.L., Ford, J.P., Saunders, R.S., Breed, C., Masursky, H., McCauley, J.F., Schaber, G., Dellwig, L., England, A., MacDonald, H., Martin–Kaye, P., and Sabins, F., 1982, Shuttle imaging radar experiment, *Science*, **218**, 996.

Elachi, C., Roth, L.E., and Schaber, G.G., 1984, Spaceborne radar subsurface imaging in hyperarid regions, *IEEE Transactions on Geoscience and Remote Sensing*, **GE–22**, 383.

Engman, E.T., 1981, Remote sensing applications in watershed modeling in catchment hydrology, *Proceedings of the International Symposium of Rainfall–Runoff Modeling*, Mississippi State, Mississippi.

European Space Agency (ESA), 1989, ERS–1: European Remote Sensing Satellite, ESA Publication BR–36, ISBN 92–9092–019–X, November.

European Space Agency (ESA) and Joint Research Centre (JRC), 1988, Imaging radar experiments in Europe: illustrated experimental results (1978–1987), ESA Publication TM–01, October.

Ford, J.P., 1980, Seasat orbital radar imagery for geologic mapping: Tennessee–Kentucky–Virginia, *American /association of Petroleum Geologists*, **64**, 2064.

Ford, J.P., Blom, R.G., Bryan, M.L., Daily, M.I., Dixon, T.H., Elachi, C., and Xenos, E.C., 1980, Seasat views North America, the Caribbean and Western Europe with imaging radar, JPL Publication 80–67, November.

Ford, J.P., Cimino, J.B., and Elachi, C., 1983, Space shuttle Columbia view the world with imaging radar: the SIR–A experiment, JPL Publication 82–95, 179pp.

Ford, J.P., Cimino, J.B., Holt, B., and Ruzek, M.R., 1986, Shuttle imaging radar views the Earth from Challenger: the SIR–B experiment, JPL Publication 86–10, March.

Froideaux, C.M., 1980, Radar, an optimum remote sensing tool for detailed plate tectonic analysis and its application to hydrocarbon exploration (an example in Irian Jaya, Indonesia), In: *Radar Geology: An Assessment*, California Institute of Technology, JPL 80–61, 457.

Guindon, B, Teillet, P.M., Goodenough, D.G., Palimaka, J.J., and Sieber, A.J., 1984, Evaluation of the crop classification performance of X, L and C–band SAR imagery, *Canadian Journal of Remote Sensing*, **10**.

Hallikainen, M., Tares, T., Hyyppa, J., and Somersalo, E., 1989, Classification of forest types by radar scatterometer, *Proceedings IGARSS' 89 / 12th Canadian Symposium on Remote Sensing*, Vancouver, Canada, pp. 2378–2381.

Hoekman, D.H., 1987, Measurements of the backscatter and attenuation properties of forest stands at X–, C–, and L–band, *Remote Sensing of Environment*, **23**, 397–416.

Hoffer, R.M., and Lee, K.S., 1989, Forest change classification using Seasat and SIR–B satellite SAR data, *Proceedings IGARSS' 89 / 12th Canadian Symposium on Remote Sensing*, Vancouver, Canada, 10–14 July, pp. 1372–1375.

Hoogeboom, P., 1982, Classification of agricultural crops in radar images, *IEEE Transactions on Geoscience and Remote Sensing*, **GE–21**, 329–336.

IEEE, 1976, Standard letter designation for radar frequency bands, IEEE Std 521–1976, Nov.30.

Ishiwada, Y., and Nemoto, Y., 1986, Present status of Japanese ERS–1 project, *Proceedings of the Second Spaceborne Imaging Radar Symposium*, Jet Propulsion Laboratory, JPL Publication 86–26, April.

Jackson, T.J., Chang, A., and Schmugge, T.J., 1981, Active radar measurements for estimating soil moisture in Oklahoma, *Photogrammetric Engineering and Remote Sensing*, **47**, 801.

Jackson, T.J., and O'Neil, P.E., 1985, Aircraft scatterometer observations of soil moisture on rangeland watersheds, *International Journal of Remote Sensing*, **6**, 1135.

Koopmanns, B.N., 1983, Side looking radar, a tool for geological surveys, In: *Remote Sensing Reviews*, edited by G.P. De Loor, (New York: Harwood Academic Publishers), 1, 19.

Leberl, F., Domik, A., Raggam, J., and Kobrick, M., 1986, Radar stereo mapping techniques and application to SIR–B images of Mt. Shasta, *IEEE Transactions on Geoscience and Remote Sensing*, **GE–24**, 473.

Le Toan, Th., Shahin, A., and Riom, J., 1980, Application of digitised radar images to pine forest inventory: first results, *Proceedings 14th International Symposium on Remote Sensing of Environment*, Costa Rica.

Li, R.Y., Ulaby, F.T., and Eyton, J.R., 1980, Crop classification with a Landsat – radar sensor combination, *Proceedings of the 1980 Symposium on Machine Processed Remotely Sensed Data*, Purdue University, West Lafayette, IN, June 2 – 6.

Lillesand, T.M., and Kiefer, R.W., 1987, *Remote sensing and image interpretation*, (New York: John Wiley and Sons).

MacDonald, H.C., and Waite, W.P., 1970, Optimum radar depression angles for geological analysis,"*Modern Geology*, **2**, 179.

MacDonald, H.C., and Waite, W.P., 1973, Imaging radars provide terrain texture and roughness parameters in semi–arid environments, *Modern Geology*, **4**, 145.

McCauley, J.F., Breed, C.S., Schaber, G.G., McHugh, W.P., Haynes, C.V., Issawi, B., Grolier, M.J., and El–Kilani, A., 1986, Paleodrainages of the eastern Sahara – the radar rivers revisited, *IEEE Transactions on Geoscience and Remote Sensing*, **GE–24**.

McCauley, J.F., Schaber, G.G., Breed, C.S., Grolier, M.J.,Haynes, C.V., Issawi, B., Elachi, C., and Blom, R., 1982, Subsurface valleys and geoarcheology of the Eastern Sahara, *Science*, **218**, 1004.

Martin–Kaye, P.H.A., and Williams, A.K., 1973, Radargeologic map of Eastern Nicaragua, *Momoria de la IX Conferencia Inter–Guyanas, Boletin de Geologia*, Publicacion Especial No.6, Caracas, Venezuela, pp.600–605.

Mehta, N.C., 1983, Separability of agricultural crops with airborne scatterometry, AgRISTARS, Report SR–L3–04428, JSC–18885, June.

Mehta, N.C., 1984, Use of multi–frequency, multi–polarisation, multi–angle airborne radars for class discrimination in a southern temperate forest, *Proceedings of the IGARSS'84 Symposium*, Strasbourg, France, pp. 149–154.

Meyer–Roux, J., 1987, The10 year research and development plan for the application of remote sensing in agricultural statistics, Commission of the European Communities, Joint Research Centre, Ispra, SP.1.87.39/EN, August.

Moe, T., Schmugge, T.J., and Jackson, T.J., 1984, Calculations of radar backscattering coefficient of vegetation covered soils, *Remote Sensing of Environment*, **15**, 119.

National Aeronautics and Space Administration (NASA), 1986, Synthetic aperture radar (SAR) Earth Observing System instrument panel report, Volume IIf.

National Aeronautics and Space Administration (NASA), and Jet Propulsion Laboratory (JPL), 1986, Shuttle imaging radar–C science plan, JPL Publication 86–29, September.

National Aeronautics and Space Administration (NASA), and Jet Propulsion Laboratory (JPL), 1989, Science plan for the Alaska SAR facility program, JPL Publication 89–14, September.

National Research Council, 1985, *A strategy for Earth science from space in the 1980's and 1990's. Part II: atmosphere and interactions for the solid earth, oceans and biota*, (Washington D.C.: National Academy of Sciences).

National Research Council, 1986a, *Global change in the geosphere – biosphere: initial priorities for an IGBP*, (Washington D.C.: National Academy of Sciences).

National Research Council, 1986b, Remote sensing of the biosphere, (Washington D.C.: National Academy of Sciences).

Ottl, H., 1986, The X–SAR system, *Proceedings of the Second Spaceborne Imaging Radar Symposium*, Jet Propulsion Laboratory, JPL Publication 86–26, April.

Paris, J.F., 1986, Probing thick vegetation canopies with a field radar scatterometer, *IEEE Transactions on Geoscience and Remote Sensing*, **GE–24**, 886–893.

Parry, D.E., and Trevett, J.W., 1979, Mapping Nigeria's vegetation from radar, *The Geographical Journal*, **145**.

Pitts, D.E., Badhwar, G.D., and Reyna, E., 1988, The use of helicopter mounted ranging scatterometer for estimation of extinction and scattering properties of forest canopies, Part II: experimental results for high density aspen, *IEEE Transactions on Geoscience and Remote Sensing*, **GE–26**, 144–152.

Richards, J.A., Sun, G.Q., and Simonett, D, 1987, L–band radar backscatter from forest stands, *IEEE Transactions on Geoscience and Remote Sensing*, **GE–25**, 487–498.

Sabins, F.F. Jr., 1983, Geologic interpretation of Space Shuttle radar images of Indonesia, *American Association of Petroleum Geologists Bulletin*, **67**, 2076.

Schmugge, T.J., 1984, radar remote sensing of soil moisture, *Proceedings of the Eighteenth International Symposium on Remote Sensing of Environment*, Paris, France, October.

Schaber, G.G., McCauley, J.F., Breed, C.S., and Olhoeft, G.R., 1986, Shuttle imaging radar: physical controls on signal penetration and subsurface backscatter in the eastern Sahara, *IEEE Trans. on Geoscience and Remote Sensing*, **GE–24**.

Shanmugan, K.S., Ulaby, F.T., Narayana, V, and Dobson, C., 1983, Identification of corn fields using multidate radar data, *Remote Sensing of Environment*, **13**, 251–264.

Sieber, A.J., 1985, Forest signatures in imaging and non–imaging radar scatterometer data, *ESA Journal*, **9**, 431–448.

Sieber, A.J., 1985, Statistical analysis of SAR images, *International Journal of Remote Sensing*, **6**, 1555–1572.

Sieber, A.J., 1986, International forest inventory: the global forest ecosystem as viewed by ERS–1, Proposal To ERS–1, Commission of The European Communities, Joint Research Centre, T.N.I.04.E2.86.105.

Sieber, A.J., and Noack, W., 1986, Results of an airborne synthetic aperture radar (SAR) experiment over a SIR–B (Shuttle Imaging Radar test site in Germany, *ESA Journal*, **10**, No.3.

Sieber, A.J., and Trevett, J.W., 1983, Comparison of multi–frequency band radars for crop classification, *IEEE Transactions on Geoscience and Remote Sensing*, **GE–21**, 285–294.

Soares, J.V., Bernard, R., and Vidal Madjar, D., 1987, Spatial and temporal behaviour of a large agricultural area as observed from airborne C–band scatterometer and thermal infra–red radiometer, *International Journal of Remote Sensing*, **7**, 981–996.

Technology Research Association Of Resources Remote Sensing System (RRSS), 1986, Observation system for Earth Resources Satellite–1, Minato–ku, Tokyo 105, Japan.

Theis, S.W, McFarland, M.J., Rosenthal, W.D., and Jones, C.L., 1982, radar remote sensing of soil moisture, Texas A&M University, Collece Station, TX, RSC–3458k–129.

Trevett, J.W., 1986, *Imaging radar for resources surveys*, (London: Chapman and Hall)

Ulaby, F.T., 1974, Radar measurements of soil moisture content, *IEEE Transactions on Antennas Propagation*, **AP–22**, 257–265.

Ulaby, F.T., Allen, C.T., Eger, G., and Kanemasu, E., 1984, Relating the radar backscattering coefficient to leaf area index, *Remote Sensing of Environment*, **14**, 113–133.

Ulaby, F.T., Barr, J., Sobti, A., and Moore, R.K., 1974, Soil moisture detection by Skylab's radar sensors, *Proceedings of URSI Commission II Specialist Meeting*, Berne, Switzerland, September.

Ulaby, F.T., and Batlivala, P.P., 1976, Optimum radar parameters for mapping soil moisture, *IEEE Transactions on Geoscience and Electronics*, **GE–14**, 81–93.

Ulaby, F.T., Batlivala, P.P., and Dobson, M.C., 1978, radar backscatter dependance on surface roughness, soil moisture and soil texture: Part 1 – Bare soil, *IEEE Transactions on Geoscience and Electronics*, **GE–16**, 286–295.

Ulaby, F.T., Bradley, G.A., and Dobson, M.C., 1979, radar backscatter dependance on surface roughness, soil moisture and soil texture: Part II – vegetation covered soils, *IEEE Transactions on Geoscience and Electronics*, **GE–17**, 33.

Ulaby, F.T., Brisco, B, and Dobson, M.C., 1983, Improved spatial mapping of rainfall events with spaceborne SAR imagery, *IEEE Transactions on Geoscience and Remote Sensing Letters*, **GE–21**, 118.

Ulaby, F.T., and Bush, T.F., 1976, Monitoring wheat growth with radar, *Photogrammetric Engineering and Remote Sensing*, **42**, 557–568.

Ulaby, F.T., Cihlar, J., and Moore, R.K., 1974, Active radar measurements of soil moisture content, *Remote Sensing Environment*, **3**, 185–203.

Ulaby, F.T., Dobson, M.C., and Brunfeldt, D.R., 1983, Improvement of moisture estimation accuracy of vegetation covered soil by combined active / passive radar remote sensing, *IEEE Transactions on Geoscience and Remote Sensing*, **GE–21**.

Ulaby, F.T, Moore, R.K., and Fung, A.K., 1986, *Radar remote sensing: active and passive*, (Artech House), a three volume set: *Volume 1: radar remote sensing fundamentals and radiometry, 1981, Volume 2: Radar remote sensing and surface scattering and emission theory, 1982, Volume 3: From theory to applications, 1986.*

Van Zyl, J.J., Zebker, H.A., and Elachi, C., 1987, Imaging radar polarisation signatures: theory and observation, *Radio Science*, **22**, 529–543.

Viksne, A., and Liston, T.C., 1970, SLR reconnaissance of Panama, *Photogrammetric Engineering*, **36**, 253–259.

Wang, J.R., Engman, E.T., Shuie, J.C., Ruzek, M., and Steinmeier, C., 1986, *IEEE Transactions on Geoscience and Remote Sensing*, **GE–24**, 510.

Wing, R.S., 1970, Cholane area – San Andreas fault zone, Calofornia: a study in SLAR, *Modern Geology*, **1**, 173.

Wing, R.S., and MacDonald, H.C., 1973, Radar geology, petroleum exploration technique, Eastern Panama and North Western Columbia, *American Association of Petroleum Geologists Bulletin*, **57**, 825.

Zoughi, R., Bredow, J., Osman, S., and Moore, R.K., 1989, Fine resolution signatures of coniferous and deciduous trees at C–band, *International Journal of Remote Sensing*, **10**, 147–169.

VEGETATION CANOPY REFLECTANCE: FACTORS OF VARIATION AND APPLICATION FOR AGRICULTURE.

F. BARET
INRA Bioclimatologie
Domaine St Paul
BP 91, 84143 Montfavet
France

1. Introduction

Since the launch of the first land applications earth observation satellite in the early 1970s, increasing attention has been paid to the interpretation of the reflectance signal collected by radiometers rather than photographic film. In the beginning, this interpretation largely relied on the experience acquired by the photointerpreters. However these technics were not optimum for the interpretation of the satellite sensor data because of the differences in scale and wavelength domain used. The transformation of the initial qualitative and subjective visual interpretation into quantitative, automatic and objective interpretation rapidily appeared as a necessity. Experimental and theoretical studies were developed to give an as robust as possible background in order to relate reflectance data to crop characteristics which could be then used by agronomists for yield or biomass assesment.

In the first part of this paper, the factors governing canopy spectral response are briefly described. We will restrict the study to the visble, near-infrared and middle-infrared spectral domains (400nm-3000nm). Sieber and Churchill, else where in this book, discuss microwave/vegetation interactions, and Bartholomé discusses the role of satellite sensor derived thermal data in crop yield estimation. A second section defines differing vegetation indices (combinations of single spectral reflectance bands) and shows how one can use them to extract some canopy characteristics. Finally the differing approaches currently used to derive crop characteristics from remote sensing data are reviewed and then future developments are discussed.

2. Factors governing canopy spectral reflectance.

Elementary interaction of any electromagnetic radiation with a canopy takes place on vegetation elements such as leaves, stems and spikes, plus soil background and water (which is important for flooded crops for example). We will firstly study the optical properties of these elements. Then, we will analyse the sensitivity of canopy reflectance on the amount and spatial orientation of canopy elements.

A. S. Belward and C. R. Valenzuela (eds.),
Remote Sensing and Geographical Information Systems for Resource Management in Developing Countries, 145–167.

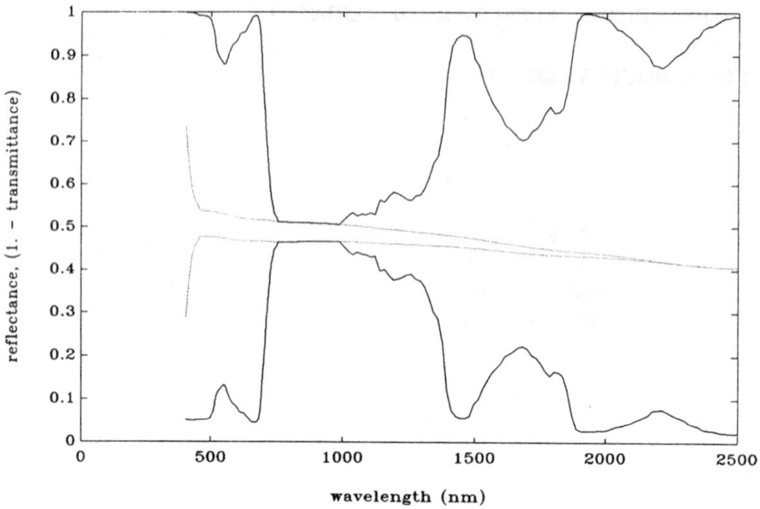

Figure 1. Typical leaf reflectance and transmittance spectra. The solid lines correspond to a typical green leaf; the dotted lines to an albino and dry leaf (without pigments and water). Simulated data from Jacquemoud and Baret (1990) PROSPECT model.

2.1. OPTICAL PROPERTIES OF CANOPY ELEMENTS.

2.1.1. *Leaves.* If one neglects emitting phenomena which are very weak in the spectral domain considered in this paper, electromagnetic theories divide any interaction into scattering and absorbing phenomena. At leaf level, the scattering is mainly due to refractive index (n) discontinuities encountered by a light beam along its path through the leaf mesophyll. It passes from 1.0 in the air to 1.33 for water and to 1.4 for cell walls and cuticule wax (Jacquemoud and Baret, 1990). The refraction index of these three main leaf constituents is not strongly wavelength dependent from 400 to 3000nm. For an ideal non absorbing leaf (i.e without pigments or water), the corresponding reflectance and transmittance spectra will be flat and very high (figure 1.).

In the near-infrared domain where there are strong absorption features, the magnitude of the reflectance and transmittance is governed by the mean number of discontinuities encountered in the leaf. For compact and thin leaves, the transmittance is greater than the reflectance. This is the case of most monocotyledons' leaves, especially if they are young. In contrast, for spongy mesophyll and thick leaves, transmittance tends towards 0.0 while reflectance reaches a saturation value close to 0.8. This can be the case for old and senescing dicotyledons. However, most green leaves exibit reflectance values very close to the transmittance values in the near-infrared.

We can divide leaf absorbing material into two categories depending on the wavelength domain. In the visible spectral domain chlorophylls and accessory pigments for which absorbption is due to changes in the spin and angular momentum of electrons and transitions between orbital states of electrons in particular atoms, chlorophyll a and b, carotenoids for example. Chlorophyll a and b pigments absorb strongly in the blue and red wavelengths, and carotenoids absorb mainly in the blue. Brown pigments appearing with leaf senscence absorb with a decreasing intensity from the blue to the short near-infrared. In the middle infrared, water absorption due to changes in the vibrational-

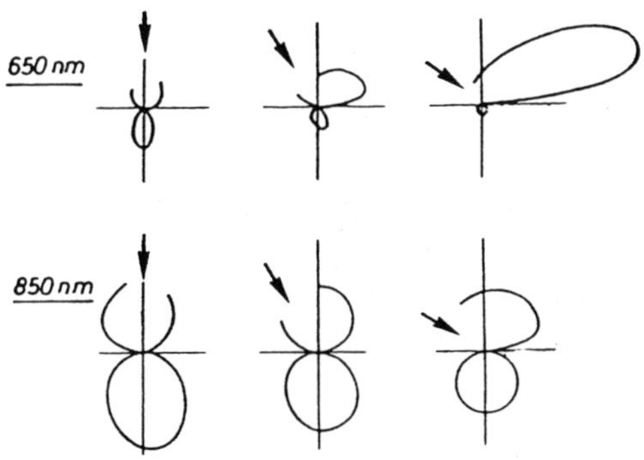

Figure 2. Directional variation of reflectance and transmittance of soybean leaves in the red (650nm) and near infrared (850nm) (from Breece and Holmes, 1971). Arrows indicate the incidence direction. reflectance and transmittance values are given in the principal plan (the plan of the incident radiation) in polar plot representation. For each view zenith angle, reflectance and transmittance are proportional to the radius. A lambertian scatterer corresponds to a perfect circular pattern in this representation.

rotational modes with two main absorption bands (around 1450 nm and 1900 nm) is the main controling factor.

Typical spectra for different leaves show off (figure 1) similar trends in reflectance and transmittance for all leaves which are caused by absorption mechanisms. Three spectral domains are classicaly defined. Firstly, the visible (400-700nm), with low reflectance and transmittance caused by the strong chlorophyll pigment absorption. Secondly the near-infrared (700-1300nm) characterized by high reflectance and transmittance levels due to the absence of strong absorbing features. Thirdly the middle infrared (1300-3000n), with variable reflectance and transmittance values linked to water absorption characteristics.

These basic properties allow us to understand differences in reflectance or transmittance of leaves from species to species or induced by various stresses; for example, a decrease in chlorophyll concentration (due to natural senescence or to mineral or water deficiencies or to deseases) leads to higher levels of reflectance and transmittance at visible wavelengths; in the same way, any decrease of leaf water content per leaf unit area increases reflectance and transmittance values in the middle-infrared; the near-infrared features are in general more stable than the visible and middle-infrared ones. We often observe a slight increase of the reflectance and decrease of the transmittance in this spectral domain from juvenile stages to senescence. This is due to the gradual changes in the mesophyll corresponding to an increase in the mean number of refractive index discontinuities.

The above described properties also explain differences between leaves and other vegetation elements such as stems, spikes and fruits, for which transmittance tends to 0.0 because of their thickness while reflectance values are again governed by pigment and water concentration per unit element area.

The phase function (describing the directional variation of reflectance and transmittance) of leaves results from two different processes (Grant, 1987); diffuse reflectance or transmittance due to internal scattering of the mesophyll; and specular reflectance which results from surface interaction and is mainly determined by leaf surface roughness depending on the state of the epicuticular wax.

These principles explain why leaves (see figure 2.) are generally close to the lambertian approximation for transmittance or near-infrared reflectance: Most of the scattering takes place inside

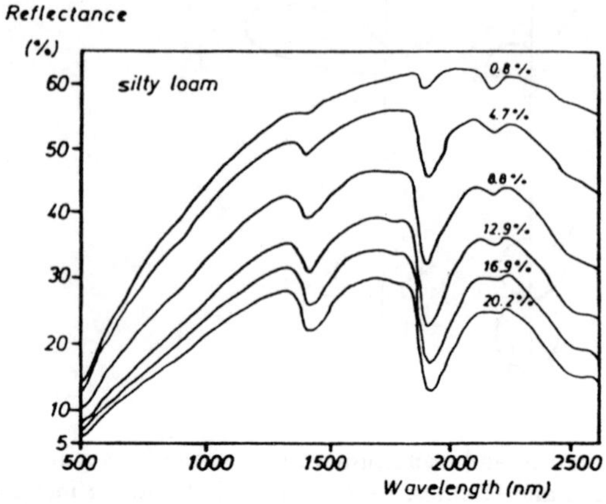

Figure 3. Typical soil spectra as function of surface moisture (from Bowers and Hanks 1965).

the leaf. In contrast, in strong absorbing domains where the light doesn't penetrate the leaf to a great extent, the specular component is very strong and also exhibits some interesting polarization features (Vanderbilt *et al.* 1985).

2.1.2. *Soil background.* Soil reflectance spectra nearly always exhibit the same features (figure 3). Reflectance increases from 400nm to around 2000nm and then decreases in the middle infrared with some deep valleys around 1450nm and 1900nm. Again this can be explained by scattering and absorption by the soil constituents. Absorption is due to particular metallic ions, by organic matter or brown pigments which mainly absorb in the visible domain. In the middle infrared, water and organic matter and carbonates are the principal absorbing elements. Scattering is determined by the grain size of soil elements relative to the wavelength and also by the macroscopic roughness which controls the shadowing and light trapping effects. From these considerations we can conclude that for a given soil type, two factors will determine its reflectance; Firstly, the roughness for which an increase generally results in lower reflectance values for all the wavelengths. Secondly, the surface moisture for which an increase also results in lower reflectance values on the whole spectra.

Soil phase function vary widely from the lambertian approximation to more anisotropic patterns. The shadowing effect linked to soil roughness is the main factor which controls the phase function for a given soil. Figure 4 shows typical directional patterns. In the perpandicular plan (the plan perpandicular to solar beam), the phase function for this soil is azimuthaly symetric and not very far from the lambertian case. But in the principal plan (the plan of the solar beam), we observe a dissymetry: reflectance values in the direction of the sun are enhanced; it corresponds to the so-called hot spot effect (or opposition effect) for which minimum shadow fraction is sensed. Of course, in the opposite direction, because of the increasing shadowed parts recorded, reflectance decreases. We can notice (figure 4) that the phase function of a soil is generally very slightly wavelength dependant.

An important property of soil background is the linear relationship observed between reflectances in one given wavelength and in any other wavelength (Richardson and Wiegand, 1977). This is termed as soil lines and can be expressed as

$$R_s(\lambda_2) = a(\lambda_1,\lambda_2) . R_s(\lambda_1) + b(\lambda_1,\lambda_2) \tag{1}$$

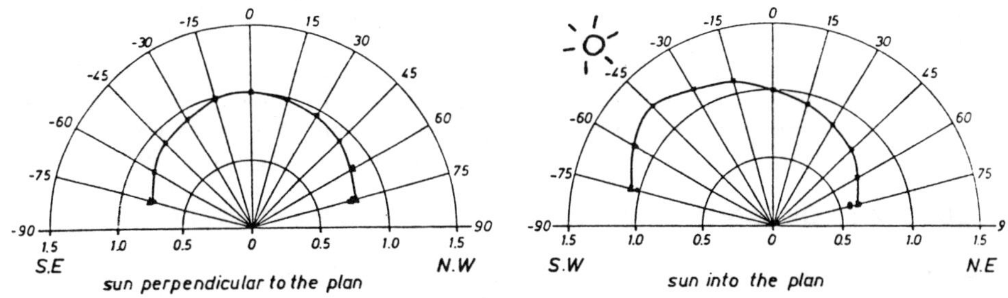

Figure 4. Phase function of a bare soil for the four MSS channels. Reflectances are normalized by nadir viewing values. Lambertian soil corresponds to the circle of radius 1.0 in this polar plot representation (From Guyot *et al.* 1980).

where λ_1 and λ_2 are any two wavelengths, $R_s(\lambda_1)$ and $R_s(\lambda_2)$ the corresponding soil reflectances, and $a(\lambda_1,\lambda_2)$ and $b(\lambda_1,\lambda_2)$ are the soil line parameters for the two wavelengths (λ_1,λ_2). Figure 5 shows the distribution of the soil line parameters for red (λ_1) and near-infrared (λ_2) wavelengths reported in the literature. This relationship also holds for fully senescent vegetation, and this allows us to define a "senescent line" rather than a soil line (Baret, 1986). This is important because it represents a lot of real backgrounds for vegetation having significant amounts of dead material at the bottom of the canopy.

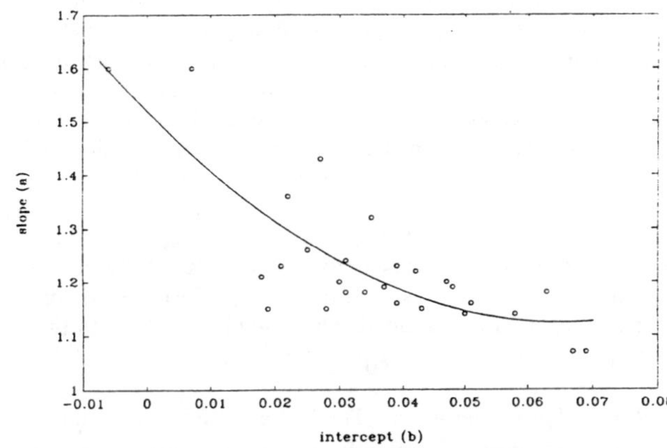

Figure 5. Distribution of soil line parameters (a, b) for red and near infrared domains reported in the literature.

2.1.3.: *Water.* Water can be an important background in agricultural terms where flood irrigation is practiced, rice crops for example. Similarly to leaves, water reflectance is the sum of a specular component, depending on surface features and governed by Fresnel's laws, and of a scattered component issued from internal interaction (volume scattering). The specular reflectance from the water body's surface is usually perceived as 'sunglint' from air and space platform observations (Cracknell *et al.* 1987). In the middle infrared, the strong absorption properties of liquid water leads to near zero reflectance. In consequence, the magnitude of the scattered component varies signicantly only in the visble and near infrared domains. It largely depends on the nature of the bottom, depth of water, and on the amount and optical properties of suspended particles. These suspended elements can be algua with absorption characteristics close to the chlorophyll pigments, and "yellow substance" which can be either mineral (suspended soil elements) or organic. For turbid water, the soil line concept remains valuable in the visible and near-infrared domains (Yamagata *et al.* 1988).

2.2. SENSITIVITY OF CANOPY SPECTRAL REFLECTANCE TO SOME FACTORS.

The spectral reflectance of a vegetation canopy will be a weighted composition of each element (soil, water, vegetation). The weighting factors depend either on external parameters such as illumination or viewing geometry or on canopy characteristics describing the spatial arrangement of the elements (mainly the leaves). We will illustrate this section by simulating canopy spectral response to differing sets of input parameters with a reflectance model. The SAIL reflectance model (Verhoef 1984, 1985) is used which allows computation of the bidirectional reflectance of a horizontaly homogeneous canopy with lambertian leaves. For simplification, we will just study a one layer canopy. This model has been partialy validated on several occasions (Badwhar *et al.* 1984, Goel 1985, Rigal and Terres 1988 for example). The model requires a minimum set of input parameters to compute the corresponding bidirectional reflectance. These can be divided into canopy parameters and external parameters. Canopy parameters include leaf area index (LAI) defined as leaf area (one side) per unit ground area); leaf inclination distribution function (LIDF) which describes the orientation of the leaves. In this simulation, the azimuth of the leaf is considered uniform and the inclination follows an ellipsoidal distribution function (distribution of inclinations equivalent to those found on an ellipsoid) characterized by the average leaf angle (θ_l) (Campbell 1986, Wang and Jarvis 1988); leaf reflectance $r(\lambda)$ where λ is the wavelength. For this simulation study, we consider two spectral domains: red for which green leaves have around 0.07 reflectance, and infrared with 0.45 reflectance; leaf transmittance $t(\lambda)$. The corresponding leaf transmittance in the red and near infrared domains is taken as 0.007 and 0.50 respectively; and finally soil reflectance which is considered for simplification as lambertian and having the same reflectance level in the red and near infrared domains. The external parameters include illumination geometry characterized by the solar zenith angle (θ_s). We consider only direct incoming radiation; viewing geometry characterized by the view zenith angle (θ_0) and the view azimut (ψ_0) defined relatively to the sun azimut ($\psi_0=0$ corresponds to the azimut of the sun).

For each spectral domain the directional distribution of the reflectance for differing LAI, soil reflectance, leaf inclination or sun position is presented. The study is restricted to the principal plane ($\psi_0=0°$ or $\psi_0=180°$) and to the perpendicular plane ($\psi_0=90°$ or $\psi_0=270°$).

2.2.1. *Leaf area index and soil optical properties effects.* The leaf area index is the most important crop parameter used to describe canopy structure. It varies from 0.0. (bare soil) to a maximum value of around 8.0 to 10.0 for annual crops, and even more than 15.0 for very dense prerennial forests. Figure 6a shows the variation of the directional reflectance for various LAI levels. For a given view direction, canopy reflectance starts from the soil background reflectance value for LAI=0.0 (no vegetation) and tends towards a limit termed "infinite reflectance". In the red domain, the reflectance rapidly decreases and reaches the infinite reflectance value for LAI close to 2.0. In contrast, in the near-infrared domain, the reflectance increases more gradually until it reaches the infinite reflectance value for LAI close to 8.0. This behaviours can be explained in terms of contrast between leaf and soil optical properties and by leaf transmittance levels. In the red domain, soil reflectance is generaly higher than leaf reflectance whilst the reverse case is true in the near-infrared. Over all cases, when LAI is sufficiently high enough to be close to the infinite value, soil no longer has an effect on canopy spectral response. In the red

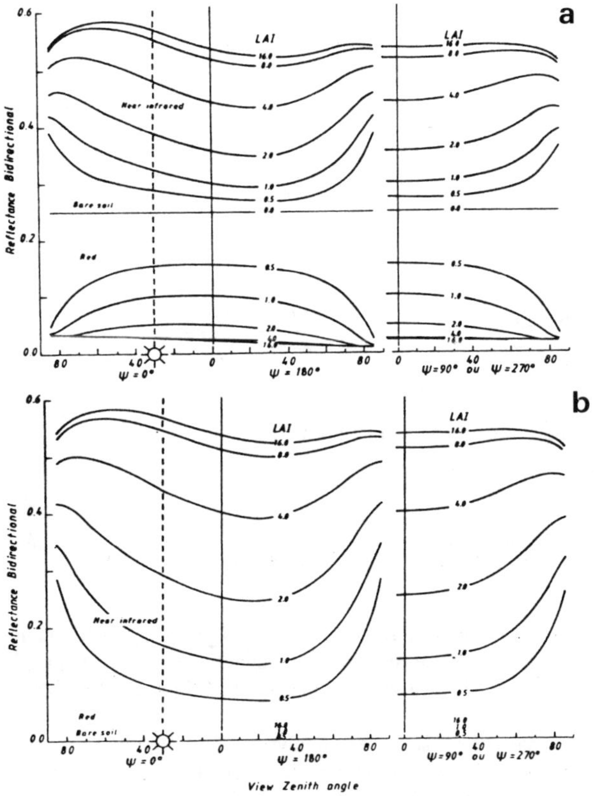

Figure 6. Bidirectional reflectance simulated by SAIL model as function of the view zenith angle (θ_0) in the principal plan and in the perpandicular plan. LIDF is spherical (θ_1=57°), solar zenith angle (θ_s) is 30°. Each curve correspond to a particular LAI ranging from 0.0 to 16.0. Top curves correspond to the near-infrared domain and bottom curves to the red domain. Soil reflectance is taken as 0.25 (figure 6a) and 0.0 (figure 6b) for both spectral domains.

domain, leaf transmittance is very weak so that almost no radiation penetrates inside the canopy after the initial interception of light by the upperest leaf layers. The saturation level, hence the infinite reflectance will be reached as soon as the canopy completly covers the soil background, a situation obtained for relatively low LAI (about 3.0). On the contrary, in the near-infrared because of the high leaf transmittance level, a number of leaf layers are required to obtain a full extinction of any incident beam. In consequence, multiple leaf layers will contribute significantly to canopy reflectance particularly where leaves have high near-infrared reflectance. These properties lead to important multiple scattering phenomena explaining the high infinite reflectance values obtained in the near-infrared spectral domain.

This simple behaviour has been formalized into semi-empirical models relating spectral reflectance to LAI (Baret 1988)

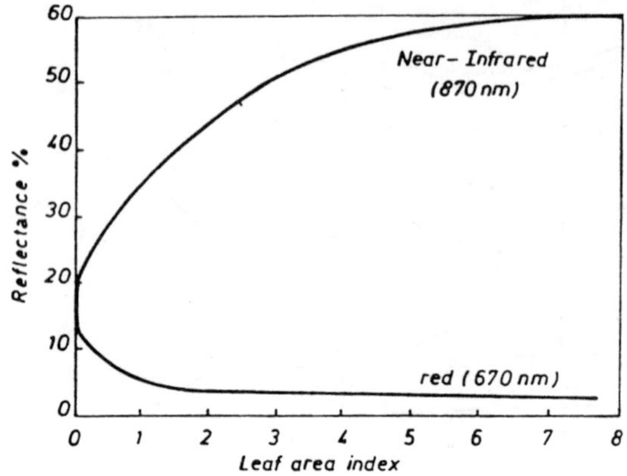

Figure 7. Red and near-infrared reflectance as function of LAI.

$$R_c(LAI,\theta_l,r,t,R_s,\theta_s,\theta_o,\psi_o)=R_\infty(\theta_l,r,t,\theta_s,\theta_o,\psi_o)+(R_s-R_\infty(\theta_l,r,t,\theta_s,\theta_o,\psi_o)).e^{-K(\theta,r,t,\theta,\theta,\psi).LAI} \quad (2)$$

where R_s is the soil background reflectance, $R_c(LAI,\theta_l,r,t,R_s,\theta_s,\theta_o,\psi_o)$ is the canopy bidirectional reflectance, $R_\infty(\theta_l,r,t,\theta_s,\theta_o,\psi_o)$ is the canopy infinite reflectance, and $K(\theta_l,r,t,\theta_s,\theta_o,\psi_o)$ is an extinction coefficient. This semi-empirical model is illustrated by figure 7 for the red and the near infrared domains.

If we now consider a particular LAI and analyse the directional distribution of canopy reflectance (figure 6.), we generally observe non lambertian properties characterized by (i) dissymetry in the principal plane due to the shadowing effects as described for soil background. For a view direction close to sun direction, a minimum of shadow is sensed. When view direction is very far from sun direction, the relative shadowed area seen increases and reflectance values are thus lower; (ii) a symetry in the perpandicular plane. Reflectance values are generally in between the values observed in the principal plane for a given viewing zenith angle. Figure 6 also shows that the directional reflectance depends on the contrast between soil and leaf optical properties. Similar processes occur when the mean optical path increases due to changes in the view direction and a corresponding increase of the LAI. This can explain why in most cases (except for errect canopies), the directional response is closer to the lambertian approximation when the reflectance tends to its infinite value for large LAI. However, this behaviour is controled by leaf angle distribution.

2.2.2 *Effect of the leaf angle distribution.* As previous simulations were computed with a spherical leaf angle distribution, we will now manipulate it by compressing or streching and look at the corresponding reflectance response. Figure 8. shows that a planophile canopy ($\theta_l=0°$) is a perfect lambertian body assuming that all it's elements are lambertian. In this particular case, the directional reflectance equals the hemispherical reflectance. The anisotropy of the reflectance increases as the canopy becomes more errect. In this case and for non saturing LAI, the reflectance is closer to the soil reflectance (soil contribution increases when leaves are more errect). For large LAI, errect canopies correspond to low reflectance values because of light trapping effects. This has for exemple been observed over dense tropical forests by Malingreau *et al.* 1989.

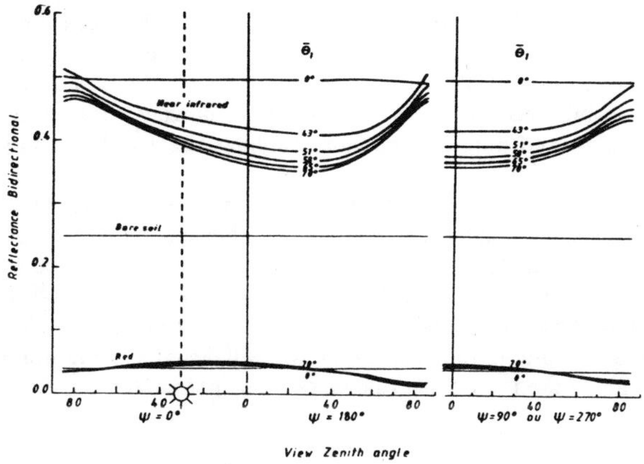

Figure 8. Idem figure 6a but for LAI =2.0 and differing LIDF characterized by the average leaf angle (θ_l). θ_l=0° and θ_l=57° correspond respectively to planophile and spherical LIDF canopies.

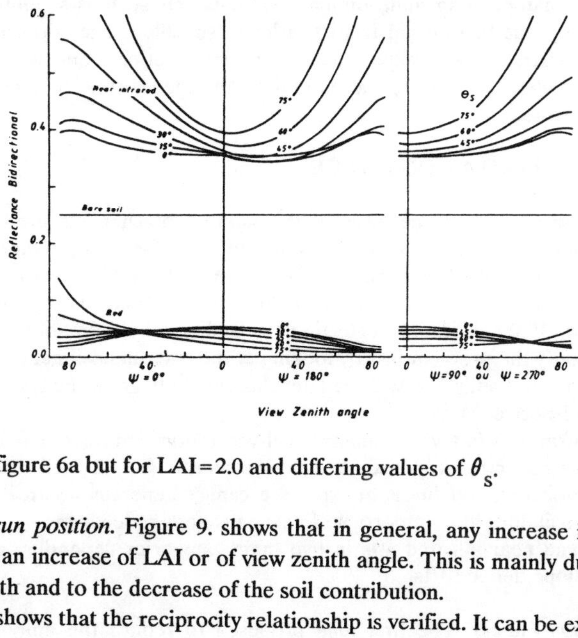

Figure 9. Idem figure 6a but for LAI=2.0 and differing values of θ_s.

2.2.3. *Effect of sun position.* Figure 9. shows that in general, any increase in solar zenith angle has a similar effect to an increase of LAI or of view zenith angle. This is mainly due to the increase in length of the optical path and to the decrease of the soil contribution.

Figure 9 shows that the reciprocity relationship is verified. It can be expressed by

$$R_c(LAI,\theta_l,r,t,R_s,\theta_s,\theta_o,\psi_o)=R_c(LAI,\theta_l,r,t,R_s,\theta_o,\theta_s,\psi_o) \tag{3}$$

where R_c and R_s have the same meaning as in equation (2). This important property expresses the fact that the source and the sensor positions can be exchanged without affecting the resulting reflectance.

During this simulation study, we have proposed a qualitative interpretation of physical processes governing the spectral response. We can distinguish four principal mechanisms; firstly the magnitude and sign of the difference between soil background optical properties and leaves reflectance and transmittance; secondly the contribution of multiple scattering which increases with the optical path and with reflectance and transmittance of leaves; thirdly the shadowing effect which determines the proportion of shadowed areas sensed; fourthly the light trapping effect which is important for errect and dense canopies.

We have also seen that canopy spectral response depends not only on vegetation parameters but also on uncontrolled variables such as soil background and irradiance geometry. In order to reduce the effect of uncontrolled variables which create noise around the relationship between canopy parameters and reflectance, differing combinations of individual wavebands have been used. At present, many remote sensing studies are based on these combinations (termed as vegetation indices) rather than on original reflectance (or radiance) data. In the next section we will describe some well known vegetation indices and show how they can be used to extract relevant canopy parameters.

3. Review of vegetation indices and their use.

Because of the difficulty in manipulating multi-dimensional data sets such as satellite images, and in order to remove disturbances due to external factors, a lot of so-called vegetation indices (VI) have been proposed. These synthesize in one unique quantity the information from two or more spectral channels. We will first try to classify these vegetation indices and then formalize general relationships with some canopy characteristics.

3.1. CLASSIFICATION OF VEGETATION INDICES.

Vegetation indices can be classified on the basis of the number of channels used. Generally, this number is dictated by the number of available and relevant channels (for example thermal data are not used to form VIs), ranging from 6 (landsat TM), to 2 (NOAA AVHRR). However a number of works report that more than 90% of the multispectral information is explained just by only two channels (Sheffield 1985, Baret *et al.* 1988). Without neglecting the additional information corresponding to extra channels, we will simplify the problem just by looking at the two dimensional case corresponding to the red and near-infrared wavebands. We will see later that the findings on the two dimensional case can be extended to the multispectral case.

All red and near-infrared VIs try to enhance soil-vegetation contrasts and to minimize soil background effects on canopy spectral response. As we have seen in § 2.1.2., soil reflectance can vary widely due to its surface moisture, roughness or type. We cannot generally control these variations, though as we saw in § 2.2.1 soil does have important effects on canopy reflectance.

We will divide red and near-infrared VIs in two main categories depending on how they are built. That is distance and slope derived VIs.

3.1.1. *The distance vegetation indices.* The first idea proposed to reduce the amount of data is to perform a principal component analysis (PCA) and to restrict the spectral information to the distance to one principal axis. Wilkinson elsewhere in this book describes the image processing methods involved in PCA. However this blind procedure has been successfully replaced by orthogonal decomposition where the axis are defined by some particular features (Jackson 1983). In the 2D case, the soil line is choosen as the first vector. This allows Richardson and Wiegand (1977) and Kauth and Thomas (1976) to define respectively the PVI (Perpendicular vegetation index for the 2D case) and the GVI (Greeness vegetation index for the nD case) as the distance to the soil line. PVI is derived fron

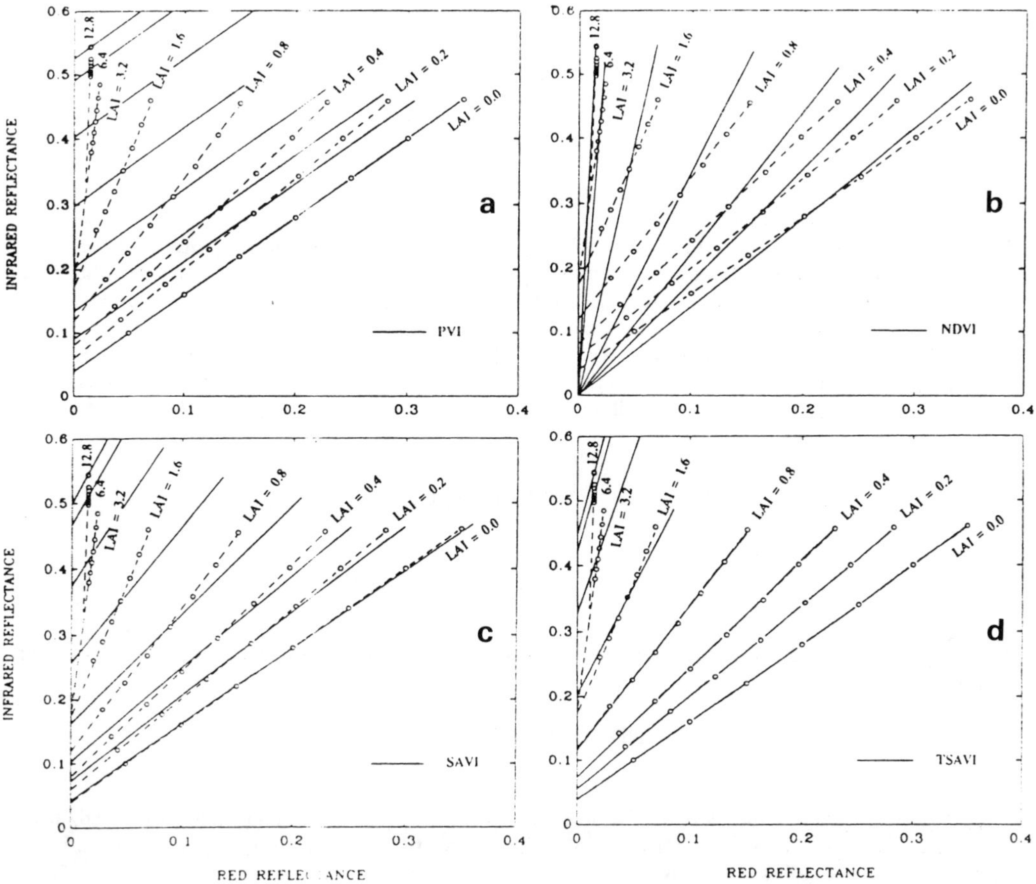

Figure 10. Graphical representation of different VI in the red near-infrared space. The dashed lines correspond to canopies with different soil backgrounds and with the sane LAI (vegetation lines). The line for which LAI=0.0 is the soil line. Open circles are the simulated reflectances. solid lines correspond to "iso-VI" (same VI values) and are drawn for each LAI across the mean vegetation point corresponding to the average soil optical properties.

the soil line equation (1) and expressed as

$$PVI=(R_{nir}-a.R_{red}-b)/\sqrt{a^2+1} \tag{4}$$

Figure 10a shows that iso-PVI lines correspond to lines parallel to the soil line. PVI value increases with LAI. We can observe that these indices initially built to remove soil background disturbances on spectral data are still sensitive to soil brightness at low and medium LAIs. Based on a more physical analysis, Clevers (1988) has elaborated a similar index called the WDVI (weighted difference vegetation index) defined by

$$\mathrm{WDVI} = R_{nir} - a.R_{red} \tag{5}$$

This index is functionally equivalent to the PVI because we can write

$$\mathrm{PVI} = (\mathrm{WDVI}-b)/\sqrt{a^2+1} \tag{6}$$

3.1.2. *The slope vegetation indices.* Most remote sensing systems don't directly measure the reflectance but the radiance. Ratioing spectral data has the theoretical advantage of removing multiplicative factor effects such as variations in irradiance conditions. But this approximation is only strictly valid if we ignore atmospheric effects which induce spectral variation of the incident and reflected radiation and also produce an additive term due to the atmospheric path radiance.

Pearson and Miller (1972) and Rouse *et al.* (1974) developed respectively the RVI (ratio vegetation index) and the NDVI (normalized difference vegetation index). These ratio vegetation indices enhance the contrast between soil and vegetation reflectance (or radiance) in the red and near-infrared space. They are defined by:

$$\mathrm{RVI} = R_{nir}/R_{red} \tag{7}$$

$$\mathrm{NDVI} = (R_{nir} - R_{red})/(R_{nir} + R_{red}) \tag{8}$$

From equation (7) and (8), we can write the mutual functional equivalence (Perry and Lautenschlager 1984)

$$\mathrm{NDVI} = (\mathrm{RVI}-1)/(\mathrm{RVI}+1) \tag{9}$$

Both VIs correspond to the slope of the line joining the origin and the data point in the red/near-infrared feature space diagram (figure 10b). We can see that these initial VIs are again very sensitive to soil variations, especially for low LAI.

Huete (1988) modified the NDVI in order to reduce soil background effects. The SAVI (soil adapted vegetation index) he proposes is based on a particular property of the reflectance features in the red near-infrared representation: for a given canopy (given LAI and LIDF), any changes in soil background reflectance affects the near-infrared reflectance proportionally to the red reflectance. This can be expressed by "vegetation lines" equivalent to the soil lines

$$R_{nir} = \alpha . R_{red} + \beta \tag{10}$$

A measure of the slope α will result in a soil independent VI. The SAVI is an approximation of this slope α (figure 10c). SAVI is expressed as

$$\mathrm{SAVI} = (1+L).(R_{nir} - R_{red})/(R_{nir} + R_{red} + L) \tag{11}$$

where L is a parameter which can be adjusted to minimize soil effects. Huete (1988) fixes L to 0.5. We can see that soil background effects are still not insignificant. Starting from this point, Baret *et al*

(1989) and Baret and Guyot (1990) improve the SAVI by taking into account the soil line parameters a and b as in (1)

$$TSAVI = a.(R_{nir}-a.R_{red}-b)/(a.R_{nir}+R_{red}-a.b+X(1+a^2)) \tag{12}$$

where X is again a parameter which is adjusted so as to minimize soil background effects. It has been set to 0.08. Figure 10d shows that soil background effects are reduced in comparison to the other three VIs. We can also see that TSAVI is equivalent to NDVI for a particular set of soil line parameters: (a,b) = (1.0, 0.0).

Of all vegetation indices, the most successful VIs in terms of removing soil back ground effects are the "slope VI" adapted to a particular soil line. The definition of all these VIs can be extended to any 2D spaces (for example near-infrared/middle-infrared space) and also to nD spaces but with an increasing complexity. We are now able to study how these VI are related to crop parameters.

3.2. CANOPY PARAMETER ASSESSMENT USING VEGETATION INDICES.

We saw in § 2.2.1. that the LAI is functionaly linked to spectral response. The LAI is an important vegetation characteristic widely used by agronomist or ecologist for example. Importance is largely due to the fact that it measures the magnitude of the leaf area development controlling most of the physiological and energy budget processes. The first part of this section establish a framework for studying relationships between VI and LAI.

The efficiency ϵ_i is the other widely used parameter which can be derived from LAI but also directly from radiometric data: Efficiency ϵ_i corresponds to the fraction of photosynthetically active radiation (400-700nm) absorbed by the canopy. This quantity allows us to evaluate the primary production through the following identity equation:

$$DM = (DM/APAR).(APAR/IPAR).(IPAR/S).S = \epsilon_c.\epsilon_i.\epsilon_s.S \tag{13}$$

where DM is instantaneous dry matter production, S is the incoming global radiation (300-3000nm), IPAR is the incoming photosynthetically active radiation (400-700nm), and APAR is the absorbed photosynthetically active radiation. ϵ_i is then defined as the fraction of APAR absorbed by the canopy (APAR/IPAR), ϵ_s is the fraction of photosynthetically active radiation contained in the global incoming radiation (IPAR/S), and ϵ_c is the efficiency of the conversion of APAR into biomass (DM/APAR). Equation (13) is often rewritten in it's integral form, relating the cumulated primary production (AGDM) to the set of efficiency coefficients

$$AGDM = \int_t DM.dt = \int_t \epsilon_c.\epsilon_i.\epsilon_s.S.dt \tag{14}$$

where t is the time. In equation (14), ϵ_s is very stable for most irradiance conditions and is usually close to 0.5 (Varlet-Grancher *et al.* 1981). ϵ_c depends on growth conditions such as different stresses, CO_2 level, temperature, on phenological stages and on vegetation type. Values of ϵ_c range between 0.0 to more than 3.0 $g.MJ^{-1}$ and can be derived from crop growth models as we will see later. The last efficiency parameter, ϵ_i, is functionaly linked to canopy structure, as indeed VIs are. Thus we can see

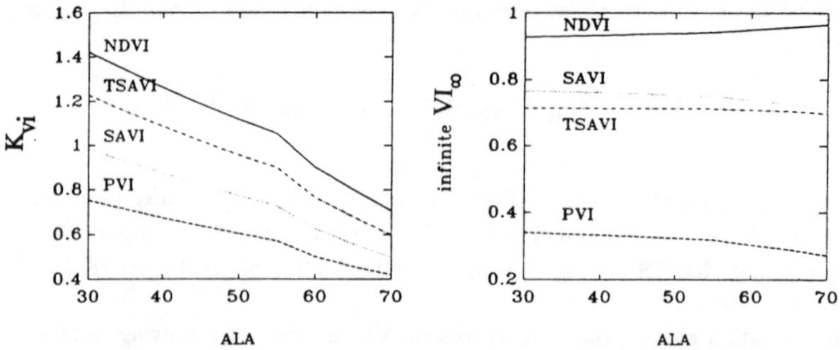

Figure 11. Extinction coefficient (K_{vi}) and asymptotic values (VI_{∞}) as a function of the average leaf inclination ($ALA = \theta_l$) for NDVI, PVI, SAVI and TSAVI. (From Baret and Guyot 1990).

that VI and ϵ_i are mutually dependant However the time scale for ϵ_i corresponds to values integrated over a day (there are no accurate methods to evaluate AGDM at time scales shorter than a day) whereas radiometric data are instantaneous. The second part of this section describes the relationships between VI and ϵ_i.

3.2.1. *LAI estimated from VIs.* A large number of studies report that vegetation indices reach a saturation level with high LAI (Hatfield *et al.* 1984, Wiegand and Hatfield 1988, Clevers 1989, Baret and Guyot 1990). This behaviour is similar to that observed with canopy reflectance as shown in § 2.2.1. and is formalized by the following equation

$$VI = VI_{\infty} + (VI_s - VI_{\infty}).e^{-K_{vi}.LAI} \tag{15}$$

where VI_{∞} is the infinite value of VI when LAI tends towards infinity, VI_s is the value of VI for the soil background, and K_{vi} is an extinction coefficient. Figure 11 shows the values of the extinction coefficients K_{vi} and asymptotic levels of VI, VI_{∞} for the most important red-near infrared VIs. This asymptotic trend is a source of important errors in estimating LAI from radiometric data at high LAI levels. The VIs are no longer sensitive to changes in the LAI, but are still sensitive to variations in some factors such as leaf oprientation, sun or viewing position, and leaf optical properties. This continued sensitivity to factors other than LAI will lead to important bias in LAI assesment through VI evaluation for LAIs beyond about 3.0.

The other major problem involved in using equation (15) to assess LAI is the necessity for *a priori* knowledge concerning the values of the parameters K_{VI}, VI_{∞}, and of course VI_s (or the parameters of the soil line (a,b) for PVI, SAVI or TSAVI). Parameters $K_{VI}$and VI_{∞} depend on canopy architecture and vary from species to species, but also with the level of stresses and with phenology. Most of the values for the parameters K_{VI} and VI_{∞} fitted to experimental data exhibit some variations even for a given species (figure 12). This noise will also introduce errors in the estimation of LAI from VIs.

3.2.2. ϵ_i *estimated from VIs.* Field measurements and theoretical results agree to relate ϵ_i coefficient (integrated over a day) to LAI through

$$\epsilon_i = \epsilon_{i\infty}.(1.0 - e^{-K_{\epsilon}.LAI}) \tag{16}$$

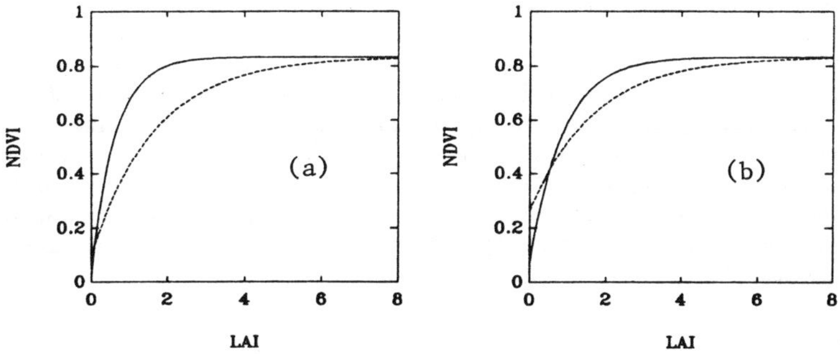

Figure 12. Experimental relationships between NDVI and LAI fitted for Corn crops sown at two differing dates (a and b)(From Baret 1989). Solid line: before anthesis; dashed line: after anthesis. We can notice the variations among sowing dates and phenologic stages in the parameter K_{vi} corresponding to the slope of the curves.

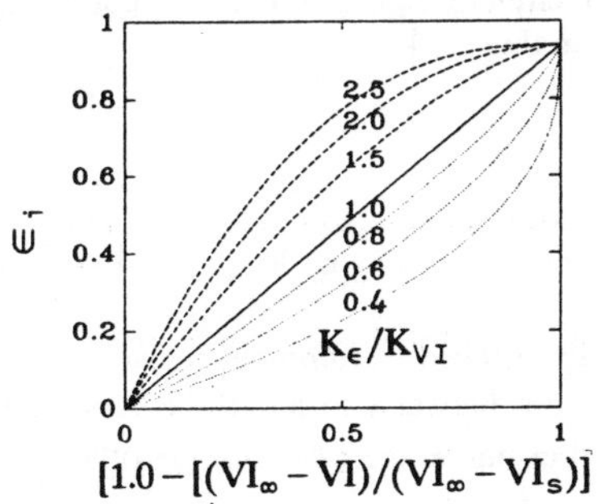

Figure 13. general relationship between ϵ_i and VI. abscissa correspond to a normalisation of VI: $[1.0-[(VI_\infty-VI)/(VI_\infty-VI_s)]$. The relationship is given for differing values of K_ϵ/K_{VI}.

Where $\epsilon_{i\infty}$ is the asymptotic value of ϵ_i reached for LAI tending towards infinity, and K_ϵ is an extinction coefficient. $\epsilon_{i\infty}$ and K_ϵ depend on canopy structure, leaf optical properties in the PAR domain and irradiance conditions through the day. $\epsilon_{i\infty}$ is generally close to 0.94 for green leaves, and K_ϵ ranges from 0.3 for very errect and sparse canopies to more than 1.5 for regular and planophile canopies. Coupling equations (16) and (15) produces a general relationship between VI and ϵ_i

$$\epsilon_i=\epsilon_{i\infty}.[1.0-[(VI_\infty-VI)/(VI_\infty-VI_s)]^{(K\ /K\ )}] \qquad (17)$$

Figure 13 illustrates this relationship for differing values of K_ϵ/K_{VI}. This ratio ranges from 0.5 to 2.5 depending on the VI used and on the canopy architecture and irradiance conditions used to estimate ϵ_i. If the ratio K_ϵ/K_{VI} is very close to one, we get a linear relationship between ϵ_i and VI as reported by some authors, Hatfield *et al.* 1984 and Sellers 1985 for example. Here again, we have to have *a priori* knowledge of the values of the parameters $\epsilon_{i\infty}$, VI_∞ and $K\epsilon/K_{VI}$ which are subjected to variations within and between species.

In this section, we have defined general relationships between VI and LAI or ϵ_i. It appears as a necessity to have *a priori* knowledge of these relationships parameters (equations (15) and (16)). Because these values vary from species to species, a preliminary classification separating the different species must be performed on the satellite scene before applying them. However it also appears that other factors induce some variations. One way to reduce this scattering around these relationships is to use more than three channels or directional viewing. There is no doubt that the increasing accuracy in estimating LAI or ϵ_i is linked with an increasing complexity of the algorithms used. It will probably implicitly tend toward the so-called inverse problem consisting in extracting canopy architecture parameters and optical properties of its elements from a larger set of multispectral multi directional set of radiometrical data by inverting reflectance models.

4. Strategies for biomass and yield estimation.

In this section we will analyse the different approaches currently used to estimate biomass or yield from reflectance data. Atmospheric effects, spatial resolution and accuracy of the satelite calibration on the calculation of VIs from satellite digital counts will not be discussed here, though these points are reviewed by Hill elsewhere in this book.

4.1. CURRENT PROCEDURES.

Two main approaches to yield estimation are adopted; those using instantaneous relationships between biomass and radiometric data, and those using multitemporal radiometric data.

4.1.1. *Instantaneous relationship between radiometric data and biomass or yield* A number of experiments show good correlations between VI and the corresponding above ground biomass or yield (Aase and Siddoway 1981, Gerberman *et al.* 1984, among others). These can be set in the form of a product of terms which are more or less independent:

$$AGDM = (AGDM/LDM).(LDM/LAI).LAI = (Lr)^{-1}.(LSA)^{-1}.f^{-1}(VI) \tag{18}$$

where AGDM is the above ground dry mass, LDM is the leaf dry mass, Lr is the fraction of total dry matter corresponding to leaves (LDM/AGDM), LSA is the leaf specific area (LDM/LAI), and f is the function corresponding to equation (15) and defined by: VI = f(LAI).

The yield can also be derived from total above ground biomass through the following relationship

$$YIELD = HI.AGDM \tag{19}$$

where HI is the harvest index which corresponds to the ratio of economicaly usable parts to total biomass. This varies widely with species, cultivar, and growth conditions (Donald and Hamblin 1976). It follows that instantaneous relationships between yield or total biomass production and VI are not direct and depend on many factors; leaf dry matter fraction (Lr) will vary widely with species, phenology and stresses, depending on the relative amount of non leaf materials such as stems, fruits and spikes. For most annual crops it starts from 1.0 at emergence and gradually decreases until 0.0 at

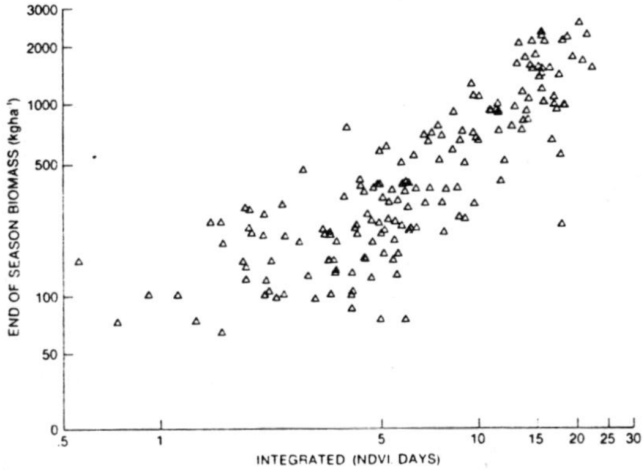

Figure 14. Relationship between net primary dry biomass produced at the end of the growing season and the NDVI integrated over the growing season in Senegal for herbaceous plants (from Prince and Tucker 1986).

maturity where all green leaves have disappeared; leaf specific area (LSA) depends mainly on species and to a lesser extent on phenology or stresses (Rawson *et al.* 1987); Last but not the least, AGDM and yield estimates from VI are directly subject to errors in extracting LAI from VIs as discussed earlier in § 3.2.1. These remarks explain the variability observed among empirical relationships found in the literature (Tucker *et al.* 1980, Baret 1988 among others). In consequence, this approach appears to be highly dependent upon the conditions in which they are obtained such as year, phenological stage, location, cultivar and treatment; and as we have seen can only be used for low amounts of vegetation (not saturating LAI). Therefore, it seems somewhat difficult to use the relationships (18) or (19) in a predictive way.

4.1.2. *Using multitemporal data.* In § 3.2 we demonstrated that VI is related to the efficiency coefficient ϵ_i which was used in equation (14) to compute biomass production. However, this approach needs frequent estimates of ϵ_i in order to perform the time integration over the growing season. Meteorological satellites such as NOAA AVHRR offers a very high temporal resolution to monitor the VI evolution. Using this source of data, Tucker *et al.* (1985) and Goward *et al.* (1986) demonstrated that NDVI integrated over the whole growing season gives a good description of the biomass produced in large areas (figure 14.)

This linear feature can be aproximated by assuming a linear relationship between ϵ_i and NDVI as found by Asrar *et al.* (1984) and time independent values of ϵ_s, ϵ_c and S. In these conditions, equation (14) can be set in the form of

$$\mathrm{AGDM} = \epsilon_c.\epsilon_s.S_d.\epsilon_{i\infty}.\ {}_{\mathrm{season}}[(VI-VI_s)/(VI_\infty-VI_s)] \quad (20)$$

where S_d is the mean daily incoming global radiation. Although this global procedure, applied to large areas seems to give interesting results, it can be refined by introducing the temporal dependency of each term. This has been partially performed at ground level by some authors and gives reliable estimates of biomass for some crops (Ajai *et al.* 1984, Wanjura and Hatfield 1985, Gallo *et al.* 1985, Guerif *et al.* 1988, Steinmetz *et al.* 1990 among others). However, only few applications using earth

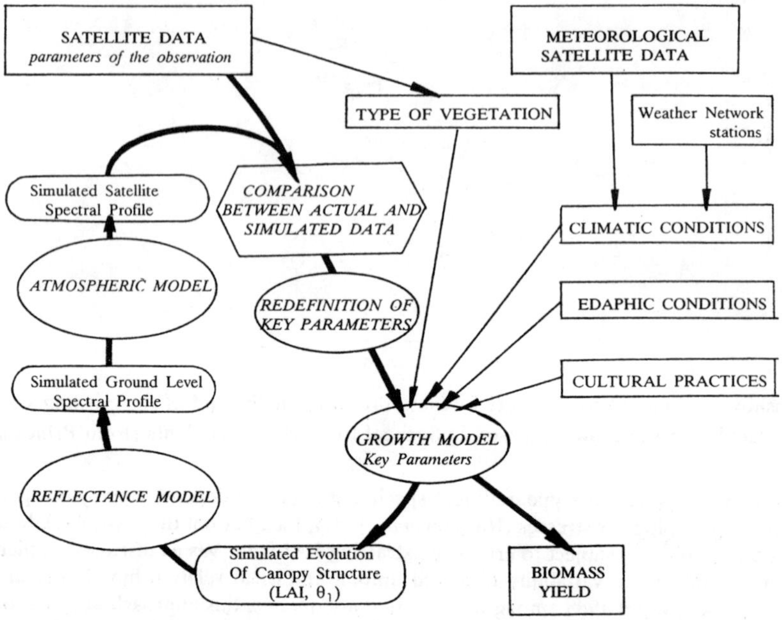

Figure 15. general procedure to derive biomass or yield from remote sensing data and other sources of information.

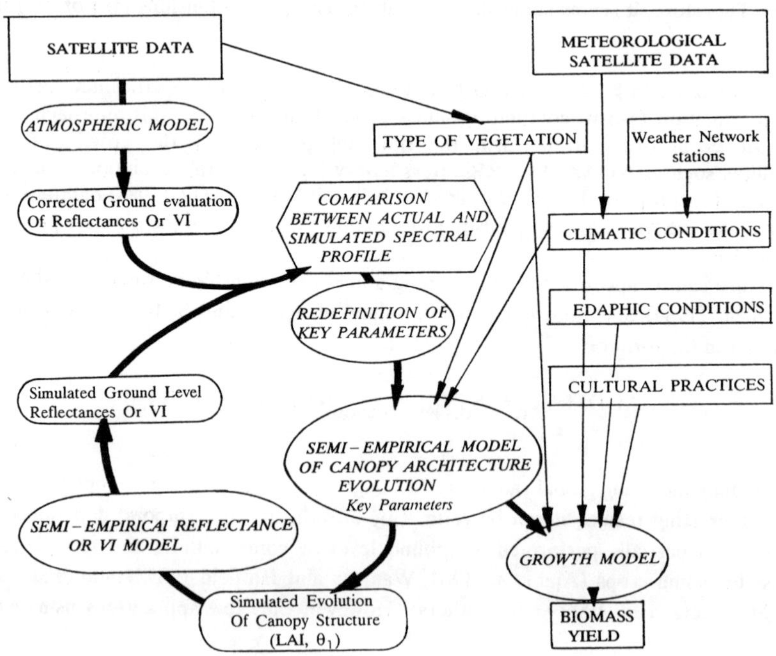

Figure 16. Simplified procedure to derive biomass and yield from remote sensing data and other sources of information.

observation satellite data have been developed: The main limitation to using this technique is the high temporal resolution it requires. Interpolation algorithms have been elaborated by different authors (Baret 1986, Clevers 1986; Maas 1988) so that it is possible to describe the whole spectral profile (time evolution of radiometric data) from a minimum of satellite measurements. This more mechanistic approach is the one actually developed to assess the economic yield of crops.

4.2. FUTURE DEVELOPMENTS: COMBINING SPECTRAL PROFILE AND GROWTH MODELS.

The definition of the parameters governing the primary production and the partition between harvested and unharvested parts of crops requires external information such as climatic conditions, edaphic characteristics, and information about cultural practices, in addition to spectral data. This multi-source information has to be integrated into a general growth model for which input data are the external factors described above and output should be the primary production and yield, and also an evolution of canopy architecture which can be transformed into a spectral profile through reflectance modelling (Baret *et al.* 1988). By adjusting key parameters, we can reduce the distance between the spectral profile observed by satellite sensors and the one simulated through growth models. Figure 15 shows the procedure proposed. This approach requires requires a lot of external information, though in part earth observation and meteorological satellites contribute to complete them. The main problems actually remaining in applying this procedure are firstly the lack of completely tested and reliable reflectance and growth models, secondly the fact that the key parameters have to be more clearly defined in the growth models, and thirdly the necessity for frequent observation.

However, the approach does have the advantage that data from different satellite configurations and also complementary data such as radar response can be used. At present, it seems that research is oriented toward a simplified version of this initial procedure. Two separate models are used, first to extract canopy architecture evolution from remote sensing data, second to compute biomass and yield. Figure 16 describes this simplified procedure. A semi-empirical model of canopy architecture evolution is used to simulate the evolution of single band reflectance or VIs through a semi-empirical model. This simulated spectral profile is then compared to the satellite data. The parameters of the model of canopy architecture evolution are optimized in order to minimize the distance between the actual (satellite) and simulated spectral profile. The resulting evolution of canopy architecture is finally used in a specific growth model to compute biomass production and yield. This simplified procedure avoids the problems of defining the key parameters of the initial procedure growth model. The use of semi-empirical models adapted to each situations brings some guaranty of their accuracy.

5. Conclusion

This brief description shows that, although we now have a good qualitative understanding of the mechanisms and factors governing canopy spectral response, applications are still limited by the difficulty in reliably assessing canopy characteristics from remote sensing data. This is because of two distinct but linked causes. Firstly the limited quantity of information contained in actual satellite data as compared to the number of factors affecting canopy reflectance. This will be improved in the future with high spectral resolution satellites, with satellite concepts such as "Polder" (Deschamps *et al.* 1990) which performs multidirectional observation and polarization measurements, and also with other types of remote sensing data such as radar and thermal infrared. Secondly the lack of reliable and tested physical or physiological, mechanistic or even semiempirical models, which take into account most of previously described factors. Research is needed to improve and test reflectance models as well as growth models and canopy architecture evolution models.

However, present applications of satellite data are more simply but empirically derived from local relationships between radiometric data and sampled ground truth. The remote sensing capabilities are then used to extend the local relationships to the whole satellite scene. These procedures will be discussed elsewhere in this book.

We have also pointed out the necessity for multitemporal data. This will have positive though indirect consequences in smoothing noise due to uncontrolled variables such as soil background variations and atmospheric effects.

All these basic studies briefly described in this paper also help in the definition of satellite sensors systems and also in evaluating the accuracy corresponding to such estimations. This is essential where remote sensing derived canopy characteristics have to be compared with more classical methods based on field sampling and statistical approaches for high spatial resolution problems. It is also of importance when Man wants to deal with global scale problems. We must evaluate the magnitude of the errors in the satellite derived parameters introduced in global models.

Aase J.K., Siddoway F.H., Millard J.P., 1984, Spring wheat leaf phytomass and yield estimates from airborne scanner and hand-held radiometer measurements. *International Journal of Remote Sensing* **5**(5) : 771-781.

Ajai, Sashikumar M.N., Aggarwal P.K., Singh A.K., Sinha S.K., 1984, A model for dry matter and grain yield estimation in wheat, *Proc. Crop Growth Conditions & Remote Sensing; 22-23 june 1984; ISRO-ICAR:611-614.*

Badwhar G.D., Verhoef W., Bunnik N.J.J., 1985
Comparative study of Suits and SAIL canopy reflectance models, *Remote Sensing of the Environ*ment **17**:179-185.

Baret F., 1986, Contribution au suivi radiométrique de cultures de céréales. *Thèse de Doctorat, Université Paris-Sud. Orsay. n° d'ordre : 98, 182 pp.*

Baret F., 1988, Un modèle simplifié de réflectance et d'absorptance d'un couvert végétal. *Proceedings of the 4th International colloquium on spectral signatures of objects in remote sensing. Aussois-Modane (France). Jan. 18-22. ESA-SP 287 : 113-120.*

Baret F., Guyot G., Begue A., Maurel P., Podaire A., 1988, Complementarity of middle infrared with visible and near infrared reflectance for monitoring wheat canopies. *Remote Sensing of the Environment.* **26**:213-225.

Baret F., 1989, La regression non linéaire: quelques exemples d'application dans le domaine de la télédétection, *in Remote Sensing From Space: Physical Aspects and Modelling, Toulouse, aug. 1988,* CEPADEUS ed.:413-420.

Baret F., Guyot G., Major D., 1989, TSAVI: A vegetation index which minimizes soil brightness effects on LAI or APAR estimaion, *in 12th Canadian Symposium on Remote Sensing and IGARSS'90, Vancouver, Canada, July 10-14 1989.*

Baret F, Guyot G., 1990, Potential and linmits of vegetation indices , submitted to *Remote Sensing of the Environment.*

Bowers S.S., Hanks R.J., 1965, Reflection of radiant energy from soils. *Soil Science*, **100**(2) : 130-138.

Breece H.T., Holmes R.H., 1971, Bidirectional scattering characteristics of healthy green soybeans and corn leaves in vivo. *Applied Optics*. **10** : 119-127.

Campbell G.S., 1986, Extinction coefficients for radiation in plant canopies calculated using ellipsoidal inclination angle distribution, *Agricultural and Forest Meteorology*. **36**:317-321.

Clevers J.P.W., 1986, The use of multispectral Photography in Agricultural Research, *Symposium on Remote Sensing Ressources and Development and Environmental Management, Enschede, Aug. 1986, ISPRS vol26(1):227-232*

Clevers J.G.P.W., 1988, The derivation of a simplified reflectance model for the estimation of leaf area index. *Remote Sensing of the Environment* **25**(1) : 53-70.

Cracknell A.P., Khattak S., Vaughan R.A., 1987, The automatic identification of areas of sunglint in AVHRR data, *proceedings of the Annual Conference on Remote Sensing Society held in Nottingham U.K., 7-11 septmber 1987 pp 403-415.*

Deschamps p.y., Herman M., Podaire A., Leroy M., Vermande P., 1990, A spatial instrument for the observation of polarization and directionality of earth reflectances: POLDER, *submitted to IGARSS'90, Washington.*

Donald C.M., Hamblin J., 1976, The biological yield and harvest index of cereals as agronomic and plant breeding criteria, *Advances in Agronomy* **31**(1):85-105.

Gallo K.P., Daughtry C.S.T., Bauer M.E., 1985, Spectral estimation of absorbed photosynthetically active radiation in corn canopies. *Remote Sensing of the Environment* **17** : 221-232.

Gerberman A.H., Cuelar J.A., Gausman H.W., 1984, Relationship of sorghum canopy variables to reflected infrared radiation for 2 wavelength and 2 wavebands. *Photogrammetric Engineering and Remote Sensing* **50**(2) : 209-214.

Goel N.S., Thomson R.L., 1987, Evaluation of canopy Reflectance models for LAI estimation through it's inversion. *IEEE Trans. on Geoscience And Remote Sensing G.E., 23(5):674-684.*

Gosse G., Varlet-Grangher C., Bonhomme R., Chartier M., Allirand J.M., Lemaire G., 1986, Production maximale de matière sèche et rayonnement solaire intercepté par un couvert. *Agronomie* **6**(1) : 47-56.

Goward S.N., Tucker C.J., Dye D.G., 1986, North american vegetation patterns observed with meteorological satellite data. *In. : Coupling of ecological studies with remote sensing. US. Man and the Biosphere Program Department of State* (Dye M.I., Crossbey D.A. Jr, Editor). Dept. of State Publ. 9504 : 96-115.

Grant L., 1987, Diffuse and specular characteristics of leaf reflectance. *Remote Sensing of the Environment* **22** : 309-322.

Grant L., Daughtry C.S.T., Vanderbilt V.C., 1987, Polarized and non-polarized reflectances of Coleus blumei. *Environmental and Experimental Botany*. **27**(2) : 139-145.

Guerif M., Delecolle R., Gu X., Guinot J.P., Japiot M., Steinmetz S., 1988, Estimation de la biomasse et du rendement de cultures de blé dur à partir des indices de végétation SPOT, *in Proc. 4th International Colloquium on Spectral Signatures of Objects in Remote Sensing, Aussois, France, 18-22 jan. 1988, ESA-SP-287: 137-141.*

Guyot G., Malet P., Baret F., 1980, Analyse des indicatrices de réflexion de l'orge et du blé, possibilité de la stéréoradiométrie. *Proc. Int. Symp. I.S.P. Hamburg. International Archives of Photogrammmetry, XXII-B7. Commission VII : 372-381.*

Hatfield J.L., Asrar G., Kanemasu E.T., 1984, Intercepted photosynthetically active radiation in wheat canopies estimated by spectral reflectance. *Remote Sensing of the Environ*ment **14** : 65-75.

Huete A.R., 1988, A soil adjusted vegetation index (SAVI): *Remote Sensing of the Environment* **25**:295-309.

Jackson R.D., 1983, Spectral indices in n-space. *Remote Sensing of the Environment* **13** : 409-421.

Jacquemoud S., Baret F., 1990, PROSPECT: A model of leaf optical properties spectra. submitted to *Remote Sensing of the Environment.*

Kauth R.J., Thomas G.S., 1976, The tasseled cap. A graphic description of the spectral temporal development of agricultural crops as seen by Landsat. *Proceedings of Machine Remotely Sensed Data. IEE Catalog n° 761103-1. MPRSD New-York.*

Maas S., Richardson A.J., Wiegand C.L., Nixon P.R., 1985, Use of plant spectral and weather data in modeling corn growth. *In Proceedings of the 19th International Symposium on Remote Sensing of the Environment ERIM. Ann. Arbor. Mi., 21-25 October.*

Monteith J.L., 1977, Climate and the efficiency of crop production in Britain. *Philosophical Transactions of the Royal Society, London*. B282 : 277-294.

Pearson R.L., Miller L.D., 1972, Remote mapping of standing crop biomass for estimation of the productivity of the short-grass prairie, Pawnee National Grasslands, Colorado. *Proceedings of the 8th International Symposium on Remote Sensing of the Environment, Ann. Arbor. : 1357-1381.*

Perry C.R., Lautenschlager L.F., 1984, Functional equivalence of spectral vegetation indices. *Remote Sensing of the Environment.* **14** : 169-182.

Prince S.D., Tucker C.J., 1986, Satellite Remote Sensing of rangelands in Botswana. II. NOAA AVHRR and herbaceous vegetation, *International Journal of Remote Sensing* **7**(11):1555-1570.

Rawson H.M., Gardner P.A., Long M.J., 1987, Sources of variation in Specific leath area in wheat grown at high temperatures, *Australian Journal on Plant Physiology* **14**:287-298.

Richardson A.J., Wiegand C.L., 1977, Distinguishing vegetation from soil-background information. *Photogrammetric Engineering and Remote Sensing,* **43** : 1541-1542.

Rigal D., Terres J.M., 1988, Modèlisation du profil spectral de cultures de maïs et de soja. Estimation de la biomasse produite, INRA Bioclimatologie, BP91, 84140 Montfavet, France, 85pp.

Rouse J.W., Haas R.H., Schell J.A., Deering D.W., Harlan J.C., 1974, Monitoring the vernal advancement of retrogradation of natural vegetation. NASA/GSFC, Type III, Final Report. Greenbelt, MD. 371 pp.

Sellers P.J., 1985, Canopy reflectance, photosynthesis and transpiration. *International Journal of Remote Sens*ing **6**(8) : 1335-1372.

Sheffield C., 1985, Selecting band combinations from multispectral data. *Photogrammetric Engineering and Remote Sensing,* **51**(6) : 681-687.

Steinmetz S., Gguerif M., Delecolle R., Baret F., 1990, Spectral estimates of the absorbed photosynthetically active radiation and light use efficiency of a winter wheat crop subjected to nitrogen and water deficiencies, *International Journal of Remote Sens*ing (under press).

Tucker C.J., Holben B.N., Elgin J.H., Mcmurtrey J.E., 1980, Relationship of spectral data to grain yield variation, *Photogrammetric Engineering. and Remote Sensing* **46**(5):657-666.

Tucker C.J., Vanpraet C.L., Sharman M.J., Van Ittersum G., 1985, Satellite remote sensing of total herbaceous biomass production in the Senegalese Sahel : 1980-1984. *Remote Sensing of the Environment.* **17** : 233-249.

Vanderbilt V.C., Grant L., Biehl L.L., Robinson B.F., 1985, Specular, diffuse, and polarized light scattered by two wheat canopies. *Applied Optics* **24** : 2408-2418.

Varlet-Grancher C., Chartier M., Gosse G., Bonhomme R., 1981, Rayonnement utile à la photosynthèse des végétaux en conditions naturelles: Caractérisation et variations, *Acta Oecologia Oecologica Plant*arum **2**(16):189-202.

Verhoef W., 1984, Light scattering by leaf layers with application to canopy reflectance modeling : the SAIL model. *Remote Sensing of the Environment* **16** : 125-141.

Verhoef W., 1985, Earth observation modeling based on layer scattering matrices. *Remote Sensing of the Environ*ment **17** : 165-178.

Wang Y.P., Jarvis P.G., 1988, Mean leaf angles for the ellipsoidal inclination angle distribution, *Agricultural and Forest Meteorology* **43**: 319-321.

Wanjura D.F., Hatfield J.L., 1981, Spectral Procedures for estimating crop biomass. *Transactions of the ASAE* **28**(3): 922-927; Published by the American Agricultural Engineers, St Joseph, Michigan.

Yamagata Y., Wiegand C., Akiyama T., Shibayama M., 1988, Water turbidity and perpandicular vegetation indices for paddy rice flood dammage analyses, *Remote Sensing of the Environment* **26**: 241-251.

REMOTE SENSING FOR VEGETATION MONITORING ON REGIONAL AND GLOBAL SCALES

A. S. BELWARD
Commission of the European Communities,
Joint Research Centre,
Institute for Remote Sensing Applications,
21020 Ispra (Varese) Italy.

1. Vegetation and the earth system

Studies on regional and global scales, such as the International Geosphere-Biosphere Programme, highlight the concept of the earth as a single system (IGBP 1988). Maintenance of this system depends on achieving equilibrium between the various components; as Turk *et al.* (1972) state " A balanced ecosystem is considered to be in a healthy condition".

It is also true that "The presence of an organism modifies the environment it is exposed to" (Monteith 1973); the activities of mankind are particularly evident in this respect. Our expanding population is paralleled by consumption of natural resources, by changes in global vegetation and soil types and by pollution of the oceans and atmosphere. In other words we have a profound impact upon the biogeochemical cycles of the earth system.

Information is needed to assess the impact of mankind's activities, to predict future conditions, and to arrive at a more complete understanding of the processes supporting the earth system . But what information should be obtained ?

Vegetation, accounting for 99% of the earth's biomass, occupies a key position with regard to global environmental concerns. The vegetative cover is a major source of global energy and moisture for the atmosphere; although only 29% of the earth's surface is terrestrial, and of this a maximum of around 20% is actually vegetated, the total surface area of leaf material in contact with the atmosphere is far greater than the total surface area of the entire planet (Verstraete and Dickinson 1986). Changes in vegetation types and rates of growth and development directly lead to changes in energy, water and material transport within the earth system. This in turn leads to environmental change. For example deforestation, conversion of natural vegetation to agricultural land and biomass burning are firmly implicated in the increasing concentration of atmospheric CO_2, and the resultant changes in the equilibrium temperature of the earth.

However, being responsive to both climate and human activity the vegetation canopy is in itself a good indicator of ecosystem condition. Vegetation is dynamic in responding and adapting to prevailing environmental conditions; dynamics manifest in changes in the distribution of vegetation types, and changes in plant growth and development. In either case monitoring such change calls for repeated observation and measurement.

The study of vegetation dynamics on regional and global scales is concerned with the community, biome and ecosystem levels of organization, not individual plants, or even populations. Repeated observation and measurement on these scales is not possible using

A. S. Belward and C. R. Valenzuela (eds.),
Remote Sensing and Geographical Information Systems for Resource Management in Developing Countries, 169–187.

conventional *in situ* measurements. However, satellite sensor remote sensing does provide a mechanism for global observation on a regular and repeatable basis. Vegetation monitoring on regional / global scales is an integral part of natural resource surveys such as tropical deforestation assessment, agricultural production forecasting, bush fire monitoring and environmental degradation monitoring. These are the themes of subsequent chapters of this book; this chapter introduces some of the fundamental considerations to using remote sensing for this type of work.

2. Candidate Remote Sensing Systems

The value of the space perspective and temporal dimension in vegetation surveys, such as those needed for regional scale environmental studies, has long been recognized (Steiner 1970). However, the information that can be obtained from satellite sensor observations will to a large extent depend on the characteristics of the system in question. It is therefore relevant to review existing remote sensing systems with reference to regional / global scale vegetation monitoring.

2.1 HIGH SPATIAL RESOLUTION SYSTEMS

The high spatial resolution remote sensing systems, such as Landsat or SPOT can only provide relatively low frequency temporal coverage of large areas, even given the pointable sensors of the SPOT system (see the earlier chapter by Hill). High spatial resolution data also means large data volume, especially where multi temporal data over very large geographical areas are concerned. Of course, this is not to say that high resolution data have no part to play in regional scale work, (see for example McGraw and Tueller (1983) on the applications of high resolution data in the management of extensive grasslands) but problems of high cost, high data volume and low frequency of cover do represent serious drawbacks to such use of these data.

2.2 LOW SPATIAL RESOLUTION SYSTEMS

Of the currently operational earth observation remote sensing systems the Advanced Very High Resolution Radiometer (AVHRR), an imaging radiometer aboard the US National Oceanic and Atmospheric Administration's (NOAA) polar-orbiting TIROS-N/NOAA series of satellites, is the best suited for regional scale vegetation monitoring. The basic system parameters are shown in table 1.

The orbital and viewing parameters are such that global cover is available on a daily basis, thus largely overcoming problems of cloud free image availability. In addition, the lower data volumes resulting from reduced spatial resolution make the analysis of long term time-series for large areas a realistic proposition (Justice 1986), and as we shall see, the spectral characteristics of the AVHRR data are reasonably well suited for vegetation dynamics studies.

AVHRR data are available in different forms in which spatial and temporal resolution vary. These are described below in terms of spatial and temporal scales relating to regional scale vegetation monitoring.

2.2.1 *High Resolution Picture Transmission* (HRPT) are real time full resolution digital data for all five channels. The data have 10 bit radiometric resolution and a resolution cell size at nadir of 1.1 by 1.1 kilometres, degrading to 2.4 by 6.9 kilometres at extreme view angles. As the system characteristics of AVHRR result in a variable ground area associated with each across track pixel,

Table 1. Characteristics of the NOAA AVHRR remote sensing system

Satellites in the NOAA series and approximate periods of operation

TIROS-N:	19 Oct 1978 to 30 Jan 1980
NOAA-6:	27 Jun 1979 to 30 Jun 1986
NOAA-7:	24 Aug 1981 to 11 Jan 1985
NOAA-8:	3 May 1983 to 6 Mar 1986
NOAA-9:	25 Feb 1985 to 7 Nov 1988
NOAA-10:	17 Nov 1986 to present
NOAA-11:	8 Nov 1988 to present

NOAA Satellite orbital parameters

Orbit type:	sun-synchronous, near polar circular
Orbital inclination:	99.092°
Altitude:	833 km
Orbital period:	102 minutes
Orbits / day:	14.2
Complete cycle:	9 days
Equator crossing times:	02.30, 07.30, 14.30 & 19.30 hrs
Ground cover (latitude):	90° south - 90° north

AVHRR viewing parameters

Total swath width:	2700 km
Scan angle:	55.4° either side of nadir
Instantaneous field of view:	1.39 mrad
Ground resolution:	1.1 x 1.1 km at nadir increasing to 2.4 x 6.9 km at maximum scan angles

AVHRR sensor characteristics

Channel 1:	0.58 - 0.68 μm (Saturation at approximately 100% albedo)
Channel 2:	0.72 - 1.10 μm (Saturation at approximately 100% albedo)
Channel 3:	3.55 - 3.93 μm (Saturation at approximately 320 K)
Channel 4:	10.3 - 11.3 μm (Saturation at approximately 320 K)
Channel 5:	11.5 - 12.5 μm (Saturation at approximately 320 K)
Quantisation level:	1024 digital counts, 10 bit for all channels

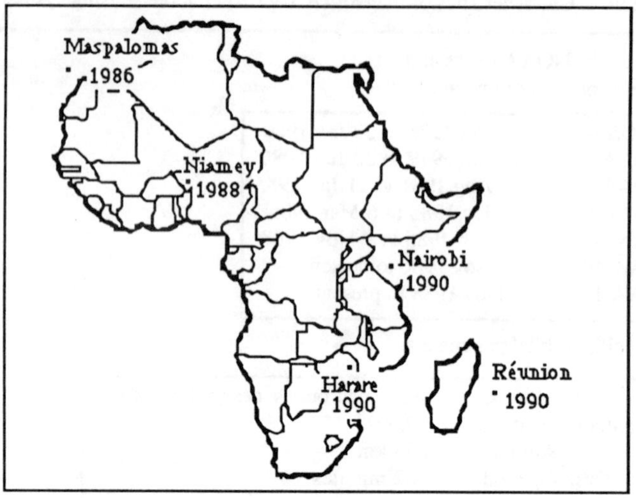

Figure 1 Distribution and date of commencement of service of ground receiving stations for Africa

and as viewing conditions and the nature of the target are subject to change, it can only be safely assumed that the individual spectral measurements provide an "integrated" view of any area at the kilometre level.

Vegetation parameters measured at this level relate to the structure and composition of plant communities, to canopy physical parameters such as biomass and to the phenology of the communities. Observations of this nature, if made on suitable temporal and geographical scales, can give a good indication of prevailing vegetation dynamics.

It is important however, to note that only those areas proximate to ground receiving stations will be covered by HRPT data, as cover is limited to the area in which the satellite is in direct line of sight of the station's antenna. This will vary depending on the position of the station and the antenna elevation used. Cover is at least 18 million km^2 around the station, so regional, if not continental and global cover is assured.

Whilst it is clear that certain environmental phenomena have to be measured in terms of days, biomass estimation for example, other factors, such as major changes in vegetation type, require monitoring over a long period. This need for an historical perspective calls for an archive of data extending back over a number of years. The length of archive available for different parts of the globe will depend on the period of operation of stations offering cover, and this is highly variable.

Figure 1 shows the distribution and date of commencement of service of ground receiving stations for Africa. In comparison the ground receiving station at the University of Dundee in Scotland has an archive of daily HRPT data extending back to 1978.

2.2.2 *Local Area Coverage* (LAC) are again full resolution 1.1 km, 10 bit data, though recorded on-board the satellite, rather than directly transmitted to receiving station. This offers considerable advantages in terms of geographical scale of cover as data for parts of the globe not covered by ground receiving stations can still be imaged.

Unfortunately the geographical extent of the cover and the temporal characteristics of LAC data acquisition are not consistent, as data coverage is limited by the on-board recording facilities.

Acquisitions are based on user requests, and requests always exceed capacity; further, requests are given a priority ranking and scientific investigations come 5th out of 6 (Kidwell, 1986). The LAC data set therefore does not form a particularly reliable basis for regional and global scale vegetation monitoring.

2.2.3. *Global Area Coverage* (GAC) are reduced spatial resolution 10 bit data again recorded on-board the satellite. The GAC data are generated by on-board sampling of the full resolution data such that for a given scan line the first four pixels out of every five are averaged, and only every third scan line is processed. This gives a spatial resolution, at nadir, of 1.1 km by 4 km, with a gap of around 3 km between pixels across the scan lines, although the data are generally treated as 4 km resolution (Kidwell 1986).

The reduced data volume means that the on-board recording facilities are able to preserve complete daily global cover. The lower data volumes also make the processing of long time-series over extensive areas, such as entire continents a far less costly exercise than would be possible from full resolution data; for comparison see Sharman (1989) who discusses the data processing requirements for construction and analysis of long time-series of HRPT data for Europe.

Figure 2 compares the volume of data from one channel from AVHRR data at each resolution for Africa over one year and 7 year periods. This clearly illustrates the implications of "fine" resolution for time series construction (it is worth pointing out that although HRPT/LAC resolution represents a 16 fold increase in data processing requirements over GAC, over 1,300 Landsat TM pixels are needed to cover the same area as a single HRPT pixel).

The GAC data set has a temporal scale with daily frequency of cover and a time horizon of around 9 years (the global GAC archive dates back to June 1981). This is the most complete global cover data set available from any imaging sensor, and as such represents a significant data source for environmental studies on regional and global scales. However, the advantages in terms of data volume, extent of cover and frequency of cover are offset by the reduced measurement scale and associated reduction in information content.

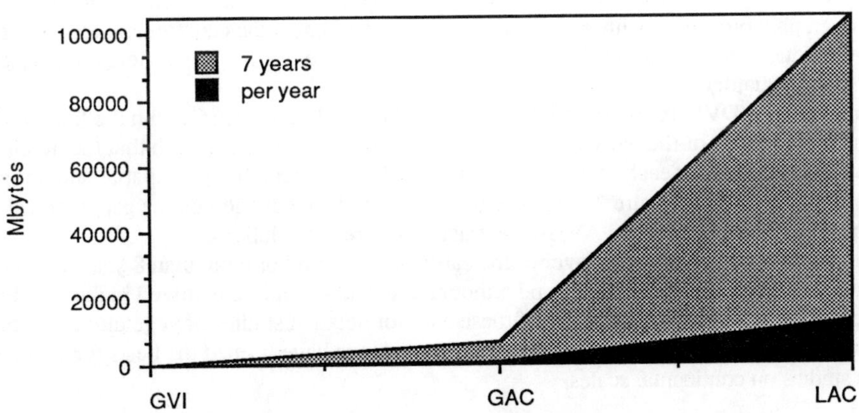

Figure 2 Effect of resolution on data volume for time series: One and Seven years Normalized Difference Vegetation Index data for Africa

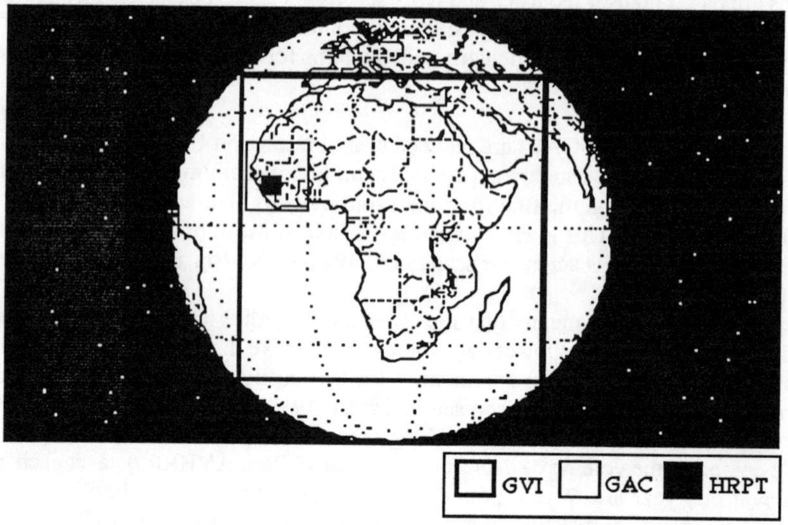

Figure 3, area covered by 512 pixels by 512 lines of AVHRR at different resolutions

2.2.4.*Global Vegetation Index* (GVI) data differ from the preceding data types, being only available as ready mapped processed 8 bit data, rather than the raw (albeit undersampled in the case of GAC), 5 channel, 10 bit data of the HRPT, LAC and GAC products.

The GVI data are, as the name suggests, already in the form of a geophysical product, namely the Normalized Difference Vegetation Index (NDVI). The significance of the NDVI is discussed by Baret in an earlier chapter. The NDVI combining channels 1 and 2 is calculated from the daytime passes of all 14 daily orbits of GAC data. These are mapped for the first generation product onto a polar stereographic projection with a resolution cell size of 15 km at the equator, and for the second generation product onto a plate carrée projection (16 km at the equator), and a Mercator projection (19.5 km at the equator).

Only one GAC NDVI pixel is selected for each GVI pixel (NOAA 1986), with the final product made from a composite of the daily NDVI arrays over a seven day period such that the maximum NDVI value is retained for each location. Thus the GVI is produced by both spatial and temporal sampling of the GAC data. Figure 3 illustrates the effects of data reduction on the geographical area covered by a 512 pixel by 512 line AVHRR extract at different resolutions.

The GVI data set offers weekly cover of the earth over a period of more than 8 years (the global GVI archive dates back to April 1982), and although the undersampling involved in the creation of the GVI products severely limits the use of these data for detailed studies of vegetation dynamics, the low data volumes and consistent historical earth coverage have formed the basis for vegetation dynamics studies on continental scales.

3. AVHRR for vegetation monitoring using visible and near-infrared wavelengths

3.1 VEGETATION INDICES FROM AVHRR DATA

Most remote sensing techniques for vegetation monitoring use data recorded through those parts of the electromagnetic spectrum which provide a strong signal from vegetation whilst contrasting with background material (Tucker and Sellers, 1986). Earlier chapters by Belward and Baret have shown that the red and near-infrared wavelengths satisfy this criterion, and that combinations of these two wavebands as vegetation indices can be used as surrogate measurements of various attributes of a vegetation canopy, such as total phytomass and leaf area index.

Calculating vegetation indices from AVHRR data involves the transformation of the original spectral channels into a new synthetic vegetation index channel. Creation of a vegetation index channel from AVHRR simply involves the arithmetic combination of channel 1 and 2 data using appropriate image processing routines (see Wilkinson, elsewhere in this book). Building up regular vegetation index images from sequential satellite overpasses adds the temporal dimension necessary for vegetation monitoring.

3.2 DEVELOPMENT OF THE AVHRR SLOPE INDICES

Townshend and Tucker (1981) exchanged the similar AVHRR channels 1 and 2 for the Landsat MSS bands 5 and 7, used in the ratio index developed by Rouse *et. al.* (1974), to create an analogous index referred to as VI_{av}

$$VI_{av} = \frac{(\text{AVHRR ch 2 - AVHRR ch 1})}{(\text{AVHRR ch 2 + AVHRR ch 1})} \tag{1}$$

where AVHRR ch2 represents the response in AVHRR channel 2 (centre wavelength 0.833 µm bandwidth 0.240 µm) and AVHRR ch1 represents the response in AVHRR channel 1(centre wavelength 0.635 µm bandwidth 0.117 µm). VI_{av}, and indeed the same ratio using data recorded at equivalent wavebands from other sensors, is now generally referred to as the NDVI. Similarly, AVHRR channels 1 and 2 can be used in the other slope indices discussed by Baret, however the NDVI remains the most frequently used at present.

The NDVI is a bounded ratio ranging in value from -1 to +1. Negative and low values are associated with cloud, bare soil, low green vegetation density and senescent vegetation. Higher values relate to photosynthetically active cover such that, by inference, the higher the value the greater the green phytomass (Sellers 1985). Holben (1986) calculated average NDVI values from GAC AVHRR channel 1 and 2 planetary albedo's for a range of different cover types, these are shown in table 2.

For computer aided display and analysis the NDVI is usually scaled to an 8 bit integer. For example the standard NOAA global NDVI image product, the GVI (see § 2.2.4.) is scaled

$$\text{GVI} = 240 - \left[\left\{ \frac{(\text{near-infrared - red})}{(\text{near-infrared + red})} \right\} + 0{,}05 \right] \times 350 \tag{2}$$

Table 2. NDVI values for different cover types as measured from NOAA-7 (Holben 1986)

Cover type	Planetary albedo Channel 1	Planetary albedo Channel 2	NDVI
Dense green-leaf vegetation	0.050	0.150	0.500
Medium green-leaf vegetation	0.080	0.110	0.140
Light green-leaf vegetation	0.100	0.120	0.090
Bare soil	0.269	0.283	0.025
Clouds (opaque)	0.227	0.228	0.002
Snow and ice	0.375	0.342	-0.046
Water	0.022	0.013	-0.257

In common with most vegetation index scaling techniques a constant is added to remove negative values and a multiplicatory constant is used to obtain the required range of values. The scaling used by NOAA, as shown in equation 2, has the additional (unusual) feature of inverting the NDVI so that when displayed the highest NDVI values appear darkest in the image.

3.3 ALTERNATIVES TO SLOPE INDICES

The distance vegetation indices, discussed by Baret, such as the Perpendicular Vegetation Index (PVI) of Richardson and Wiegand (1977), or the Greeness Vegetation Index and Soil Brightness Index (SBI) of Kauth and Thomas (1976) do have the advantage that they they begin to take into account background soil type. However, their calculation from AVHRR data is less easy than the ratio based indices. For example calculation of the PVI first requires *a priori* identification of non-vegetated areas so the precise equation of the soil line can be calculated (see Baret). Then axis translation and rotation, from red / near-infrared to SBI / PVI feature space, is needed to obtain the synthetic vegetation index channel. This is achieved using standard formulae for axis rotation

$$\begin{aligned} SBI &= (\text{red} - R0)\,\text{Cos}\,\alpha^\circ + (\text{near-infrared} - IR0)\,\text{Sin}\,\alpha^\circ \\ PVI &= -\,(\text{red} - R0)\,\text{Sin}\,\alpha^\circ + (\text{near-infrared} - IR0)\,\text{Cos}\,\alpha^\circ \end{aligned} \tag{3}$$

where red = AVHRR channel 1, near-infrared = AVHRR channel 2, R0 = red value at the near-infrared intercept, IR0 = near-infrared value at the red intercept,and α° = the slope of the soil line.

3.4 PRACTICAL ASPECTS TO THE USE OF AVHRR DERIVED VEGETATION INDICES

3.4.1 *The distance vegetation index* approach to regional vegetation studies has been limited, mainly because of the problems associated with the identification of areas of bare soil. This is particularly pronounced in AVHRR based work as it is very difficult (and in most biomes impossible) to find areas of bare soil at the scale of the AVHRR pixel. Note that this limitation also applies to the soil adjusted slope indices discussed by Baret. Additional limitations arise because even though the

PVI, unlike the NDVI, is generally considered to be independent of soil background (soil variation is concentrated in the transform parallel to the soil line) the concept of a 'universal soil line' has been questioned; Huete *et al.* (1984) for example, report deviations from linearity when a wide spectral range of soils was considered.

3.4.2 *The NDVI* is the vegetation index most frequently used for regional scale vegetation monitoring using AVHRR data. This can in part be explained by it's effectiveness as a surrogate measure of biophysical parameters, partly by the fact that calculation of the NDVI, unlike PVI, requires no *a priori* information concerning the imaged scene and partly by the fact that the calculation of the NDVI is such that there is some normalization for variations in viewing conditions (Justice *et al.* 1985).

There are two main drawbacks to consider when using the NDVI. Firstly, the index tends to produce a saturation response to very high green leaf densities (Tucker 1977). Secondly, the NDVI is not independent of substrate, and at low vegetation covers soil background reflectance can exert a considerable influence on sensor response (Otterman 1981, Huete *et al.* 1984). In addition to these general points there are specific points to consider when using NDVI derived from AVHRR

First the effects of cloud cover. The effect of cloud cover on the detection of vegetation using optical wavelength remote sensing systems is axiomatic; clouds obscure the ground. This can be a significant problem. For example, throughout the West African growing seasons of 1987- '88 and '88 - '89 there were no completely cloud free days (J. Vogt, personal communication). However, as shown in table 2, NDVI values for clouds and vegetation contrast sharply, with clouds having low or negative values, thus cloud non-cloud regions within an image can at least be separated.

Second, the effect of variable viewing conditions, including geometry and atmosphere. The AVHRR covers a swath width of more than 2700 km so there is considerable variation in sun-target-sensor geometry along the scan lines, furthermore these relationships will change with latitude and season. Add to this the fact that atmospheric conditions are variable and also affect radiance values (Duggin and Piwinski 1984) and one begins to see that the changes in the viewing conditions make direct comparisons between radiance values difficult (see the earlier chapter by Hill). However, although variable, it has been shown by Holben and Fraser (1984) that variations such as off nadir viewing and atmospheric aerosol content only cause reductions in NDVI values. This is an important point and has lead to the development of a technique referred to as maximum value compositing which minimizes the effects of variations in viewing conditions and at the same time reduces the deleterious effects of cloud contamination. This technique is described in detail by Holben (1986).

Maximum value composites (MVC's) are created from registered daily NDVI data over a variable time period (periods varying from 7 days (NOAA 1986) to a month (Justice *et al.* 1985) are reported in the literature), such that for each location the pixel with the highest NDVI value for the time period is preserved. This process will tend to select near-nadir pixels, those imaged through the clearest atmospheres and those least contaminated by cloud (Holben 1986). Some contamination by thin cloud and sub-pixel clouds remains, as of course does contamination where cloud is present for the entire compositing period.

Whilst offering considerable advantages there are also problems associated with the MVCs. Holben (1986) points out that although composites made over longer time periods are more effective in reducing cloud and variable viewing condition influences, care has to be taken that the time period selected allows adequate representation of the development curves of the vegetation target. It is quite possible for the MVC process to smooth out vegetation response curves if the compositing period is too long. Nevertheless the advantages of the MVC approach are such that it

:d for the study of regional vegetation dynamics, specific examples of which are given /.

4. AVHRR data for vegetation monitoring using thermal wavelengths

The thermal channels of the AVHRR provide data responding to vegetation factors other than the plant pigment / structure controlled responses of the near-infrared and red wavelengths. As shown in table 1 channels 3, 4 and 5 are all sensitive to emitted radiation, and variations in this have been successfully used to discriminate between different land cover classes. For example, Kerber and Schutt (1986) showed how spectral separability between forest, urban and agriculture/grass land cover classes using AVHRR data was dominated by emitted radiation, particularly in channel 3, Malingreau *et al.* (1989) have shown clear divisions between major eco-biological zones of West Africa using channel 3 data, and the data have been extensively used for identifying disturbance in tropical forests, as discussed in the later chapter by Malingreau.

A major advantage to using channel 3 for regional vegetation monitoring in tropical areas is that, of the AVHRR channels, channel 3 is the least affected by water vapour and in addition offers reasonable penetration through (thin) smoke and haze. In contrast, atmospheric moisture, smoke and haze can in tropical areas limit the use of vegetation indices based on channels 1 and 2 for vegetation monitoring, and even atmospheric transmissivity within channels 4 and 5 can be significantly reduced over tropical areas by the water column (Malingreau *et al.* 1989).

However, care must be taken in interpreting channel 3 data. In chapter 1 we saw there can be both emitted and reflected components of any signal recorded through the electromagnetic spectrum at the wavelengths used for channel 3. The proportion of any channel 3 signal attributable to reflected or emitted energy will depend on the temperature and physical characteristics of the target plus viewing conditions. High values will in general be recorded from targets with high surface temperatures, though they can also result from highly reflective targets, for example sunlit cloud edges can saturate the channel 3 sensors even though the cloud surface is cold.

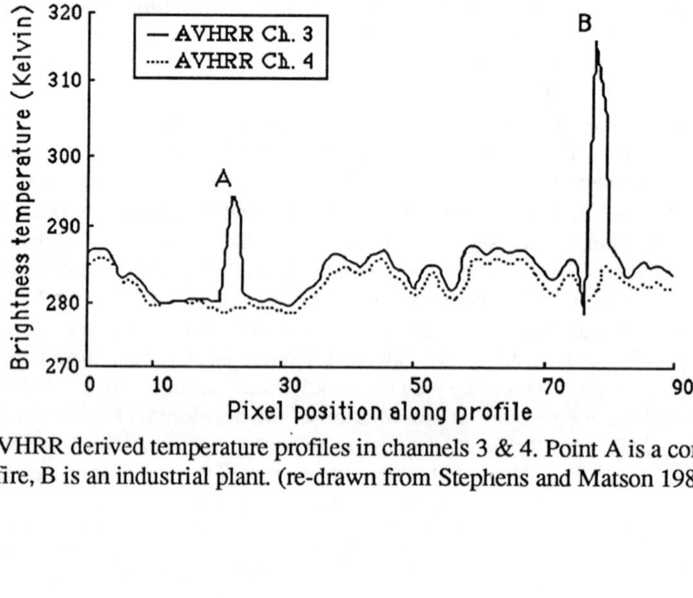

Figure 4 AVHRR derived temperature profiles in channels 3 & 4. Point A is a controlled forest fire, B is an industrial plant. (re-drawn from Stephens and Matson 1986)

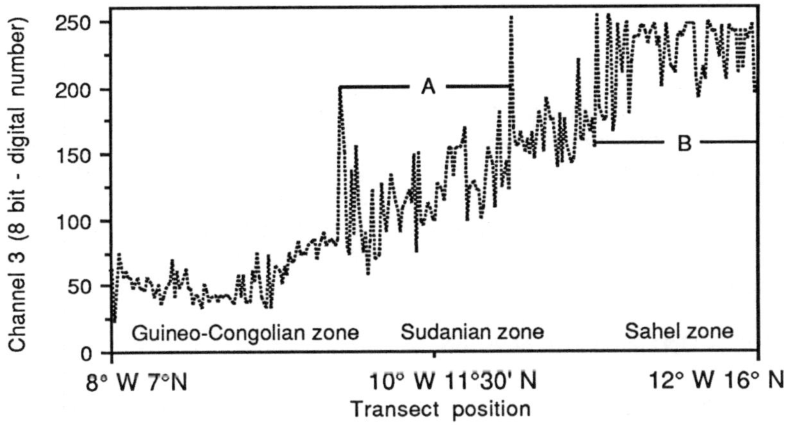

Figure 5, AVHRR channel 3 temperature profile across West Africa from 12 December 1987

Planck's radiation law (see Bird, chapter 1) shows how at high temperatures maximum radiant emittance shifts to shorter wavelengths. The significance of this for regional vegetation monitoring is shown in figures 4 and 5.

Figure 4 (from Stephens and Matson 1987) shows AVHRR derived temperature profiles in channels 3 & 4; background earth surface temperatures show a temperature difference of around 1 to 2 K between AVHRR channel 3 and 4, but for elevated heat sources the differences are far greater. At point A the profile crosses a small controlled forest burn and at point B a phosphorous plant; temperature values for channel 3 are 16.2 K and 33.9 K higher than channel 4 at A and B respectively. Thus we can see that channel 3 data are of considerable value for detecting elevated heat sources.

Figure 5 shows a channel 3 temperature profile across West Africa from 12 December 1987. A general warming from the tropical forests of coastal West Africa to the Sahel -Sahara zone in the north is evident. Scattered around this trend are discrete "hot spots". In the Sudanian zone these are quite distinct from the cool background temperature (points A) and are indicative of bush fire events. In the Sahel zone the fire events are far less clear (points B), though December is in fact the peak period for fire activity in this region. This is because the background temperature is that much higher; the reflective bare sandy soils and high surface temperatures come close to saturating the sensor (as shown in table 1 channel 3 saturates at 320 K / 47° C) and as the dry season advances this is indeed the case. Thus although of considerable value in the study of vegetation dynamics, there are strong seasonal / latitudinal limits to such use of channel 3 data. A detailed discussion concerning the use of channel 3 for biomass burning and vegetation dynamics studies can be found by Grégoire elsewhere in this book.

5. Scale considerations in applying the AVHRR approach to regional vegetation monitoring

The relationship between geographical scale of interest, frequency of observation required, time horizon over which data are needed and ground resolution necessary to record vegetation dynamics varies enormously. The preceding sections have shown that the AVHRR is a source of data available over a range of spatial (both in terms of geographical extent of cover and ground resolution) and temporal (in terms of frequency of observation and length of archive) resolutions reasonably well suited for vegetation monitoring on regional and continental scales. Here the relationship between data availability and information needs when applying the AVHRR approach to vegetation monitoring is considered.

5.1 GLOBAL AND CONTINENTAL SCALES

Studies at these scales are at present restricted to analysis of the NDVI from either GAC at reduced resolution or the GVI archives. At the time of writing there are no time series of full resolution 10 bit 5 channel data available on a continental basis, though such work is in progress (JRC, 1990).

Whilst NDVI time series from GAC data are available for selected continents, it is the GVI data which provide a global view of past and present vegetation dynamics. Given the coarse resolution of these data observations relating to detailed events in the vegetation canopy are not possible. However, Justice *et al.* (1985), Malingreau (1986) and Townshend and Justice (1986) have, among others, shown that these data are of considerable value for differentiating between the dynamics of major biomes; as Malingreau (1986) states "At a conceptual level there is no apparent incompatibility between the spatial scale of the AVHRR GVI observations and that of large-scale environmental phenomena."

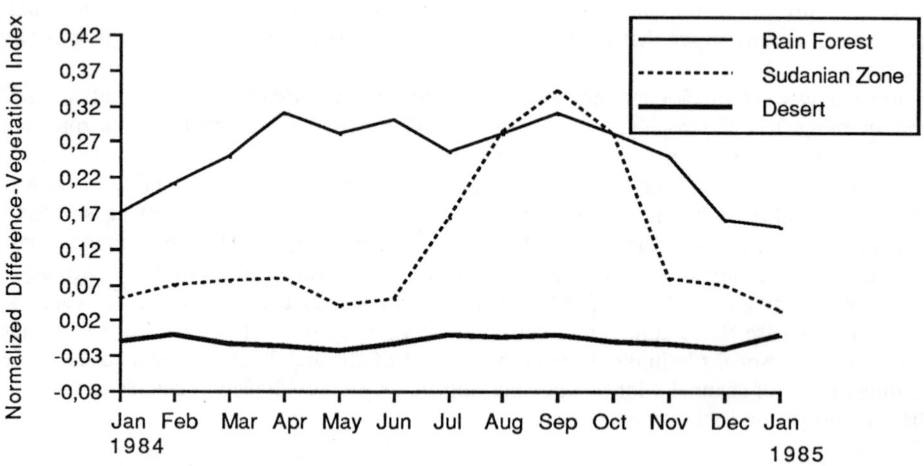

Figure 6 NDVI development curves for three African cover types (re-drawn from Townshend and Justice 1986)

An example of the type of data available from the GVI archive can be seen in figure 6. This is from Townshend and Justice (1986) who relate temporal NDVI MVC profiles to the phenology of major vegetation types of Africa, and show that such profiles can be used to detect differences in vegetation dynamics from year to year. Figure 6 shows NDVI profiles for three contrasting cover types. As one would expect, the NDVI profile obtained from the Sahara desert is almost flat with consistently low values. In contrast, the profile obtained from the Sudanian zone to the south is far more variable, with the response of the vegetation to the single rainy season (starting in May) being particularly marked. Further south still, the profile from the tropical rain forest displays a far less marked seasonal response, reflecting the abundant rainfall for this region.

5.2 REGIONAL SCALES

For vegetation monitoring on regional scales the choice of AVHRR data is much wider. Here, care must be taken that the scale of observation chosen matches the vegetation phenomena of interest. It is both expensive and superfluous to process fine resolution data for monitoring environmental phenomena occurring on coarse scales. For example figure 7, from Malingreau and Belward (1989), shows NDVI development curves for an agricultural region of the Ganges plain where winter wheat and summer rice are grown as successive crops. The general pattern of crop development and length of growing season can be detected equally well at either ground resolution; so if a general picture is all that is required the extra investment involved in using GAC data is unnecessary.

However, there are aspects of regional scale vegetation monitoring which call for more precise ground location than is possible with the GVI product. Examples include those instances where reference to ground calibration is required, such as agricultural production forecasting (see Bartholomé elsewhere in this book) or grassland phytomass estimation, and situations where the intrinsic scale of the environmental phenomenon of interest can be finer than the AVHRR ground resolution, such as bush fire monitoring. The latter two are discussed below to illustrate some of the spatial and temporal limitations of AVHRR for vegetation monitoring.

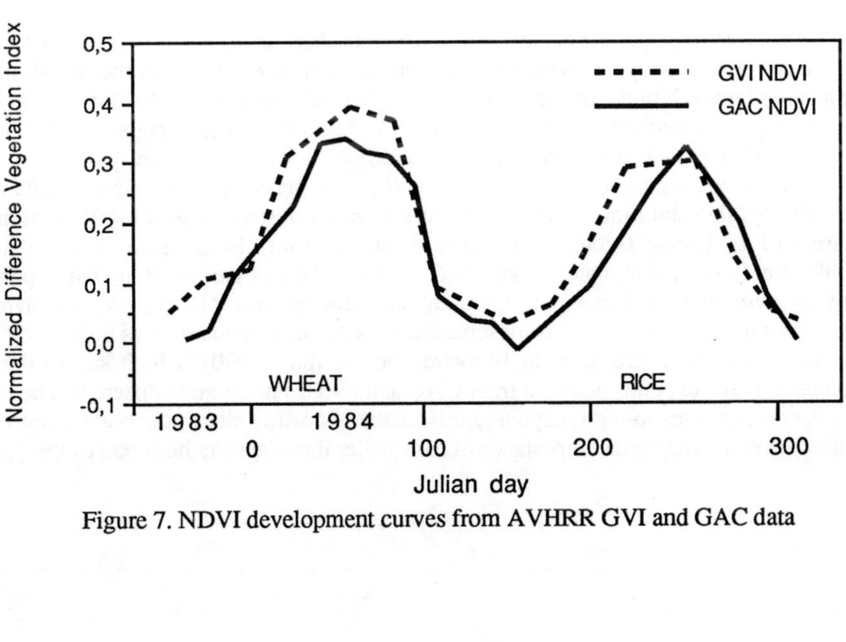

Figure 7. NDVI development curves from AVHRR GVI and GAC data

5.2.1. *Grassland phytomass estimation* is an essential element of range management. The rangelands are a key global resource covering in excess of three billion hectares, around twice the area devoted to arable agriculture (Myers, 1984). Successful management of this resource requires that the animal population is kept in balance with the plant productivity of the rangeland, thus close monitoring of the temporal and spatial development of the grasslands is needed.

The general relationships between increasing amounts of range phytomass and increasing values of AVHRR derived NDVI are well established, though largely remain qualitative at present (see for example Tucker *et al.* (1985) and the special edition of the *International Journal of Remote Sensing* (Justice 1986) covering the monitoring of the grasslands of semi-arid Africa using AVHRR data). The lack of sufficient ground data (Justice and Hierneaux 1986) and difficulty in precise location of ground sampling points in the imagery (Hierneaux and Justice 1986) are identified as key limitations. This is emphasized by the fact that where good ground data have been combined with high resolution imagery (Landsat MSS) reliable estimates of rangeland phytomass have been made from vegetation indices (Prince and Astle 1986, Taylor *et al.* 1986).To overcome this problem of precise location of ground sample points, the use of Landsat MSS in a calibration role for the low resolution AVHRR has been suggested (Prince and Tucker 1986, Taylor *et al.* 1986, D'Souza and Hiederer 1989). Though the high cost of Landsat MSS data, plus the likely effects of cloud cover in reducing Landsat MSS image availability tend to limit operational effectiveness of such an approach. Accurate positioning of AVHRR data with respect to ground sampling sites can thus be seen as key limitation to the quantitative use of these data.

5.2.2 *Bush fire monitoring* using AVHRR data illustrates other limitations of the AVHRR as a source of data for vegetation monitoring. Fire is a major agent of change in tropical vegetation communities and monitoring of burning patterns is seen as a key input to vegetation dynamics studies on regional and global scales, let alone the impact of biomass burning on atmospheric chemistry.

Recent work (for example Malingreau *et al.* 1985, Grégoire *et al.* 1988 and Langaas and Muirhead 1988), although identifying problems associated with a sensor which saturates at 47° C, has demonstrated the use of AVHRR channel 3 data for quasi - operational fire monitoring in the tropics.

A drawback affecting operational fire monitoring is that, as seen in §2.2, AVHRR (full resolution) data are not routinely available on a global daily basis. Further, the length of full resolution archive for much of the tropics is limited, yet historical data are needed for the important study of spatio-temporal burning patterns. It is the lower resolution data, principally GAC, which provide the guaranteed global daily data, and the necessary historical perspective.

However, the on-board sampling scheme used in GAC data generation is far from optimum; the average of all 15 full resolution pixels has been shown by Justice *et al.* (1989) to be more representative of LAC/HRPT data, and Belward and Lambin (1990) have shown that the undersampling limits the spatial representivity of GAC data. These points are of limited importance in the study of environmental phenomena operating on scales far greater than the resolution of the AVHRR but limit the use of GAC for the observation of point phenomena such as fires.

This is illustrated in figure 8, from Belward and Lambin (1990), which shows that the distribution pattern of fire pixels detected from GAC and HRPT can be quite different. The pixels selected in the on-board sampling completely misses the major fire site in the bottom left of the figure whilst over emphasizing the importance of the smaller fire events in the centre of the figure.

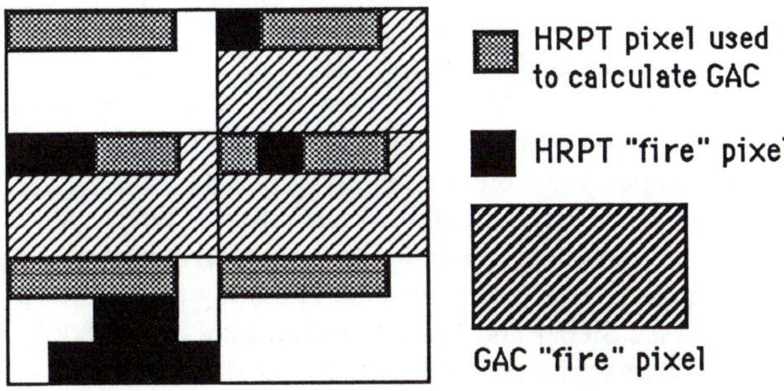

Figure 8. Contemporaneous HRPT and GAC AVHRR channel 3 data showing how the GAC undersampling affects the use of these data for the detection of bush fires (re- drawn from Belward and Lambin 1990)

Of course the detection of fire events using GAC data will depend on the spatial characteristics of the fires with reference to the sampling strategy employed in the GAC data generation; in the West African savannah, running fires, characterized by intense but narrow burning fronts, are often the case (Grégoire 1990) and can therefore be missed in the sampling process; in comparison the burning patterns in the forests of Kalimantan and North Borneo are largely picked up by the GAC data (Malingreau *et al.* 1985).

The limitations to the use of GAC data due to sampling effects extend not just to fire monitoring, but more generally to the detection of small structures, linear features and features with pronounced periodicity, particularly when their orientation is in the across-track direction.

6. Conclusions and future trends

Regional, and even global scale vegetation monitoring is possible using data from the AVHRR sensor. Furthermore, this instrument will remain the basic source of data for such studies over the next few years. There is still much research to be done with regard to sensor calibration, development of atmospheric correction routines (particularly to account for aerosols) and algorithms for the detection of sub-pixel clouds and cloud shadows before quantitative study of global vegetation dynamics is possible. However, past work has shown that ecologically significant parameters such as primary productivity, phenological development, surface temperatures and fire events can all be measured with varying degrees of precision, and applied to key global environmental issues such as deforestation, desertification and atmospheric pollution.

The successful application of the AVHRR data to vegetation monitoring over extensive areas, and the pressing need for such information, has lead to the proposal of improved AVHRR "type" instruments for inclusion on NASA's earth Observing System (EOS) polar platforms to be launched in 1997-98 and for the European Space Agency polar platform (see the earlier chapter by Hill). Both these systems will offer a choice of wavebands within a fixed range, both aim to offer global coverage on a near daily basis. They will also offer more precise sensor calibration and greater spatial and spectral resolution than the AVHRR system. Improved spatial resolution is expected to help in accurate ground location, whilst increased spectral range will allow the collection of data valuable for more specific characterization of vegetation factors such as plant pigment ratios (blue, green wavelengths) and moisture content and hydration state (middle-infrared wavelengths).

At the time of writing the exact specifications have yet to be finalized, but in any case the ground segment of such systems will have to be capable of handling increased data transfer, archiving and processing loads compared with those from the AVHRR; for reference 5 channel 10 bit HRPT cover of Africa amounts to around 400 Mbytes of data per day.

Although both the improved spectral and spatial resolution of the proposed systems will allow more precise measurement of many factors relating to vegetation dynamics the problem of areas of persistent cloud cover remains. The future systems operating in the microwave region of the electromagnetic spectrum (see Sieber and Churchill elsewhere in this book) such as ERS-1 and RADARSAT will overcome this problem, and will provide data complementary to that from the optical wavelengths.

Vegetation monitoring on regional and global scales is already a reality with the AVHRR; the end of the century is going to see the availability of a suite of instruments providing data over a range of spatial, spectral and temporal scales. Technological progress is well advanced, research relating to the development of the ecological concepts needed for analysis of vegetation dynamics over this range of scales must now follow.

References

Belward, A. S., and Lambin, E.,1990, Limitations to the identification of spatial structures from AVHRR data. *International Journal of Remote Sensing,* **11**, 921 - 927.

D'Souza, G., and Hiederer, R., 1989, Monitoring rangeland vegetation in the Sahel using Landsat MSS and NOAA AVHRR, *Proceedings of the 4th AVHRR data users' conference, held at Rothenburg ob der Tauber, Federal Republic of Germany, on 5th - 8th September 1989.* pp 111-116.

Duggin, M. J., and Piwinski, D. 1984, Recorded radiance indices for vegetation monitoring using NOAA AVHRR data; atmospheric and other effects in multitemporal data sets, *Applied optics*, **23**, 2620-2623.

Grégoire, J. M., 1990, Effect of the dry season on the vegetation canopy of some river basins of West Africa as deduced from NOAA - AVHRR data, *Hydrological Sciences Journal* (in press).

Grégoire, J. M., Flasse, S., and Malingreau, J-P., 1988, Evaluation de l'action des feux de brousse, de novembre 1987 à février 1988, dans la région frontalière Guinée-Sierra Léone. Rapport S.P.I. 88.39, Centre Commun de Recherche - Etablissement d' Ispra.

Hiernaux, P. H. Y., and Justice, C. O., 1986, Suivi du développement végétal au cours de l'été 1984 dans le Sahel Malien, *International Journal of Remote Sensing,* **7**, 1515-1531.

Holben, B. N., and Fraser R. S., 1984, Red and near-infrared sensor response to off-nadir viewing, *International Journal of Remote Sensing,* **5**, 145-160.

Holben, B. N., Characteristics of maximum-value composite images from temporal AVHRR data, *International Journal of Remote Sensing,* **7**, 1417 - 1434.

Huete, A. R., Post, D. F., and Jackson, R. D., 1984, Soil Spectral Effects on 4-Space Vegetation Discrimination, *Remote sensing of environment*, **15**, 155-165.

IGBP, 1988, International Geosphere - Biosphere Programme: A study of Global Change, IGBP report number 4. A plan for action. Report prepared for discussion at the first meeting of the Scientific Advisory Council for the IGBP, Stockholm, Sweden on 24th to 28th October, 1988. (IGBP secretariat, Royal Swedish Academy of Sciences, Box 50005, S-10405, Stockholm, Sweden).

JRC., 1990, Commission of the European Communities Joint Research Centre-Ispra annual report for 1989, (in press).

Justice, C.O., Markham, B.L., Townshend, J.R.G., and Kennard, R.L.,1989, Spatial degradation of satellite data, *International Journal of Remote Sensing*, **10**, 1539-1561.

Justice, C. O., 1986, Editorial, *International Journal of Remote Sensing*, **7**, 1385-1390.

Justice, C. O., and Hiernaux, P. H. Y.,1986, Monitoring the grasslands of the Sahel using NOAA AVHRR data: Niger 1983, *International Journal of Remote Sensing*, **7**, 1475-1497.

Justice, C. O., Townshend, J. R. G., Holben, B. N., and Tucker, C. J., 1985, Analysis of the phenology of global vegetation using meteorological satellite data, *International Journal of Remote Sensing*, **6**, 1271-1318.

Kauth, R. J. and Thomas, G.S., 1976, The tasselled cap - a graphic description of the spectral - temporal development of agricultural crops as seen by Landsat, *Proceedings of the 2nd International Symposium on Machine Processing of Remotely Sensed Data* , pp 4B41 - 4B51.

Kerber, A. G. and Schutt, J. B., 1986, Utility of AVHRR channels 3 and 4 in land-cover mapping, *Photogrammetric Engineering and Remote Sensing*, **52**, 1877-1883.

Kidwell, K. B., 1986., NOAA polar orbiter data user's guide, NOAA NESDIS, National Climate Data Center, Washington, D.C.

Langaas, S. and Muirhead, K., 1988, Monitoring bush fires in West Africa by weather satellites. *Proceedings of the 22nd International Symposium on Remote Sensing of Environment held in Abidjan, Cote d'Ivoire, on October 20 - 26, 1988* , pp. 253 - 268.

Malingreau, J.P., Stevens, G., and Fellows, C., 1985, 1982-83 forest fires of Kalimantan and North Borneo. Satellite observations for detection and monitoring, *Ambio*, **14**, 314-321.

Malingreau, J.P., 1986, Global vegetation dynamicss: satellite observations over Asia, *International Journal of Remote Sensing*, **7**, 1121-1146.

Malingreau, J.P., Tucker, C. J., and Laporte, N., 1989, AVHRR for monitoring global tropical deforestation. *International Journal of Remote Sensing*, **10**, 855 - 867.

Malingreau, J.P., and Belward, A. S., 1989, Vegetation monitoring using AVHRR data at different resolutions. *Proceedings of the 4th AVHRR data users' conference, held at Rothenburg ob der Tauber, Federal Republic of Germany, on 5th - 8th September 1989*. pp 141-144.

McGraw, J. F., and Tueller, P. T., 1983, Landsat computer-aided analysis techniques for range vegetation mapping, *Journal of Range Management*, **36**, 627-631.

Monteith, J. L., 1973, *Principles of Environmental physics* (London: Edward Arnold).

Myers, N., 1984, *GAIA, an atlas of planet management* (New York: Doubleday).

NOAA, 1986, Global Vegetation Index User's Guide, National Environmental Satellite Data and Information Service, National Oceanic and Atmospheric Administration, Washington D. C.

Otterman, J., 1981, Reflection from soil with sparse vegetation, *Advanced space research*, **1**, 115-119.

Prince, S.D., and Astle, W.L., 1986, Satellite remote sensing of rangelands in Botswana. I. Landsat MSS and Herbaceous vegetation, *International Journal of Remote Sensing*, **7**, 1555-1570.

Prince, S.D., and Tucker C.J., 1986, Satellite remote sensing of rangelands in Botswana. II. NOAA - AVHRR and Herbaceous vegetation, *International Journal of Remote Sensing*, **7**, 1555-1570.

Richardson, A.J., and Wiegand, C.L., 1977, Distinguishing vegetation from soil background, *Photogrammetric Engineering and Remote Sensing*, **43,** 1541 - 1552.

Rouse, J.W. Jr., Haas, R.H., Schell, J.A., Deering, D.W., and Harlan, J.C., 1974, Monitoring the vernal advancement and retrogradation (green wave effect) of natural vegetation. NASA type III, Final report, RSC 1978-4, Goddard Space Flight Center, Greenbelt, Maryland.

Sellers, P.J., 1985, Canopy reflectance, photosynthesis and transpiration, *International Journal of Remote Sensing*, **6**, 1335-1372.

Sharman, M., 1989, Monitoring agriculture in Europe using AVHRR: management and use of a large multi-temporal dataset. *Proceedings of the 4th AVHRR data users' conference, held at Rothenburg ob der Tauber, Federal Republic of Germany, 5th- 8th September 1989*. pp 175-178.

Steiner, D., 1970, Time dimension for crop surveys from space, *Photogrammetric engineering* , **36**, 187- 194.

Stephens G., and Matson, M., 1987, Regional and Global fire detection using AVHRR data, *proceedings of the 21st International symposium on Remote Sensing of Environment, Ann Arbor, Michigan, October 26-30*, pp 447-457.

Taylor J.C., Belward A.S., Hewett, D.G., and Wyatt, B.K., 1986, Mapping of vegetation and soils and estimation of biomass in the Sahel. Final report to the Commission of the European Communities on: Monitoring of pastureland ecosystems in the Sahel and Mapping of Cloud cover and rainfall.

Townshend, J.R.G., and Justice, C.O., 1986, Analysis of the dynamics of African vegetation using the normalized difference vegetation index. *International Journal of Remote Sensing*, **7**, 1435-1445.

Townshend, J.R.G., and Tucker, C. J., 1981, Utility of AVHRR of NOAA 6 and 7 for vegetation mapping. In *Matching remote sensing technologies and their applications* edited by J. A. Allen (Reading: Remote Sensing Society).

Tucker, C.J., and Sellers, P.J., 1986, Satellite remote sensing of primary production,*International Journal of Remote Sensing*, **7**, 1395-1416.

Tucker, C.J., Vanpraet, C.L., Sharman M.J., and Van Itersum, G., 1985, Satellite remote sensing of total herbaceous biomass production in the Senegalese Sahel: 1980 - 1984, *Remote sensing of environment*, **17**, 233-249.

Tucker, C.J., 1977, Asymptotic nature of grass canopy spectral reflectance, *Applied optics*, **16**, 1151-1156.

Turk, A., Turk, J. and Wittes, J. T., 1972, Ecology, pollution, environment, (Philadelphia: W. B. Saunders).

Verstraete, M. M. and Dickinson, R. E., 1986, Modeling surface processes in atmospheric general circulation models, *Annales Geophysicae*, **4/B**, 357-364.

REMOTE SENSING AND AGRICULTURAL PRODUCTION MONITORING IN SAHELIAN COUNTRIES

E. BARTHOLOMÉ
Institute for remote Sensing Applications
Joint Research Centre,
Commission of the European Communities
I-21020 Ispra (Varese), Italy

1. Introduction

Agricultural production monitoring has for many years been seen as an important field of investigation for the remote sensing research community. On the one hand, considerable efforts have been made to define the estimation of crop production. On the other hand, these results are not used to any great extent in the operational monitoring and early warning systems set up by national and international agencies. This is partly due to the lack of relevance of the scientific experiments and results to the end-user needs, and partly due to the cost of proposed techniques with regard to available budgets.

It is thus important to try and define the user's requirements and constraints before looking for possible technical solutions involving remote sensing, and to ensure that any system proposed is adapted to specific human and physical environments.

1.1.WHO ARE THE END-USERS

Potential end-users of information regarding crop production can be found at levels from the field to the continent. Direct observations are surely faster, less expensive and more efficient than remote sensing and other "high tech" methods for monitoring the crop status at the lowest levels, that is from every field to local and sub-regional levels. More expensive monitoring procedures such as remote sensing should be limited to spatial levels for which there is no reasonably feasible alternative. End-users of crop production information obtained with such techniques are thus decision makers who are responsible for management tasks from regional to international levels. In the context of Sahelian Africa, they are either officials from Countries Governments, or members of the International Donor Community (Reilly, 1986).

1.2. NATURE OF THE INFORMATION NEEDED

The need for information is thus both at the national level (export/import management) and at the regional level. In both cases, two different time constraints can be considered. For food management between surplus and deficit areas, timely information is usually more important than high accuracy (Spiers *et al.*1986). In addition the information provided can be restricted to foodcrops of strategic importance. This is typical of Early Warning Systems. In contrast, for rural management and economic planning it is useful to have at one's disposal precise and reliable information regarding all the aspects of agricultural production. For these applications, the time factor is less stringent. This is the domain of Statistical Surveys.

A. S. Belward and C. R. Valenzuela (eds.),
Remote Sensing and Geographical Information Systems for Resource Management in Developing Countries, 189–214.

1.3. OPERATIONAL CONSTRAINTS

Financial resources are usually limited for survey administrations of the Sahelian Governments. Yet these administrations should be the primary information producers. Furthermore the personnel availability tends to be reduced because of "structural adjustments" in relation to more severe economic policies. This situation makes it difficult to improve observation procedures using traditional techniques.

Donor agencies also have a need for reliable information for their cooperation policy . They are generally available for financing the introduction of up-to-date tools in the national services, but they are more reluctant to subsidize the running costs involved, the view being that these should be supported by national budgets at the end of the installation phase. This aspect should not be underestimated when proposals are made to introduce new methodologies or techniques which have proved to be efficient in economical and technical contexts completely different from the one of developing countries.

1.4. AGRICULTURAL LANDSCAPE CONSTRAINTS

In the West African Sahelian Countries, the greatest part of the agricultural production is obtained in traditional farming systems.In these systems fallowing remains the main technique for maintaining fertility. While fallow land can account for 2/3 to 3/4 of the agricultural domain in Sudanian areas, it tends to disappear in other regions (especially in the Sahelian zone) where the increase of cropped area is the only actual answer to demographic growth and yield reduction linked with worsening climatic conditions. The main tool used is still the hoe, mechanization as not yet reached a majority of farmers in most regions, or else remains confined to donkey-powered carts and scratch ploughs. The average level of chemical fertilizer input is very low: for instance 1.5 kg/ha in Niger (Diaper 1987) and 10 kg/ha in Senegal (Direction de l'Agriculture, 1989). This means that in this country about 120,000 ha, that is 5% of the cultivated land, receive an appropriate amount of fertilizer. The average field size is rather small compared with the pixel size, even for Spot and Landsat TM data. Yields are low , the amount of vegetation can even be lower in cultivated fields than elsewhere, giving a crop spectral response very similar to bare soil. Moreover the tree stratum is usually important in the agricultural domain of the Sudano-Sahelian regions, for example shea-butter trees (*Vitellaria paradoxa*),apple-ring trees (*Acacia albida*) baobab trees (*Adansonia digitata*). The within-field growing pattern is often heterogeneous because of variable soil quality and water availability which induce an irregular germination of seeds. In addition, the inter-field growing differences are reduced by the strong dependency of vegetation growth upon water provided by rainfall during the rainy season. The Sudano-Sahelian agricultural landscape introduces many constraints on the use of remotely sensed imagery, the "spectral signature" is not specific to a given crop species, but describes the scene complexity.

The usefulness of remote sensing techniques for crop area estimation and yield prediction has been demonstrated in the specific context of improved agricultural practices (Berg and Grégoire 1982, Spiers *et al.* 1986, Mc Murtrey *et al.* 1988). These trials dealt with areas managed not by private farmers or rural communities, but by public administrations with important investment means at their disposal. As a result the agriculture tends to be strongly modernized in these areas: systematic land partition in large fields (>1 ha), irrigation, mechanization, use of fertilizers and selected seeds, etc....Nevertheless this improved agricultural sector at present accounts only for a limited and even sometimes marginal part of total cultivated areas for crop production.

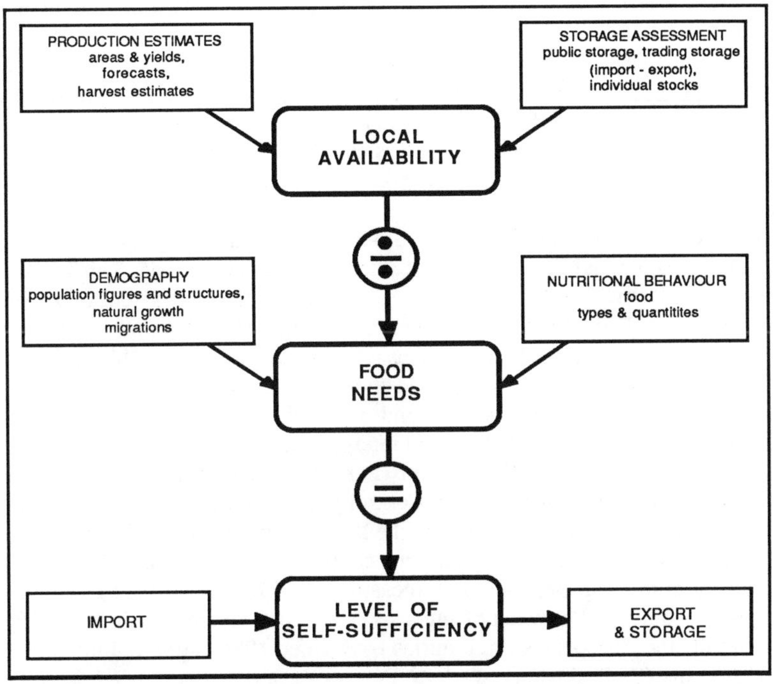

Figure 1: general flowchart of the different parameters to be taken into account in a food availability warning system (Bartholomé 1989b)

1.5. REMOTE SENSING AND OTHER INFORMATIONS

Remote sensing is able to provide only a small part of the information needed for agricultural production monitoring. This is the case for agricultural statistics where information about the structure of the production systems is useful among others for the rural planner: land property, level of mechanization, amount and movements of inputs/outputs, household structures, etc...This is also the case for early warning systems (Reilly 1986) where the basic question is to know if the population of a given region will have sufficient food, and for how long these reserves will last. As shown in figure 1, the answer involves not only information about foodcrop production, but also knowledge about demography, food consumption behaviour, and foodcrop trade and storage, factors which cannot be measured from space.

1.6.CROP MONITORING SYSTEMS IN WESTERN SAHELIAN COUNTRIES.

To try and cope with the need for information expressed by decision-makers, several operational systems running in parallel have been set up to monitor crop production and food availability in the Sahelian countries of West Africa. The most widely known are reviewed below.

FEWS (Famine Early Warning System) is funded by the cooperation agency of the United States of America. Its aim is to collect all types of useful data in the countries concerned (Mauritania, Mali, Burkina, Niger, Chad, Sudan and Ethiopia),transferring the data to the U.S.A. for analysis, and reporting to American authorities and national governments (Reilly, 1986). Until now, qualitative interpretations of AVHRR GAC-derived NDVI's (see Belward in this book for further details) is the main remote sensing input in this system. The objective is to provide information about location of "at-risk populations", with a quantitative assessment of per capita food availability (FEWS bulletins 1987, 1988, 1989).

FAO has extensive experience in information production including regular food status reports: GIEWS (Global Information and Early Warning System) has run since 1985. This is based mainly on the aggregation in Rome FAO headquarters of existing information collected on-site by numerous correspondents. Efforts are made in the frame of the ARTEMIS project to complement these data with remote sensing-derived observations (Hielkema *et al.* 1986). This project is oriented towards the routine production of pre-processed thematic outputs such as NDVI composites, rainfall estimates, crop moisture availability and anomaly maps. These products should be distributed to each country for further analysis and interpretation.

The CILSS project DIAgnostic PERmanent (Permanent Diagnosis, CILSS is the french acronym for Interstate Committee for Drought Control in the Sahel) is run with financial support from the European Development Fund (EDF) and the Italian Cooperation Administration. It aims to improve traditional statistical surveys for agriculture, cattle breeding, etc... Efforts are also made to use the collected data for crop production forecasting: in 1989, the first figures of "harvest forecast" were published by the Permanent Diagnosis project on November 2th.

In addition to this, the European Communities have since 1986 supported the installation of Early Warning Systems (SAP, Système d'Alerte Précoce) at the national level with the help of Non-Governmental Organizations (Kennes, 1988). The objective here is to collect at the lowest administrative level agrometeorological data, information about food prices on markets, and nutritional indicators.

The AGRHYMET Program is a CILSS activity aimed at operational agrometeorological and hydrological monitoring for the member states: Cape Verde, Mauritania, Senegal, Gambia, Guinea-Bissau, Mali, Burkina Faso, Niger and Chad. It produces decadal and monthly bulletins describing potential crop status and yield based on rainfall data processing from the synoptic network with agrometeorological models. In addition, qualitative analyses are performed on locally received and processed AVHRR-HRPT data.

From this brief overview it can be concluded that remote sensing is beginning to be used on an operational basis for crop monitoring, but almost only at the international level. This involves routine processing of imagery and computation of indicators for crop yield and crop status. However, it should be borne in mind that systematic quantitative interpretation is not currently carried out from these data, and that no monitoring of cropped area is made using remote sensing techniques. These points illustrate the difficulties encountered until now in the Sudano-Sahelian context with high resolution satellite imagery, as mentioned earlier. The cost of such operations is still considered to be a limiting factor. It was estimated to be about 0.6 US$/km^2 in Sudan (Spiers *et al.* 1986) for a semi operational campaign involving the use of a limited number of Landsat MSS images.

There is a clear requirement for timely production estimates and for improved agricultural statistics both at national and regional levels. Remote sensing is still barely used to fulfil these requests. The following pages describe the current research and development status of remote sensing and associated techniques, and prospects for applications in the near future.

2 Production of agricultural statistics

2.1 CURRENT TECHNIQUES

Agricultural surveys are usually organized by national administrations. They all follow very similar sampling procedures. Everywhere in the CILSS Countries the method used is multi-level sampling. The stratification is always based on socio-demographic criteria (for example village size, number of farms per village), sometimes with complementary criteria (degree of modernization of the techniques used). The first step is the selection of sample-villages. The second step is the selection of farms in every sampled village. The third step is a selection of fields in each chosen farm. All the selected fields are measured, their content is recorded, and yield plots are placed on a sub-sample (DIAPER, 1987).

The statistical concepts behind such an approach are well tried and tested, and demographic information is readily accessible. On the other hand, population figures are not always very accurate. A comparison of several sources for Mali shows erratic variations between years, and suggests that the problem may be greater at lower levels of Administration units, and when extrapolation is made from old statistics (table 1). In this last case the strong influence of internal and international migrations is emphasized.

As no precise farm enumeration, sampling is made on "Compounds". A compound ("concession" or "carré" in French) is a physically defined spatial unit in a village, usually bordered by a fence or a wall and containing housing and other buildings for several families. Some of these families may not be active in farming activities. However doubts are expressed on the fact that statistics regarding compounds should be considered as better than population figures, as the compound is the basic unit for tax collection.

	1 1976	2 1976	3 1983	4 1983	5 1987	6 1987	7 1988	8 1990
KAYES	872750	863100	965975	1015200	929483	1136542	1113000	1122162
KOULIKORO	932237	1338200	1124135	1761600	1159406	2075226	1724000	1314876
SIKASSO	1098068	1085900	1356454	1277300	1834510	1430305	1376000	1429336
SEGOU	1082224	1070400	1270500	1262800	1680314	1376199	1396000	1459589
MOPTI	1129041	1116500	1345745	1312200	1409917	1468922	1326000	1326687
TIMBUKTU	490456	484500	572400	567300	522115	626249	476000	448424
GAO	370903	366900	438000	428900	290359	479315	403000	392467
BAMAKO	419239		675000		804172			731565
COUNTRY	6394918	6325500	7748209	7625300	8630276	8592758	7814000	8225106

Table 1: comparison of population statistics for Mali according to several sources: DNSI census 1976 (atlas PIRT), 2/ DNSI census 1976 (centrales DIAPER), 3 DNO estimates (Bulletin CNAVS), 4: DNSI estimates (centrales DIAPER), 5/ PADEM survey/OSCE estimate (bull. OSCE), 6/ extrapolation from DNSI 1985 data, 7: FEWS estimate (FEWS bull Nov. 1987 & Dec 1988), 8: FEWS bull. Jan. 1990 from 1987 census.

This method is not well suited for crop production statistics, because the number of farms by compound can vary widely, and also not all the farmers live in compounds. This is the case for instance for urban population (civil servants, merchants, politicians, religious chiefs,...) doing farming as a second activity.

For the third sampling step, all the fields of the farm are supposed to be visited, but usually the farmer do not declare all of them, either because they are too far from the village to be easily visited, or because they are cultivated not by themselves, but by the wifes, the youngs, or in common with other people. this evidently introduces area underestimates (PADEM 1985).

In addition, field surveying is considered to be heavy and tedious work. As a result, the sampling level is kept very low. In Mali for instance, on the average one field is measured for approximately 270 ha of cultivated surface. In Senegal, 1% of the 250000 compounds are visited by about 250 enumerators. As each compound farms about 10 ha, each enumerator has to make every year field size measurements over 100 ha (Direction de l'Agriculture, 1989). As the sampling procedure is repeated every year, the selected fields change accordingly and thus complete field surveying has to be carried out every year.

The currently used surveys thus show several weaknesses for which solutions should be found. Any new approach should try 1) to make field survey easier to the enumerator by establishing permanent observation sites, 2) to improve the accuracy of crop production estimates, either by increasing the sample size or by reducing its variability 3) to improve the sampling frame by taking into account information regarding agricultural land rather than socio-demographic criteria. These improvements should be made without increasing the current running costs, and ideally should result in their reduction.

2.2 IMPROVEMENT OF AGRICULTURAL SURVEY PROCEDURES

The alternative approach proposed here is in a phase of pre-operational evaluation in the frame of the EDF-funded CILSS project "surveillance des Ressources Naturelles Renouvelables au Sahel" (monitoring of natural renewable resources in the Sahel). This uses methods successfully tested in regions of intensely mechanized agriculture (Holko and Sigman 1984, Fournier and Meyer-Roux 1986, Italian Ministry of Agriculture, undated) and adapted to specific agricultural conditions of the area (Bartholomé 1985, 1986). It is based on the hierarchical structure of geographical space. As shown in figure 2, three different levels have been selected: the agricultural region, the land-facet and the segment.

2.2.1 *The agricultural region* is the space within which the structure of agricultural landscapes is homogeneous. This homogeneity takes into account both the natural conditions and the production techniques used by the populations. This implies that types of crops or crop associations and corresponding surface proportions tend to be stable throughout the region. From a statistical point of view, this means that the landscape structure is stationary throughout the region, and that any sample is representative of the whole region. Thus within the limits of such a region the sample size will be smaller for a given accuracy level than in any artificially defined area such as administrative units.

As the above-mentioned criteria are discernible in the landscape, the agricultural regions can be identified and delineated with medium resolution imagery such as Landsat MSS data. The relatively low cost and limited data volume for computation allows systematic coverage of that part of the national territory devoted to cultivation. Regional landscape features tend to be stable, so mapping

mapping will, at this level, not need frequent updating, but will remain valid for several years (typically 5 - 10 years according to regions).

Within each agricultural region *the agricultural domain* can be easily drawn either by visual or digital classification of remotely sensed imagery (SODETEG 1988). The agricultural domain is defined as that part of the territory which is actually devoted to cultivation. It also comprises fallow lands, and some non agricultural areas which cannot be separated either because of their small size compared to the pixel size (such as small roads, footpaths,) or because of spectral confusion (sterile surfaces appear like arable land on dry season images). As the agricultural domain can for different reasons (low population pressure, soils not suitable for agriculture) account for a small part of the territory, it is important to focus the attention on it instead of wasting time and energy in making observations on areas where no crop will be found.

After that first step the territory to be investigated is reduced to the agricultural domain and is ready for sampling thanks to the delineation of homogeneous regions.

2.2.2 *The land-facet* is a physiographic unit considered in this instance for its agricultural characteristics (Derruau 1976, Lebeau 1979). The land-facets are thus defined exclusively in the agricultural domain. Specific production types and cultural practices are associated with environmental potential. In some cases this potential is the result of long-term agricultural practices, such as the rings of permanent fields around the huts which show a typically improved soil with respect to other areas of the cultivated domain, because of regular manuring.

Land-facets can be classified and mapped with available standard 1/50,000 air photographs or with high resolution satellite imagery (Landsat TM and SPOT). This step is useful mainly to ensure that ground observations are distributed proportionally according to the different types of crop production conditions. It is useful for crop yield forecasting because it allows the water storage conditions specific to each land-facet type to be taken into account.

2.2.3 *The segment* is a piece of territory several tens of hectares in size and bordered by easily recognizable limits such as hill ridges, tracks, rivers, or defined as a geometric block of land, in which the cadastral map is established. The segments are kept fixed for several years, possibly five to ten years. Field boundaries have just to be checked and updated. The cadastral map is used by enumerators to collect observations on every field. Thus, although ground survey has still to be carried out by enumerators, the work load is considerably lighter as field areas no longer need to be measured. Given this, it is possible for the enumerator to make more observations. The sample size can be increased for a same workload, and thus accuracy is improved.

2.2.4. *A summary* of general survey strategy can is as follows: 1) delineation of homogeneous agricultural regions, 2) delineation within each agricultural region of the agricultural domain, 3) partition of the agricultural domain in sampling units homogeneous in size, 4) selection of a number of sampling units, 5) delineation and area estimation of the different land-facets in selected sampling units, 6) delineation of possible segments in the different land-facets of the selected sampling units, 7) segment selection proportionally to each type of land-facet 8) cadastral mapping of the selected segments 9) field survey 10) extrapolation by land-facet and by homogeneous area 11) data redistribution by administrative unit.

2.3 CADASTRAL MAPPING

The cadastral map can be established by classical topographical methods (SODETEG 1989), or by air survey. Unfortunately systematic air photo coverage on sufficiently large scales and appropriate times is not available in Africa. Moreover very few Countries are equipped to carry out such air photo campaigns by themselves.

An alternative procedure is based on the use of small format photography. Graham *et al.* (1985) identify low cost, portability and ease of use as advantages over conventional air photography. To this, Bartholomé *et al.* (1988) reporting on an air photo campaign in Ségou, Mali, with an ultra light aircraft equipped with a 70 mm Hasselblad camera (see also Grégoire and Zeyen 1986), showed this system to be a very efficient way of taking air photos during the rainy season as compared to traditional aircraft, because of the possibility of staying on stand-by in the working area, and thus profiting of any cloud-free period. On the other hand traditional aircraft offers decisive advantage for extensive surveys because it allows acquisition of photos over large areas in a short time.

The mapping accuracy with photomosaics assembled from enlarged 70 mm air photos does not reach photogrammetric standards but is sufficient for agricultural surveys. An accurate map scale can be defined by measurement of a ground base, or by precise knowledge of the above-ground altitude during the flight. This can be achieved for instance with laser rangefinders (Grégoire and Zeyen 1986).

Field area can be obtained from the the cadastral map by several methods, ranging from automatic map digitization with microcomputers to more traditional point sampling with a grid superimposed onto the map.

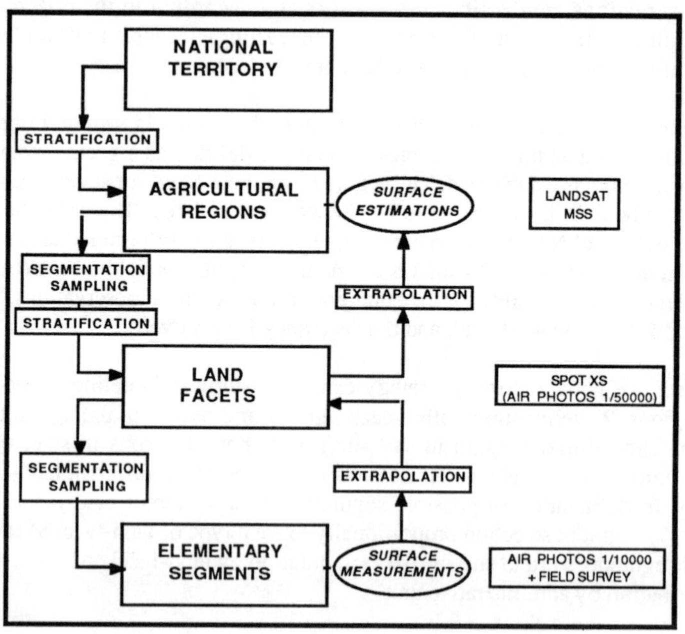

Figure 2: flowchart of a stratification procedure (after Bartholomé 1986)

3 Crop area estimates

3.1 DIRECT CROP AREA ESTIMATIONS BY SATELLITE REMOTE SENSING

In European and American Countries, Remotely sensed imagery is used to extrapolate the sample collected by enumerators to large ares through regression techniques between ground observations and corresponding spectral responses (Holko and Sigman 1984). Such an approach requires two basic conditions to be met: 1) sufficient ground observations must be made to allow the computation of statistically significant relationships, 2) the crop in question must have good spectral contrast in relation the other cover types on the image.

The first condition is far from satisfied at present, primarily because resources needed for the intensive ground survey are lacking. There are also indirect technical reasons, such as the lack of appropriate cadastral map or air photographs needed to establish the relationship between ground observations and remotely sensed imagery.

Specific attention has been devoted to the second aspect, especially since the launch of the SPOT 1 satellite. However, in the West African context, research concerning spectral discrimination between crops and other cover types remains generally inconclusive. Even though discrimination is possible for specific crops in specific areas , the results could be hardly generalized outside the reference test-sites (Revillon 1987, Imbernon *et al.* 1988, Belem *et al.* 1987, Lambin 1988). This is illustrated by figure 3 which shows red and near-infra-red reflectance values for different cover types in the Ségou region in Mali.

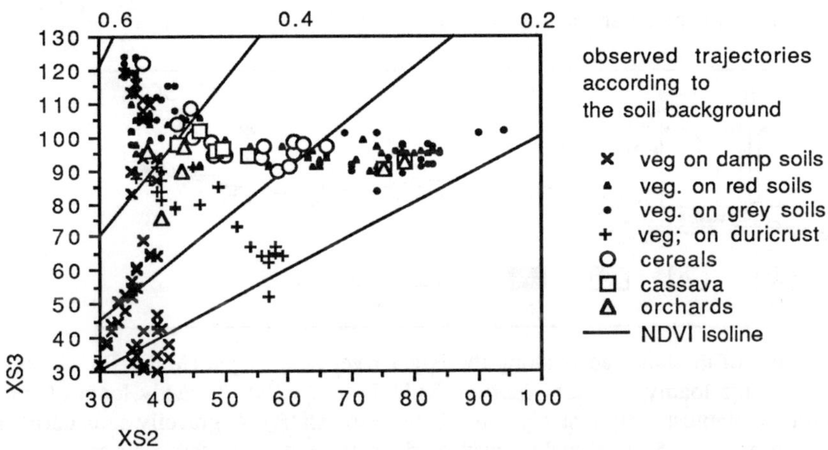

Figure 3: scattergram of different types of land cover in the red (XS2) and near-infrared (XS3) channels of the SPOT image of Oct. 9th, 1986, around Ségou (Mali). Cover types have been defined by photo-interpretation of corresponding air photographs. For high NDVI values all vegetation types are mixed, while for low NDVI values the differentiation is due mainly to the background reflectance properties (Bartholomé 1989a).

Note the considerable overlap in the spectral response of different crops. Differences are mainly due to phytomass amount and spectral properties of soil background. In some instances spectral differences are attributed to variations in crop management calendars (Ndikumana *et al.* 1987). This problem is exacerbated by the limitations of a traditional farming system; working capacity within these systems is often so low that the same crop is sown over a time period of a month or more. Crop species cannot thus be safely determined from temporal reflectance curves or remotely sensed estimation of phytomass amount.

Crop classification and surface estimates from high resolution satellites imagery cannot yet be considered as an operational technique in the specific context of Sahelian agriculture.

3.2 CROP AREA ESTIMATION BY AIR SURVEY

Air campaigns have been proposed as an alternative to field surveys by Ottichilo *et al.* (1985) for example. The procedure is based on parallel transects along which vertical photos are taken with a small format camera. Crop area estimates are then derived from each photo by counting the number of points of a sampling grid which fall into cultivated fields. The efficiency of the method was demonstrated for regions with simple cropping systems dominated by one crop (maïze in Kenya for Ottichilo or Millet in Niger). However, drawbacks to the methods exist for regions of mixed cropping. Figure 4 illustrates such a situation, again from the Ségou region. Analysis of large scale near-infrared colour air photographs showed that crop discrimination depended very much on differences in crop management as well as phenology (Bartholomé *et al.* 1988). Thus for example rice was clearly identified because it is specific to irrigated areas, while millet could not be separated from sorghum because of their similar planting pattern.

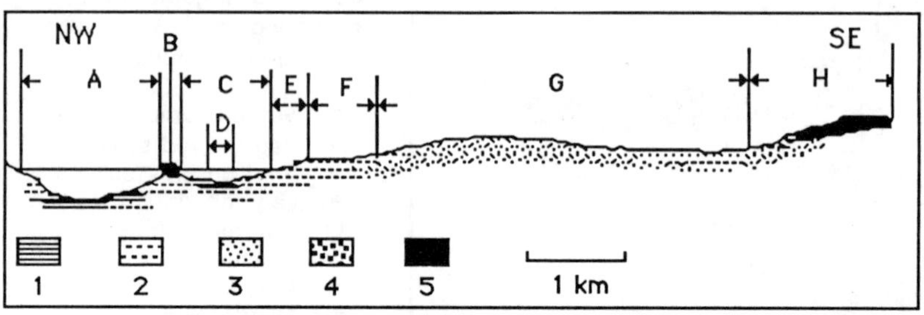

Figure 4: organization of the land-facets along the Niger river near Ségou (Mali). Profile total length: ± 10 km. 1: fine loamy alluvium (partly Typic Tropaqualfs); 2: sandy loam (Typic Tropaqualfs &Plinthic Haplustalfs); 3: sandy soils (Ultic Haplustalfs); 4: gravelly soils derived from duricrust decomposition; 5: lateritic duricrust; A: river, B: man-made dam; C: rice growing area (erected rice, *Oryza sativa*); D: floating rice (*O. glaberrima* and *O. sativa*); E: fringe area: cassava (*Manihot esculenta*), orchards of mango trees (*Mangifera indica*); F: permanent cropping of millet (*Pennisetum americanum*), sorghum (*Sorghum bicolor*), hungry rice (*Digitaria exilis*), cow pea (*Vigna unguiculata*), Bambara ground nuts (*Voandzeia subterranea*) and orchards (citrus); G: rainfed crops, mainly millet, sorghum and peanuts (*Arachis hypogaea*) with short fallows, shea-butter trees (*Vitellaria paradoxa*), apple-ring trees (*Acacia albida*), H: semi-natural vegetation (bush and grass). (Bartholomé 1989a).

In a systematic survey at regional or national level using air survey, only broad classes can be defined so as to ensure consistency of interpretation without ground reference. For rainfed crops for instance, two basic classes should be considered as they can safely be recognized: tall cereals (millet and sorghum, possibly with cow pea), and peas (including peanuts and ground nuts).

The air survey approach is thus excellent for measuring total cropped area and for estimating main cropping classes. This level of information is sufficient for crop production forecasting, but not for agricultural statistics where a precise description is needed for each type of crop.

Moreover, the operational application of this technique for repeated annual surveys at the national level is probably limited because it is very demanding, among other things, in terms of aircraft availability and of qualified manpower for photo-interpretation.

3.3 RESEARCH PERSPECTIVES

We have already seen that remote sensing is not at present able to provide all the information needed for agricultural statistics. It is worth mentioning however, that even if consistent high resolution satellite data for the whole region were available, ground processing systems could not cope with the volume of data.

However, an alternative approach could be to monitor by remote sensing methods a small series of sites scattered throughout the territory. These sites should be selected to be statistically representative of the agricultural domain. Such an approach can satisfy the information needs at the sub-continental level, but probably not at the national level, and surely not at the regional level. The objective should be limited to an assessment of overall cropped and harvestable areas on the selected sites.

Even at that rather low level of information, some critical problems have still to be taken into account. Neither dry season nor rainy season imagery alone can give an appropriate answer to the question of the cropped area. The former allows us to recognize bare soils, only a part of which will be used for cultivation. Bare soil also comprises fallow, sterile eroded soils, infrastructures, etc. Rainy season images alone do not allow systematic discrimination of crops from other vegetation. A combination of informations from both seasons should allow us to define crops as vegetation growing on surfaces recognized as potentially cultivated land on dry season images. Preliminary observations show that the best suited period lasts for a maximum of one month, around the time of full crop development. This period falls on the second half of the rainy season, and is characterized by frequent cloudiness. No other constraint than the period should be imposed to ensure maximum probability for obtaining good images. This means that the probability of using TM data is low, and that any cloud-free SPOT image should be accepted, even with a large lateral viewing angle. The superimposition of such data with dry season imagery makes it necessary to control the image geometry, and especially the parallax effects introduced by the viewing angle. A digital elevation model may thus be necessary to correct possible geometrical discrepancies, especially in hilly regions. In most areas sufficiently detailed topographic maps needed to create the elevation model by simple digitization of contour lines do not exist. A possible alternative is to process satellite stereoscopic imagery. Observation sites would have to be maintained in the same place for several years to spread the costs involved by the establishment of the digital elevation model.

JRC are taking steps to evaluate the technical aspects of such an approach, which in any case is bound to be rather expensive. Its justification on economic grounds will have to be carefully assessed.

An alternative solution which could be considered is farmer strategy modelling. This approach involves the understanding of the farmers response in terms of cultivation according to a limited number of parameters which can somehow be quantified: land-facets, farming techniques, socio-cultural behaviour, recent history of crop production, market prices, current weather conditions. Such an approach uses remote sensing only to define the spatial frame (land use, land facets, land potentialities maps). In contrast much effort is needed to understand the causal factors of farmer strategies. As far as these can be modelled, forecasts could be done of production types and in a specific region according to the above-mentioned variables and the current weather situation.

4 rainfed cereal yield monitoring and estimation

4.1 AGROMETEOROLOGICAL AND STATISTICAL METHODS

Two different approaches are classically used to forecast crop yields. The first one is based on empirical analysis of historical yield record variations according to climatic conditions. It allows the yield trend to be taken into account, which in most regions of the world tends to be positive thanks to the improvement of agricultural practices. The yearly departures from the trend are linked to specific climatic events. Such an approach can hardly be applied to Sahelo-Sudanian regions because yield records are not sufficiently accurate or reliable, or do not cover a sufficiently long time period.

The second approach is based on the establishment of agrometeorological models. these aim at simulating the dry matter production during a given period according to the influence of various climatic and environmental parameters. A very thorough knowledge of physiological processes is needed for such an approach. However, despite their relative perfection, they have been developed more as research tools than for operational crop assessments, particularly because they require the use of very sophisticated and often unconventional parameters (Frère and Popov, 1986). As a result, operational approaches are based on simplified models which relate crop potential yields to plant water availability. This approach is usually referred to as the "water requirement satisfaction index" method (Frère and Popov 1986, Lidon and Forest 1983). This involves previous knowledge of crop coefficients K_{cr} which relate the maximum plant water requirement ET_m to potential evapotranspiration PET, and are specific to the crop and dependent upon the phenological stage

$$K_{cr} = \frac{ET_m}{PET} \tag{1}$$

where ET_m is the evapotranspiration of a desease-free crop grown in large fields under optimum soil water and fertility conditions and achieving its full production potential for the given growing environment (Frère and Popov 1986); PET is the potential evapotranspiration measured either on grass-covered lysimeters (Penman method) or on evaporating pans. Crop coefficients are empirically determined by in-field measurements.

FAO proposed a simplified approach which is based on previous knowledge of crop type and vegetative cycle length, crop coefficients for each phenological stage, soil water availability, and climatological normals of potential evapotranspiration (Frère and Popov 1979, 1986). This so-called "FAO index" is based on the computation of weekly or decadal (10-day period) differences between available water from rainfall and soil moisture, and maximum crop water requirement. Crop water requirement is simply obtained by multiplying for a given week or decade the crop coefficient specific to the crop and the phenological stage by the normal value of potential evapotranspiration. The satisfaction index is then integrated throughout the season. It can be expressed as

$$WRSI = (1 - \frac{WD}{WR_{tot}}) * 100 \tag{2}$$

where WRSI is the water requirement satisfaction index for a given decade. WD is the total water deficit (in mm) since the start of the crop growth until the current decade, and WR_{tot} is the total crop water requirement for the whole growing season.

An important improvement to the FAO approach has been provided by the the "IRAT/CIRAD model". This model replaces the potential evapotranspiration with the actual evapotranspiration ET. This parameter is computed with crop-specific models taking into account rainfall and soil water (Lidon and Forest 1983, Ganzin, 1989). In this case the water requirement satisfaction index "ISBE" is the ratio between actual and maximum evapotranspiration

$$ISBE = \frac{ET}{ET_m} \tag{3}$$

with ET_m derived as for equation (1).

The final index is then derived from a combination of the water requirement satisfaction index on the whole vegetative cycle and the so-called "critical period".

None of these models provide a direct assessment of crop yield, they are only able to assess the water stress level, which will limit the production. Many other factor have a strong influence on yields: cultural practices, fertilizers, crop varieties for example. Nevertheless, empirical relationships can be established between observed yields and water satisfaction indices for specific regions or environments. These indices are thus calibrated in terms of "potential yields". Because of the spatial variability of non meteorological factors, potential yields have a spatially-limited significance.

The quality of estimates is a function of the quality of agrometeorological stations and of point observation networks which are used to determine the model parameters. Spatial generalization of potential yields is obtained by purely mathematical interpolation methods, it is thus sensitive to the quality and representativeness of each node. In particular observations have to be collected so that they allow a statistically significant picture of the average crop status in the region around the node to be drawn. Information availability at the national level relies on the actual communication network efficiency.

4.2 EVAPOTRANSPIRATION MAPPING WITH REMOTE SENSING DATA

In comparison with the classical agrometeorological approach, remote sensing offers at least three theoretical advantages: 1) it allows us to get measurements on the whole area of interest, and not on point sites; 2) it allows direct perception of the spatial distribution of phenomena; and 3) the observation quality is supposed to be homogeneous throughout the observed space. By contrast it contrast it does not allow direct measurement of evapotranspiration or yield, but only emitted or reflected radiation.

Measurements in the thermal infrared bands allow surface temperature estimates to be derived. This physical parameter is usually not measured with high accuracy because of lack of knowledge about surface emissivities. Surface temperature is not in itself a direct indicator of vegetation productivity, but is a variable of the energy balance equation (see Bird, chapter 1). The combination of satellite-derived surface temperature with ground-measured air temperature allows an efficient measurement of water availability according to vegetation type. Seguin (1985) states that water availability variations can account for a temperature difference range of 20° C, while vegetation types can produce temperature differences of 5° C. Jackson *et al.* (1977) established an empirical linear relationship between actual evapotranspiration and temperature differences

$$ET - R_n = a + b\,(T_s - T_a) \tag{4}$$

where ET is the actual evapotranspiration of the crop cover, R_n is the net radiation, T_s is the surface temperature, T_a is the air temperature, a and b are empirically defined parameters. This relationship has been verified in the United States, in Europe and in Africa (Jackson *et al.* 1977, Galli de Paratesi and Reiniger 1982, anonynous 1986, Søgaard 1988).

Surface temperature can easily be estimated from satellites providing two thermal channels, such as AVHRR T_4 and T_5 channels (see Belward, in this book). The most widely used algorithm is the "split window"

$$T_s = T_4 + a\,(T_4 - b\,T_5) \tag{5}$$

where a and b are empirically established parameters. While it is usually considered to provide good accuracy on sea surfaces, the figures on land are less safe because of uncertainties regarding the actual emissivity, which can change with time in relation with vegetation state, vegetation cycle and soil humidity.

These perturbations can have a large effect on the T_4 - T_5 difference. To overcome this problem, Søgaard (1988) proposed to estimate T_4 - T_5 difference on several water surfaces around the region of interest and to compute the tendency surface from these anchor points.

Air temperature is available only through meteorological stations, which can be a major limitation to the application of the model. It was suggested by anonymous (1984) and Søgaard (1988) that Jackson's model can be simplified into

$$ET - R_n = a + b\,T_s \tag{6}$$

where net radiation R_n is a function of the incoming solar radiation S_i, albedo a, and thermal infrared net radiation L_n (Seguin 1987)

$$R_n = S_i\,(1 - a) + L_n \tag{7}$$

S_i is a function of the solar constant (1375 J $s^{-1}m^{-2}$) and of astronomic equations describing the apparent sun movement for any point of the earth.

The albedo can be derived from the visible channels of satellite imagery (Raschke 1987). To ensure good precision in the data careful attention should be paid to interfering factors such as optical characteristics of the atmosphere and satellite view angle.

According to Søgaard (1988) the thermal infrared net radiation L_n can be entirely estimated from remote sensing data. L_n is basically a function of surface temperature (computed with split window techniques) and water vapour pressure. This variable cannot be directly derived from satellite imagery, it is substituted with the precipitable water content of the air column which can be derived from the T_4 - T_5 difference (Dalù 1987, Hansen and Søgaard 1987).

If on the one hand maps of actual evapotranspiration can be established with remotely sensed imagery, on the other hand a precise validation and parameter adjustment is difficult because no appropriate field data are available. Furthermore comparison with actual evapotranspiration obtained at agrometeorological stations is of little significance because these values are specific to crop species and even variety, while the satellite pixel mixes a wide range of cover types.

Extensive field data collection should thus be pursued to calibrate the satellite-derived actual evapotranspiration maps, and to assess their significance with respect with crop evapotranspiration in order to give them meaning in terms of potential crop yield at the regional level.

4.2 YIELD ASSESSMENT WITH REMOTELY SENSED VEGETATION INDICES

4.2.1 *Theoretical model.* Monteith (1972) proposed to relate phytomass production to intercepted photosynthetically active radiation in a model which is now widely accepted, and which can be written as follows

$$Y = \int_{t_1}^{t_2} PAR\ .\ APAR\ .\ \varepsilon\ dt \tag{8}$$

with Y the total phytomass production, PAR the photosynthetically active part of the solar radiation (in Joules), APAR that part of photosynthetically active radiation which is absorbed by the vegetation (%), ε the efficiency coefficient expressing the amount of dry phytomass produced by unit of absorbed energy (g/J). A detailed discussion of this model can be found in Baret, earlier in this book.

Vegetation indices are combinations of reflectances in the visible and near-infrared channels. The most used is the normalized difference vegetation index, "NDVI"

$$NDVI = \frac{IR - R}{IR + R} \tag{9}$$

where IR and R stand for near-infrared and red, respectively. The reflectance in every single channel is the result of combined vegetation and soil background reflectance weighted by the relative amount of both elements. More details about vegetation indices can be found in Baret, in this book. While they are usually associated with leaf area indices, vegetation indices have also been both theoretically (Sellers 1985, 1987, Tucker and Sellers 1986, Choudhury 1987) and empirically (Daughtry *et al.* 1983, Asrar *et al.* 1984) related to the amount of absorbed photosynthetically active radiation (figures 5 and 6). While theoretical models suggest that the relationship is linear only in very specific conditions of soil background reflectance, the observations indicate that a linear fit between both variables is sufficient

$$APAR = a + b\,NDVI \tag{10}$$

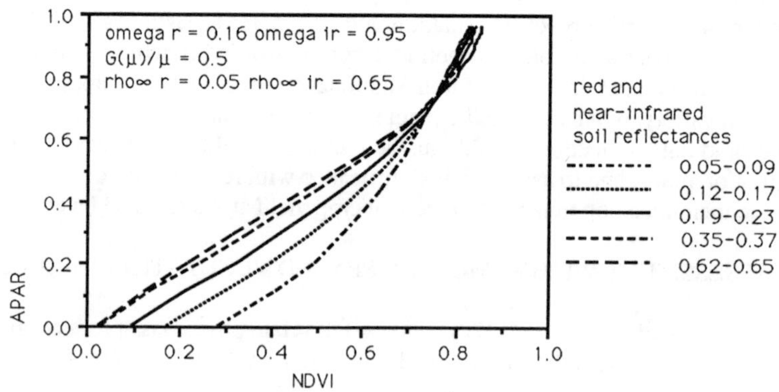

Figure 5: theoretical relationship between NDVI and APAR for different soil reflectances, using Kumar's model (1988).

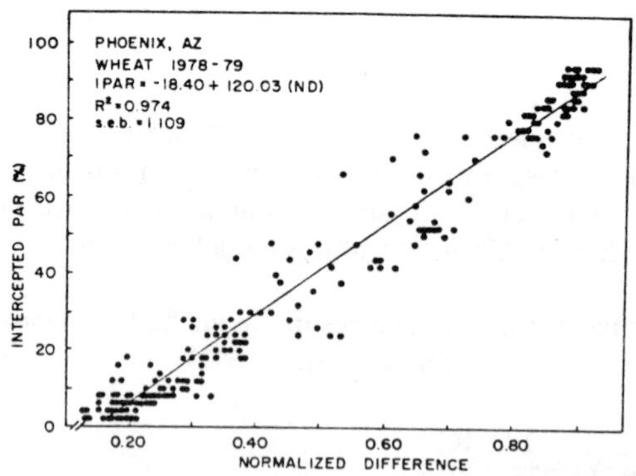

Figure 6: observed relationship between NDVI and APAR on wheat experimental fields (Hatfield *et al.* 1984).

The theoretical relationship between vegetation indices and phytomass production is thus not directly established, but is based on the integration of several partial relationships: absorbed light and phytomass production through the photosynthetic process, foliage surface and absorbed light, and vegetation indices and foliage surfaces (leaf area index).

Monteith's model gives an estimate of total phytomass production, while what is requested is grain yield P. This can be obtained thanks to the grain-to-straw ratio G, so that the equation (8) becomes

$$P = \int_{t_1}^{t_2} PAR \,.\, (a + bNDVI) \,.\, \varepsilon \, dt \qquad (11)$$

It should be stressed that, to the contrary of the previously mentioned agrometeorological models, this one allows direct computation of yields from remote sensing data, and other parameters. Moreover, phytomass production is linked to the total amount of energy captured by the vegetation cover over a certain time period. It can be seen from equation (11) that integrated NDVI is related to biomass amount, while instantaneous NDVI is linked to biomass productivity at a given instant.

Monteith's model has been successfully verified in experimental conditions (Steven *et al.* 1983, Asrar *et al.* 1985, Weiser *et al.* 1986, Daughtry 1988), it has only be used as a justification to empirical studies over grassland (Tucker *et al.* 1983, De Wispelaere and Peyre de Fabergues 1985, Justice *et al.* 1986) and crop productions (Johnson *et al.* 1987, Mc Murtrey *et al.* 1988, Bartholomé 1989b, Rassmussen 1990). Almost all these studies were spatially limited because it is the only way to collect ground information, and thus to establish satellite to ground empirical relationships. However, the end-user request for inter-regional and international information cannot be satisfied by these results.

4.2.2 *empirical applications.* To try and overcome this problem, Bartholomé (1989b) analyzed the possibility of relating weekly maximized vegetation indices from AVHRR GAC data averaged for about 85 administrative districts in four countries (Senegal, Mali, Burkina Faso, Niger) and official yield figures published for the same administrative units.

Figure 7 shows that regional NDVI averages were able to describe about 2/3 of the yield variability in 1984. The remaining unexplained third is probably due to poor quality of yield data and inappropriate NDVI measurements which mix signals both from agricultural and non agricultural vegetation.

While yield figures are usually subject to a priori criticism, it should be noted that significant effort as been made with the aid of several donors to improve the collection of agricultural statistics since 1982-1984 depending on the country. Furthermore, yield estimates are less sensitive than surface estimates to defaults in the sampling frame, in that they are an average and not an extrapolation. Yet, the limited number of observations means we cannot consider official yield figures as providing a very accurate picture of the situation. As a result of this it was found that considering average yields for both millet and sorghum improved relationships with NDVI's because of the increase in the number of observations.

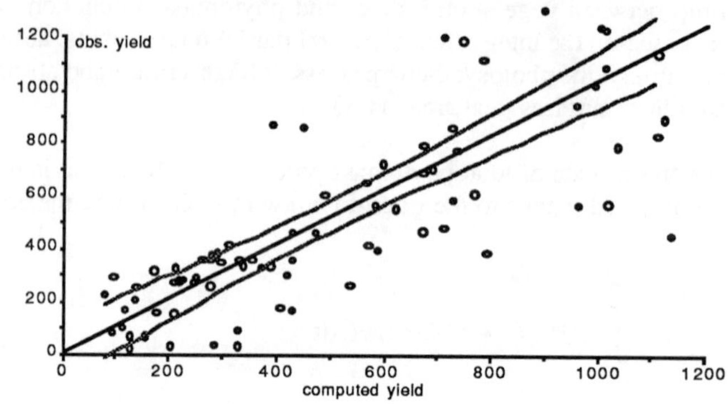

Figure 7: observed millet and sorghum yields against estimated yields for 85 administrative units of senegal, Mali, Burkina Faso and Niger in 1984. Yields were estimated with NDVI regional averages integrated until Sept 3d. ($R^2 = 0.64$). (Bartholomé 1989b).

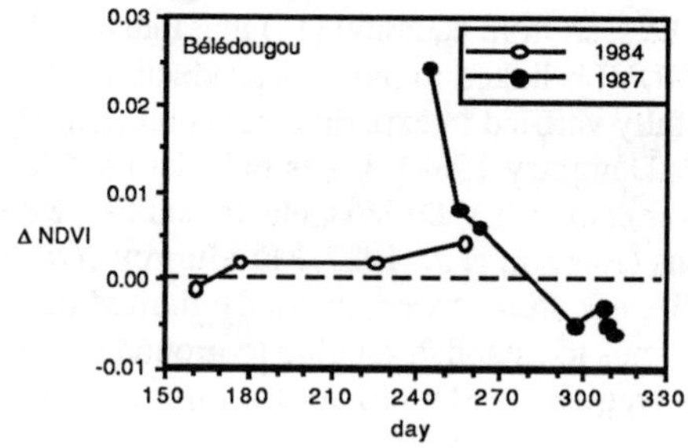

Figure 8: temporal variation of the NDVI difference between agricultural and non agricultural domains in the Bélédougou region (Mali, 30 km N of Bamako). The non agricultural domain has higher NDVI values during the rainy season, but the tendency changes at the start of the dry season.

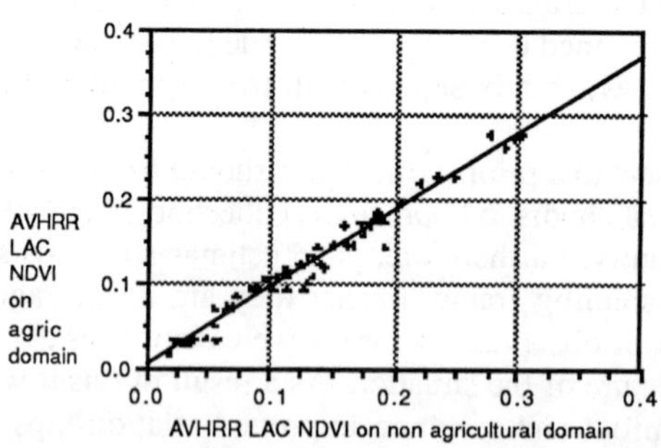

Figure 9: regression line between agricultural domain and non agricultural domain NDVIs . Data were collected on six sites between 12° N and 15° N in Mali from AVHRR LAC and HRPT data.

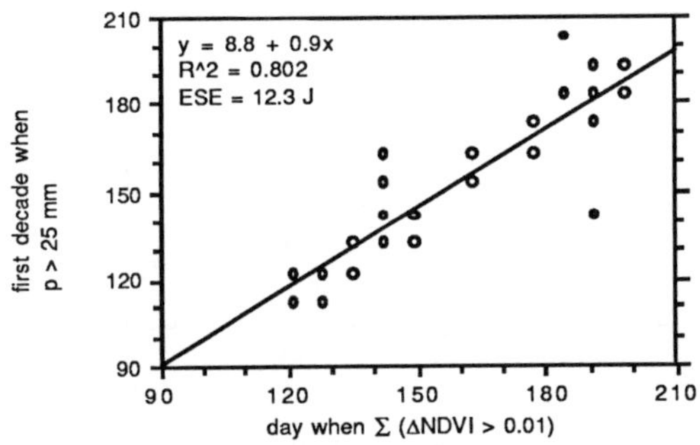

Figure 10: relationship between the start of NDVI curve ("radiometric emergence") and the start of "useful" rains.

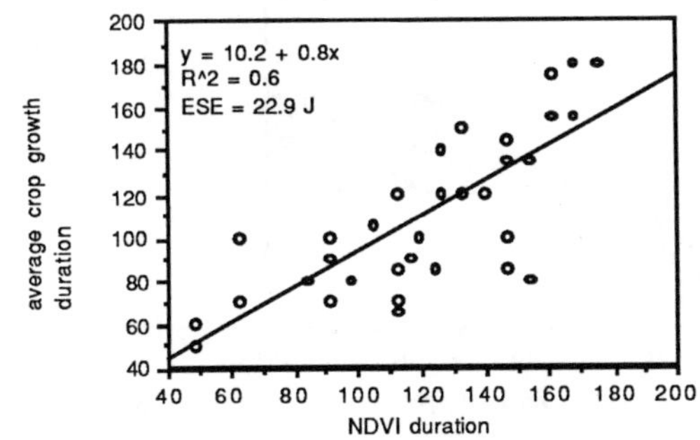

Figure 11: relationship between duration of radiometric season and average crop growth duration.

A comparison of AVHRR LAC NDVI's from agricultural and non agricultural domains in Mali showed that while both kinds of cover types behave roughly in the same way under water stress conditions, differences are clearly visible in growth cycles: crop surfaces start greening after natural vegetation and reach senescence later (figure 8). This was also observed in Burkina Faso (Rassmussen 1990). Moreover, the agricultural domain shows lower NDVI values than natural vegetation on the average (figure 9).

Temporal analysis of NDVI curves allowed definition of three important periods of the vegetative cycle. The start of the regionally averaged NDVI curve allows to define correctly the start date of useful rains (figure 10), which correspond with the period of first successful sowings for rainfed crops, but not with emergence, which happens later. The regional NDVI maximum is two to three weeks earlier than expected for measurements on crops only. In this case, NDVI maximum is associated with the booting-flowering transition, which is important information for agrometeorological models. NDVI duration was also closely associated with average crop cycle length (figure 11). This allows us to define harvest time. Regional NDVI averages can thus be a good indicator of agrometeorological season, but more spatially precise observations are needed to monitor specific crop growth cycle.

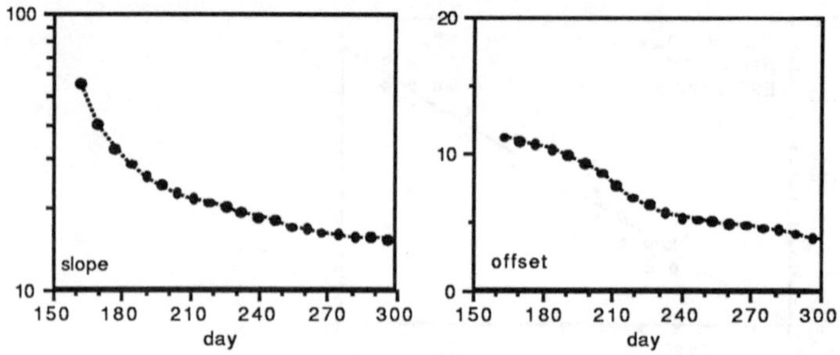

Figure 12: temporal evolution of slope and offset values of the yield versus integrated NDVI relationship for 85 administrative regions of Senegal, Mali, Burkina Faso and Niger.

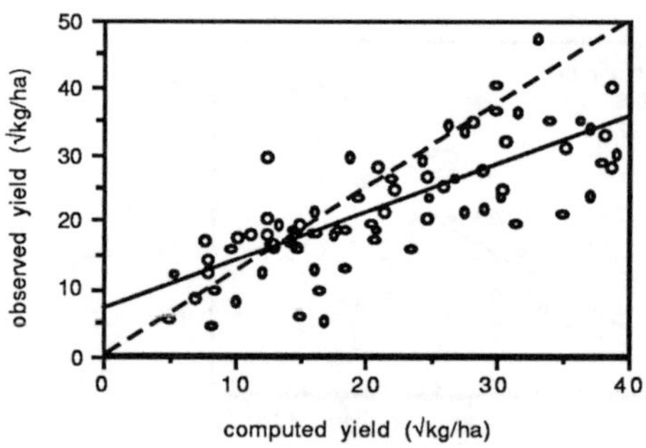

Figure 13: comparison between model-computed and observed yields. Plain line is the regression line, dashed line is the 1:1 line.

The temporal analysis of integrated NDVI versus yield relationships showed that the regression parameters change progressively with time, in a way which can be empirically modelled (figure 13). Provided that the relationship can be demonstrated to be stable from year to year, it could then be possible to compute the relevant slope and offset of the regression for any time of the growing season, and thus to derive corresponding quantitative yield forecasts.

4.2.3 *Deterministic approach.* The theoretical model was also tentatively applied to remote sensing data in the same experiment (Bartholomé 1989b). This involves previous knowledge of the model's parameters.

In experimental field conditions the APAR - NDVI relationship can easily be empirically determined by ground measurements. This is not more possible with remote sensing data, because a sufficient number of measurements must be made to allow the calculation of average values of the parameters for a sufficient number of pixels. A solution has been proposed, and tested with SPOT data, by Bartholomé (1989a). Vegetation cover percentages are estimated with large scale air photographs for a series of fields. This allows leaf area indices to be derived, which in turn are related to APAR. APAR values were then related to SPOT NDVI's to show, thanks to Kumar's thanks to Kumar's model (1988), that the converging point of the different trajectories determined by the soil background type (figure 1) correspond to a specific APAR value of about 80% (to be further confirmed) derived from existing relationships between cover percentage, leaf area index and APAR. As the APAR - NDVI relationship can be accepted as linear, the calibration is obtained thanks to two points: the soil line NDVI (see Baret in this book) is associated with APAR = 0 (no vegetation), and the converging point where NDVI is associated with an APAR value of about 80%.

Global radiation was considered as constant for the entire zone under study and for the whole growing season, as only two main growth periods can be derived from NDVI temporal curves: the vegetative and the reproductive stages. This was checked with global radiation data derived from three meteorological stations in the region. It is worth mentioning that global radiation can be derived from satellite data (Brakke and Kanemasu 1981, Dedieu *et al.* 1983) at least on a monthly basis, but this approach was not considered here as it would have involved heavy data processing work.

Very little information is available regarding the efficiency coefficients. Usual values for plants such as millet and sorghum can be of around 2.5 g/MJ in experimental conditions (Gosse *et al.* 1986), but significantly different values have been reported for millet in Niger: 1.14 g/MJ (Ong and Monteith 1984); 3.37 g/MJ (Imbernon *et al.* 1990).

Grain-to-straw ratios observed in experimental conditions and in normal farming conditions can vary from 1/10 to 1/2 for millet and sorghum.

Figure 13 shows that good agreement can be obtained between observed yields and estimations calculated with Monteith's model at the district level for four West African countries. However, these results are sensitive to the selected parameter values. In this case where a low efficiency factor and a high grain-to-straw ratio were considered, underestimations are observed for low yields and overestimations for high yields. Spatial analysis showed that low values are observed in Northern Sahelian areas where farming is limited to very small areas with above average water storage capacity. Crops are thus in better growing conditions than natural vegetation. High yields instead are observed in the Sudano-Guinean fringe of Mali and Burkina Faso. In these regions natural vegetation starts growing earlier than crops. Trees and grass more completely covers the ground than cultivated species. As a result average NDVI's overestimate crop productions.

A significant improvement of remotely sensed assessment could thus be obtained by using NDVI averages computed only on the agricultural domain instead of on the whole territory. More work has still to be done to collect information regarding grain-to-straw ratios and efficiency factors before operational application can be envisaged for the modellistic approach.

5. Conclusions

The current state of the art shows that, while remote sensing techniques can be used in an operational way to improve agricultural statistical surveys, direct crop area estimates seem hardly feasible in Sahelian Countries.

Work has still to be done to know if early estimates of global harvestable areas can be obtained with satellite imagery in this specific environmental context.

Low spatial resolution satellite data are currently used by several organizations to complement the information collected by ground networks. Imagery exploitation remains qualitative until now. A gap has still to be filled to allow accurate interpretation of physical parameter images and maps in maps in terms of quantitative yields and crop production estimates, which is the information actually needed by the end-users.

To attain this objective research should involve both empirical and modellistic approaches. In this respect, field data collection is of the outmost importance. This should include the establishment of all necessary parameters, such as crop coefficients for agrometeorological models, efficiency factors, grain-to-straw ratios. Meanwhile representative yield figures should be collected with a sampling scheme adapted to the use of remote sensing imagery in order to create good reference data bases.

6 references

anonymous,1984, Suivi du bilan hydrique à l'aide de la télédétection par satellite application au Sénégal. Projet CEE- DG VIII "caractérisation par les techniques de télédétection de la dynamique de la désertification à la périphérie du Sahara", rapport intérimaire, 1 décembre 1984, CNES, INRA, IRAT, ISRA, DMNS.

anonymous, 1986, Suivi du bilan hydrique à l'aide de la télédétection par satellite - application au Sénégal. Contrat CCR Ispra n° 2404-84-07 ED ISPF - CNES/INRA/IRAT/ISRA/DNMS, Avignon.

Asrar, G., Fuchs, M., Kanemasu, E. T., Hatfield, J. L., 1984, Estimating absorbed photosynthetic radiation and leaf area index from spectral reflectance in wheat. *Agronomy Journal,* **76,** 300-306.

Asrar, G., Kanemasu, E.T., Jackson, R.D., Pinter, P.J., 1985, Estimation of total above-ground phytomass production using remotely sensed data. *Remote Sensing of Enivronment,***17**, 211-220.

Bartholomé, E., 1985, L'apport de la télédétection dans l'estimation des productions agricoles en pays sahéliens - éléments d'approche méthodologique. Internal report.

Bartholomé, E., 1986, Remote sensing as an aid for area estimation of rainfed crop in sahelian countries. *Proceedings XXth ERIM Int. Symp on R.S. of Env., Nairobi 4-10 Dec. 1986.* pp 1119-1128.

Bartholomé, E., 1989a, Calibration of vegetation indices for APAR measurements using intrinsic satellite image information. *Proc. of RSS annual conference, Univ of Bristol 13-15 Sept. 1989.* pp 47-52.

Bartholomé, E., 1989b, Contribution aux méthodes de suivi par télédétection satellitaire des rendements régionaux des cultures céréalières pluviales en Afrique soudano-sahélienne. Dissertation présentée pour l'obtention du grade de Docteur en Sciences, Université Catholique de Louvain, Louvain-la-Neuve, Belgique, Institut pour les applications de la télédétection, CCR -Ispra, Italie.

Bartholomé, E., Grégoire, J.-M., Zeyen, R., 1988, Small format air photo from ultralight aircraft as an aid for data collection of agricultural statistics in Sahelian countries *Proceedings IGARSS '88 Symposium, Edinburgh, 13-16 Sept 1988 -Ref: ESA SP-284 (IEEE 88CH2497-6).* pp 269-270.

Bélem, C., Berger, M., Kilian, J., Lainé, G., Morant, P., 1987, Evaluation et suivi des systèmes de cultures vivriers de la zone cotonière Ouest du Burkina Faso à partir de données satellitaires SPOT - étude du parcellaire agricole. Short note.

Berg, A., Grégoire, J.-M., 1982, Use of remote sensing techniques for rice production forecasting in West Africa (Mali and Guinea: Niger-Bani Project). *Proc. EARSeL-ESA Symp. Igls, Austria, 20-21 April 1982 ESA SP-175.* pp 161-168.

Brakke, T. W., Kanemasu, E. T., 1981, Insolation estimation from satellite measurements of reflected radiation. *Remote sensing of Environment,* **11,** 157-169.

Choudhury, B.J., 1987, Relationships between vegetation indices, radiation absorption, and net photosynthesis evaluated by a sensitivity analysis. *Remote Sensing of Enivronment,* **22,** 209-

Dalù, G.,1987, Satellite remote sening of atmospheric water vapor. *International Journal of Remote Sensing,* **7,** 1089-1097.

Daughtry, C. S. T., 1988, Estimating absorbed radiation and phytomass from multispectral reflectance of corn and soybeans. *Proc. IGARSS'88 Symp., Edinburgh 13-16 Sept 88 - ESA Publication Division ESA SP-284.* pp 821-824.

Daughtry, C. S. T. ,Gallo, K. P., Bauer, M. E., 1983, Spectral estimates of solar radiation intercepted by corn canopies. *Agronomy Journal.* **75,** 527-531.

Dedieu, G., Deschamps, P.Y., Kerr, Y.,1983, Solar irradiance at the surface from METEOSAT visible data Proc. *Symposium on Machine Processing of R.S. data* pp 389-391.

Derruau, M., 1976, *Géographie humaine.* (Paris: Collection U - Editions Armand-Collin).

de Wispelaere, G., Peyre de Fabergues, B., 1985, Action de recherche méthodologique sur l'évaluation des ressources fourragères par télédétection dans la région du Sud-Tamesna (Niger). Rapport de 1ère année IEMVT-CIRAD, Maisons-Alfort, Paris.

DIAPER, 1987, Centrales d'informations - Niger- Banque de données de base sur l'agriculture vivrière et l'élevage. Projet FED - CILSS Diagnostic Permanent, Ouagadougou report n° D 12/87.

Direction de l'Agriculture, 1989, Résultats définitifs de la campagne agricole 1989. Direction de l'Agriculture, Division des statistiques agricoles, Dakar.

FEWS, 1987, 1988, 1989 Famine Early Warning System Country Reports. Prepared for the Africa Bureau of the USAID by Price, Williams & Ass., Inc.

Fournier, P., Meyer-Roux, J., 1985, Préparation des expériences à réaliser après le lancement des satellites de la deuxième génération (TM 1984,SPOT 1985) pour l'étude des régions défavorisées. Contrat CCR-Ispra 2272-83-12 ED ISPF, Rapport final. Minist Agric/SCEES, Paris.

Frère, M., and Popov, G. F., 1979, Agometeorological crop monitoring and forecasting, FAO plant production and protection paper n° 17.

Frère, M., Popov, G. F., 1986, Early agrometeorological crop yield assessment. FAO plant production and protection paper n° 73.

Galli de Paratesi, S., Reiniger, P., (Edit.) 1982, Heat Capacity Mapping Mission Investigation n° 25 (TELLUS Project). Final report to NASA - JRC Ispra Public. Serv.

Ganzin, N., 1989, Diagnostic hydrique des cultures - utilisation du logiciel DHC de suivi opérationnel des campagnes. Centre AGRHYMET, Cellule DHC, Niamey.

Grégoire, J.-M., Zeyen, R., 1986, An evaluation of ultralight aircraft capabiltiy for remote sensing applications in West Africa. *International Journal of Remote Sensing*, **7**, 1075-1081.

Gosse, G. ,Varlet-Grancher, C., Bonhomme, R., Chartier, M., Allirand, J.-M., Lemaire G., 1986, Production maximale de matière sèche et rayonnement solaire intercepté par un couvert végétal. *Agronomie*, **6**, 47-56.

Graham, R. W., Read, R. E., Kure, J., 1985, Small format microlight surveys. *ITC journal*, **1**, 14-20.

Hansen, B., Søgaard, H., 1987, Monitoring vegetation index and biomass production in Southern Greenland based on NOAA-AVHRR data. *Proceedings XXI ERIM Symposium Ann Arbor*. pp 883-892.

Hielkema, J. U., Howard, J. A., Tucker, C. J., Van Ingen Schenau, H. A., 1986, The FAO/NASA/NLR ARTEMIS system: an integrated concept for environmental monitoring by satellite in support of food/feed security and desert locust surveillance. *Proceedings XX ERIM Symposium, Nairobi 4-10 Dec. 1986*, pp 147-160.

Holko, M., Sigman, R. S., 1984, The role of Landsat data in improving U. S. crop statistics. *proc. XVIIIth ERIM Symp. Paris Oct. 1-5, 1984*, pp 307-320

Imbernon, J. ,Leprieur, C., Garin, J.-P., 1988, Caractérisation des surfaces cultivées (zone soudano-sahélienne) à l'aide de l'imagerie SPOT - Etude d'exploitations caractérisées par des parcellaires de faible dimension, étude de faisabilité. -Rapport LERTS, Toulouse -IRAT, Montpellier.

Imbernon, J., Lagouarde, J.P., Kerr, Y., Bègue, A., 1990, Suivi agroclimatique des cultures en zone sahélienne à partir des données NOAA de Maspalomas. Programme CCE/DG VIII "caractérisation par les techniques de télédétection de la dynamique de la désertification à la périphérie Sud du Sahara" - rapport de deuxième phase, IRAT Montpellier, INRA Avignon, LERTS Toulouse.

Italian Ministry of Agriculture and Forestry, undated, Remote Sensing applications in agriculture - crop inventories, yield forecasting, water resources. Feasibility study performed by I.T.A. consortium.

Jackson, R. D., Reginato, R. J., Idso, S. B. ,1977, Wheat canopy temperature: a practival tool for evaluating water requirements. *Water Resource research*, **13**, 651-656.

Johnson, G. E., van Dijk, A., Sakamoto, C. M., 1987, The use of AVHRR data in operational agricultural assessment in Africa. *Geocarto International*, **1**, 41-60.

Justice, C., (Ed.) 1986, Monitoring the grasslands of semi-arid Africa using NOAA-AVHRR data. *International Journal of Remote sensing,* **7**, (special issue).

Kennes, W., 1988 EC support for early warning systems. Commission of the european Communities, 26 Oct. 1988, Brussels.

Kumar, M., 1988, Crop canopy spectral reflectance. *International Journal of Remote sensing,* **9**, 285-294.

Lambin, E., 1988, Apport de la télédétection satellitaire pour l'étude des systèmes agraires et la gestion des terroirs en Afrique Occidentale. Dissertation thèse docteur en Sciences, Univ. Catholique de louvain-la Neuve.

Lebeau, R., 1979, *Les grands types de structures agraires dans le monde.* (Paris: Editions Masson, 3d ed.).

Lidon, B., Forest, F., 1983, Valorisation agricole des ressources pluviométriques - synthèse de l'atelier IRAT - CIEH tenu à Ouagadougou du 27/11 au 4/12/82 . IRAT, Montpellier.

Mc Murtrey III, J.E., Hart, G.F., Tucker, C. J., Ibrahim, M., 1988, Comparison of sorghum production potential with AVHRR normalized difference vegetation index (NDVI) changes within and between seasons for eastern Sudan. *Proceedings of XXII ERIM symposium, Abidjan 20-26 Oct. 1988.* pp 541-550.

Monteith, J. L., 1972, Solar radiation and productivity in tropical ecosystems. *Journal of Applied ecology,* **9**, 747-766.

Ndikumana, J., Philippe, E., Joannes, H., 1987, Caractérisation des signatures spectrales de quelques cultures tropicales par mesures radiométriques au sol. *Africa Pixel,* **1**, 45-57.

Ong, C. K., Monteith, J. L., 1984, Response of pearl millet to light and temperature. *Proc. of the Int. Symp. Agrometetorology of sorghum & millet in the semi-arid tropics, ICRISAT centre, Patancheru - India, 15-20 Nov. 1982.* pp 129-141.

Ottichilo, W. K., Peden, D. G., Agatsiva, J. L., Mwendwa, H., 1985, Maize harvest forecast for 1984 in Bugoma, Elgeyo Marakwet, Kakamega, Nakuru, Nandi, Trans Nzoia, Uasin Gishu and West Pokot districts. KREMU report, Ministry of Planning, Nairobi-Kenya n° 115.

PADEM, 1985, Manuel de collecte des données Projet Amélioration des Enquêtes de Ménage. PNUD/Bamako PADEM/18.

Raschke, E. , 1987, Radiation budget at ground. ISPRA COURSE on applications of remote sensing to agrometeorology, JRC ISPRA April 6-10, 1987 ARS/87/13.

Rasmussen, M. S., 1990, Une méthode simple pour estimer les rendements et les productions en mil dans le Nord du Burkina Faso à l'aide d'indices de végétation NDVI dérivés des données NOAA AVHRR. Atelier sur l'utilisation de la télédétection en statistiques agricoles au Sénégal, CSE-DA, Dakar, 2-4 mai 1990 (to be published in proceedings).

Reilly, D., 1986, Famine and early warning the FEWS project. *Proceedings XX ERIM Symposium, Nairobi 4-10 Dec. 1986.* pp 143-146.

Revillon, P. Y., 1987, Evaluation des données SPOT pour l'inventaire et le suivi des cultures en Afrique soudanienne, exemple du périmètre agricole de Bané (Burkina Faso). *Actes du colloque SPOT 1, utilisation des images, bilan, résultats Paris nov. 1987.* pp 407-414.

Seguin, B., 1985, Utilisation des satellites pour la prévision des rendements - application à la zone sahélienne. Short note.

Seguin, B., 1987, Use of surface temperature and estimation of evapotranspiration. Ispra Courses applications of remote sensing to agrometeorology April 6-10 1987 ARS/87/12.

Sellers, P. J. , 1985, Canopy reflectance, photosynthesis and trasnspiration. *International Journal of Remote Sensing,* **6,** 1335-1372.

Sellers, P.J., 1987, Canopy reflectance, photosynthesis and transpiration. II. The role of biophysics in the linearity of their interdependence. *Remote Sensing of Environment,* **21,** 143-183.

SODETEG, 1988, Surveillance des ressources naturelles renouvelables. rapport du contrat 2888-85-12 ED ISP F - CCR/CCE. SODETEG - T.A.I., Valbonne.

SODETEG, 1989, Surveillance des ressources naturelles renouvelables au Sahel - Rapport final avenant 2 contrat CILSS n° 5105.33.94.238 SODETEG - T.A.I., Valbonne.

Søgaard, H., 1988, Comparison between satellite-derived evapotranspiration and normalized difference vegetation index in the Sahelian zone. *Proceedings of XXII ERIM symposium, Abidjan 20-26 Oct. 1988,* pp 349-367.

Spiers, B., Haley, K., David, F., Trayfors, W., Olsson, J., 1986, A timely response for information needs in Sudan. *Proceed. XX ERIM Symposium, Nairobi 4-10 Dec. 1986,* pp 81-85.

Steven, M. D., Biscoe, P. V. ,Jaggard, K. W., Paruntu, J., 1986, Foliage cover and radiation interception. *Field Crops Research,* **13,** 75-87.

Tucker, C. J. ,Vanpraet, C., Boerwinker, E., Gaston, A., 1983, Satellite remote sensing of total dry matter production in the senegalese Sahel. *Remote Sensing of Envoronment,* **13,** 461-474.

Tucker, C. J., Sellers, P. J., 1986, Satellite remote sensing of primary production.*Intenrational journal of remote sensing,* **7,** 1395-1416.

Weiser, R.L., Asrar, G., Miller, G.P., Kanemasu, E.T., 1986, Assessing grassland biophysical characteristics from spectral measurements. *Remote Sensing of environment,* **20,** 141-152.

RAINFALL ESTIMATION IN AFRICA USING REMOTE SENSING TECHNIQUES

R. BONIFACIO
Department of Meteorology,
University of Reading,
Reading, RG6 2AU
United Kingdom.

1. Introduction

Rainfall is the most variable meteorological parameter and one of the most, if not the most, relevant to the welfare of human populations. This is so both in terms of it's beneficial effects, rainfall is a key limiting factor in crop and rangeland production, and deleterious effects, such as floods and as a cause of soil erosion. Increasingly dense and accurate rainfall observations are needed to meet the needs of a growing human population. However, during the last decade the availability and reliability of rainfall data from raingauge networks has been generally declining, a phenomenon which is worst in tropical and subtropical regions - precisely where rainfall monitoring needs are more acute.

Rainfall gauges can tell us the amount of rainfall which fell at that point with an absolute accuracy of 5% (HMSO 1981) but information about rainfall distribution in the surrounding area will depend on the nature of the rain producing systems.

In areas such as the African Sahel the spatial and temporal variability of rainfall imply that interpolating between raingauge values to obtain an estimate of the rainfall at a particular point may give rise to serious errors. Figure 1(a) shows that the rate at which the correlation between station rainfall in Niger decreases with distance is such that, for sites more than about 20 km or so from a gauge, the best estimate of a daily fall may be derived from climatology (e.g. mean rainfall / rainday) and not from the fall recorded at the next gauge. Figure 1(b) shows results obtained from a 10x10 km square grid of raingauges with 2km linear spacing at Ibecetene (Niger). We can see that rainfall amounts per event vary by a factor of nearly 3 within such a small area. Summing totals over 10 days or so will only partially smooth out this variability. The implications of this are discussed below.

This implies that for seasonally or semi - arid regions of Africa, characterized by high spatial and temporal variability in rainfall, proper monitoring would require an impractically large number of gauges.

The use of satellite sensor data, due to its wide and continuous spatial coverage, offers the possibility of monitoring rainfall in areas with deficient raingauge coverage, and thus providing rainfall estimates needed for environmental monitoring, hydrology and water management, crop and rangeland production and climatological studies. Satellite rainfall estimates do not aim to be a replacement for raingauge data, but a complement to it, by giving a better picture of the rainfall spatial distribution. In the case of Africa, most studies focus on the Sahelian zone (Senegal to

A. S. Belward and C. R. Valenzuela (eds.),
Remote Sensing and Geographical Information Systems for Resource Management in Developing Countries, 215–233.

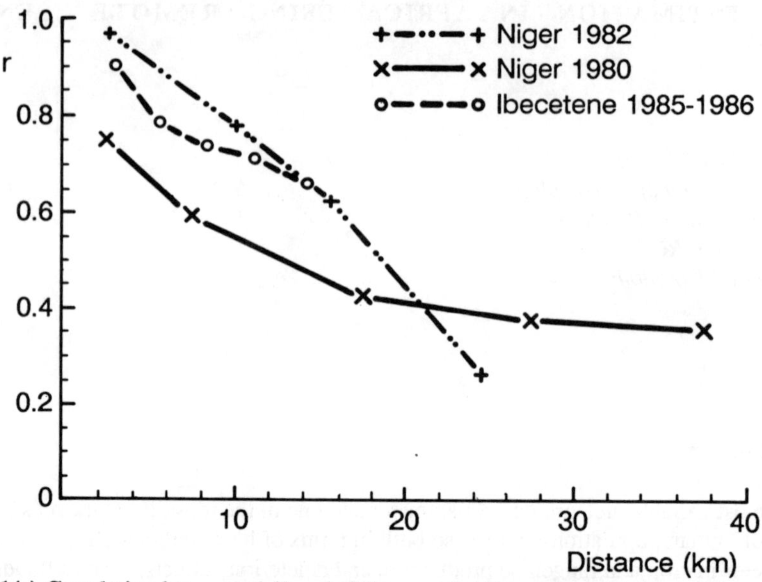

Figure 1(a) Correlation between daily rainfalls recorded at guages with increasing separation. Only occasions when rain fell at both stations are included.

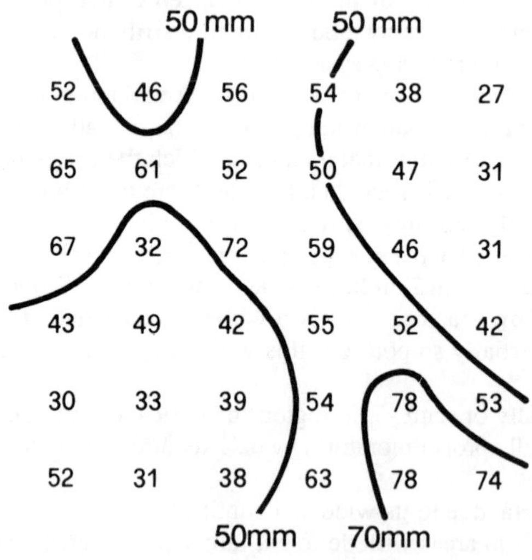

Figure 1 (b) Rainfall event over a rainguage grid at Ibecetene in Niger, July 1985.

Somalia) as this is an area where the need for accurate rainfall monitoring is particularly acute due to the recurrent drought episodes. In addition, because the area is relatively homogeneous, in terms of topography and geography, and because part of the rainfall arises from organized systems of cumulonimbus cloud cells (Bolton 1984) it is amenable to treatment by simpler rainfall estimation techniques.

2.Overview

Within the electromagnetic spectrum the regions that offer possibilities for rainfall estimation are the microwave (MW) region, the visible (VIS) region and the thermal infra red (TIR) region and many techniques using MW or TIR or/and VIS data have been developed for this purpose. None of these techniques gives a direct measurement of rainfall amounts, but each can give indications of the probable rainfall in particular weather conditions. Each has its own merits, and drawbacks, which are to some extent related to the type of rainfall producing system that is being monitored.

Rainfall may be produced by the forced uplifting of moist stable air which gives rise to thick stratiform cloud and rainfall which is extensive in space and time. Alternatively rain may fall in the form of showers or thunderstorms which are convective and give rise to rain which varies rapidly with time and distance. In the former case the cloud tops may not extend high in the atmosphere, but, in contrast, most rainfall of convective origin is given by clouds which extend up to the tropopause.

2.1. MICROWAVE

In the MW region of the spectrum information about rainfall is derived from physical interaction (absorption and scattering) with rain itself, giving direct evidence of rain areas. MW remote sensing systems can be divided into passive and active. Passive systems measure the MW radiation naturally emitted by the target, while active systems (weather radar) emit a MW radiation beam and measure the return signal (echo) from the target. We will not refer here to active systems as their limited spatial range puts them outside the scope of this work.

The most suitable frequencies for the passive MW remote sensing of rainfall are 19.35 GHz, 37 GHz and 85.5 GHz. There are however difficulties in discriminating between rain areas and the background surface, especially over water bodies or wet land surfaces, and / or, at rainrates of less than 10 - 20 mm/hr (Burke *et al.* 1982). Better results are obtained using a combination of frequencies and polarizations. The very poor spatial resolution of MW radiometers (>15 km) limits its applications, and an imaging frequency of 4 times a day (MW radiometers fly in polar orbiting satellites) poses additional problems in cases where the lifetime of the rain producing event is less than the interval between consecutive images (e.g. African arid zones).

Still, passive MW rainfall monitoring techniques yield great promise, and an increasing number have been developed over the past years, though they are still a long way from operational use (see Barrett and Martin 1981, Lovejoy and Austin 1980 and Bailey *et al.* 1986). In this respect it is probably worth mentioning the TRMM (Tropical Rainfall Measuring Mission) project, a satellite sensor specially designed for rainfall measurements in the tropical regions, especially over the oceans (Simpson *et al.* 1988). This is a multi-spectral sensor (radar, passive MW, TIR and VIS) on board a low altitude (300 km) polar orbiting satellite with a 30 degree inclination orbit allowing for increased time and space coverage. Such multi spectral capabilities and coverage will provide remote sensed estimates of rainfall with the highest accuracy yet; the launch is planned for 1995, and the operational life time is designed to be for about three years.

2.2. THERMAL INFRA RED AND VISIBLE

In these wavebands information about distribution and intensity of rainfall is indirect. Observation of clouds in the VIS channel will yield cloud top reflectivity, while the TIR channel will provide us with cloud top temperatures. It is from either, or both, of these parameters that information about rainfall amounts is derived.

2.2.1 Visible. The principle is that high reflectivity is linked to high liquid water content and large cloud thickness, which implies higher probability of rain. However, one can find thin bright clouds, and cloud thickness alone is not the only factor determining the rainfall falling from it, other factors include cloud temperature, drop size distribution, concentration, and evaporation (Barrett 1981).

2.2.2 Thermal Infra Red. Low (TIR derived) cloud top temperatures indicate high cloud tops, which correspond to large cloud thickness, which implies greater probability of rain and tells us something about cloud type. However, not all high, cold cloud is associated with rainfall and significant rain may fall from warmer topped clouds, especially near coastal areas. Also the underlying terrain and the type of weather system, of which the rain bearing cloud is part, may cause the resulting rainfall to vary widely.

2.2.3 The techniques. During the last decades a large number of satellite sensor rainfall monitoring techniques using TIR and VIS data have been developed, with varying degrees of complexity (with or without the use of other data such as climatological records and raingauge data). No technique, so far, is able to deal effectively with all types of climate and weather situations.

Current techniques can be grouped into manual and interactive techniques, and objective techniques. Special reference will be made to techniques developed by the Remote Sensing Unit at the University of Bristol and to the technique developed by the TAMSAT (Tropical Agricultural Meteorology using SATellites) group at the University of Reading as this is, as yet, the only technique in fully operational use.

3. Manual and interactive rainfall monitoring techniques

These were the pioneering techniques for rainfall estimates. Examples are two techniques developed at the University of Bristol -the Bristol Method and the BIAS (Bristol/NOAA InterActive Scheme) technique. They are particularly suitable to low and middle level stratiform cloud in middle latitudes which are difficult to tackle by objective techniques.

The Bristol Method uses high quality gridded VIS and TIR imagery which is interpreted visually to decide which type of cloud is present, and to determine the areal extent of each cloud type in each grid-cell. Additional parameters used are, a terrain factor to account for terrain topography influence on rainfall, and a synoptic factor to account for the type and characteristics of the rain cloud system. The rainfall estimates are derived from a combination of these parameters via local regressions.

This method is very labour intensive, requiring attention from skilled personnel throughout the process for the interpretation of the imagery and merging with synoptic weather and topographical data.

The BIAS technique is a further development of the Bristol Method. The technique works on, and outputs at, full pixel resolution instead of at grid cell resolution, and the translation from cloud type / duration indices to rainfall estimates is made via global regressions, which are applicable to new areas of study. BIAS has been partially automated and implemented in an interactive image processing system, thus it is still, to some extent, labour intensive and requiring sophisticated machinery. This method has been particularly successful when used as an interpolation scheme between raingauges, and when applied to large scale synoptic rainfall events which give rise to relatively homogeneous rainfall at the ground.

4. Objective rainfall monitoring techniques

Objective techniques are particularly suited to the monitoring of convective rainfall arising from convective cloud systems with life times much shorter than stratiform clouds. Convective systems give most of their rainfall in their growth stage. Once they have reached their maximum horizontal extent they generally give little precipitation. Therefore proper monitoring of this type of rainfall requires a high frequency of satellite sensor observation; a large number of satellite images precludes the use of manual and interactive techniques.

Examples of objective techniques are CIST, Convective Infrared Satellite Technique (Griffith *et al.* 1981), the NESDIS convective precipitation technique (Scofield and Oliver 1977), the PERMIT (Polar orbiter Effective Rainfall Monitoring Technique) (Barrett *et al.* 1986) and ADMIT (Agricultural Drought Monitoring Integrated Technique) (Barrett and D'Souza 1985 and 1986) techniques, both from the Remote Sensing Unit at the University of Bristol, and the Cold Cloud Statistics Technique (Milford and Dugdale, 1990) from the TAMSAT group at the University of Reading.

The CIST and NESDIS techniques were applied to the USA plains for the estimation and prediction of rain from convective storms. The others were applied to, or specifically designed for, application to seasonally or semi arid - Africa.

4.1. PERMIT AND ADMIT

PERMIT uses four TIR images a day, either from polar orbiting, or from geostationary satellites. Daily rain cloud areas are identified as those below an appropriate temperature threshold and summed up for periods of 10 days or more. These sums are then translated into rainfall estimates using climatological mean rainfall per rainday statistics. The estimates can be adjusted by synoptic station data if available.

ADMIT is a bispectral technique using VIS and TIR (TIR only at night time) data to differentiate between dry and wet areas; rain is most likely to fall from bright (thick) and cold (high altitude top) clouds. Daily, ten daily (or more) rainfall estimates are derived from the rain/no rain map in a similar way to PERMIT.

These two techniques have similar performance levels, though ADMIT generally gives marginally better results than PERMIT. However, the reduced daily number of images used may not be enough for a proper assessment of rainfall in arid tropical regions, like the African Sahel, where storms may have life times as cold cloud of less than 3 to 4 hours.

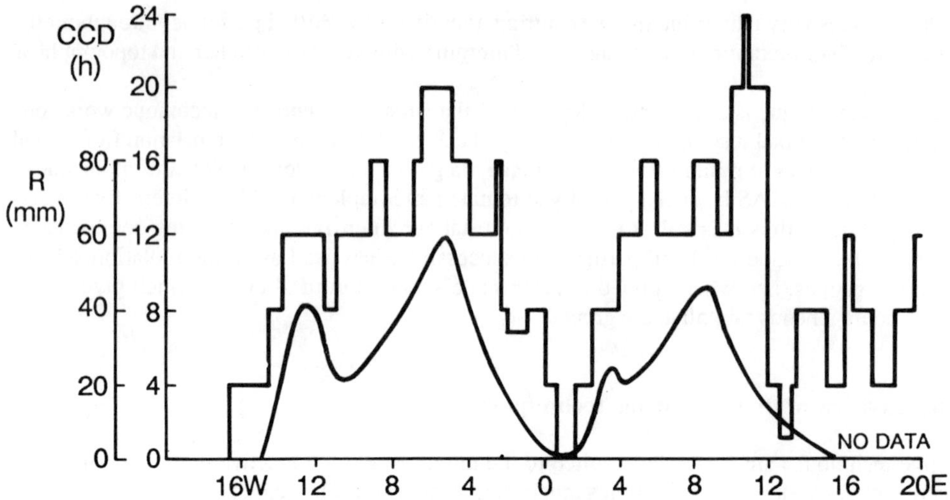

Figure 2 Rainfall (lower curve) and CCD (upper curve) along 15°N for the period 11 to 20 July 1986.

This fact accounts for the use of a much higher temperature threshold (about -30° C) than those used by the techniques based on frequent geostationary satellite imagery (-50 / -60° C). Rainfall estimates, so derived, should be used as averages over distances covered by the moving rain clouds during the image intervals.

4.2. THE COLD CLOUD STATISTICS TECHNIQUE

4.2.1. Principles and outline This technique was developed specifically for sub-saharan West Africa. In these regions the most important rain mainly arises from organized systems of cumulonimbus cloud cells, orientated roughly North - South, with a length of one to several hundred kilometers which characterize the rain bearing system, called a squall line or disturbance line (Bolton 1984). The cold tops of these clouds appear in the TIR as areas with temperatures typically less than -50/-60° C. Squall lines may have an active lifetime of only a few hours, with rain rate falling by a factor of 20 during the first two hours, and much of the rain (70% in Niger) falling at night. Thus we need to use as many images per day as possible to correctly monitor the rainfall, and we may assume that VIS data gives probably little additional information.

If we know the total duration of cold cloud below a given temperature threshold for say, 10 days, this will be a good indication of the length of time of squall line activity over a given pixel. A relationship between cold cloud duration (CCD) and measured rainfall will give a measure of the average rainfall over that pixel. The conversion factor will depend on the temperature threshold used, the stage of the wet season and the location of the pixel. The threshold used must be high enough to include all the precipitating cloud, but low enough to exclude warmer clouds not associated with heavy rainfall. To derive a CCD map each TIR image is scanned and all pixels with temperature below the threshold are assigned a unit score. At the end of the required period the units are added for each pixel; The ability of the CCD technique to provide useful information on rainfall is illustrated in figure 2, where the good correspondence between the maxima and minima of the two parameters is clear.

4.2.2. Details of operation Within the relatively simple scheme outlined above there are many details to be sorted out, such as the most appropriate temperature threshold, the regression model to use and others.

The need to use a large number of images per day has been referred to above. Available data suggest that for intervals between images larger than 2 hours the information derived (CCD) becomes significantly degraded (Milford 1990). An hourly interval that allows for eventual interruptions without loss of information is adequate, but a dedicated system can as easily use half hourly data.

The satellite rainfall estimates can have any time resolution above one day. Bearing in mind the noisy nature of rainfall, the longer that time period the less error will be involved in the estimate when compared to raingauge data. The error due to the inability of the raingauge to represent average rainfall over an area the size of a METEOSAT pixel is discussed later. In practice the chosen time resolution for single pixel data is ten days. Similarly, increasing the area over which the estimates are made will increase its accuracy. It appears that daily rainfall estimates may be useful over areas of the order of 10,000 km^2 or more.

Once a large number of data pairs (CCD and rainfall for the same period and location) are available, contingency tables are built for different temperature thresholds and the threshold which best discriminates between days with, or without, significant rainfall is chosen; the amount of rainfall considered as significant is that above a limit defined according to users requirements, see table 1(a) and (b). Following this, rainfall totals are regressed against CCD; the simplest possible regression model is to use a single threshold and a linear relationship.

More sophisticated processing can be done, or additional data used, e.g. minimum temperature of the cloud, visible and/or water vapour channel radiance (Jacobs 1987, Turpeinen 1987, Negri and Adler 1987, Snijders 1990). However, the improvements were for very small data samples, or were not large enough to justify the additional difficulties and problems in the operational implementation of the technique. Due to the skewness of the rainfall data the regression is made excluding zeroes and between the median of data pairs grouped into classes according to the CCD (see table 2(a) and (b) for a data set) which improves the estimates and gives more weight to the larger falls.

A relationship thus derived should be tested with independent data sets, applying it in subsequent years and to different areas, to see how well the relationship holds from year to year (essential for the technique to work in the operational, pre - calibrated mode) and away from the original area of calibration (see tables 3 and 4). Comparisons between estimated and observed rainfall will show those areas within the original regions where estimates are consistently above or below the observed values and so, provided we have enough data, we can reduce the area for which each calibration is valid. In each region the chosen threshold varies and the coefficients will depend on time (month) and location (latitude) (see figures 3 and 4); currently used zones are shown in figure 5.

T °C	- 40 °C		- 50 °C		- 60 °C	
CLOUD DURATION	NO CLOUD	CLOUD	NO CLOUD	CLOUD	NO CLOUD	CLOUD
NO RAIN	4	18	7	15	11	11
RAIN	7	295	16	286	30	272

Table 1(a) Decadal contingency tables, using three temperature thresholds. For -50° C occasions with rain and no cloud are almost equal in number to those with cloud and no rain. The threshold used is therefore -50°c.

T °C	- 40 °C		- 50 °C		- 60 °C	
CLOUD DURATION	< 5h	> 5h	< 5h	> 5h	< 5h	> 5h
< 10 mm RAIN	4	18	7	15	11	11
> 10 mm RAIN	7	295	16	286	30	272

Table 1(b) As 1(a) but for discriminating between rainfalls above and below 10mm. A threshold of -60° C is indicated.

TRESHOLD -50C COLD CLOUD DURATION (hours)

ACTUAL RAINFALL (mm)	0.0 - 0.0	0.5 - 2.0	2.5 - 4.5	5.0 - 7.0	7.5 - 9.5	10.0 - 12.0	12.5 - 14.5	15.0 - 17.0	17.5 - 19.5	20.0 - 22.0	22.5 - 24.5	25.0 - 27.0	27.5 +
0 - 0	10	12	3	2	3	3	4	1	0	0	0	0	0
1 - 10	15	16	31	36	31	24	17	11	3	2	0	0	0
11 - 20	1	2	9	15	16	21	30	15	16	4	4	2	1
21 - 30	1	1	11	9	15	21	18	22	15	4	4	3	1
31 - 40	0	1	2	9	7	17	6	12	14	7	4	1	1
41 - 60	0	1	1	11	14	18	20	16	23	16	5	5	2
61 - 80	0	0	1	4	1	7	6	8	11	6	4	5	0
81 - 100	0	0	0	1	1	2	3	5	5	2	4	4	5
101 - 120	0	0	0	0	0	0	0	3	2	4	1	2	1
121 - 140	0	0	0	1	0	0	0	1	0	0	0	0	0
141 - 160	0	0	0	0	0	0	0	0	0	1	0	0	1
160+	0	0	0	0	0	0	0	0	0	0	0	0	0

Table 2(a) Contingency table of rainfall against CCD. Data for Niger, July 1985, 1986, 1987.

QUARTILE TABLE AT -50C

CCD RANGE	N	RAINFALL (mm) 25%	MEDIAN	75%
0.0 - 0.0	27	0.0	0.8	2.0
0.5 - 2.0	33	0.0	1.5	6.2
2.5 - 4.5	58	3.3	7.9	21.3
5.0 - 7.0	90	4.1	13.3	35.4
7.5 - 9.5	90	6.4	16.2	31.6
10.0 - 12.0	113	11.1	24.3	39.5
12.5 - 14.5	104	12.6	20.6	43.3
15.0 - 17.0	94	19.1	29.3	50.7
17.5 - 19.5	69	24.9	36.0	55.3
20.0 - 22.0	47	33.2	46.3	67.4
22.5 - 24.5	26	26.6	42.3	72.8
25.0 - 27.0	24	38.2	62.6	88.1
27.5 +	12	36.6	62.8	92.0

Table 2(b) Interpolated quartile values for the same data.

Year	Number of obs	Exact category	1 cat out	2&3 cats out	>3 cats out	% <1 cat out
1985	327	69	106	132	20	54
1986	309	68	106	109	26	56
1987	318	60	121	109	28	57

Table 3 Comparison of actual and estimated 10 day rainfall
Niger, July 1985, 1986, 1987

Country	Number of obs	Exact category	1 cat out	2&3 cats out	>3 cats out	% <1 cat out	> 1 cat out	
							too high	too low
Gambia	21	5	7	6	3	57	1	8
Senegal	12	3	6	3	0	75	1	2
Mali	15	4	3	8	0	47	8	0
B. Faso	6	1	3	2	0	67	2	0
Tchad	23	7	9	5	2	70	4	3
Sudan	13	2	3	4	4	38	8	0
Niger	63	18	21	20	4	62	17	7
All	153	40	52	48	13	60	41	20

Table 4. Comparison of actual and estimated dekad rainfall, 13° N - 15° N
July 1987, using calibration from Niger only, July 1985

CCD (hrs)	number of stations	Rainfall statistics median	25%	75%
0.0	5	2	0	9
0.0 - 4.5	64	11	5	24
5.0 - 9.5	100	20	11	36
10.0 - 14.5	55	41	18	57
15.0 - 19.5	7	48	36	84

Table 5a) Median and quartiles of the distribution of rainfall for grouped CCDs between 13° and 15° N, 10 day periods, July 1985

Median	Lower quartile	Upper quartile	Number of events
1.0	0.6	1.4	1
6.7	3.8	10.2	1
9.9	4.9	14.4	1
17.5	12.2	28.3	2
17.9	12.1	20.3	1
21.2	15.3	25.0	2
32.5	27.4	35.1	2
34.7	29.6	43.5	2
37.2	28.5	42.4	3
51.1	41.1	59.2	3
53.9	50.7	58.9	2
56.2	46.6	66.1	1
72.1	65.7	76.5	2
86.0	67.9	100.1	4

Table 5b) Rainfall distribution on a rainguage grid (Niger) for 10 day totals, July, August 1895 & 1986, showing quartiles in order of ascending median

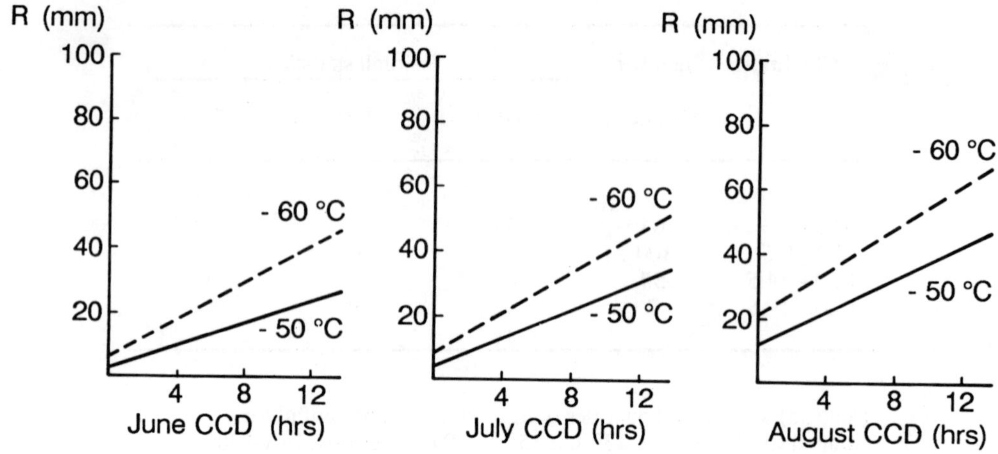

Figure 3 Differences between months in regressions of rainfall on CDD for decads, excluding periods with no cold clouds, Niger 1985.

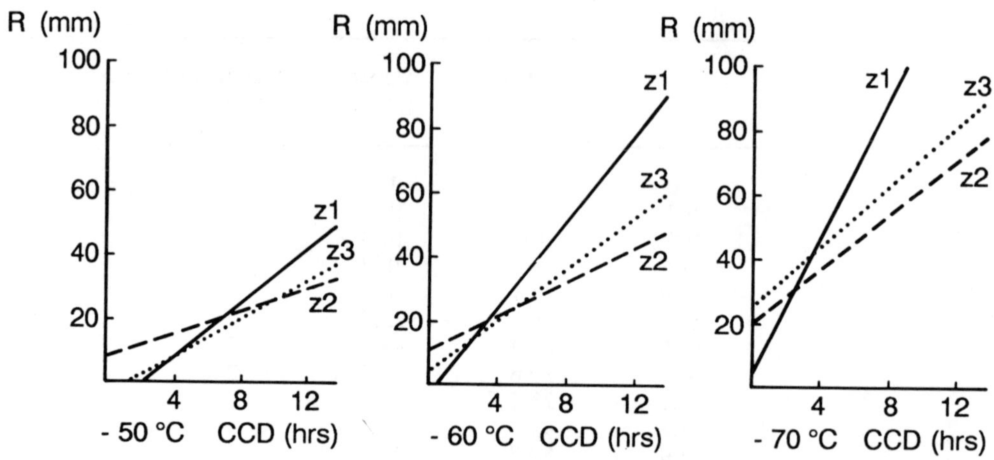

Figure 4 Differences between latitudes in regressions of rainfall on CCD for decades, excluding periods with no cold cloud, Niger 1985. Z1 is north of 15°N, Z2 is between 13°N and 15°N and Z3 is south of 13°N.

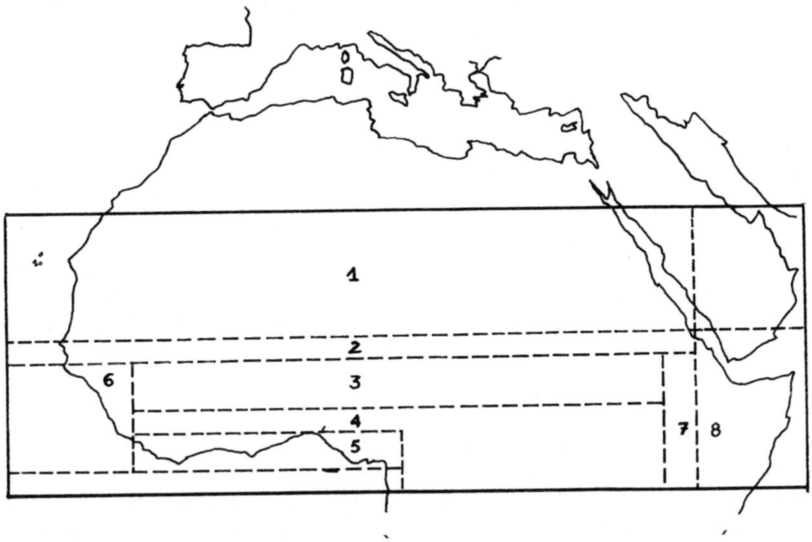

Figure 5 Calibration zones recommended based on 1988 TAMSAT rainfall estimates (METEOSAT map projection).

Expansion of this technique to other areas must be preceded by careful study of the local rain bearing weather systems, to distinguish homogeneous zones, where a single calibration can be applied, from those zones where the technique is either not applicable or needs to be combined with local weather knowledge. Some studies over Kenya showed that areas where the technique looks promising may be bordered by areas where it fails as the origin of the rain changes from convective to low stratiform cloud. Application of the technique to Ethiopia will require knowledge of the local orographic influences on rainfall. Any comparisons with ground data (e.g. assessment of the technique's performance) must be judged against the performance of the raingauge network, especially in areas of such rainfall variability (see figure 1(b)). Table 5(a) gives the relationship between CCD and rainfall statistics (median and lower and upper quartile) and table 5(b) the quartile ranges of point measurements associated with measured (not estimated) area rainfall. It is seen that about half of the inter quartile range of the rainfall estimates is due to the poor representativity of the raingauge data. For areas up to about 500 km^2 relationships between area average rainfall and point measurements have been calculated (Flitcroft 1989) for the case where the whole area has been affected by a single storm.

4.2.3. Going operational A technique that operates successfully in pre-calibrated mode is suitable for operational deployment, provided one develops the necessary hardware. An ideal operational system runs without human interaction and is robust and highly reliable - it should not fail in case of interruption in the transmission and should restart by itself after a power failure.

Such a system has been developed at the University of Bradford, UK (Hyatt *et al.* 1988), using the algorithms developed by TAMSAT as outlined above. It comprises a dish (1.8m to 2.5m) and a high performance, low noise, amplifier linked to a main receiver. This is linked to an IBM PC AT or good quality clone and a colour printer. Its other main characteristic is its low price (at the time of writing - May 1990 - this is about US$ 30,000 for a duplicate system) which puts it within the reach of a multitude of (new) users; simple basic training of an operator is required.

The system processes on line all the available daily TIR images, building the CCD image along the specified period of time (10 days). With the calibration built in, it produces the rainfall estimates, output to a colour printer, at the end of that time. So far the system has been mounted in a number of African countries : Morocco, Algeria (Tamanrasset and Algiers), Tunisia, Sudan (Khartoum) and Botswana (Gaborone); the ones yielding operational output in the form of rainfall estimates are the Tamanrasset and Sudan stations, the others being used mainly to study CCD relations to rainfall and weather patterns (there is little hope of success for this technique in the Mediterranean coastal areas).

4.2.4. Applications and future prospects The adequacy of the accuracy of the rainfall estimates thus produced depends on the end user's requirements and the decisions he will make using the products. As an example, rainfall data for hydrological monitoring have much more stringent frequency and accuracy requirements than those for drought monitoring. The first operational application of the Reading technique was within the ARTEMIS (Africa Real Time Environmental Monitoring using Imaging Satellites) system from FAO (United Nations Food and Agriculture Organization) an operational remote sensing system operating in near real time that supports environmental monitoring using data from both geostationary and polar orbiting satellites, via products like rainfall estimates, number of raindays, vegetation indices (van Ingen Schenau and Venema 1986, van Ingen Schenau *et al.* 1988). The main objective of the system is to support FAO's Early Warning function and in Desert Locust control and survey. The rainfall estimates in the former case allow the identification of areas affected by dry spells where crops could be at risk and, in the latter, the survey of areas with out of season rainfall that may serve as locust breeding grounds. In either case the need is to detect areas with rainfalls above and below about 15 mm. With a properly chosen threshold, identification of areas with 2 or 3 consecutive dry decades is straightforward even with estimates of wide uncertainty. This covers the most obvious application of a reliable rainfall estimation technique in this case enhanced by the coupling with other independent data from other satellites.

Another application with great promise deals with the application of the technique to hydrological (river level) monitoring. The principle is that the more one increases the area over which the CCD is averaged, the more accurate the results will be. So if we look at a large enough area, like a large river catchment basin (10000 km^2 or more), one can expect to obtain acceptable daily rainfall estimates over that area, something which is not possible for smaller areas. When these estimates were applied to a hydrological model on the river Bafing it fared better than when using data from existing raingauge networks (figure 6) (Hardy *et al.* 1989). More detailed work is currently being done in the TAMSAT group with the Blue Nile catchment basin with good prospects for seasonal flood forecasting.

Other research going on at the moment, building up on previous research (Justice *et al.* 1990) is trying to use satellite rainfall estimates to predict end of season biomass in the arid zones of the African Sahel (areas where rainfall is the main determining factor in crop/rangeland production). In

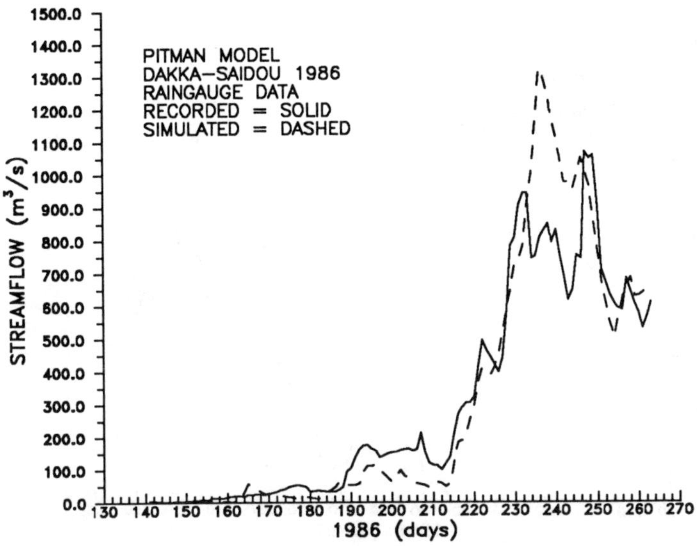

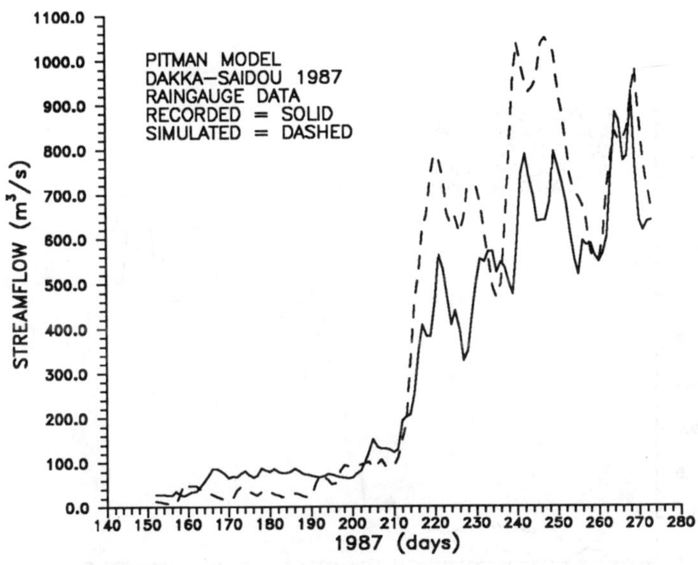

Figure 6 (a) Recorded and modelled streamflow at Dakka - Saidou (Guinea) using raingauge data

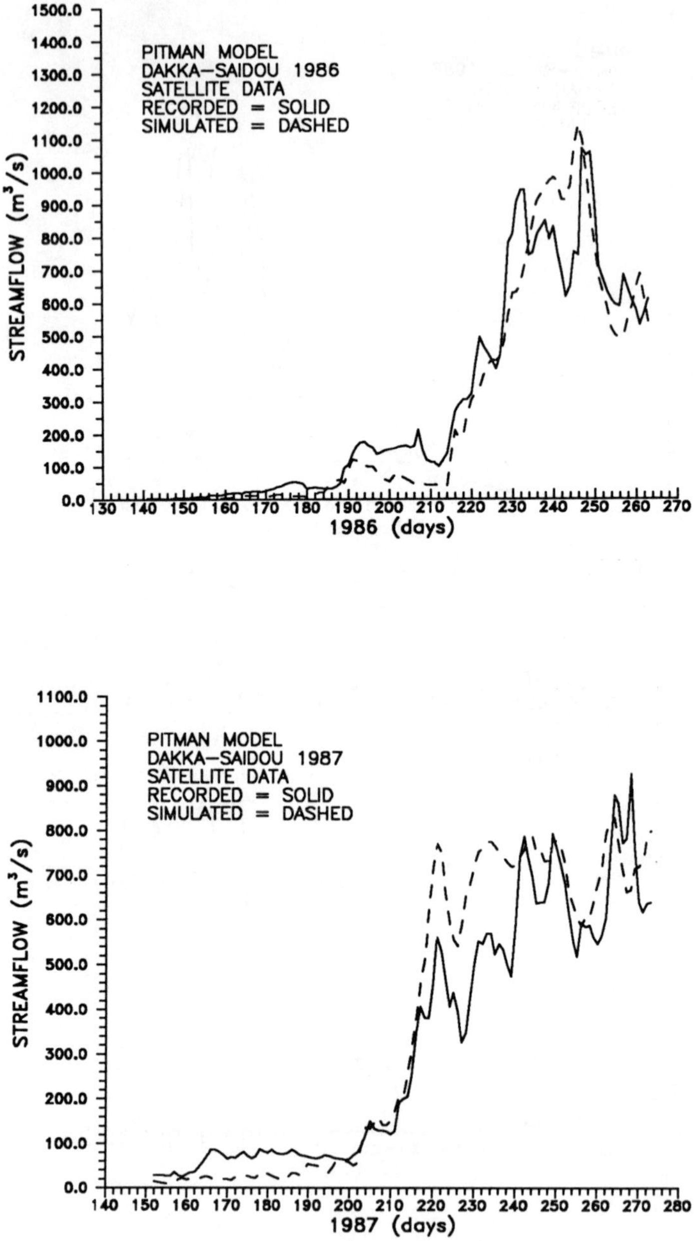

Figure 6 (b) Recorded and modelled streamflow at Dakka - Saidou (Guinea) using satellite rainfall estimates.

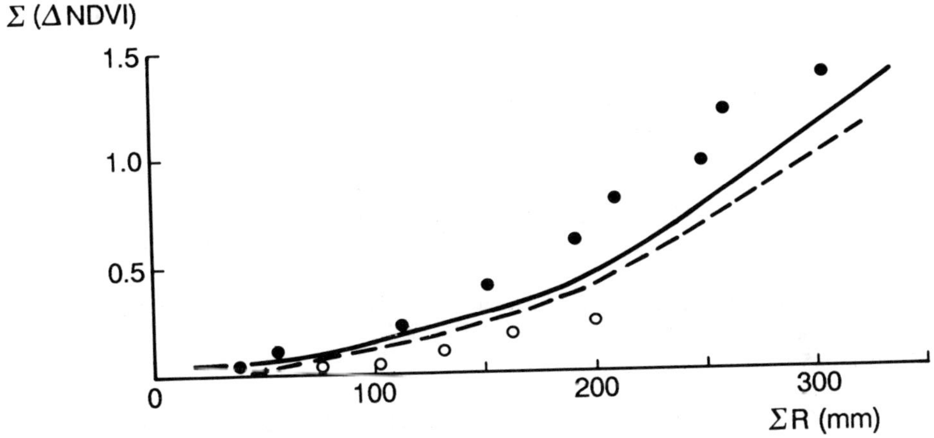

Figure 7 Comparison of accumulated NDVI (above a dry season minimum) and accumulated rainfall estimated from METEOSAT. Data are means for areas approximately 50 km across centered on 15°N :

O 1985, centered on 6°E
● 1986, centered on 6°E
--- 1985, centered on 0°, 3°, 6° W (averaged)
--- 1986, centered on 0°, 3°, 6° W (averaged)

these areas the rainfall might be used as a predictor of the Normalized Difference Vegetation Index (NDVI). This index is related to vegetation productivity, see the chapter by Baret earlier in this book.

The evolution in time and the corresponding integrated NDVI is related to end of season biomass production at scales of 25 km or so. Figure 7 shows the relationship between accumulated NDVI and rainfall.

Good and varied use of the satellite rainfall estimates is therefore possible, with the proviso that one is aware of the limitations, weaknesses and strengths of the method. Also a great deal of work still needs to be done with the objectives of enlarging the area of application of the method and of refining it by incorporating other factors such as size, rate of growth, speed of travel of the cloud, its minimum temperature and sharpness of the front.

On the other hand future satellites will offer new possibilities and increased abilities for rainfall monitoring, but it will be some years before anything reaches operational status. Also of fundamental importance are the continuing advances in computer technology and the ever decreasing hardware prices that allow more powerful research to be done and simpler and cheaper systems to be set up. The need for rainfall monitoring will always be there.

References

Bailey, J.O., E.C. Barrett, C. Kidd, 1986. Satellite passive microwave imagery and its potential for rainfall estimation over land, *Journal British Interplanetary Society*, **39**, 527 - 534.

Barrett, E.C., 1989, Satellite remote sensing of rainfall. In *Applications of remote sensing in agriculture* edited by F. Toselli, (Dordrecht: Kluwer).

Barrett, E.C. and D.W. Martin, 1981. *The Use of Satellite Data in Rainfall Monitoring*, (London: Academic Press).

Barrett, E.C. and G. D'Souza, 1985, METEOSAT dry day, drought and rainfall products for Africa : An introduction. *Proceedings of the 5th METEOSAT Scientific Users Meeting*, Rome, Italy, 10pp.

Barrett, E.C. and G. D'Souza, 1986, *The Development of an objective range of dry day products for drought monitoring over Africa*, Final Report to ESA, contract 6137/84/D.HS(SC).

Barrett, E.C., C. Power, M. Beaumont, T. Richards, 1986, *Cloud cover and rainfall in the Western Sahel*, Final Report to NERC, contract F6/C1/19, Remote Sensing Unit, University of Bristol, England, 85 pp.

Bolton, D., 1984, Generation and propagation of African squall lines. *Quarterly Journal of the Royal Meteorological Society*, **110**, 695 - 721.

Burke, H., K. Hardy, N. Tripp, 1982 : Detection of rainfall rates utilizing space borne microwave radiometers, *Remote Sensing of Environment*, **12**, 169 - 180.

Chadwick, A.F., G. Dugdale, A.F. Traore, J.R. Milford, 1986: Operational rainfall mapping over the Sahel, *Proceedings of the 6th METEOSAT Scientific Users Meeting*, held in Amsterdam, Nov. 1986, (Darmstadt: EUMETSAT).

Flitcroft, I., J.R. Milford, G. Dugdale, 1989, Relating point to area average rainfall in semi arid West Africa and the implications for rainfall estimates derived from satellite data. *Journal of Applied Meteorology*, **28**, 252 - 266.

Griffith, C.G., W.L. Woodley, J. Augustine, 1981 : Satellite rain estimation in the US High Plains. *Journal of Applied Meteorology*, **20**, 53 - 66.

Hardy, S., G. Dugdale, J.R. Milford, J.V. Sutcliffe, 1989, The use of satellite derived rainfall estimates as inputs to flow prediction in the river Senegal. In *New directions for surface water modeling* edited by M. L. Kavvas (Wallingford: IAHS Press) pp 23 - 30.

Her Majesty's Stationery Office, 1981, Handbook of meteorological instruments, Vol 5. The measurements of precipitation and evaporation. HMSO, London.

Hyatt, S., J. Stephenson, J. Williams, A. Zahran, 1988. A PDUS for rainfall estimation over the Sudan. *Proceedings of the 7th METEOSAT Scientific Users Meeting*, Madrid, held in Sept 1988, (Darmstadt: EUMETSAT) pp 413-416.

Jacobs, C.M., 1987. Preliminary report on the applicability of the METEOSAT system in rainfall mapping over Zambia. MARS Project Report, Staring Centrum, Wageningen Agr. University.

Kruger, L., G. Strubing, G. Schulz, 1982. Satellite data as a basis for the estimation of rainfall and runoff, *Advances in Hydrometry*, IAHS Publ. **134**, 273 - 284.

Lovejoy, S. and G. Austin, 1980 : The estimation of rain from satellite borne microwave radiometers. *Quarterly Journal of the Royal Meteorological Society*, **106**, 255 - 276.

McDougall, V., M. Saunby, G. Dugdale, J.R. Milford, 1988 : Relationships between rainfall and cloud top temperature in tropical Africa - Seasonal and regional effects, *Proceedings of the 7th METEOSAT Scientific Users Meeting*, Madrid, Sept. 1988, (Darmstadt: EUMETSAT). pp 297-303.

Milford, J.R., 1990. Potential and limits of rainfall estimates using cold cloud statistics, *Proceedings of the NOAA/USAID International Workshop on the Use of Satellites for Estimating Precipitation*, held in Washington D.C., USA. December 1988 (in press)

Milford, J.R. and G. Dugdale, 1990, Estimation of rainfall using geostationary satellite data, in Applications of Remote Sensing in Agriculture, *Proceedings of the 48th Easter School in Agricultural Science, University of Nottingham, April 1989* (London: Butterworth) (in press)

Moses, J.F. and E.C. Barrett, 1986. Interactive procedures for estimating precipitation from satellite imagery and surface characteristics, *Proceedings of the International Workshop on Hydrologic Applications of Space Technology*, held in Cocoa Beach, Florida, August 1985, IAHS Publ. 60, (Wallingford: IAHS press) pp 18 - 23.

Negri, A.J., R.F. Adler, 1987 : Infra red and visible satellite rain estimation. Part II: A cloud definition approach. *Journal of Climate and Applied Climatology*, **27**, 1565 - 1576.

Scofield, R. and V.J. Oliver, 1977 : A scheme for estimating convective rainfall from satellite imagery. NOAA Tech. Memorandum, NESS 86, Washington D.C., 47 pp.

Snijders, F., 1990, Rainfall monitoring based on METEOSAT data. *International Journal of Remote Sensing*, in press.

Turpeinen, O.M., 1987 : Validation of the ESOC precipitation index. *Proceedings of the 6th METEOSAT Scientific Users Meeting*, Amsterdam, November 1986, (Darmstadt: EUMETSAT)., volume II

van Ingen Schenau, H. and J.C. Venema, 1986 : ARTEMIS : An integrated approach to environmental monitoring using data from METEOSAT and NOAA imaging satellites. *Proceedings of the 6th METEOSAT Scientific Users Meeting*, Amsterdam, November 1986, (Darmstadt: EUMETSAT), volume I.

van Ingen Schenau, H., J.C. Venema, J.U. Hielkema, 1988 : ARTEMIS environmental monitoring of the African Continent : First Results. *Proceedings of the 7th METEOSAT Scientific Users Meeting*, Madrid, Sept 1988, (Darmstadt: EUMETSAT) pp 285-295.

WATERSHED DEGRADATION - USE OF THERMAL DATA AND VEGETATION INDICES AS INDICATORS OF ENVIRONMENTAL CHANGES - HYDROLOGICAL IMPLICATIONS OF CHANGES IN LAND SURFACE COVER

J-M. GREGOIRE
Commission of the European Communities,
Joint Research Centre,
Institute of Remote Sensing Applications,
21020, Ispra (Varese) Italy.

1. The case for watershed monitoring

Among natural resources supporting human needs and economic development, water is obviously of special importance; both in terms of quantity and quality. This is particularly true in the developing countries for at least three reasons: firstly climatic, limited amounts of rainfall in most of the areas between the tropics; secondly anthropogenic, high population growth and migrations; and thirdly economic, lack of financial resources for water management including quality control.

A distinction has to be made between precipitation and water resources. In fact for a given territory, whatever is its extent, precipitation can be considered as potential water resources for human activities and natural needs. Only part of the precipitation (figure 1) will contribute to the water reserves in the environment (in the vegetation cover, the soil, the deep ground water tables and the surface water bodies) and form the actual water resources. Effective support to human needs and economic development will depend largely on the space and time relationships between potential and actual resources.

Such relationships are driven, to a large extent, by the surface characteristics of the environment. Some of these are permanent on the human time scale, such as topography or morphopedology, others may be affected by seasonal or inter-annual changes. In this last category, vegetation cover status is by far the most important factor controlling the transformation between potential and actual water resources.

Any changes affecting these surface characteristics, due to natural reasons or anthropomorphic charge, will have direct consequences on the water resources effectively usable by human activity and economic development. Thus, any water resources management plan must take into account changes in the surface characteristics of the geographical domain receiving the precipitation: the watershed.

A. S. Belward and C. R. Valenzuela (eds.),
Remote Sensing and Geographical Information Systems for Resource Management in Developing Countries, 235–252.

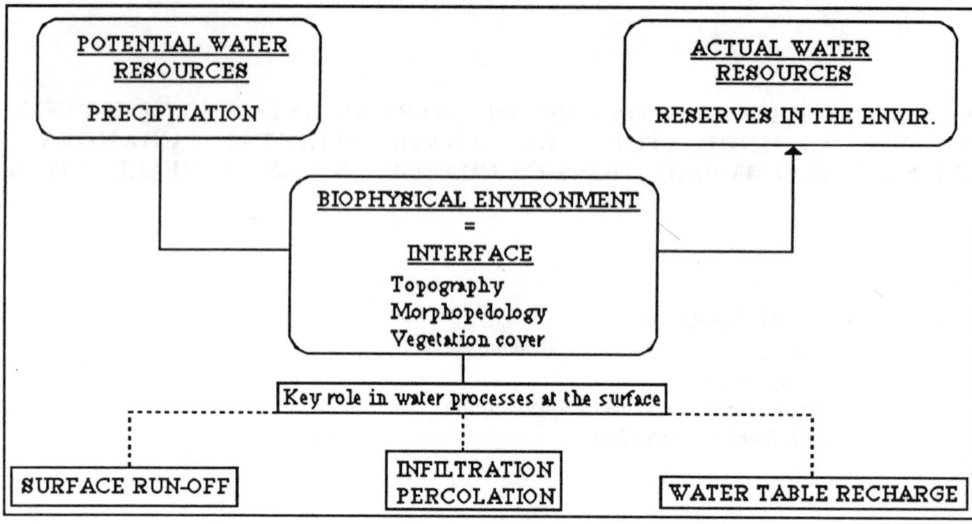

Figure 1. Link between potential and actual water resources

2. Space and time domain of interest for watershed monitoring

Quite often important river basins for water resources cover several countries. This is true for example in South-America (Amazon), Europe (Rhine, Danube), Africa (Niger, Nile), and in South-East Asia (Mekhong). In such a situation, a tool for monitoring change should be able to provide information at regional, if not continental levels.

In the time domain, annual distribution and frequency of rainfall show high variability according to region. Figure 2 (Balek 1983) illustrates this point. The equatorial rainfall, figure 2a, shows two annual peaks, no pronounced dry season and a high regularity. Monsoonal rainfall, figure 2b, among the heaviest on the earth, is concentrated within 4 to 5 months in summer and shows a very high variability year by year. The rainfall regime in semi-arid regions, such as the subtropical parts of Africa, are characterized by highly irregular rainfall distribution with a marked seasonality, figure 2c.

Surface dynamics on the river basins will, to a certain extent, follow the temporal pattern of the rainfall. Thus, a monitoring tool should be able to provide information on a time base varying from one to 10 days.

Both spatial (regional view) and temporal constraints (high frequency) make remote sensing the only tool for quantitative monitoring of surface dynamics. Among the space platforms actually available, meteorological satellites such as METEOSAT and NOAA certainly represents the most effective tool (Table 1).

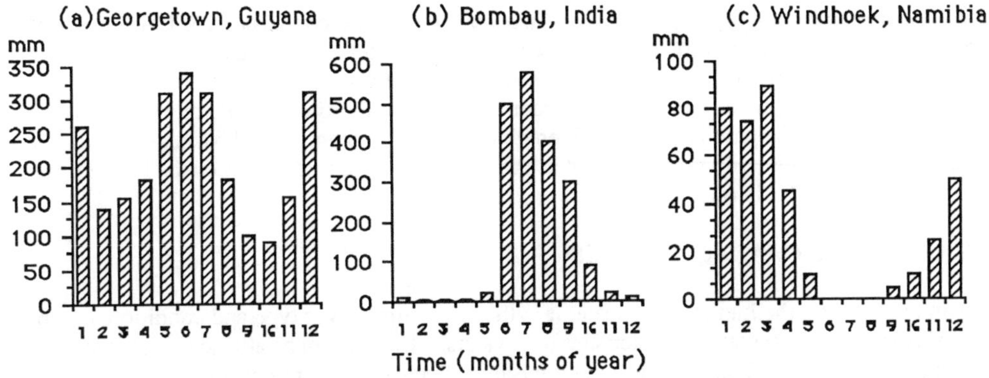

Figure 2. Monthly rainfall distribution typical of tropical forest regions (a), monsoonal regions (b), and dry tropical areas (c). (adapted from Balek, 1983)

3. Requested information for large watershed monitoring

As already mentioned, the information required concerns the surface characteristics which play a major role in the relationship between potential and actual water resources. This means that, within the "soil-vegetation-atmosphere" system, the "vegetation cover" and "morphopedological substratum" sub-systems are the two domains of interest for which biophysical parameters have to be defined and quantified.

3.1. PERMANENT CHARACTERISTICS

On the human time scale, most of the morphopedological characteristics of a watershed can be considered as constant, in terms of impact on the water processes at the surface. In fact, in the absence of vegetation cover, processes such as infiltration, surface run-off, flood flow and percolation directly depend on (a) the intensity of the precipitation, (b) the topography, (c) the size of the watershed, (d) the type of morphopedological substratum, for example variations in permeability and thickness are of particular importance, (e) the drainage pattern. These physical characteristics are usually extracted from existing topographical and soil maps. Obviously, part of these documents can be derived or updated using remote sensing data.

Table 1. Main characteristics of Landsat, SPOT, NOAA and METEOSAT platforms. (see Hill and Belward elsewhere in this book for more complete descriptions)

Satellite	Spectral domain	Visiting period	Resolution	Swath
Landsat-TM	visible + thermal	18 days	30 + 120 m	180 km
SPOT	visible	2 to 26 days	20 + 10 m	60 km
NOAA	visible + thermal	24 hours	1 + 4 km	2700 km
METEOSAT	visible + thermal	geostationnary	25 km	full disk

3.1.1.Parameterization of morphometric characters. Following conclusions of Ebisemiju (1979) and Kohl (1986), morphometric parameters can be reduced to four key factors in order to describe the main characteristics of a watershed. These are, firstly, the mean length of all stream channels (km), secondly, the mean length of 1st order channels in the basin (km), thirdly,the drainage density, defined as the ratio between total length of first order channels and basin area (km^{-1}), and finally, the maximum basin relief, defined as the difference in altitude between the highest and the lowest point of the basin (m).

3.2. CHANGING CHARACTERISTICS

Again considering the human time scale, it is the vegetation cover type and condition which are probably the most important characteristics of a watershed which can be affected by changes.

One has to consider both seasonal and inter-annual changes; their impact on water processes at the surface are quite different. In the first category, seasonal changes, the modifications affecting the vegetation cover are usually the result of an adaptation process to natural conditions of the watershed, local and regional climate, rainfall pattern, soils distribution, for example. Such an adjustment optimizes the relationship between potential and actual water resources for given environmental conditions on a watershed.

In the second category, inter-annual changes, the modifications affecting the vegetation cover are no longer the result of a more or less stable equilibrium, but the consequence of forcing phenomena such as changes in regional climate or land-use practices. In this case, the origin of the changes can be considered as being external to the system and thus the equilibrium will be broken on the watershed for a certain period of time, until a new equilibrium is found in the relationship between potential and actual water resources. In this dynamic situation human activity plays a major role.

What are the vegetation cover characteristics which have to be taken into account for the understanding of water processes at the surface ?

Vegetation has a direct impact on interception of precipitation, evapotranspiration and indirectly on infiltration rate, percolation and surface run-off (Figure 3). Vegetation cover acts upon these processes mainly via its structure and physiology (Eagleson 1978). In terms of structure, three characteristics are of special interest because all three are accessible to a certain extent by remote sensing, (1) the overall morphology of the vegetation cover, (2) the density of coverage, (3) the biomass.

Surface albedo and aerodynamic properties are affected by these factors and thus the available energy usable for evapotranspiration processes in the system.

In terms of physiology, the main process is the stomata control which allows vegetation to regulate its transpiration rate according to water availability in the soil (root zone) and atmospheric demand, up to a certain limit which is species-dependent.

3.2.1.Parameterization of changing characteristics. Remote sensing radiometric data can be used to derive at least four physical parameters linked to changing characteristics of watersheds surfaces. Firstly the surface temperature (Ts), this brings information on the water status at the surface and can be used to evaluate evapotranspiration rates. Secondly the radiance in middle-infrared wavelengths which brings information on the overall state on the vegetation cover, soil surface humidity and presence of "hot spots" in the environment (for instance bush fires). Thirdly and fourthly, the Normalized Difference Vegetation Index (NDVI) and albedo, both derived from

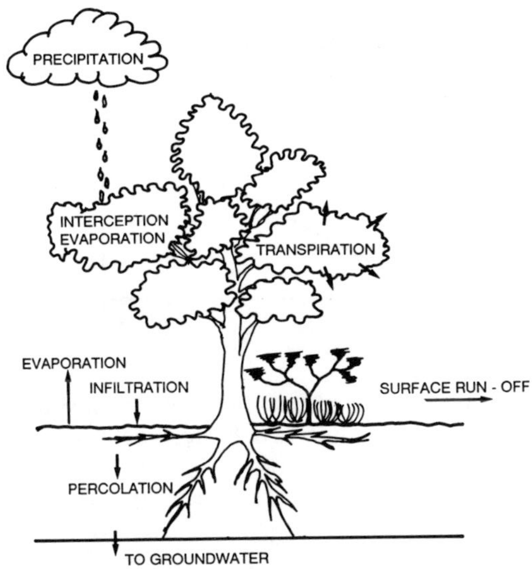

Figure 3. Water processes at the surface, in presence of vegetation (adapted from Verstraete and Dickinson 1986).

visible and near-infrared data; these are linked to green vegetation density, vigor and biomass. Bird and Baret, in earlier chapters, discuss the above in more detail.

3.3. BUSH FIRE INTENSITY

In certain cases, monitoring of changes in surface characteristics of the watershed can go beyond simple diagnostics, as discussed up to now, and can tackle the processes at the origin of the changes. Not all processes can be approached with a remote sensing. Moreover, bush fire dynamics, seen as a major agent of change in the tropical vegetation canopy, can be efficiently monitored by meteorological satellite data. The use of these data for detection and monitoring of active fires (Matson *et al.* 1984, Stephens and Matson 1987, Grégoire *et al.* 1988, Langaas and Muirhead 1988, Citeau *et al.* 1988, Belward and Lambin 1990) and evaluation of burned areas (Frederiksen *et al.* 1989) has been amply demonstrated, and this in various regions of the inter-tropical belt (Sadowski and Westover 1986, Malingreau and Tucker 1988).

3.3.1. Parameterization of bush fire intensity. The basic information of interest for bush fire intensity covers three aspects. First the total number of fire events observed on a given watershed for a certain period of time (usually the dry season in the tropical areas). Second, the percentage of these fire events which have occurred at the end of the dry season, (the consequences on the environment of the late fires is much more important than the ones due to early fire events). Third the total area, within the watershed, affected by burning during a given period of time (here again during the dry season).

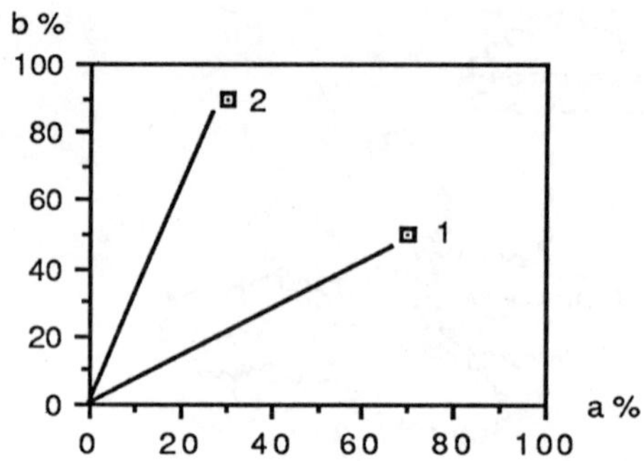

Figure 4. Parameterization of bush fire intensity for two watersheds
(1 & 2 are the watersheds, a & b are as defined below)

These aspects can be combined in a single index of fire intensity (FI)

$$FI = \sqrt{(a^2 + b^2)} \qquad (1)$$

where "a" is the ratio between "total number of fire events" (pixels with a saturated DN value in AVHRR channel 3) for a given dry season and the number of pixels in the watershed, "b" is the percentage of fire events observed in the second half of the dry season.

Figure 4 shows how such an evaluation of bush fire intensity can be represented graphically. For a given period of time or dry season, watershed 1 has been affected by more fire events than watershed 2, in proportion to their respective areas. But the latter shows a higher percentage of fires at the end of the dry season, the period during which fire impact on the vegetation cover is maximum. The final result is an equal "bush fire intensity" on the two watersheds, but as we can see from the graphs the spatial and temporal fire dynamics are quite different. The index FI allows us to balance the number of fire events by its temporal pattern.

4. Data management

Remote sensing based methodology for watershed monitoring relies on two main sources of information, maps which give access to permanent attributes of the watersheds, and satellite images which allow changes affecting some of the surface characteristics, mainly vegetation cover, to be quantified. The overall information available for analysis and interpretation can be divided into three data bases.

4.1. THE GRAPHIC DATA BASE

This includes polygon files used as overlays to define watersheds boundaries on the remote sensing images. The geometric characteristics of the polygons are linked to the type of imagery used.

A single polygon file may represent a single watershed or aggregation of many watersheds depending on scale of observation.

4.2. THE SATELLITE IMAGERY DATA BASE

Monitoring changes in land surface characteristics relies on time series of images. Here again, the approach is usually multi-level. For a given set of watersheds, several image types can be assembled in a time series. The choice of the images is based on spectral domain of interest, frequency of acquisition and spatial resolution. Pre-processing of the satellite data involves three operations.

Firstly geometric correction of the documents; all the images have to be geometrically corrected to a master which can be either a cartographic document (topographical map for instance) or the first image of the time series.

Secondly, geometric registration; most of the time geometrically corrected images are not registered, translations in x (longitude) and y (latitude) directions have to be performed in order to reach a perfect registration. At this stage the master is usually the first image of the series which is used as reference.

Thirdly the data require radiometric corrections; being spread over several months, or even years, the images will have a radiometric content affected by two types of distortions: one which is linked to changes in sensors characteristics. For instance the AVHRR sensor on board NOAA-11 does not have the same pre-launch calibration characteristics as that on board NOAA-9. A time series built on images of these two platforms will have to take into account such modifications. The second one is linked, for a given sensor, to calibration deterioration during time after the launch. Kaufman and Holben (1990) have quantified this process for the AVHRR sensor on board NOAA-7 and 9 satellites, and suggested some correction procedures. Hill and Wilkinson, in earlier chapters, review the above preprocessing operations in more detail.

4.3. THE BIOPHYSICAL PARAMETERS DATA BASE

Merging data base I (watershed boundaries) and II (time series of satellite imagery) one can derive the information requested to quantify changes in the surface characteristics for a given set of watersheds, shown in figure 5. The numerical data include Surface temperature, Ts (°C), Normalised Difference Vegetation Index, NDVI (dimensionless), Digital count in AVHRR channel 3, DN C3 (dimensionless) and the Fire intensity index, FI (dimensionless).

5. Application to some West-African river basins

Among the large river basins of West Africa, five watersheds have been selected in order to illustrate the approach introduced in the previous sections. They are located along a north-south transect crossing the three main ecological zones of West Africa, south of the Sahara: the sahelian, sudanese and guinean domains (see figure 6).

The Faleme is one of the main tributaries of the Senegal river which, flowing from the northern edge of the Fouta Djallon in Guinea, crosses the contact zone between the sudanese and sahelian domains. Most of the basin, about 75 %, is located in the latter. According to the UNESCO map of vegetation of Africa (White 1986), the Faleme watershed is in the "undifferentiated sudanian woodland" area.

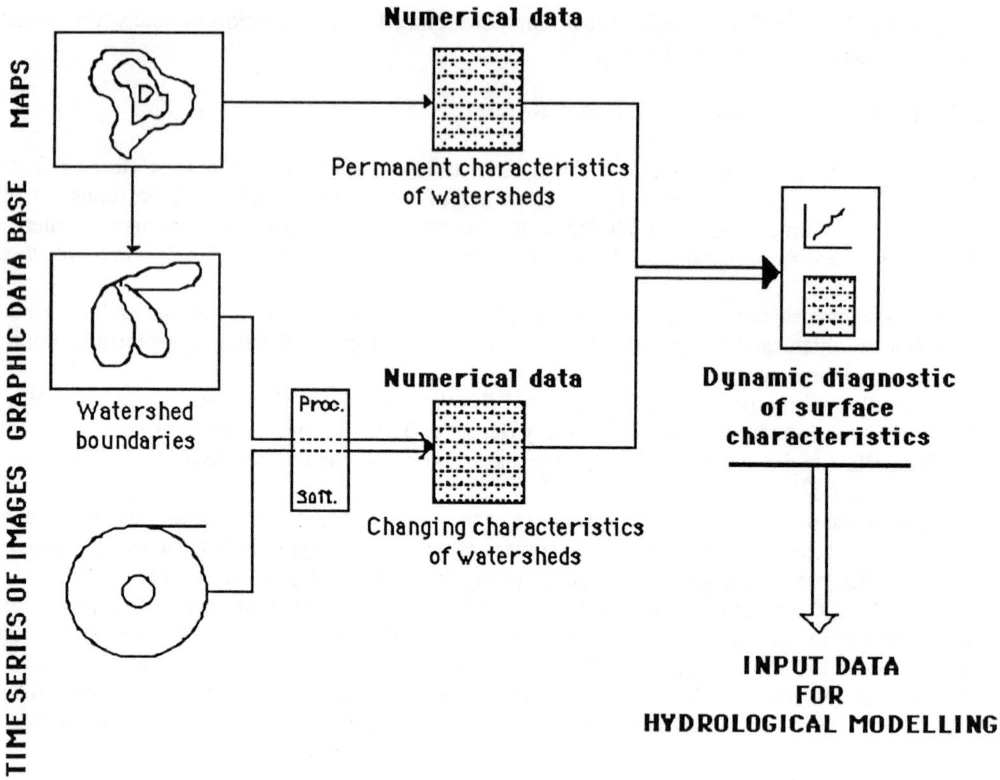

Figure 5. Flowchart showing how the data bases link to provide data for hydrological modelling.

The Niandan is one of the five tributaries of the Niger river in Guinea; it is located entirely in the sudanese domain. The climax vegetation cover is known as "mosaic of lowland rain forest and secondary grassland".

The Feredougouba and Bogho watersheds are located on the border between Guinea and Ivory Coast; tributaries of the Sassandra river, they are representative of the contact zone between the sudanese and guinean domains. The climax vegetation cover is the same than the one defined for the Niandan basin.

The Nzo watershed, also a tributary of the Sassandra river, is situated in the guinean domain of Ivory Coast; the climax vegetation cover is defined as "mosaic of guineo lowland rain forest and guineo rain forest".

5.1. DATA PROCESSING

Following the procedure described in § 4. and illustrated in figure 5, changes in surface characteristics have been monitored during the 1987/88 and the beginning of the 1988/89 dry seasons. Obviously during this time of the year the cloud cover limitations are minimum. Though

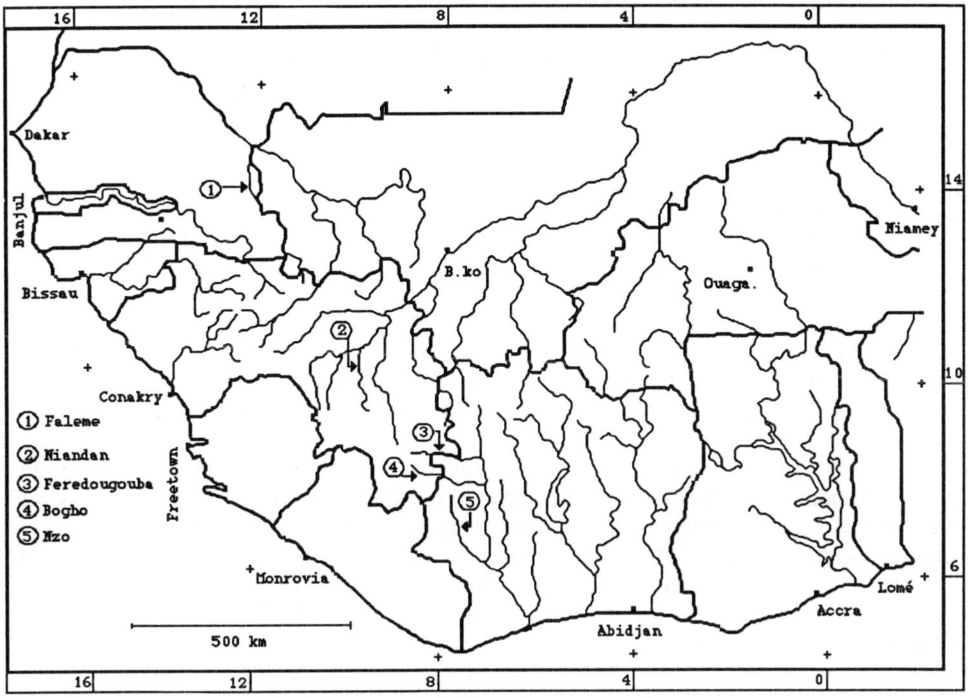

Figure 6. Location of the five watersheds used in the study

low cloud cover is an undoubted advantage, the principal reasons for the choice of this period are methodological; in fact, the effect of the dry season on the environment enhances the variability at the surface of the watersheds, makes the differentiation between ecological conditions and the detection of fragile ecosystems easier, and moreover, bush fire activity is minimal during the rainy season.

5.1.1. Graphic data base . Watersheds boundaries were defined on topographical maps at 1 to 1000.000 scale (Carte Internationale du Monde; sheets: Abidjan-NB 29/30; Bobo-Dioulasso NC-29/30; Conakry NC-28/29; Bamako ND-28/29) and then vector digitized on a digitizer tablet. Each graphic file was then geometrically corrected, and registered to the first HRPT image of the time series, using manually defined ground control points. The final result was 5 polygon files usable as graphic overlays for data extraction. These are shown below in table 2. For each river basin, the corresponding polygon covers ≈ 95% of the total area of the watershed. A maximum data "loss" of 5% occurs in the processing chain. Assuming an HRPT pixel of 1 km^2, the study areas range from 6500 to 29000 km^2.

Table 2. Polygon files corresponding to the watersheds

Polygon file name	Number of HRPT pixels in the polygon
Faleme	29345
Niandan	13012
Feredougouba	6585
Bogho	8319
Nzo	8095

5.1.2. Time series of images. 12 HRPT orbits were selected between November 1987 and January 1989. After geometric correction, based on satellite navigation parameters (Saradeth and Falk 1989), three 512 x 512 pixels sub-windows have been defined and the corresponding 12 scenes registered.

Finally, digital numbers were converted into radiance, except for AVHRR channel 3, and biophysical parameters using a software package developed by Vogt (1989). The output files are 5 channel images, where channel 1 is surface temperature in °C (derived from brightness temperatures in AVHRR channels 4 and 5 via a split window algorithm), channel 2 is the NDVI (derived from AVHRR channels 1 and 2), channel 3 is the band averaged spectral albedo of AVHRR channel 1, channel 4 is the band averaged spectral albedo of AVHRR channel, channel 5 is the original AVHRR channel 3.

5.1.3.Biophysical parameters data base. As mentioned in § 4.3. surface characteristics of the five watersheds were defined, for the 12 selected HRPT passages, on the basis of the following parameters; the mean value of NDVI, surface temperature and digital count in AVHRR band 3, the standard deviation in the same three derived channels, and the fire intensity index FI

5.2. ANALYSIS AND INTERPRETATION OF THE RESULTS

Temporal dynamics of the biophysical parameters, derived from the HRPT images, have been interpreted along three guide lines; modalities of general warming of the environment on the watersheds and dynamics of the vegetation cover; temporal pattern of bush fire activity; differentiation between river basins.

5.2.1. General warming and dynamics of the vegetation cover. The first effect of the dry season on the environment is a progressive warming due to several processes. For example, the decrease of the green vegetation density and of the evapotranspiration process, increase of bare soils surfaces and air temperature near the soil surface, decrease of soil surface humidity.

Such an evolution can be characterized using the temporal behaviour of the AVHRR channel 3 digital counts. Malingreau and Tucker (1988) have shown the potential of this channel for forest clearing detection in the Amazon basin. In fact, its spectral characteristics (3.55-3.93 μm) makes it sensitive to radiation emitted by the surface (linked to the radiative temperature) and to a lesser extent to reflected radiation. This double sensitivity make channel 3 well suited for characterizing the effect of the dry season on the vegetation canopy. Figure 7 shows the distribution histograms in AVHRR channel 3, surface temperature and NDVI on the five river basins, at three dates in the

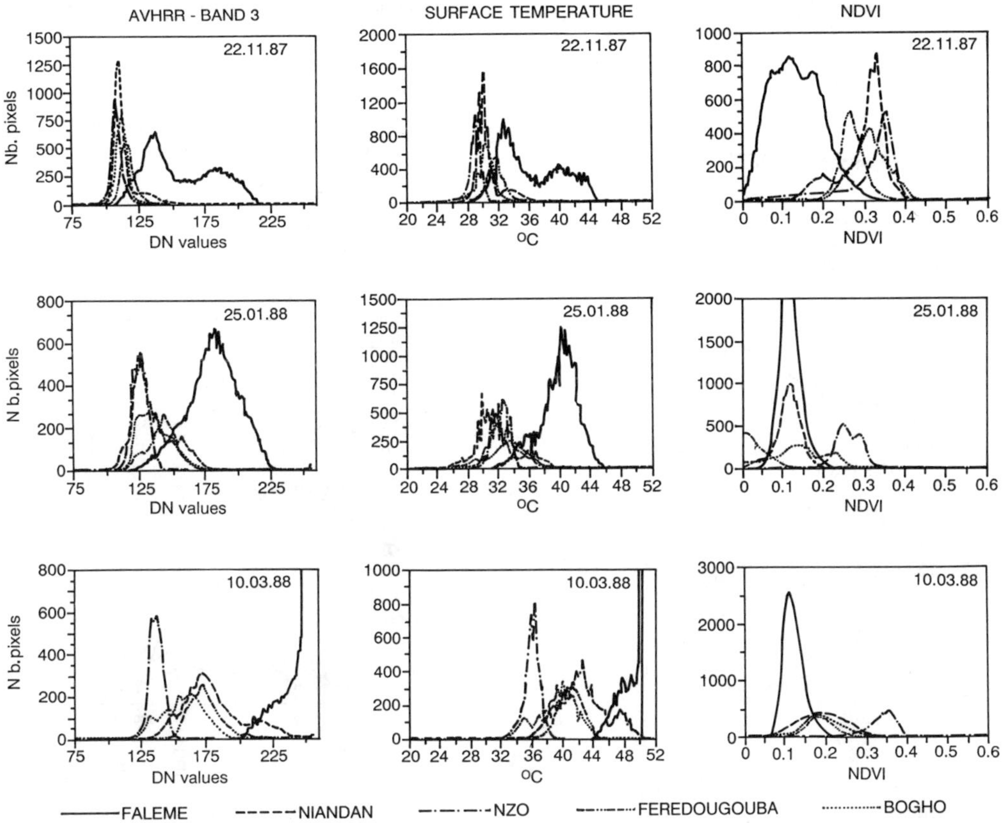

Figure 7. Histograms showing distribution of AVHRR channel 3, surface temperature and NDVI values for five West African river basins at three different dates.

1987/1988 dry season. These are the beginning (22.11.87), middle (25.01.88) and last third (10.03.88) of the dry season. Interpretation of these data leads to the following observations.

AVHRR channel 3 digital counts and surface temperature data clearly show the progressive warming of the surface on the basins; a trend which is coupled to a decrease in NDVI values.

Looking at the temporal evolution of mean values for thermal and visible data, shown in figure 8, three features appear. First, the expected general trend, increase in the thermal and decrease in NDVI, is apparent in both channel 3 and surface temperature data, but in the sahelian domain (Faleme) and soudanese (Niandan) the channel 3 values seem to be more closely correlated to NDVI evolution than surface temperatures. In the guinean domain (Nzo) and in the contact zone between guinean and soudanese domains (Feredougouba and Bogho), there are no significant differences between channel 3 and surface temperatures temporal dynamics.

Second, in the Faleme watershed the channel 3 data for the march image are saturated on most of the surface of the basin; this is certainly not due to fires but more probably to very high air and

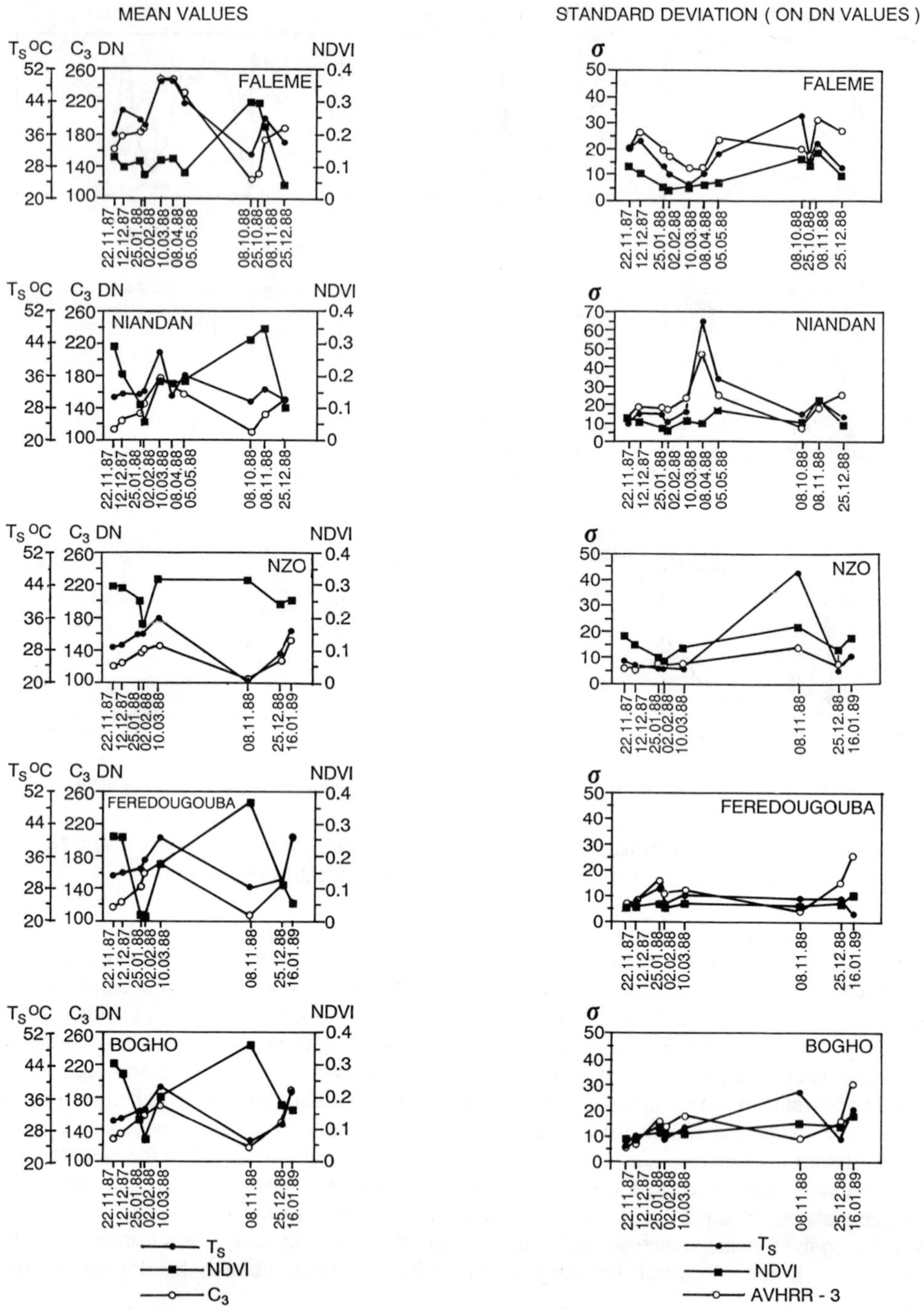

Figure 8. Temporal evolution of mean values and standard deviations of AVHRR channel 3, surface temperature and NDVI

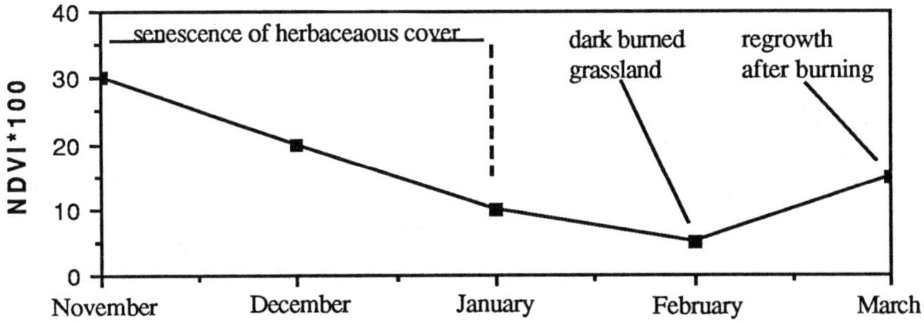

Figure 9. Possible effects of burned areas and post-fire regrowth on NDVI mean values

soil temperatures. This indicates that thermal data derived from AVHRR imagery are of very limited use in the sahelian domain after February.

Third, for the 1987/88 dry season (November 87 to April 88), thermal data show a continuing increase while NDVI values show a decrease from November 87 to February 88 and then a significant increase (≈ 0.1 NDVI unit), at least in the soudanese and guinean domains (Niandan, Feredougouba, Bogho and Nzo). This phenomenon might have a double origin. 1) a decrease in incoming solar radiation due to atmospheric effects, as mentioned by Bartholomé (1989). 2) a re-growth of herbaceous vegetation after extensive bush fire events; such a phenomenon, well known in most of the soudanese zone of West Africa, can cause a significant increase in NDVI values right after the intense fire activity observed in January/February. If this hypothesis is verified, the temporal evolution of NDVI values observed from November 87 to March 88 could be interpreted as shown in figure 9. From November to January, the progressive senescence of the herbaceous vegetation cover causes a decrease in NDVI mean value over a given watershed. Then, due to intense bush fire activity, the percentage of dark burned areas becomes high enough to force the NDVI mean value to a minimum. Finally, post-fire re-growth produces an increase of the index.

Spectral heterogeneity, as derived from the shape of the histograms, also reveals some interesting features. For instance the Faleme basin, situated in the transition zone between the soudanese and sahelian domains, shows a bimodal histogram at the beginning of the dry season in both the AVHRR channel 3 data and surface temperature, unimodal distribution in January and saturated values at the end of the dry season. Although we are dealing here with statistical distributions, and not spatial ones, the evolution observed corresponds to an "homogenization" process on the ground. The spectral characteristics of the sudanese part of the watershed, visible on the November data, are no longer detectable in January. The same trend can be observed in the NDVI data.

As already mentioned by several authors (Seguin *et al.* 1989, Becker 1989, Yokoyama 1989 for example), surface temperatures and/or channel 3 data appear to be complementary to vegetation indices and NOAA-AVHRR imagery can be considered an optimum tool as it combines the two types of information usable on regional scales.

Such data could form a sound basis for the detection of fundamental inter-annual changes affecting a given river basin. A decrease of spectral heterogeneity in time, from year to year, could be an indicator of such changes; mean values data could then provide indications on the sense of the changes.

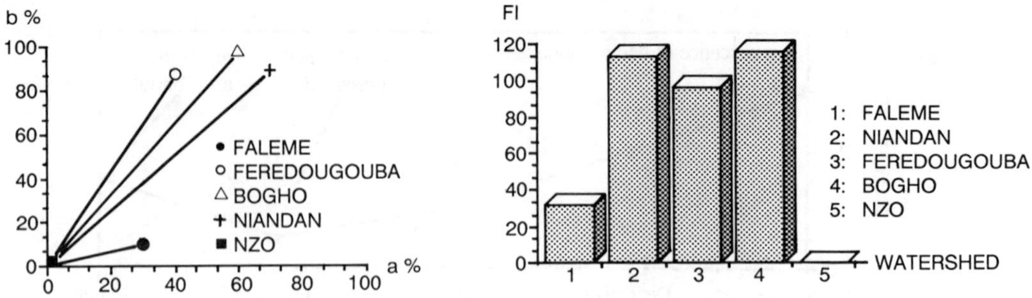

Figure 10. Comparison of Fire Index (FI) values for the different watersheds.
FI, a and b are as defined in § 3.3.1, equation (1)

5.2.2. Bush fire intensity. In terms of fire events observed on the five NOAA-AVHRR orbits from November 87 to March 88, the classification of the peak fire intensity for the selected river basins is as follows: Faleme (90 fire pixels), Niandan (86), Bogho (47), Feredougouba (23) and Nzo (2). Note that on the Faleme watershed the channel 3 data for the March image are saturated over most of the surface of the basin; this is certainly not due to fires but more probably to very high air and soil temperatures; the March data have thus not been counted for this basin, but were used for the other four.

Such fire activity corresponds, more or less, to what one would expect according to the ecological domains under consideration.

If, instead of considering the overall fire activity, one takes into account the spatial dimension and the temporal pattern of the phenomena then the results are slightly different. To do so, the Fire Index, defined in § 3.3, has been calculated for the five river basins for the period November 1987 to March 1988. Figure 10 shows the results.

The highest "Fire Index" is observed on the river basins located in the so called transition or contact zones, that is, the Feredougouba and Bogho watersheds situated at the border between the guinean and soudanese domains, and the Niandan watershed, which is situated completely in the soudanese domain, where the vegetation cover is a mosaic of clear forest or woodland and shrub savanna. The river basins situated in "extreme" environments show a much lower Fire Index, this is the case for the Faleme watershed in the sahelian domain, and the Nzo watershed in the guinean domain, the latter even showing a fire index of zero.

This is the result of a combined effect between spatial dimension of the fire activity and temporal pattern; weighted by the total area of the basin, the absolute number of fire events has obviously not the same meaning; the temporal effect is well illustrated by the Fire Indices of Niandan and Bogho watersheds. The latter, with an absolute number of fire events being half that observed on the Niandan, ends with more or less the same Fire Index value because most of its fire events ($\approx$98 %) have occurred in the second half of the dry season.

Obviously, these results are completely dependent upon the initial hypothesis used to define the Fire Index. In terms of impact on the vegetation cover, and subsequently on the water processes at the surface, the late fires occurring at the end of the dry season have a much higher effect than the one observed at the beginning. Such an hypothesis could be incorrect or mis-applied. The main problem is certainly the definition of what is a late fire; at which stage in the dry season, one considers that a fire is such for a given ecological environment ? A possible answer could be given by the temporal pattern of the NDVI mean value on the watersheds. On the basis of figure 9 data, an NDVI value of 0.15 could be a good threshold for the soudanese domain and the guineo-soudanese contact zone.

In terms of effect on the environment in a general sense, the transfer of particles and gas into the low atmosphere might be higher for early than late fires, though again this is open to question. Moreover, due to differences in fuel loading, the same number of fire events will not produce the same atmospheric transfers in a shrub savanna and in an open forest environment.

All this shows that the "Fire Index" proposed here is far from being the optimum tool to quantify the impact of bush fires. More research work is needed to parameterize the phenomena in terms of 1) remote sensing derived information 2) possible input for modelling of hydro-climatological dynamics. In this line, the Institute of Remote Sensing Applications of the J.R.C. is planning, in collaboration with African partners, an intensive field campaign in southern Guinea for January 1991.

5.2.3. Differentiation between watersheds. Turning back to our general objective, the hydrological implications of environmental changes on large river basins, one can build scenarios based on the parameters derived from remote sensing data.

In a seasonal time frame, the watersheds can be classified according to their specific warming dynamics, senescence of the vegetation (mainly herbaceous cover) and bush fire impacts at the end of the dry season. Such diagnostics may then be used as input data in the interception/runoff modules of hydrological models for the simulation of water processes at the surface during the rainy season.

In an inter-annual time frame, changes in surface characteristics both in terms of quantitative modifications (for example density of vegetation cover and fire intensity) and qualitative ones (spectral heterogeneity in the thermal and visible domains, or temporal patterns for example) can be used to detect fundamental changes affecting the watersheds. For instance, the burning calendar of a given basin is highly correlated to its ecological domain (Grégoire, 1990); any modification of this calendar, from one year to an other, is certainly a good indicator of deep changes in the watershed general conditions.

Finally, evolution analysis might provide useful indications for management and protection actions. Priority river basins can be selected within a regional management plan, such as the one promoted by the Commission of the EEC in Guinea (AGROTECHNIK 1989).

6. Perspectives in monitoring of environmental conditions of large river basins

The methodological aspects proposed here, and illustrated for some river basins of West Africa, have shown the potential of low resolution satellite data for monitoring the environmental conditions of large watersheds. They have also shown that efforts have to be made in at least two domains.

First, the building of databases on surface parameters. Such databases should have three characteristics: 1) although based on satellite derived information, they must contain data compatible with modelling in the hydro-climatological field. This means that remote sensing specialists and hydro-climatologists should join in common research projects in order to fill the gap between satellite data and required information for modelling (Verstraete and Dickinson 1986): 2) they must give information significant at a regional level. In fact, it is generally agreed that modelling of water processes at the surface is well controlled for very small watersheds (several tenth square kilometers) (Girard 1975) and for very large ones (several millions square kilometers); for the intermediate levels of perception (several thousand square kilometers), the approach is far from clear mainly because of a lack of input data at a proper scale (Solomon *et al.* 1988, Rango 1989); 3) their information content must be presented in a multilevel structure, the final layer being referred to watersheds, intended as landscape units, and not to satellite imagery as is usually the case; orbit numbers are not of interest for hydrological modelling ! Brivio *et al.* (1990) will make a proposal in this sense to the International Symposium on Remote Sensing and Water Resources to be held in The Netherlands, August 1990.

Second, the understanding of the complex interactions between land surface characteristics, meso scale meteorology and water processes at the surface, once again at a regional scale. If (!) suitable data bases are available, research activities could lead to a better comprehension of these processes and in turn allow the data bases to be modified in the right way. In terms of remote sensing this means a better definition of the temporal/spatial/spectral characteristics of future sensors such as MERIS (Anonymous, 1988) or MODIS (Townshend and Justice 1988).

Acknowledgements

I wish to thank my colleague L. Chiesa for his help in processing the data.

References

Anonymous, 1988, Feasibility study of a Medium Resolution Imaging Spectrometer (MERIS). Report of the ESTEC Contract no. 7366/87/NL/JS, Sira Ltd & Aerospatiale.

AGROTECHNIK, 1989, Comparaison de l'état du couvert végétal en 1973 et 1986 sur le Haut-Bassin du Niger - Zone des Hauts-Bassins Versants du Haut-Bakoy et du Tinkisso. Rapport final Projet No. 5604.30.94.381, République de Guinée et FED.

Balek, J., 1983, *Hydrology and water resources in tropical regions* (Amsterdam-Oxford-New York: ELSEVIER).

Bartholomé, E., 1989, Contribution aux méthodes de suivi par télédétection satellitaire des rendements régionaux des cultures céréalières pluviales en Afrique de l'Ouest, Thèse de Doctorat à l'Université Catholique de Louvain, Belgique.

Becker, F., 1989, Remote sensing of vegetation using thermal infrared bands. In *Ispra-Courses "Remote Sensing for Vegetation Monitoring"*, Joint Research Center, Ispra Establishment, Italy.

Belward, A.S., and Lambin, E., 1990, Limitations to the identification of spatial structures from AVHRR dat, *International Journal of Remote Sensing*, **11**, 921 - 927.

Brivio, P.A., Grégoire, J-M., and Zilioli, E., Use of low resolution satellite data for watershed modelling: Proposal of a methodology for West African conditions. accepted as a Paper for the *International Symposium on Remote Sensing and Water Resources, 20th-24th August 1990, Enschede, The Netherlands.*

Citeau, J., Demarcq, H., Mahé, G. et Franc, J., 1988, Une nouvelle station est née... *Veille Climatique Satellitaire.*, **25**, 23-29.

Eagleson, P.S., 1978, Climate, soil and vegetation - IV.The expected value of annual evapotranspiration,*Water Resources Research,* **14**, 5, 731-739.

Ebisemiju, F.S., 1979, An objective criterion for the selection of representative basins, *Water Resources Research,* **15**,1, 148-157.

Frederiksen, P., Langaas, S., and Mbaye, M., NOOA AVHRR and GIS-based monitoring of fire activity in Senegal - A provisional methodology and potential applications, *Proceedings of Third International Symposium on Fire Ecology; Freiburg University;16th-20th May 1989.*

Girard, G., 1975, Les modèles hydrologiques pour l'évaluation de la lame écoulée en zone sahélienne et leurs contraintes, *Cahier ORSTOM, série Hydrologie,* XII, **3**, 189-221.

Grégoire, J-M., Kohl, H-G., Space technology and prediction of the availability of flood water in the Niger river, *Proceedings of the Symposium "Hydrologic Applications of Space Technology "* IAHS Publication , **160**, 287-293.

Grégoire, J-M., Flasse, S., et Malingreau, J-P., 1988, Evaluation de l'action des feux de brousse, de novembre 1987 à février 1988, dans la région frontalière Guinée-Sierra Leone. Exploitation des images NOAA-AVHRR. Rapport CCR S.P.I. 88.39, Centre Commun de Recherche - Etablissement d'Ispra, Italie.

Grégoire, J-M., 1990, Effect of the dry season on the vegetation canopy of some river basins of West Africa as deduced from NOAA-AVHRR data, *Hydrological Sciences Journal,* **35**, 323 - 338.

Kaufman, T.J. and Holben, B.N., 1990, Calibration of the AVHRR visible and near-IR bands by atmospheric scattering, ocean glint and desert reflection,*Journal of Climatology and Applied Meteorology,* in press.

Kohl, H.G., Integrating Landsat MSS satellite and conventional geomorphologic data to define a hydro-ecological zoning of the Upper Niger basin with special reference to the spatial and temporal dynamics of the retention capabilities, *Proceedings of the XXth Symposium on Remote Sensing of Environment, Nairobi, Kenya ,4th-10th December 1986.* pp 699-703.

Langaas, S., and Muirhead, K., Monitoring bushfires in West Africa by weather satellites, *Proceedings of the 22nd International Symposium on Remote Sensing of Environment, Abidjan, Côte d'Ivoire, 20th-26th October 1988.* pp. 253-266.

Malingreau, J.P., and Tucker, C.J.,1988, Large-scale deforestation in the southeastern Amazon Basin of Brazil, *AMBIO,* **17**, 1, p. 49.

Matson, M., Schneider, S.R., Aldrige, B., and Satchell, B.,1984. Fire detection using the NOAA series satellites. Technical Report NESDIS 7, USDC-NOAA, NESDIS, Washington.

Rango, A., 1989, Remote Sensing and Large-Scale Global Processes. *Proceedings of Third Scientific Assembly of the International Association of Hydrological Sciences (IAHS), Baltimore, USA, May 1989*, edited by A. Rango (Wallingford: IAHS Publications).

Sadowski, F.G., and Westover, D.E., Monitoring the fire-danger hazard of Nebraska rangelands with AVHRR data, *Proceedings of Tenth Canadian Symposium on Remote Sensing, Edmonton,Alberta, Canada, May 1986*. pp 355-363.

Saradeth, S., and Falk, A., Orbit parameter recovery and image rectification techniques for NOAA AVHRR data, *Proceedings of IGARSS'89/12th Canadian Symposium. on Remote Sensing, held at Vancouver, Canada, on 10th-14th July 1989*. pp 1640-1643.

Seguin, B., Assad, E., Freteaud, J.P., Imbernon, J., Kerr, Y., and Lagouarde, J.P.,1989, Use of meteorological satellites for water balance monitoring in Sahelian regions, *International Journal of Remote Sensing*, **10**, 1101-1117.

Solomon, S.I., Pietroniro, A., Wishart, W., Pieyns, S., Guillot, B., and Carn, M., 1988, Application de la télédétection à l'évaluation des ressources en eau du Sahel, *Veille Climatique Satellitaire*, **24**, 26-37.

Stephens, G., and Matson, M., Regional and global fire detection using AVHRR data, *Proceedings of the Symposium on Remote Sensing of Environment, Ann Arbor, Michigan, 26th-30th October 1987*. pp 447-457.

Townshend, J.R.G., and Justice, C.O., 1988, Selecting the spatial resolution of satellite sensors required for global monitoring of land transformations, *International Journal of Remote Sensing*, **9**, 187-236.

Verstraete, M.M., and Dickinson, R.E., 1986, Modeling surface processes in atmospheric general circulation models, *Annales Geophysicae*, **4B**, 357-364.

Vogt, J., 1989, User guide to the TAN and PYS programs. Internal Technical Note, Joint Research Center - ISPRA Establishment, Italy.

White, F., 1986, La végétation de l'Afrique. Mémoire accompagnant la carte de végétation de l'Afrique Unesco / AETFAT/UNSO. Rapport ORSTOM - UNESCO. Unesco ISBN: 92-3-201955-8 ; ORSTOM ISBN: 2-7099-0832-8.

Yokoyama, R., Zhou, C.M., Tanba, S., and Ihara, H., Relation between the ground surface temperature by NOAA AVHRR and the environmental factors, *Proceedings of IGARSS'89/12th Canadian Symposium. on Remote Sensing, held at Vancouver, Canada, on 10th-14th July 1989. pp 1335-1138.*

REMOTE SENSING FOR TROPICAL FOREST MONITORING: AN OVERVIEW

J.P. MALINGREAU
Commission of the European Communities,
Joint Research Centre,
Institute for remote Sensing Applications
I-21020 Ispra (Varese), Italy

1. Introduction

The exploitation of forest resources has always been associated with human history. Food, timber and fiber extracted from the forest were basic products which allowed pre-agricultural man to survive. Later, agricultural expansion itself often occurred at the expense of forest land. A growing demand for timber during the colonial period led to large scale commercial exploitation of the most valuable forest tracts of some countries. While shifting cultivation has always been a feature of tropical forest land, the transformation of large areas of forest for agriculture has recently assumed new proportions. Under the pressure of a fast growing population and of the sustained demand for tropical timber, the last tracts of tropical forests are being seriously threatened.

This phenomenon is now so important that concern has been expressed not only for the local consequences of the destruction of the forest cover but also for its repercussions at the global level. Tropical deforestation has thus become a major ecological issue of the end of this century (Woodwell *et al.* 1983, Malingreau and Tucker, 1988).

Central to any attempt to better evaluate the impact of deforestation is an adequate understanding of the processes involved. This includes an assessment of the social, economic and physical driving forces controlling the situation. It also includes improved measurements of the actual transformations taking place, over sometimes very extensive and difficult territory, as well as constant monitoring of the rates of change of such transformations. The overall order of magnitude and the spatial distribution of the process have to be known if one is to properly assess possible impacts; rates of change have to be evaluated to determine the seriousness of trends and chart possible future outcomes.

The publicity recently given to deforestation issues has its origins in both the controversies concerning its impacts and in the inaccuracies concerning its magnitude. Indeed, it is very difficult to measure such a dispersed and complex process, the more so when there are still uncertainties in the actual definition of "deforestation" itself (FAO 1981). One fact can however be advanced with certainty: the destruction of the tropical rain-forest has accelerated in the last decade and, while it is always risky to extrapolate, all indicators point to further destruction of the last remnants of this unique biome.

The topic is highly controversial in some countries because forests are important economic assets, classically considered in isolation from conservation issues. Today, conservation concerns are increasingly, and sometimes vociferously, heard because current trends are cause for worry, and because the non economic value of the tropical forest is now ranked in high priority by a large number of people (Fearnside 1985). Therefore it appears urgent to improve our understanding of deforestation dynamics; it is the purpose of this chapter to review how remote sensing techniques

A. S. Belward and C. R. Valenzuela (eds.),
Remote Sensing and Geographical Information Systems for Resource Management in Developing Countries, 253–278.

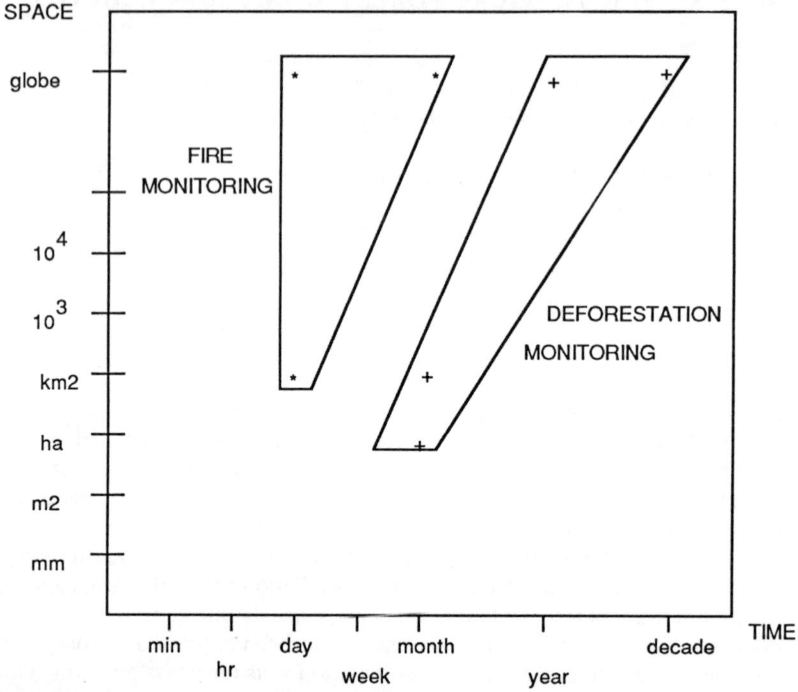

Figure 1. Stommel diagram showing the space time domain for tropical deforestation and fire monitoring. Any monitoring system has to be capable of covering the entire domain; thus historical archives, extending back over decades, are needed alongside daily measurement over areas extending from a few metres to the entire globe.

can assist in this task. Being concerned with deforestation issues in general, we will not dwell too much on the detail analysis of specific situations; instead, the aim is to demonstrate that technical advances currently allow efficient monitoring of significant changes in the forest canopy over large areas of the tropical belt.

2. Scales

While tropical deforestation starts with the cutting of a single tree by an individual farmer or logging operator, it can affect areas of varying size, from a local site to a whole mountain slope or watershed, from a country to whole subcontinents. The consequences of deforestation can be felt locally and in widely different parts of the environment. Therefore, deforestation impacts, such as land and water degradation, species extinction, microclimate change or even the increasingly familiar "global change" must be considered on a wide range of scales and the success of their analysis rests partly upon the right selection of the scale of investigation. This applies not only to the space domain but also to the time domain. The so-called Stommel diagram is a useful approach to locate a particular problem in the space-time domain (figure 1).

Table 1. A scale framework for assessing data needs in tropical deforestation monitoring

Scale	Unit	Parameter
Local	Field Farm	Cover change Clearing practices Soil fertility Shifting cultivation -return period
Regional	Watershed Country or countries Habitat	Composition of cover mosaic Soil & geomorphology Surface energy interactions Dynamics of vegetation mixes Fire dynamics
Global	Intertropical -belt Biome	Forest Non-forest transition Active deforestation fronts Fire dynamics

Of more concern to us in the present context is the requirement for matching the scale of remote sensing observations with the scale of deforestation processes. As already mentioned, these processes are varied and take place in a wide range of environmental conditions. We will find it useful to first distinguish, somewhat artificially, between three scales of interest in the forest monitoring exercise: local (or micro), regional (or meso) and global scales. These three levels are summarized in table 1. However, still missing in such summary table is an adequate exposition of the requirements for an observation system. These are now discussed.

2.1 THE LOCAL SCALE

Tropical forest monitoring on a local scale is mainly concerned with identification of features which can be directly related to human activity. Such features are thus characteristic of the clearing process itself, whether in the framework of shifting cultivation, planned colonization or logging activities. It is obvious that the detailed information required can only come from images from high resolution instruments which can distinguish between the forest canopy and the openings associated with the clearing. This type of information is needed for the planning or management of individual projects, or for research purposes. This is the case for example when assessments of the current land use of an area is required, or for natural reserve management, or when precise information is needed on the shifting cultivation return period (percentage of area used at any time and number of years under or between cropping). In many situations, especially in the early stages of deforestation, we are dealing with a somewhat diffuse process which can only be resolved if means of observation of a certain resolution are available. This is verified in the case of the slow movement of a deforestation front (see §2.2 and 2.3) where rates of advance can only be measured if data at a resolution of say a few hundred meters are available. This in turn places heavy requirements in terms of geometric accuracy since the registration of one image to another must be exact enough to guarantee that the changes which are measured are actually caused by a change in land cover and not by errors of positioning.

Local observations must also identify the features associated with the deforestation process. It is often verified that the distinction between forest and non-forest cover is not in all cases straightforward, with all kinds of secondary re-growth or tree crops crowding the confusion matrix (Achard and Blasco 1990). Two approaches can be adopted here; the first one is to emphasize the radiometric analysis in view of the identification of each individual cover class, and the second is to aim for a relative classification of such features. The first approach relates to conventional image classification techniques, such as those presented by Wilkinson, chapter 5; the second approach is based on "change detection" techniques which are solely aimed at assessing if, and where, a radiometrically discernible change has taken place in the forest canopy. These methods don't require full scene classification, but only a measure of the radiometric difference between two pixels and a decision rule to assess if such change is significant in terms of forest non-forest cover. While the approach may appear simple it is often fraught with difficulties because there is usually a seasonal signal superimposed upon the long term land cover dynamics. Seasonal effects, especially in semi-evergreen and deciduous formations, must therefore be taken into account in the interpretation. This can be done by selecting "windows" of time for observation which are phenologically comparable for each year of observation. These "bio-windows" should, as a rule, be selected at the period of maximum contrast between forest and non-forest features. Knowledge of the local vegetation seasonal dynamics is most useful in this respect (Malingreau 1986, Justice *et al.* 1985).

Current remote sensing techniques using data from the Landsat TM and SPOT HRV instruments are capable of providing the required information on the deforestation process at local scales. The cost of the data, coupled with the limited coverage per scene, mean that correct scene selection is vital. While it is clear that remote sensing analysis is highly contingent upon "available data sets", it is worth starting first with a clear user requirement plan along the lines briefly espoused in this section. In a second stage, the full archive of data must be searched and the possibility of programming the satellite acquisition (for SPOT) must be exploited. It is only as a last resort that the analyst should fall back on the use of "any" data. Indeed, this last approach is likely to lead to less than optimal results and maybe to serious errors and disappointments.

2.2 THE REGIONAL SCALE

Regional scale investigations are typically concerned with the assessment of deforestation patterns over large areas which range from the watersheds of major rivers to individual countries or even groups of countries. The analysis is usually geared towards an evaluation of changes in the forest canopy within the broader framework of watershed protection, national development, or regional planning. The approach is therefore more preoccupied with the identification of patterns or trends than with the resolution of the details of a particular situation. It is at this scale that national data of current forest cover and deforestation rates are produced. Given the importance of such data in the controversies currently surrounding the whole deforestation issue we are justified in devoting some time to the topic. Indeed, for all the talk about monitoring the earth's resources from orbiting satellites, it is still difficult to produce overall assessments of the ongoing deforestation process on regional to global scales. Some countries are better equipped than other to carry out such regular surveys, yet on these scales, data are often unreliable; worse still, their level of reliability is difficult to assess. Figure 2 shows just how widely estimates from different sources can vary (Repetto 1990). Famous case studies have been well reported but many grey areas remain in the world deforestation map.

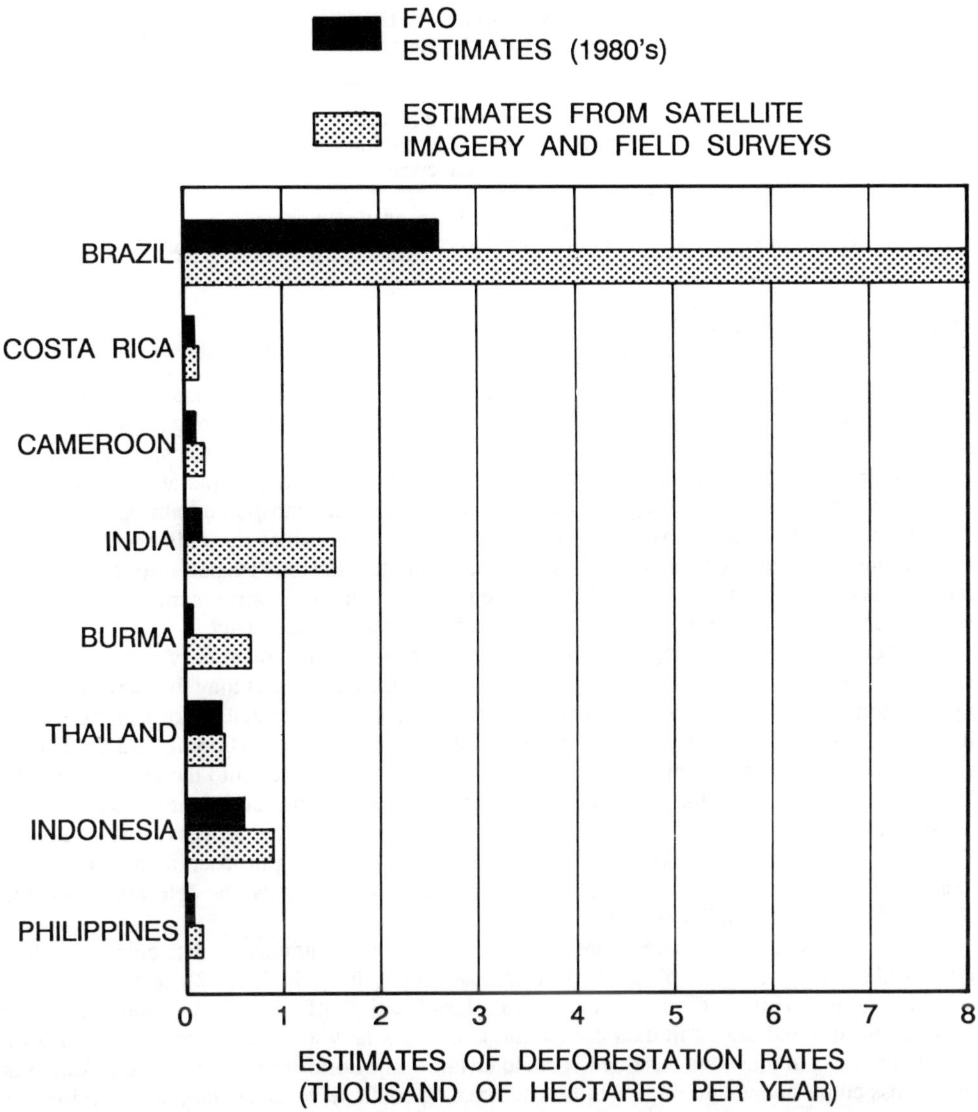

Figure 2. Estimates of deforestation rates from FAO and satellite image based surveys (after Repetto 1990)

While we are slowly realizing that significant changes in the forest cover of the tropics have occurred in the last ten years, we have also made some progress in our capabilities to survey the world and better understand the mechanics of its ecosystems. Space technology has contributed much to such understanding. Nevertheless, it is urgent that we to apply ongoing developments to a larger variety of situations than has been the case so far.

Table 2. Requirements for effective global deforestation assessment using satellite sensor observations

Global coverage
Appropriate spectral bands
Appropriate spatial resolution
Adequate repeat cycle
Assurance of continuity of cover
Availability of appropriate interpretation models
Committed user community prepared to carry out operational monitoring

2.3 THE GLOBAL SCALE

Global scale issues to be examined in the framework of tropical deforestation relate to the assessment of the planetary dimension of the processes. There are indeed strong presumptions that upon reaching a certain threshold the radical transformation of the original forest cover may have major environmental implications. Changes in the forest albedo and it's radiation balance may affect the hydrological cycle through a reduction of the latent heat transport into the atmosphere via the evaporation and evapotranspiration processes; figure 3 describes the radiation balance for tropical forest and degraded pastureland. Note in particular that the total evapotranspiration E, from the forest is far greater than that from the degraded land and that the inverse is true for the sensible heating flux H. In other words forests are considerably cooler than open land.

The equatorial belt is a region where the enormous amounts of energy transported and transformed drive the global climate. Any change in such local processes may thus have important repercussions at scales far beyond the area itself. Another factor of potentially global significance is the burning which usually accompanies deforestation activities. Biomass burning is a direct phase transformation of biological material into gases, which, when injected into the atmosphere may contribute to climate warming. There are still difficulties in estimating such parameters and it is not known for example, if the carbon dioxide released through the burning process represents 7% or 30% of the anthropogenic emissions of that gas! For our immediate purpose, biomass burning is considered as an important indicator of tropical deforestation which can be detected by satellite (Woodwell *et al.* 1983, Malingreau *et al.* 1985).

Global deforestation assessments are usually conducted by accumulating data obtained at local or national level (see for example the FAO 1980 assessment shown in figure 2); such exercises are usually fraught with difficulties related to classification of vegetation, methodological incompatibilities and biases in data compilation. There is now a growing possibility of using satellite data as a unique data base which could ensure some uniformity of scale and information type across entire biomes. In other words, satellite data could serve as a common denominator to which a large number of individual inventories could be referred. Needless to say, multistage sampling (Justice *et al.* 1989, Nelson and Holben 1986) in which data obtained at various scales and resolutions are progressively "nested" into more generalized data sets is finding here a natural field of application.

The adaptation of satellite data to global inventories of terrestrial features is based upon a series of requirements related to the sensor-platform-analysis system which are listed in table 2.

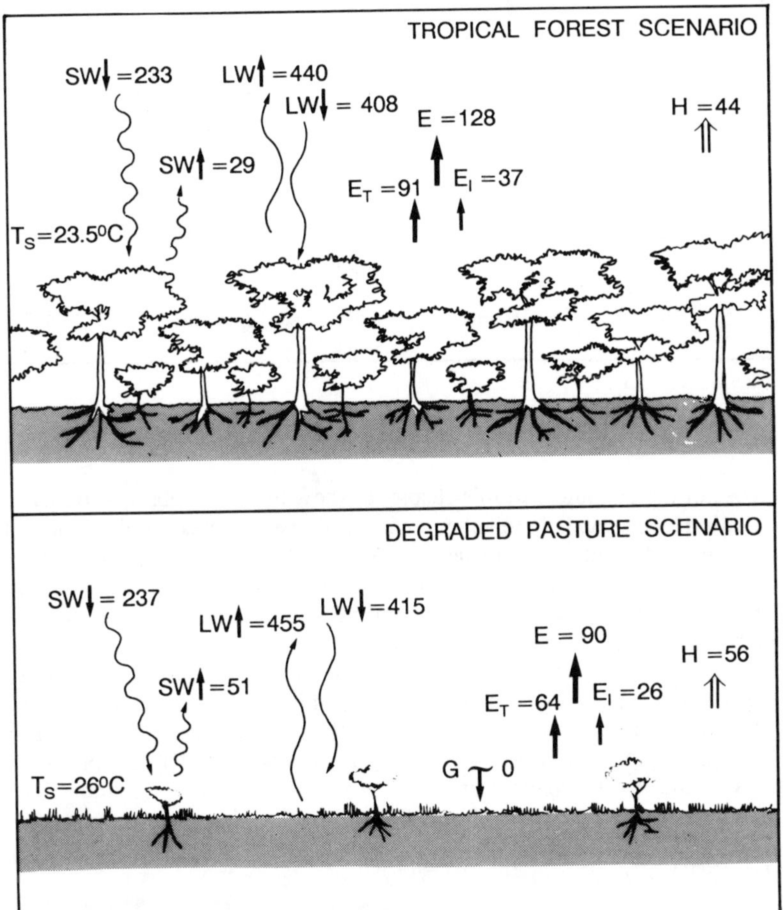

SW↓ = downward short wave (solar) radiation at the surface;
SW↑ = upward short wave (reflected) radiation at the surface;
LW↑ = upward longwave (thermal) radiation at the surface;
LW↓ = downward longwave radiation at the surface;
E = total evapotranspiration (latent heating flux);
E_T = transpiration + soil evaporation;
E_I = interception loss (evaporation of water intercepted by the vegetation);
H = sensible heating flux
G = soil heat flux;
TS = surface air temperature

Figure 3. Annual surface energy balance for different cover types in the Amazon Basin, as derived from a climate model simulation. Units are in Watts/m^2. The forest albedo (SW↑ / SW↓) at 12.5% is lower than that for the pasture, 21.5%. The forest thus absorbs more solar radiation (SW↓ – SW↑). The surface temperature (TS) is lower for forest than pasture, causing lower thermal losses (LW↑ - LW↓) by the forest surface. Thus we have a situation where forest is "cooler" than non-forest (after Nobre 1990).

Table 3 AVHRR data availability

Data	Source	Resolution	Repeat cycle	Area covered	Channel
LAC	Request	1.1 km	daily	selected	all
HRPT	Local	1.1 km	daily	in range	all
GAC	NOAA	4 km	daily since '81	world	all
GVI	NOAA	16 km	daily since '82	world	NDVI

3. The AVHRR as an observation instrument for tropical forest monitoring

It is the purpose of the remainder of this chapter to show that earth observation technologies at medium resolution levels can help in better coping with the complex problems of scale outlined above. In particular, data from the Advanced Very High Resolution Radiometer (AVHRR) have proven to be the source of much useful information on world vegetation distribution and dynamics. Belward, in chapter 9 gives a general introduction to the use of AVHRR data for regional / global scale vegetation monitoring. The use of AVHRR data for tropical deforestation monitoring has barely started and yet, being accessible to a growing number of users, these data are likely to become an increasingly important source of information for earth observation in the future. Similar medium resolution sensors such as the MODIS and the MERIS sensors to be carried by the future space platforms are likely to follow (see Hill, chapter 4), and future developments in the use of space-borne microwave sensors will provide complementary information (see Churchill and Sieber, chapter 7).

3.1 THE ADVANTAGES

The various AVHRR data types and general characteristics have been described elsewhere in this book. However, it is relevant to list here the main characteristics of the NOAA AVHRR system which make it suited for tropical forest monitoring.

3.1.1. *Consistent data sets* of AVHRR data exist for the entire globe since the early eighties and can be exploited for historical analyses (see table 3). At the lowest sampling resolution levels (Global Vegetation Index - GVI- and Global Area Coverage - GAC) the data set are truly global, (see Belward, chapter 9).

3.1.2. *The visible and near-infrared channels* of the instrument can be combined in vegetation indices related to canopy green biomass. Time series of such indices can be assembled and the signal can be associated with phenological changes in the vegetation. Such time-series can also provide a base for vegetation classification (Lloyd 1989).

3.1.3. *The thermal channels* of the AVHRR can detect thermal changes associated with transformations in the forest canopy. This characteristic has been much exploited in current work (Malingreau *et al.* 1989, Schutt and Kerber 1986)

3.1.4. *Channel 3* is appropriate for detecting elevated heat sources such as the ones associated with fires and biomass burning (Stephens and Matson 1987).

3.1.5. *Full resolution data* availability is growing. The number of ground receiving stations operating throughout the tropics to receive the HRPT data is increasing (see figure 4). Complete full 1 km resolution coverage of the tropical belt should be possible by assembling the products derived from those individual stations.

3.1.6. *Frequency of coverage* (daily) is an advantage for overcoming poor atmospheric conditions of the tropics (Holben and Fraser 1984).

3.1.7. *The resolution* of the instrument (1.1 km) is suitable for regional observations. Indeed it appears to offer a good compromise with respect to information content and data load. Lower resolution samples (i.e 4 and 15 km) are suitable for global coverage.(Belward and Lambin 1990, Malingreau and Belward 1989, Justice *et al.* 1989, Townshend and Justice 1988).

3.1.8. *Cost* is low. The data are very cheap since the main cost is the price of the support.

3.2. THE DISADVANTAGES

The above characteristics make the NOAA-AVHRR configuration a unique instrument for answering some of the problems related to the assessment of global tropical deforestation. There are however some drawbacks which should be carefully considered.

3.2.1. *The swath width* of (2700 km) leads to enormous geometric distortion which make pre-processing heavy; satellite positioning is usually not accurate enough for high levels of mapping accuracy without recourse to specialist software (Sharman 1989).

3.2.2. *Data load* from a system providing a daily global data set in five channels represent large data loads which are logistically heavy to handle.(Belward and Malingreau 1989)

3.2.3. *Atmospheric conditions* across the scene are likely to vary given the size of the area under observation ; atmospheric corrections are still difficult to apply (Hill, chapter 6).

3.2.4. *Sensor calibration* and sensor drift must be carefully evaluated if precise radiometric measurements are required (Hill, chapter 6).

3.2.5. *Scale of observation* leads to conceptual problems as the integration of the components of a landscape influencing the spectral measurement at a 1km pixel resolution are not easy to understand (Townshend and Justice 1989).

3.2.6 *A catalog* of AVHRR data holdings or acquisitions by the wide number of institutions and stations distributed around the world does not currently exist. While one can question the

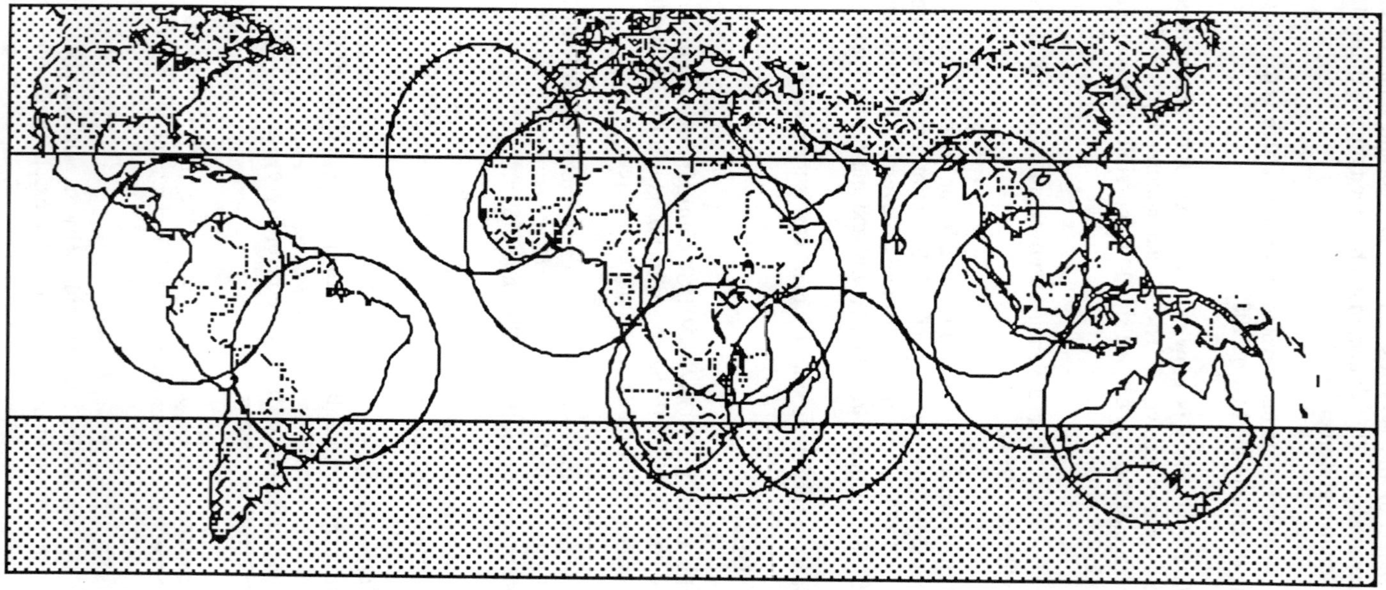

Figure 4. Ground cover of AVHRR HRPT stations operating in the tropical belt. Almost complete cover is now available, but note that the length of archive available from each station is different. Some, for example the East African stations are only just entering an operational phase so the availability of historical data is still limited.

usefulness of such a catalog when the AVHRR system provides daily acquisition of the full orbit for the area of coverage by a station, such a compilation remains useful for selecting "good scenes" for analysis. Experience has shown, however, that cloudiness rating in AVHRR catalogs provide little information on the state of the atmosphere which can be affected by smoke, haze and dust. As a rule, especially if powerful processing systems are available, it is easier to order all acquisitions from a selected period over the area of interest and to select the scenes using a quick look screening approach (visual display of full or sampled images).

4. Elements of image analysis for forest monitoring using AVHRR

This section describes the various approaches which can be adopted in analyzing AVHRR data for tropical forest monitoring. The nature of the AVHRR data, and the various pre-processing routines such as radiometric, atmospheric and geometric corrections, cloud screening, applied before the start of the analysis are reviewed in the earlier chapters by Belward, Hill and Wilkinson.

Various types of AVHRR derived information can be exploited for discriminating between relevant deforestation features. Three categories of parameters can be identified: spectral, temporal and circumstantial. These are now reviewed in turn. It is important to realize that the parameters used to identify forest and non-forest cover vary with location. Thus a particular discriminant may work in one part of the tropics and not in others. The discriminants reviewed below should, therefore, be viewed as part of a "package". There is no one discriminant that can be used throughout the globe, though all have successfully been used to separate forest from non-forest at some time in some part of the tropical belt.

4.1 SPECTRAL DISCRIMINANTS

Spectral contrast between the evergreen forest and the more seasonal replacement vegetation are usually at their peak during the last portion of the dry season. At this time the forest is still photosynthetically active, and evapotranspiration is still prominent. The canopy is highly structured with a large volume of standing green biomass. In contrast at this time of the year, seasonal vegetation types such as savannah, crops or secondary re-growth have little green matter in the canopy and are essentially reduced to a mat of senescent or dormant vegetation.

Figure 5 is a synthesis of some of the spectral characteristics of forest non-forest classes in the Amazon Basin. The figure is a plot of changes in AVHRR response [all 5 channels and the NDVI (Normalized Difference Vegetation Index - see Baret, chapter 8 -)], for a ranch opening in the forest of the Mato Grosso. Somewhat atypically - see §4.1.2 and §4.1.3 below - channel 1 (red 0.58 - 0.68 μm) shows good contrast between forest and ranch whilst channel 2 (near-infrared 0.72 - 1.1 μm) does not. In channel 1 high reflectance from the ranch contrasts sharply with the low reflectance of the surrounding forest. The senescent grass of the range is characterized by high red reflectance in comparison with the surrounding forest. The near-infrared reflectance from the dry grass and forest is similar. The lack of contrast in channel 2 leads to poor separation of forest and ranch in the NDVI, though a slight dip over the ranch is evident. Clear separation of the two cover types is possible with the thermal channels of the AVHRR; channel 3 (3.55 - 3.93μm), channel 4 (10.3 - 11.3 μm) and channel 5 (11.5 - 12.5 μm) all reveal the ranch as a "heat island" surrounded by the cooler forest. Indeed, as we saw in §2.3. figure 3, the latent heat, dissipated in the evapotranspiration process by the evergreen forest, warms the lower air layers or the soil surface of degraded pasture lands.

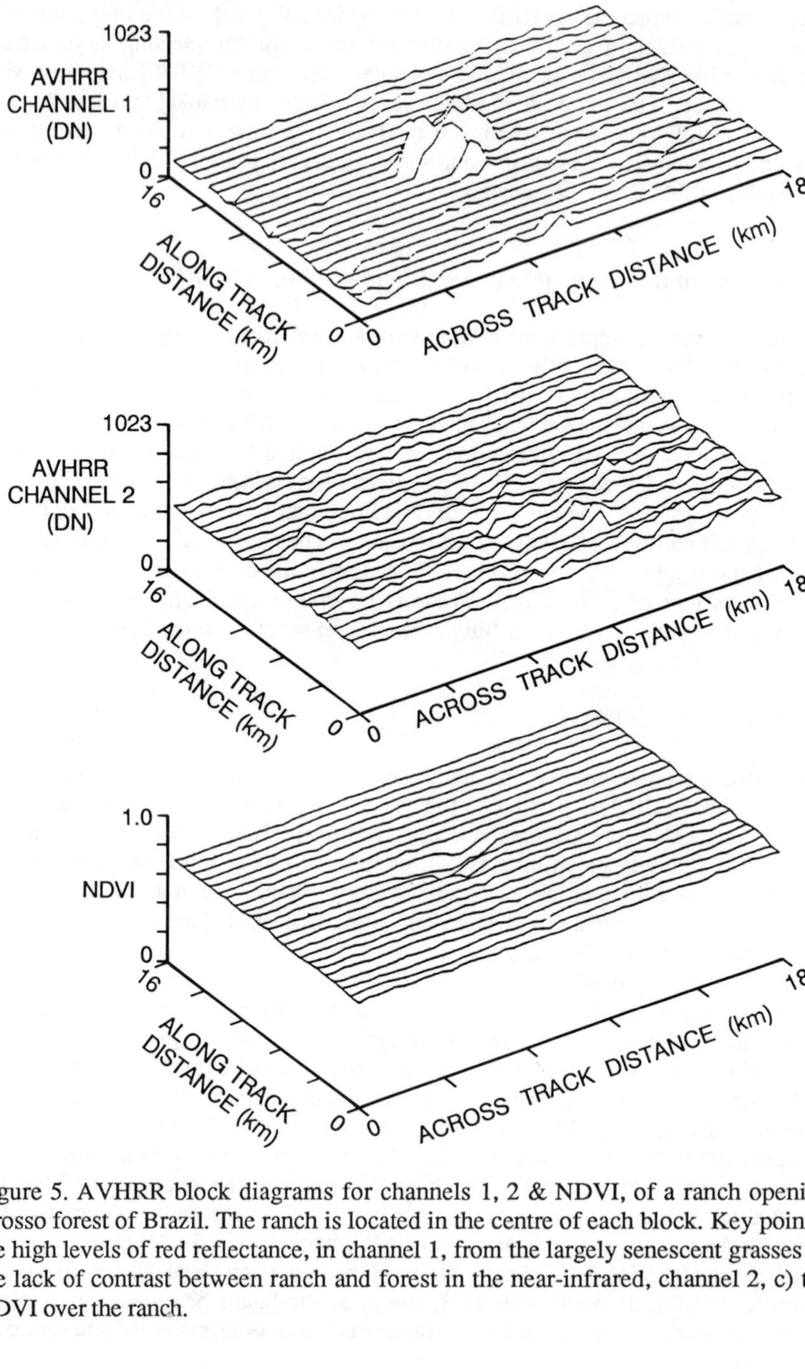

Figure 5. AVHRR block diagrams for channels 1, 2 & NDVI, of a ranch opening in the Mato Grosso forest of Brazil. The ranch is located in the centre of each block. Key points to note are a) the high levels of red reflectance, in channel 1, from the largely senescent grasses of the ranch, b) the lack of contrast between ranch and forest in the near-infrared, channel 2, c) the slight dip in NDVI over the ranch.

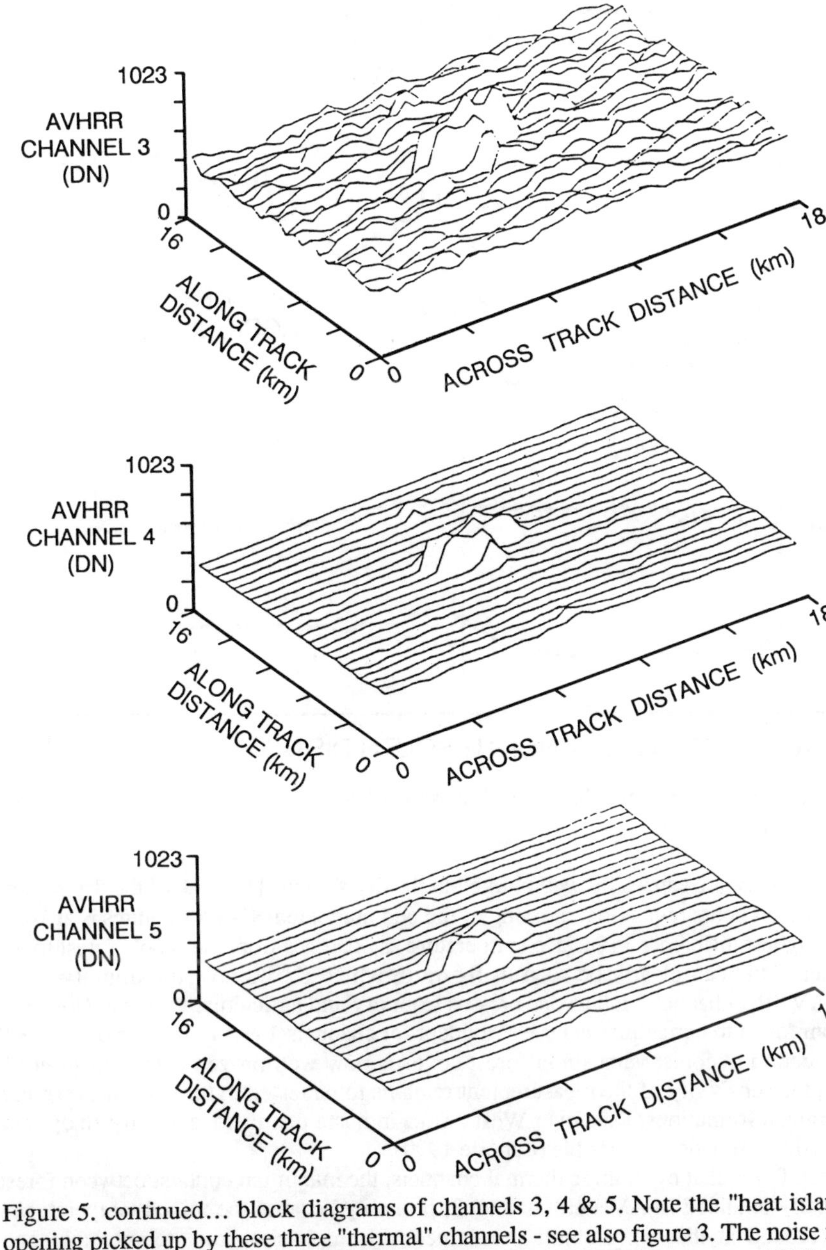

Figure 5. continued... block diagrams of channels 3, 4 & 5. Note the "heat island" effect of the opening picked up by these three "thermal" channels - see also figure 3. The noise in channel 3 data is obvious when compared with the other channels (LAC data from August 1985).

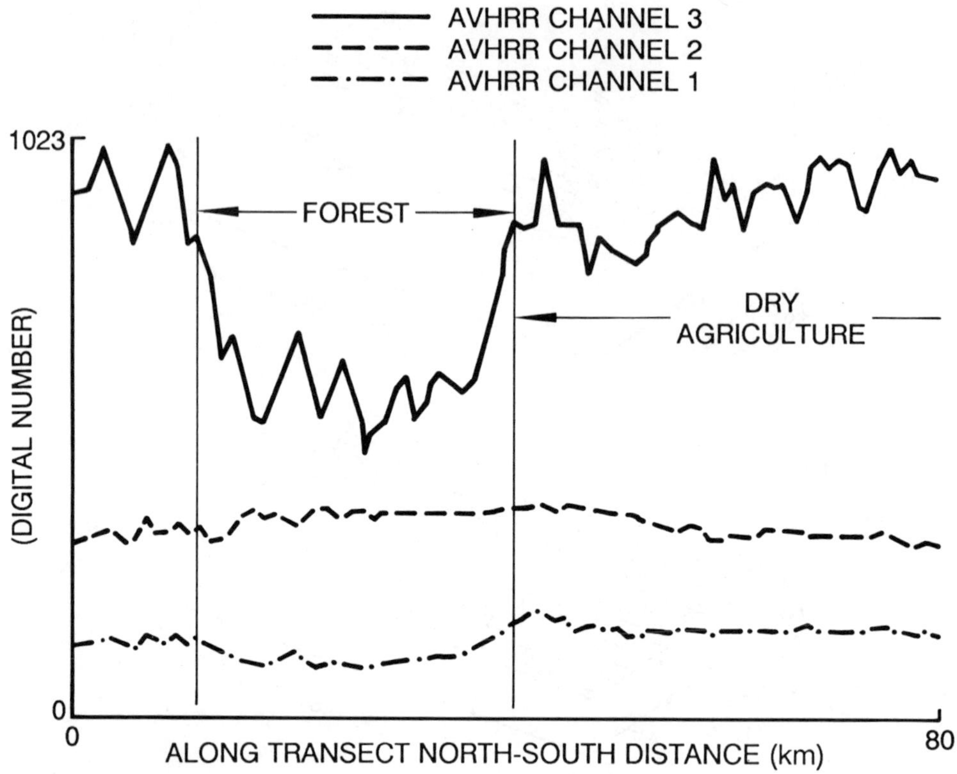

Figure 6. AVHRR transect for channels 1, 2 & 3, across the Khao - Tai range, Eastern Thailand (HRPT data from April 5th 1983)

4.1.1. *Thermal Contrasts* are especially well marked in the dry season. The peak of the dry season, as well as the afternoon hours (equatorial crossing of the odd numbered NOAA satellites is at 14.30 hrs), provide the highest difference in surface temperature as one goes from the "cool" forest to the "warm" open area. This can clearly be seen in figure 6, which shows how the digital values recorded by the AVHRR channel 3 fall as one crosses from dryland agriculture to forest. Note how the transition from forest to agriculture is far less evident for channels 1 and 2. The use of AVHRR channel 3 for detection of forest versus non-forest cover is now well proven (Malingreau *et al.* 1989). However, the sensitivity of the measurement remains to be tested in the case of evergreen versus semi-evergreen formations. Data over West Africa indicate that there is usually an optimal period when such discrimination is possible (Laporte 1990).

It has also been found that of all three thermal channels, the maximum contrast between forest non-forest is usually obtained in AVHRR channel 3, probably because the mixed temperature - albedo signal recorded at that wavelength simultaneously emphasizes the temperature and reflectance increase as one goes from a green to senescent / non vegetation cover (Malingreau *et al.* 1989). Channel 3 presents the advantage of having good atmospheric transmissivity, leading to the production of acceptable data sets in the less than ideal situations of the tropics, even though channel 3 was badly affected by noise throughout the lifetimes of NOAAs 7 and 9. The other thermal channels of the AVHRR (channels 4 and 5) can also be used and they present the advantage of being unaffected by the noise.

Using thermal data, the forest non-forest discrimination is made by determining a brightness temperature threshold which separates the two cover types. Two approaches are possible in applying the method.

The first one is applicable in areas where deforestation is taking, or has taken, place in fairly well defined geometric patterns. The objective of the "thresholding" is to find values which best reveal such a pattern; the image is then classified accordingly, with all values above the threshold being classified as non-forest and all below the threshold as forest. The approach is relatively simple to apply and is not demanding in terms of image analysis. It appears to work satisfactorily in situations of good thermal contrast for regional assessments (Malingreau and Tucker 1990). However, a major drawback is that the method can only be applied locally because of the spatial variability of surface thermal patterns. In addition, the same thermal thresholds cannot be used from one day to the other because of the possibility of intervening meteorological events. Finally, because of the limited radiometric range sometimes involved, a small change in threshold values can mean important variations in forest /non-forest proportions. The accuracy of the classification depends upon the forest non-forest brightness temperature contrasts and upon the shape of the clearings. Long interfaces will tend to over emphasize the deforested area (large perimeter to surface ratio). Features smaller than the resolution of the AVHRR will not be detected unless they follow some clearly visible geometric pattern, roads for example.

A second discrimination method is based upon analysis of distribution histograms of data recorded through thermal channels. "Training areas" or samples, taken in each of the features to be discriminated, are used to produce histograms of pixel count vs digital number. Separation is then made on the basis of a statistical rule of distance. A point can be considered as belonging to a population if, for example, it falls within the mean, plus or minus one standard deviation of the distribution. It is also possible to use other statistical rules, such as the distance between populations. This type of threshold calculation is automatic and does not necessarily allow for local adjustments.

As mentioned in the introduction to this section not all discriminants work in all places at all times, this is certainly true for the thermal channels. Figure 7, in contrast to figure 6, shows how the transition from woodland savannah to forest is less well defined in channel 3 than the forest dryland agriculture transition. In the case of the savannah forest transition it is channel 2 which more clearly defines the boundary (see §4.1.3).

4.1.2. *The visible channel* of the AVHRR (channel 1) occasionally shows the expected contrast between the green photosynthetically active canopy absorbing in the red wavelength band and the highly reflective dry vegetation or bare soil (viz figure 5a). In general however, experience with AVHRR data, has been rather disappointing in this respect. Atmospheric contamination, mainly due to scattering is relatively important in this waveband since the signal reflected by vegetation is usually rather low (Malingreau *et al.* 1989).

4.1.3 *The near-infrared channel*, on the other hand, is of direct use for discriminating between the closed primary forest and the secondary vegetation. The generally unambiguous drop in reflectance in channel 2 of the AVHRR over a closed tropical forest canopy is the sign of a highly structured formation which, despite the high overall reflectance level, can trap the near-infrared (NIR) radiation in an effective manner (Malingreau *et al.* 1989). This property can be used in two specific

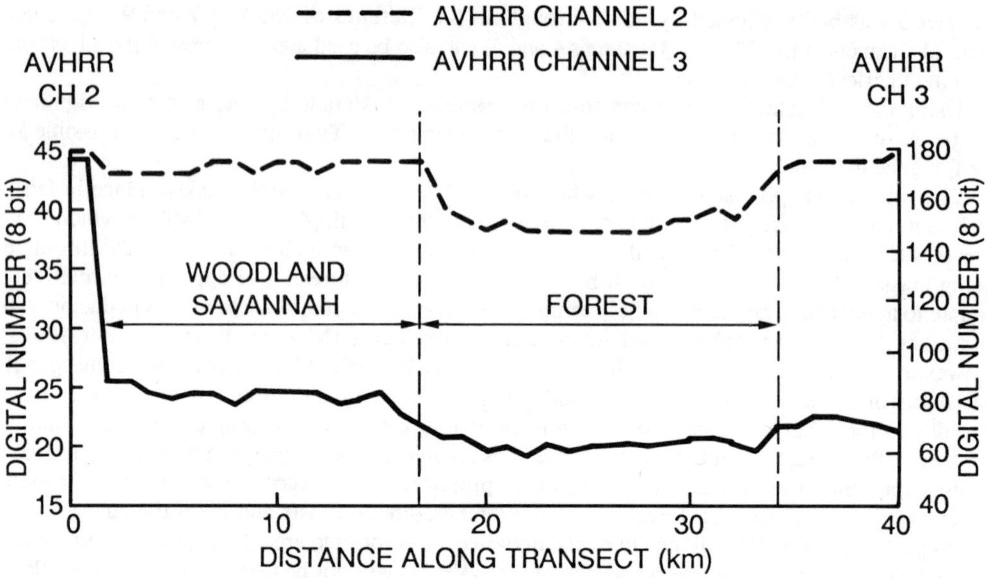

Figure 7. AVHRR transect for channels 2 & 3, across the forest savannah transition in Guinea, West Africa (HRPT data from January 13th 1987)

instances. Firstly in identifying the blocks of primary closed forest in an area covered by secondary formations; as already mentioned, the thermal contrast between such evergreen vegetation types are not easy to discern, see example from West Africa shown in figure 7. Secondly the NIR contrast can be used in identifying areas of secondary re-growth or dense plantations in the middle of the primary closed forest. Figure 8 shows an example of this from Zaire, note the sharp increase in channel 2 as the transect crosses the more open canopy of the areas of secondary re-growth, once again this contrast is not evident in the channel 3 data. As always, in cases of limited spectral contrast, a geometric arrangement of the features helps in delineation. Diffuse situations are more difficult to handle.

It is evident from the discussion in the earlier chapter by Baret that caution should be exercised when using the near-infrared channel since it can be strongly influenced by soil background effects. In addition, since the visible channel is essentially invariable, the NDVI provides only marginal additional information on the closed forest non-forest contrasts. The advantage in using NDVI is that a maximization approach can be applied to produce maximum value composites (MVCs), and thus cleaner data sets. Belward, chapter 9, reviews the concept of the maximum value composite.

4.1.4. *Relationships between channels.* There is often an inverse relationship between channel 3 and the NDVI. This is illustrated in figure 9 (Malingreau and Tucker 1987). However, the relationship has been little exploited in the study of natural vegetation. Crop related work has shown that insights into the functioning of the canopy (evapotranspiration) can be obtained using that approach. The study of seasonality of forest types could be further investigated using similar reasoning.

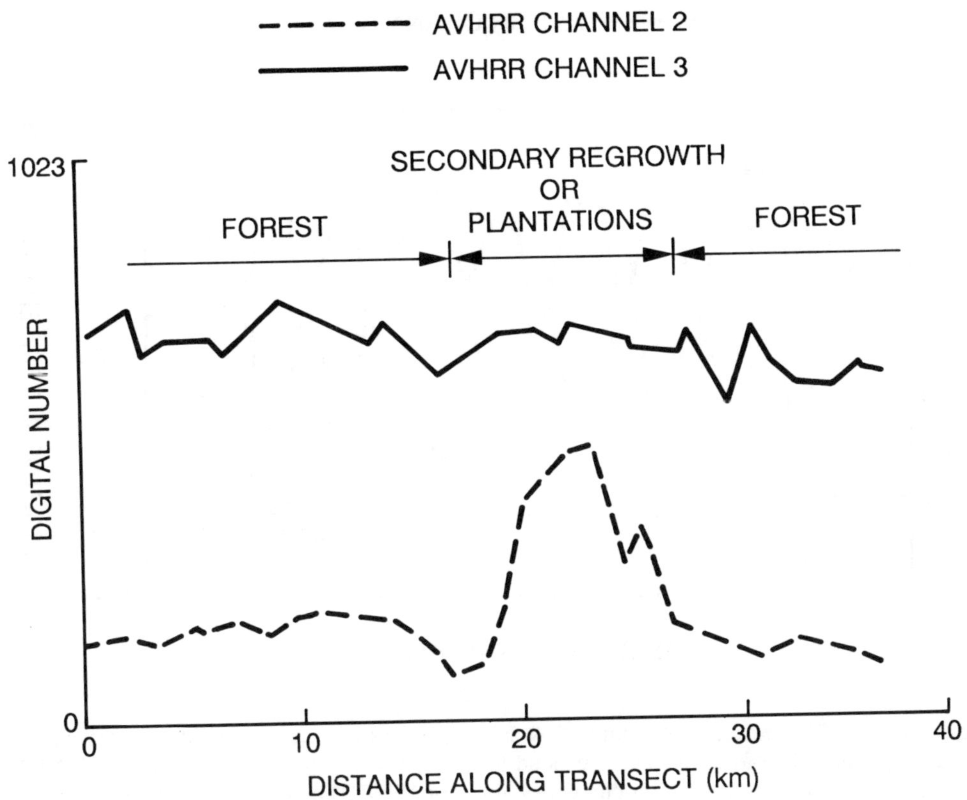

Figure 8. AVHRR transect for channels 2 & 3, across the forest secondary formation boundary in Zaire (LAC data from January 1988)

Empirical relationships are also worth exploring; for example the curves from channel 3 and channel 4 in a transect across the Burmese seasonal dipterocarp forests, figure 10, always cross each other at the forest non-forest boundary. Interestingly, the NDVI values show distinct increases for the Arakan and Kayah ranges, but no difference for the central Pegu range. Proven relationships between channels may be useful for normalizing the measurement of canopy

4.2. TEMPORAL DISCRIMINANTS

The analysis of time series of vegetation index data is potentially a powerful tool for discriminating between vegetation types and classify the plant cover according to their phenological character (Justice *et al.* 1985, Viovy *et al.* 1988). This type of measurement could in fact add a new dimension to the assessment of forest cover.

The evergreen rain-forest can be taken as a starting point for this analysis. As noted by several authors, the evergreen character of this formation is not strictly verified on temporal series of vegetation index (Tucker *et al.* 1985, Malingreau 1986). Both the GVI and the GAC products present a periodicity in signal variation which seem to be related to some canopy process. Multi-

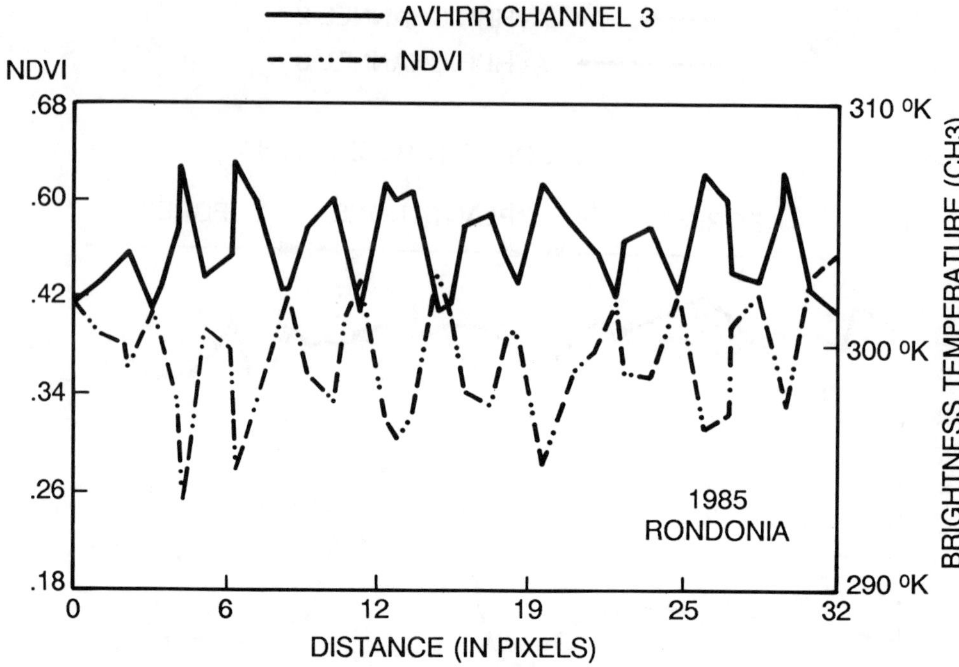

Figure 9. AVHRR transect for channel 3 & NDVI, across a series of forest clearing sites in Rondonia, Brazil. The forest areas correspond to the points with low channel 3 and high NDVI values, non-forest areas have high channel 3 and low NDVI. The curves illustrate the inverse relationship between channel 3 and NDVI (LAC data from August 1985)

year curves can even show surprising inter-annual variations which show climatic influences on systems which are essentially located in non-seasonal climatic zones, figure 11 is an example. The Central Congo Basin is non-seasonal climatically speaking, but a distinct periodicity can be seen in the NDVI temporal curve, distinct peaks appear each year around April - May and around October - November. Our understanding of such curves is still limited.

As the seasonal character of the vegetation increases, the amplitude of the NDVI increases. This is shown by the curves from different vegetation types of the Amazon basin, figure 12. Note how around April the NDVI temporal curves from the seasonal (Cerrado), semi-seasonal (Rondonia A) and "non-seasonal" tropical forest (Xingu) are highly similar; as we progress towards August the curves separate in such a way as the greater the seasonality of the cover type the greater the drop in NDVI. This amplitude is mainly related to rainfall patterns. The differential effect of the dry season on different types of vegetation appears to be a powerful discrimination tool; it has however not been systematically exploited for tropical vegetation classification.

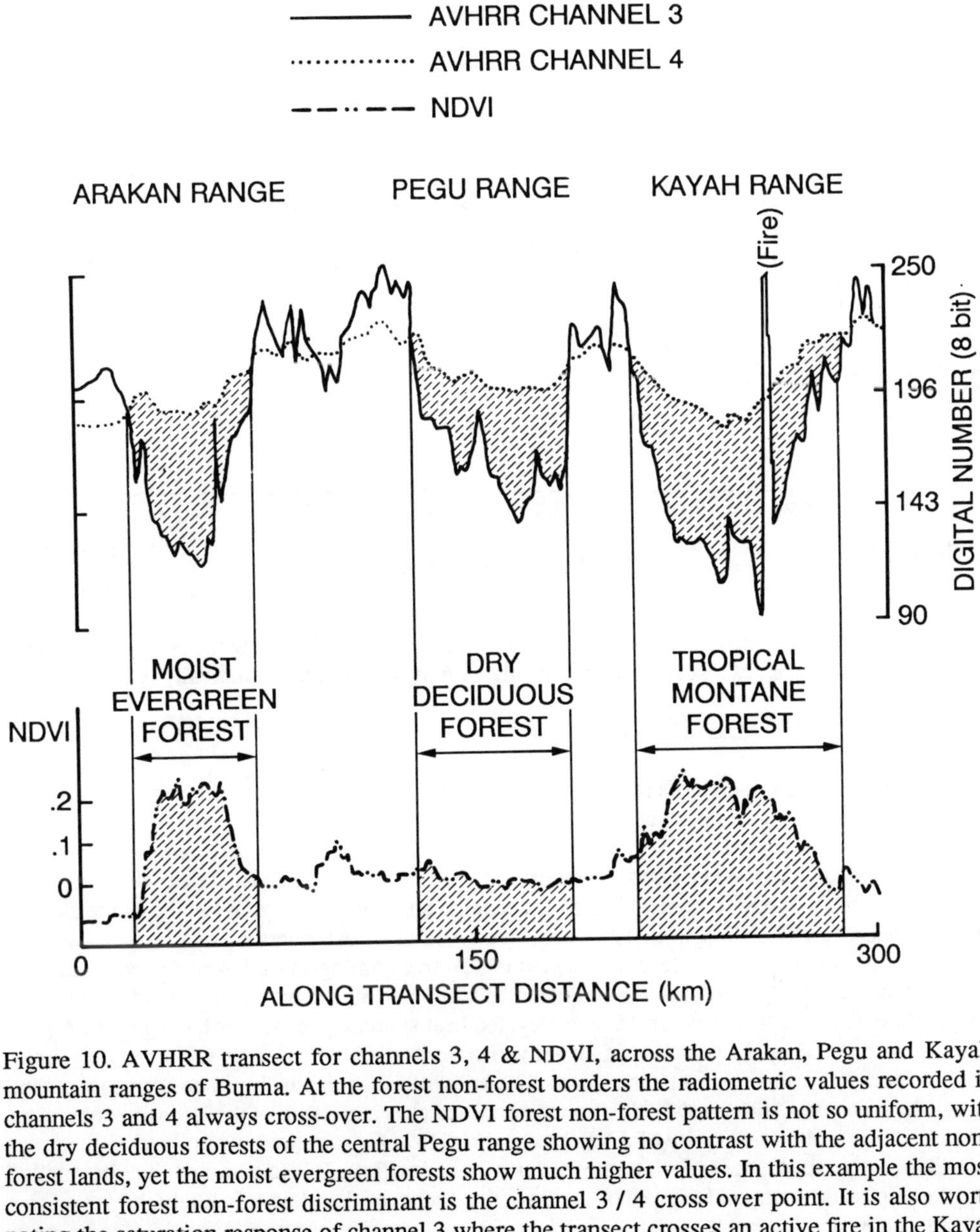

Figure 10. AVHRR transect for channels 3, 4 & NDVI, across the Arakan, Pegu and Kayah mountain ranges of Burma. At the forest non-forest borders the radiometric values recorded in channels 3 and 4 always cross-over. The NDVI forest non-forest pattern is not so uniform, with the dry deciduous forests of the central Pegu range showing no contrast with the adjacent non-forest lands, yet the moist evergreen forests show much higher values. In this example the most consistent forest non-forest discriminant is the channel 3 / 4 cross over point. It is also worth noting the saturation response of channel 3 where the transect crosses an active fire in the Kayah range. The fire goes undetected in channel 4.

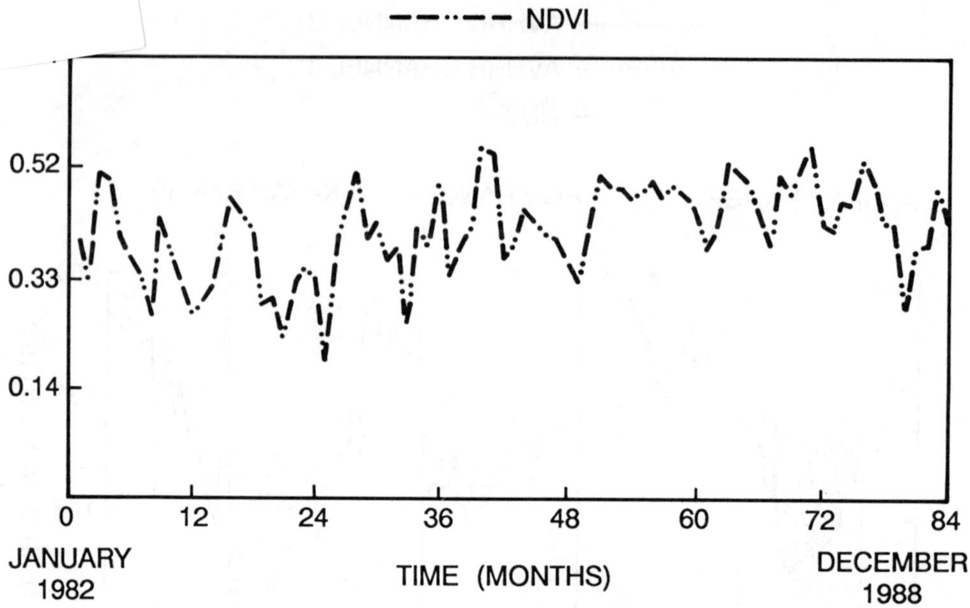

Figure 11. NDVI time series for the tropical rain forest of Zaire. Note how the NDVI shows clear seasonal variation, with two peaks per year, yet the region is invariable,climatically (GAC data)

4.3. CIRCUMSTANTIAL EVIDENCE

Various circumstantial indicators of deforestation activities can be used to better identify and then focus on "active deforestation areas".

4.3.1. *The presence of fire* is a good indicator of deforestation activity associated with traditional shifting cultivation or agricultural activities. Indeed, the burning is a necessary phase in the clearing of felled and dry vegetation. It is also a common weed and bush control practice and a way to mobilize nutrients at the soil surface before planting. Active burning as well as recent burn scars can be detected in channel 3, and used to delineate "disturbance areas" where active deforestation is taking place. Channel 3 is very sensitive to elevated heat sources (see Belward, chapter 9). A pixel can thus be marked by the presence of an active fire in its area of coverage. Obviously, in such a case the actual fire area cannot be measured. The only information which can be derived from the data is the number of "fire pixels" which on regional scales is a measure of the intensity of burning. Because of the limited sensitivity of AVHRR channel 3 (47° C saturation), warm backgrounds can also saturate the pixel and confusion may occur with active fires. It is to be noted that the satellite 14.30 overpass is a temporal sample of the total fire events of that day. It is likely, however, that this afternoon pass coincide with the peak burning hours. Night imagery could also be used, which would overcome problems associated with the daytime reflectance, emittance mix, though many fires have burned out by this time (Langaas and Muirhead 1988). These observations all point to the fact that an overriding concern with the AVHRR approach to fire monitoring is the need to

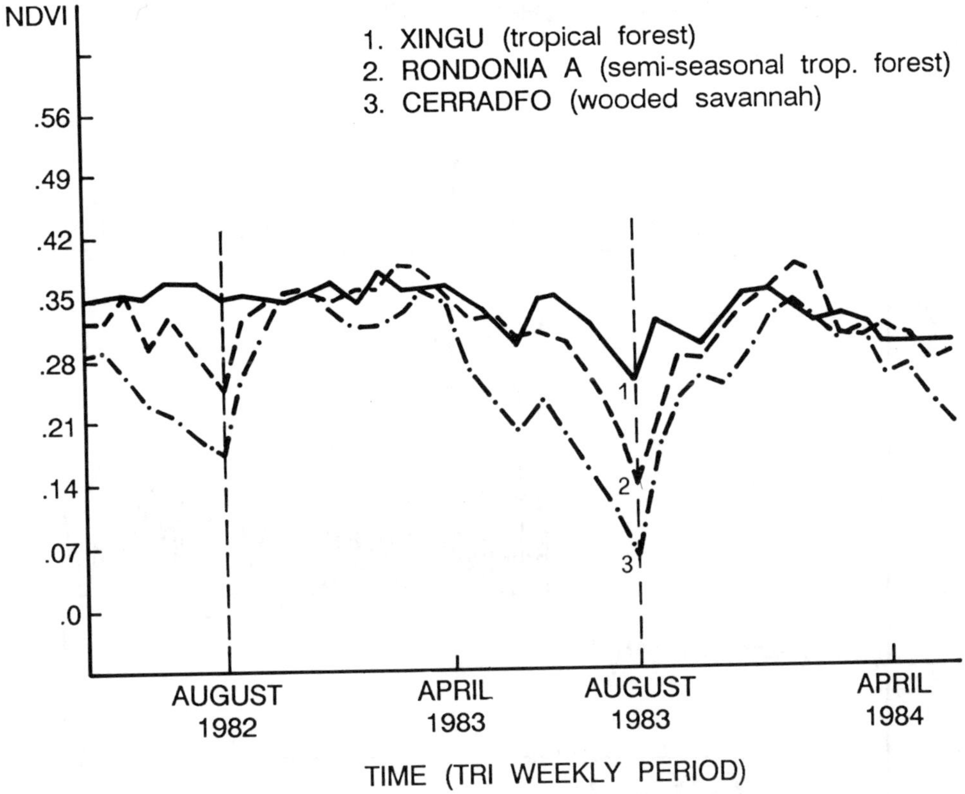

Figure 12. NDVI time series for selected sites in the Amazon Basin. Note how the drop in NDVI, occuring in August each year is more marked as the seasonal nature of the cover type increases (GVI data)

evaluate and correct for biases related to fire size, duration, and diurnal distribution. This evaluation task is considerable given the full range of environmental conditions in which fire is found. Yet, for the first time with the AVHRR data we are able to acquire a spatially comprehensive fire view of the tropical belt. For example, figure 13 shows a North South transect from the forest to the savannah of central Africa, the transition zone between the two is clearly marked by the presence of saturated pixels corresponding to the fires.

The evolution of the spatial distribution of these fires can also be interpreted to infer the type of deforestation taking place. For example regular, linear fire features can be indicative of roads, orientation in a specific direction, or regular shapes may indicate organized forest clearance (Malingreau and Tucker 1990). This type of analysis is easier to conduct in newly opened areas than in the more traditional shifting cultivation where patterns are more diffuse. Inter-annual fire counts can be compared to assess trends and rates of changes. Research remains to be done, however, to better correlate burning as detected by the AVHRR and deforestation rates.

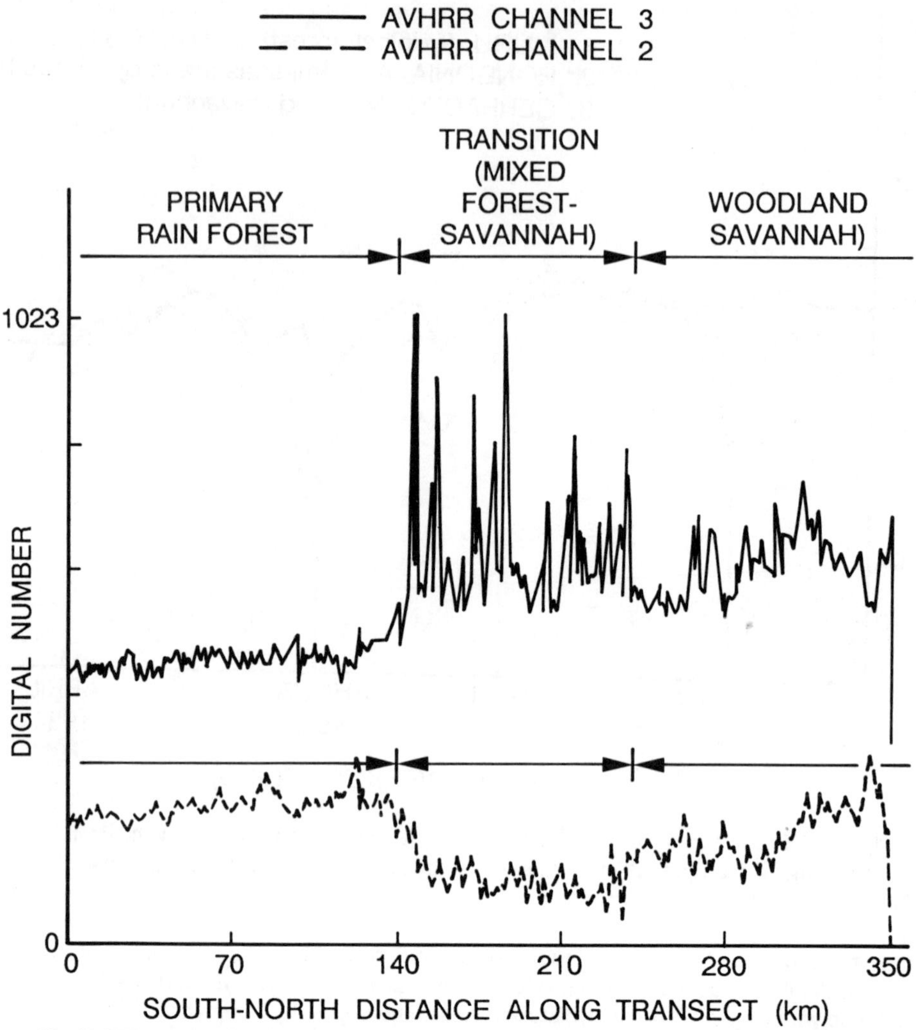

Figure 13. AVHRR transect for channels 2 & 3, across the forest savannah transition of central Africa (LAC data)

4.3.2. *Burn scars* have particular spectral characteristics which can lead to their identification and mapping using the thermal and near-infrared channels of the AVHRR. At first, because of their high post burn temperature (hot soil and ash, sometimes still smoldering), the fresh scars will show higher temperatures than the unburned areas. These can be detected using the thermal channels of the AVHRR (Malingreau and Laporte 1990). Later after the burn, the black surface being a good energy absorber, will usually reach maximum afternoon temperature and will still be detected by contrast with the cooler surroundings. This measurement has limitations, however, because as we have seen, sensor saturation occurs around 47° C which can be a normal afternoon soil temperature in unburned areas.

The low near-infrared albedo of burn scars can also be used as a discriminant and the method is relatively straightforward using channel 2 of the AVHRR. The spatial variability of the pixel values in this channel may be related to the local efficiency of burn and the amount of black carbon material left on the ground, this can be seen in the marked drop in channel 2 values as the transect, shown in figure 13, crosses the fire affected transition zone.

The whole topic of fire detection and biomass burning assessment has attracted strong interest both from managers and scientists. Indeed, for the first time an operational satellite based instrument can provide estimates of burning over large areas, an operation which was hitherto difficult to carry-out. Recent concerns for human impacts upon atmospheric chemistry and the possible implications in terms of global climatic changes has also stimulated the analysis of satellite data in view of producing global fire accounting. The possibility of using the AVHRR data at 4 km GAC resolution is being examined in this respect (JRC 1990).

4.3.3. *The presence of roads.* The opening of access roads is a prime drive for population movement and deforestation activities. Road building has actually been one of the main issues surrounding the upsurge in deforestation in the Amazon Basin (Fearnside 1985, Malingreau and Tucker, 1988). Logging roads are also an important means of access to the difficult hinterlands and it has often been observed that such a crude communication network can rapidly lead to migration and settlement. In a spatial perspective, the detection and monitoring of the progression of road building can focus the attention to areas most likely to undergo transformations due to land clearing activities. Roads can usually be detected on the higher resolution instruments carried by the Landsat and SPOT satellites. As a mean of large scale monitoring this approach is however limited. The detection of road building on AVHRR imagery is restrained to the major undertakings because of the resolution limitation.

4.3.4. *The presence of smoke* can reveal unusual events in the forest canopy. While the relationship between smoke and deforestation is not straightforward, the presence of this unusual atmospheric contamination over larger area can be used as an indicator that "something is going on" in the forest canopy. This was one of the first and main indicators of the the large scale fire which damaged extensive portion of the East Kalimantan forest during the El Nino episode of 1982-83 (Malingreau *et al.* 1985).

5. Conclusions

Remote sensing techniques have been applied to the forestry sector since the early days of air-photography. High resolution satellite data have also been routinely applied to classification and mapping problems. Confronted with a globalisation of deforestation issues in the tropics, attempts have been made to use the lower resolution but high frequency cover data from the AVHRR on the the NOAA/TIROS-N satellite series. It has been shown that, in many instances, the AVHRR 1 km data allow adequate monitoring of the closed tropical forest cover at regional to continental scales. The approach relies upon the identification of thermal or albedo contrast patterns between the forest and non-forest covers. Using fire and biomass burning as an indicator of deforestation activities the analysis can lead to an evaluation of the dynamics of deforestation fronts. With little additional development the AVHRR approach can be made operational and lead to a monitoring system which could support the regular updating of deforestation assessments. In order to optimize the methodology, it should be supplemented by other layers of information supplied for example, by an associated geographical information system. Such an addition could lead to improvements in modelling and understanding spatio-temporal patterns of deforestation in the tropics.

Acknowledgements

The assistance of Frédéric Achard and Alan Belward in finalizing the draft version of this paper is gratefully acknowledged.

References

Achard, F., and Blasco, F., 1990, Analysis of vegetation seasonal evolution and mapping of forest cover in West-Africa with the use of NOAA AVHRR HRPT data. to be published in *Photogrammetric Engineering and Remote Sensing* in a Special Remote Sensing of Tropical Moist Forests Issue (November 1990).

Belward, A. S., and Malingreau, J. P., A comparison of AVHRR GAC and HRPT data for regional environmental monitoring, Presented at *IGARSS '89, 12th Canadian Symposium on Remote Sensing, University of British Columbia, Vancouver, Canada, 10th to 14th July 1989.*

Belward, A. S., and Lambin, E.,1990, Limitations to the identification of spatial structures from AVHRR data. *International Journal of Remote Sensing* **11**, 921-927.

FAO, 1981, Tropical Forest Resources Assessment. FAO, Rome, 3 volumes.

Fearnside, P., 1985, Environmental change and deforestation in the Brazilian Amazon. in : *Changes in the Amazon Basin. Man's Impact on Forests and Rivers.* edited by J. Herming (Manchester : Manchester University Press).

Holben, B. N., and Fraser R. S., 1984, Red and near-infrared sensor response to off-nadir viewing, *International Journal of Remote Sensing,* **5**, 145-160.

JRC., 1990, Commission of the European Communities Joint Research Centre-Ispra annual report for 1989, (in press).

Justice, C. O., Townshend, J. R. G., Holben, B.N. and Tucker, C.J., 1985, Analysis of the phenology of global vegetation using meteorological satellite data., *International Journal of Remote Sensing,* **6** 1271-1318.

Justice, C. O., Markham, B. L., Townshend, J. R. G., and Kennard, R. L., 1989, Spatial degradation of satellite data, *International Journal of Remote Sensing,* **10**, 1539-1561.

Langaas, S. and Muirhead, K., 1988, Monitoring bush fires in West Africa by weather satellites. *Proceedings of the 22nd International Symposium on Remote Sensing of Environment held in Abidjan, Cote d'Ivoire, on October 20 - 26, 1988* , pp. 253 - 268.

Laporte, N., 1990, Etude de la dynamique spatio-temporelle de la végétation tropicale: utilisation de donnée satellitaires NOAA-AVHRR sur l'Afrique de l' Ouest. Ph.D. thesis, University of Paul Sabatier, Toulouse, France.

Lloyd, D., 1989, Mapping phytophenology using shortwave vegetation index imagery, *Presented at 4th AVHRR data users' conference, Reichsstadthalle, Rothenburg ob der Tauber, Federal Republic of Germany, 5th - 8th September 1989*. pp 131 - 136.

Malingreau, J.P., Stevens, G., and Fellows, C., 1985, 1982-83 forest fires of Kalimantan and North Borneo. Satellite observations for detection and monitoring, *Ambio*, **14**, 314-321.

Malingreau, J.P., 1986, Global Vegetation dynamics : satellite observations over Asia. *International Journal of Remote Sensing*, **7**, 1121-1146.

Malingreau, J.P. and Tucker, C.J., 1987, The contribution of AVHRR Data for understanding Global Processes : Large scale deforestation in the Amazon Basin. Proceedings IGARSS, Ann Arbor, Michigan, pp 443-448.

Malingreau, J.P., and Tucker, C.J., 1988, Large-Scale Deforestation in the Southeastern Amazon Basin of Brazil., *Ambio*, **17**, 49-55.

Malingreau, J.P., Tucker, C. J., and Laporte, N., 1989, AVHRR for monitoring global tropical deforestation. *International Journal of Remote Sensing*, **10**, 855 - 867.

Malingreau J-P., and Belward A. S., 1989, Vegetation monitoring using AVHRR data at different resolutions. *Presented at 4th AVHRR data users' conference, Reichsstadthalle, Rothenburg ob der Tauber, Federal Republic of Germany, 5th - 8th September 1989*. pp 141-144.

Malingreau, J.P., and Tucker, C. J., 1990, Ranching in the Amazon Basin Large -scale changes observed by AVHRR, *International Journal of Remote Sensing*, **11**, 187-189.

Malingreau, J.P. and Laporte, N. 1990, Exceptional Fire Events in the Tropics : Southern Guinea January 1987. *International Journal of Remote Sensing* (in press).

Nelson, R. and Holben, B.N., 1986, Identifying deforestation in Brazil using multi-resolution satellite data, *International Journal of Remote Sensing*, **7**, 429-448

Nobre, C., 1990, Climate model simulation of the annual surface energy balance for Amazonia, INPE Space News; the newsletter of the Institute for Space Research (São José dos Campos: INPE) **1**, 2.

Repetto, R., 1990, Deforestation in the Tropics, *Scientific American*, **262**, 18-24.

Sharman, M., 1989, Monitoring agriculture in Europe using AVHRR: management and use of a large multi-temporal data set, *Proceedings of the 4th AVHRR data users' conference, held at Reichsstadthalle, Rothenburg ob der Tauber, Federal Republic of Germany, on 5th - 8th September 1989*. pp 175-178.

Schutt, J. and Kerber, A., 1986, Utility of AVHRR channels 3 and 4 in land cover mapping, *Photogrammetric Engineering and Remote Sensing* , **52**, 1877-18883.

Stephens G., and Matson, M., 1987, Regional and Global fire detection using AVHRR data, *proceedings of the 21st International symposium on Remote Sensing of Environment, Ann Arbor, Michigan, October 26-30*, pp 447-457.

Townshend, J.R.G., and Justice, C.O., 1989, The spatial variation of land cover at very coarse scales *Remote Sensing for Operational Applications, Proceedings of the 15th Annual Conference of the Remote Sensing Society, Bristol, 13 to 15 September 1989* (Nottingham : The Remote Sensing Society) pp 415-420.

Townshend, J.R.G., and Justice, C.O., 1988, Selecting the spatial resolution of satellite sensors required for global monitoring of land transformations. *International Journal of Remote Sensing*, **9**, 187-236.

Tucker, C.J., Townshend, J.R.G. and Goff T.E., 1985, African land-cover classification using satellite data. *Science* **227** : 4685, 369-375.

Viovy, N., Achard, F., Podaire, A. and Lavenu, F., 1988, Multitemporal Characterization and monitoring of vegetation in Africa using Satellite Data, ISLSCP Symposium, Niamey May 1988.

Woodwell, G.W., Hobbie, J.E., Hourhton, R.A., Helillo, J.M., Moore, B., Peterson, B.J. and Shaver, G.R., 1983, Global deforestation : contribution to atmospheric carbon dioxide. *Science* **222**, 1081-1086.

BASIC PRINCIPLES OF GEOGRAPHIC INFORMATION SYSTEMS

Carlos R. Valenzuela
I.T.C.
P.O. Box 6, Boulevard 1945
7500 AA Enschede
The Netherlands

1. Introduction

The demand for the storage, analysis and display of complex and voluminous environmental data has led, in recent years, to the use of computers for data handling and the creation of sophisticated information systems. (Tomlinson, 1976). Effective use of large spatial volumes is dependant upon the existence of efficient systems that can transform these data into usable information (Marble and Peuquet, 1983).

Geographic information systems (GIS)-a new technology-are becoming essential tools for analyzing and graphically transferring knowledge about the world. Burrough (1986), defined a GIS as "a powerful set of tools for collecting, storing, retrieving at will, transforming and displaying spatial data from the real world for a particular set of purposes". GIS's are changing spatial data collection procedures and analytical processes. Advances in computer technology and techniques have made it possible to integrate a wide range of information (Gribb, 1984). Technological advances have increased input techniques, storage, analysis and retrieval capabilities. Furthermore, there has been a reduction in costs and an increase in accessibility, so that a larger user community has been developing (Moellering, 1982).

GIS are used to assist decision makers by indicating various alternatives in development and conservation planning and modelling the potential outcomes of a series of scenarios as illustrated in Figure 1. They provide planners with a readily accessible source of objective earth science related facts and, an inexpensive, rapid, and flexible tool for combining these facts with various products to create decision alternatives (van Driel, 1975; Star and Estes, 1989). Basic knowledge on the location, quantity and availability of natural resources is thus indispensable for more rational planning.

Computerized mapping and spatial analysis has been developed simultaneously in several related fields. The present status would not have been achieved without a close interaction among the various fields. These fields are:

- cadastral and topographic mapping
- thematic cartography
- civil engineering

A. S. Belward and C. R. Valenzuela (eds.),
Remote Sensing and Geographical Information Systems for Resource Management in Developing Countries, 279–295.

- geography
- mathematical studies of spatial variation
- soil science
- surveying and photogrammetry
- rural and urban planning
- utility networks
- remote sensing and image processing
- Artificial intelligence
- computer science

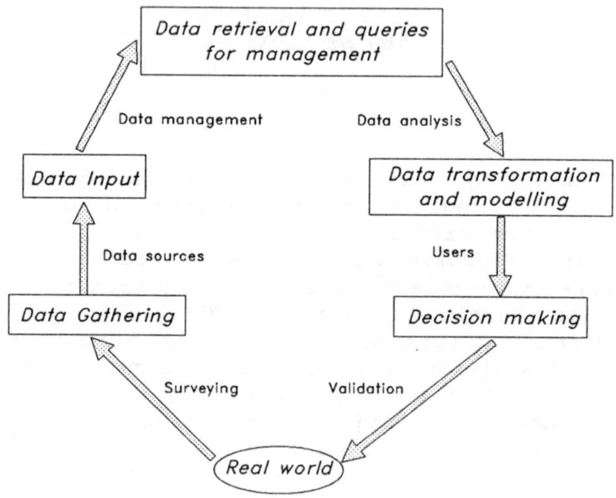

Figure 1. Geographic information systems as a management tool

GIS are rapidly becoming a standard tool for management of resources and at present it is difficult to think of a resource planning or mapping agency without a geographic information system or contemplating the implementation of such a system. Effective utilization of large spatial data volumes is dependant upon the existence of an efficient geographic handling and processing system that will transform these data into usable information (Meijerink et at., 1980).

Typical GIS applications can be grouped into the following:

- automatic cartography
- subdivision design (cut/fill, street and parcel layout)
- cadastral mapping
- highway mapping
- utility facility mapping and management
- geodesic mapping

- event mapping (accident, crime, fire, facility)
- census and related statistical mapping
- management of well log data
- land use planning and management
- environmental impact studies
- natural resources mapping and management
- land information systems
- marketing and retailing
- vehicle routing and scheduling
- urban and regional planning
- route selection of highways, pipelines and similar data
- surveying and engineering
- mineral exploration

Currently, very good software for analyzing geographic information is readily available and the computer hardware available now is quite powerful and rapidly increasing in capabilities. Visualization technology has provided high resolution, colour display capabilities, interactive editing and a wide range of other features.

Extremely versatile software systems exist, These systems deal rapidly with raster as well as point, line and polygon data, and can interconvert among all these data types.

Some of these modern software systems are really software "toolboxes" for the analysis and manipulation of geographic information: they contain a large set of tools in the form of discrete software packages which perform certain discrete functions very efficiently. These tools can be used in various combinations to accomplish any particular application "job". So, rather than devising many special purpose software systems to meet the needs of diverse users, the approach is to create modules (tools) which may be combined in several ways to satisfy the needs of a variety of users. These systems offer maximum flexibility in meeting the needs of a wide variety of users, and, by adding new tools, can keep up with rapidly changing needs in a dynamic field (Dangermond,1986).

Current types of geographic information systems are grouped into:

- engineering mapping systems (CAD/CAM) for applications in photogrammetry, topographic basemaps, road engineering
- property of parcel information systems
- generalized thematic and statistical mapping systems for natural resource management, forest inventories, vegetation, geology, soils, census mapping
- bibliographic systems about geographic documents
- geographic base file systems associated with street networks
- image processing systems

A digital geographic information system is an information system which has as its primary source of input a database composed of data referenced by geographic coordinates and in which a major part of the processing is done with a digital computer (Kennedy and Meyers, 1977). Geographic information systems consist of four basic components; data gathering and input, geographic databases, data analysis and modelling and data visualization and presentation. Kennedy and Guin (1975) considered automated spatial information systems as essentially data driven while models which use the data are important to support the decision making activities of those who use the system, a large portion of the investment will be in obtaining, converting and storing new data.

Bryan and Zobrist (1976) proposed four criteria that should be satisfied to make a GIS useful. These criteria are: (1) Point and area location must be provided, (2) it should be possible to perform variable aggregations or subsettings, (3) there should be representation of spatial arrangements of the data, and (4) the system should be able to interface with mathematical and statistical analysis programs.

Peuquet (1977) indicated that a GIS should incorporate a variety of analytical techniques, which combined, suit the needs of the various users of the system. It should be open ended, in order to allow the incorporation of new analytical techniques to accommodate changes in user needs (Tomlinson, 1976). From several reviews, it is evident that the objectives of GIS have become more ambitious as faster computer systems evolve and peripherals become more sophisticated (Miller, 1975; Power, 1975; Jerie et al., 1980, Moberg and Laefsky, 1982; Camarata, 1984; Orloff, 1985; Burrough, 1986, Valenzuela, 1988; Aronoff, 1989). Figure 2 illustrates schematically the various components of an operational GIS.

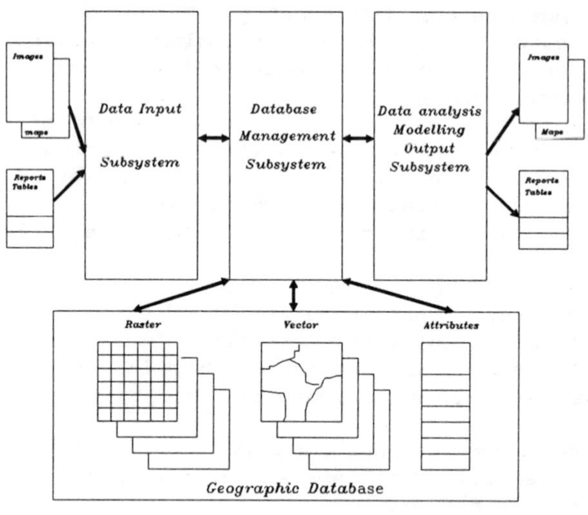

Figure 2. Schematic representation of a GIS

2.Cartography and Maps

Maps have been used since the earliest civilizations to portray information about the earth surface. The earliest specimen that is indisputably a portrayal of land features thus far discovered is a clay tablet dated about 2300 BC. certain land drawings found in Egypt and printings discovered in early tombs are nearly as old.

It appears that there were few attempts by Babylonians and Egyptians to show the form and extent of the earth as a whole. Not until the time of the Greek, geographers did speculations and conclusions as to the nature of the earth. The Greeks were outstanding among peoples of the ancient world for their pursuit and development of geographic knowledge.

Land surveying and map making were integral part of the Roman Government. As with other fields of science and technology, with the decline of the Roman Empire, progress in geography and cartography was abruptly curtailed.

A map is a kind of information system. A map is a representation, normally to scale and on a flat medium, of a selection of material or abstract features on, or in relation to, the surface of the earth. The term map is often used in mathematics to convey the notion of transferring information from one form to another, just as cartographers transfer the information from the surface of the earth to a sheet of paper. It is also used loosely to refer to any visual display of information, particularly, if it is abstract, generalized or schematic.

The terms, map, chart, and plat are used somewhat interchangeable. The connotations of use, however, are distinctive: chart for navigation purposes (nautical and aeronautical), plats (in a property boundary sense) for land-line references and ownership, and maps for general reference.

Cartography is the art and science of making maps and charts. As such, it is allied with geography in its concern with the broader aspects of the earth and its life.

2.1.TYPES OF MAPS

Nautical charts are maps of coastal and marine areas, providing information for navigation.

Aeronautical charts provide essential data for the pilot and air navigator. They are, in effect, small-scale topographic maps on which current information on aids to navigation have been superimposed.

Topographic maps are graphic representations of natural and man-made features of parts of the earth surface plotted to scale. They provide a complete inventory of the terrain and important information for all activities involving the use and development of the land. The provide the bases for specialized maps, and data for compilation of generalized maps of smaller scale.

Thematic maps are specialized maps containing information about a single subject or theme. They constitute a tool to communicate geographical concepts such as land use, population densities, crop yields, soils, landform, distribution of pH values over an experimental field, etc. Thematic maps may be quantitative (effective depth of soil). They can be expressed as **choropleth maps** or areas of equal value separated by boundaries, examples are land use maps, soil maps, census maps or administrative boundary maps. **Isopleth maps** portray quantitative data modelled by a continuous surface that is capable of mathematical description. The variations are then shown by isolines or contour-lines connecting points of equal value-. They are used for phenomena which vary smoothly across the map, such as temperature, precipitation, or population density.

The scale of a map is the ratio between distances on the map and corresponding distances in the real world. For instance, if a map has a scale of 1:25,000 then 1 cm on the map represents 25,000 cm or 0.25 km on the earth surface.

Large scale maps show great detail, the representative fraction is large : 1:10,000.

Small scale maps depict only large features, the representative fraction is small : 1:100,000.

The scale of the map is related to the information level of the map. Large scale maps have greater detail of information than smaller scale maps.

2.2.MAP PROJECTIONS

The basic problem of map projection is the representation of a curved surface in a plane, or the problem of representing the earth surface on a flat map. This may be readily surmised by trying to apply an orange peel to a flat surface. to achieve a continuous contact between the two surfaces, the orange peel would have to be distorted by stretching, shrinking or tearing and this, while being a gross oversimplification of the problem of map projections, is nevertheless a good illustration of the impossibility of a perfect solution.

All representations of a curved surface on a plane involve "stretching" or "shrinking" resulting in distortions or "tearing" resulting in interruptions. Three principal cartographic criteria are applied to the evaluation of map projection properties:

(a) Equidistance - Correct representation of distances, projections preserve distance to places from one or at most two points

(b) Comformality (orthomorphism) - Correct representation of shapes and show directions (bearings) correctly.

(c) Equivalency - Correct representation of areas

3.Data gathering and input

Methods for efficient data gathering have been and are being developed-designed for their use in geographic information systems, and providing accuracy and reliability consistent with the application level of the information. The process usually involves the interpretation of aerospace images, intensive and well-designed field observations and the support of all available existing data sources. Remote sensing products are an important data source and are also used for monitoring and updating procedures. Image processing techniques provide the tools to transform more efficiently the remote sensing data into information.

Intensive field work and efficient sampling procedures are essential to ensure the validity and accuracy of the data incorporated into the system for further manipulation, analysis and modelling procedures.

A geographic entity is defined by two types of data, ie, geometric and non-geometric. The geometric data relate to the geographic location of the entity. The non-geometric data relate recorded information (usually descriptive) to a geographic entity. The transformation of these geometric data into a format compatible with digital computers requires the use of a digitizing procedure. Attribute data are usually entered from a keyboard.

Data capture or digitization is a sequence of tasks for encoding the position identifiers of spatially-oriented data. In its narrowest form, digitizing can be viewed as determining X and Y coordinate values to describe the location of points, lines, areas or surfaces as these are depicted in one or more maps. In a broader sense, however, digitizing is the task of creating an acceptable, error-free, machine-readable data file from a variety of sources (Figure 3).

The transformation of data from maps (analogue) into a computer compatible form (digital) is one of the most time-consuming and costly tasks in the creation of a GIS. Data input requires editing operations to verify the digital data against the original map and subsequently correct the errors induced by both the human analyst and the hardware and software used.

4.Geographic database

A spatial database describes a collection of entities, some of which have a permanent location on some global, dimensioned space (Figure 4). Normally, there is a mixture of geometric (spatial) and non-geometric (a-spatial) entity types which may change in time. Geometric entity types have the basic topographic properties of location, dimension and shape. Non-geometric entities are not recognized as having permanent locations with respect to other entities. The spatial data describe the location and topology of points, lines, polygon and surface features, while the a-spatial data describe the characteristics of these features.

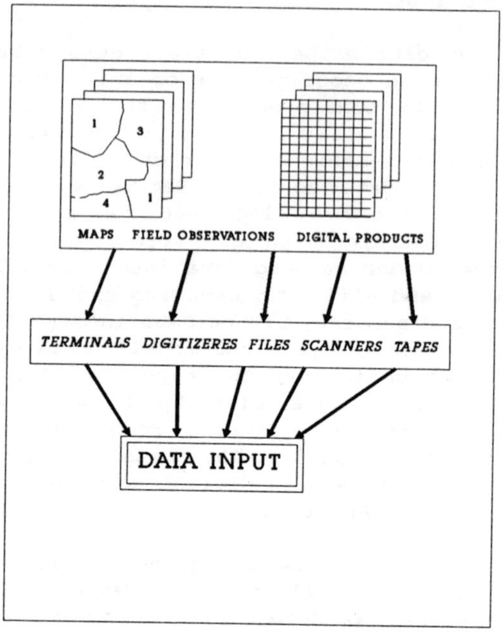

Figure 3. Data input

4.1.GRAPHICS DATABASE.

Geographic information includes attributes, a spatial location, extension or configuration and are dynamic (change in time). These geometric attributes can be metrical or topological. Metrical includes position, shape and size that can be expressed by spatial coordinates. Topological attributes are those describing characteristics such as connectivity and adjacency that are invariant under distortions and changes of scale.

Geographic entities are conventionally divided into points, lines, polygons and surfaces. Polygons are perhaps the most frequently encoded features in geographic information systems (Deucker, 1974; Werner, 1974; Peucker and Chrisman, 1975; Nagy and Wagle, 1978; Burrough, 1986; Aronoff, 1989). Essentially, the computer representation of point, line, polygon and surface entities takes one of two forms, ie, gridded (cellular, raster) data or vector data (Phillips, 1974; Baxter, 1976; Peuquet, 1977; Deucker, 1979; Nagy and Wegle, 1978; Burrough, 1986; Aronoff, 1989).

A raster data structure consists of a matrix of cells of uniform size, each one referenced by a unique positional index (line and column number).

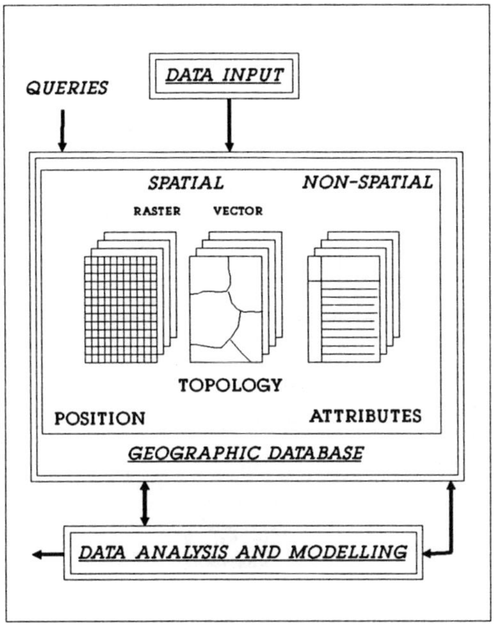

Figure 4. Geographic database

It contains a number or code representing the type of value of the attribute being mapped, which can be ordinal (scalar or vector) or nominal. A major disadvantage of the cellular organization is its wasteful use of computer storage for spatially sparse data, although this can be overcome by using some data compaction techniques (Burrough, 1986).

The vector data structure represents point entities directly by their coordinates, line entities by a chain of uniformly or non-uniformly spaced coordinates, and area or polygon entities by their boundaries.

Remote sensing products are also part of the graphics database, which makes it possible to overlay these data with GIS-derived data. Integrating image processing techniques with conventional GIS manipulation procedures provides possibilities for optimizing the transformation of data into information (information extraction) and improving the data quality for modelling or rule base operations.

Figure 5 illustrates a flow diagram of the use of remote sensing products in a GIS.

4.2. ATTRIBUTE DATABASE

In addition to the spatial representation of a feature, each entity usually has a number of other important attributes. These attributes may be either nominal, such as soil series name, crop type, or scalar, such as depth of water table or contour elevation.

A database system is essentially "a computerized record keeping system, that is, a system whose overall purpose is to maintain data and to make that data available on demand". The database itself can be regarded as a kind of electronic filing cabinet (Date, 1986).

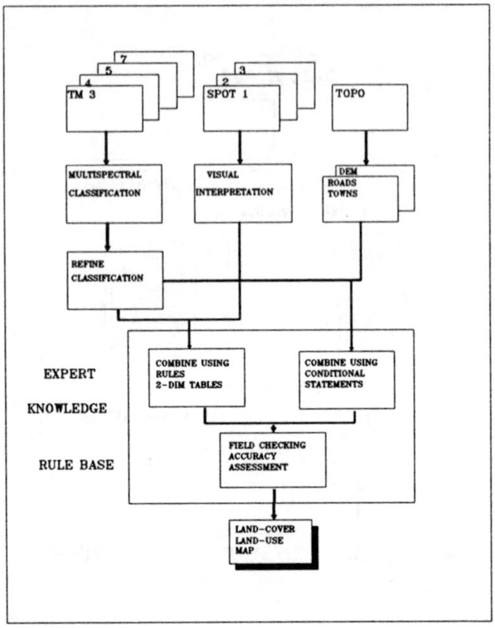

Figure 5. Integration of Remote sensing and GIS

Computerized database systems consist of four major components:

- data, which are stored in one or more databases forming the database system
- hardware, the equipment that makes it physically possible to input, store, retrieve and visualize the data
- software, which provides users with a set of statements to access and analyze the stored data
- users, a person or group of persons who use the stored data

Databases provide facilities to perform several operations commonly required in manipulations of attribute data. These operations include:

- adding new datasets to the database
- inserting new data into existing datasets
- retrieving data from existing datasets
- updating and/or transforming data from existing datasets
- deleting data from existing datasets
- removing datasets from the database

The data model, on which the database system is based, consists of both the types of data structure used to represent data entities and associations between data entities in the database, and the set of primitive operations (commonly referred to as data language) that can be used to manipulate the data stored in these structures. There are three principal data models proposed for generalized databases: relational, hierarchical and network.

5.Data analysis, modelling and cartographic manipulations

One of the most important characteristics of geographic information systems is their capability for data analysis and spatial modelling.

Conventional GIS analysis and manipulation capabilities include map overlaying, reclassification procedures, proximity analysis, optimum corridor and other cartographic modelling techniques.

Reclassifying map categories involve operations that reassigned thematic values to the categories of an existing map as a function of the initial value, the position, the size or the shape of the spatial configuration associated with each category, for instance, a soils map reclassified into a permeability map.

Overlaying maps result in the creation of a new map where the values assigned to every location on that map are computed as a function of independent values associated with that location on two or more existing maps.

Determining distance and conductivity include operations for measuring cartographic distance and result in the creation of new maps in which the distance and route between point can be expressed as a simple euclidian distance or as a function of absolute and/or relative barriers.

These capabilities are used either by themselves or in conjunction with other simulation or statistical modelling. These include land evaluation, agro-ecologic, crop production, hydrologic, land degradation, erosion modelling, risk and hazard assessment and others. Image processing techniques are also an important tool in a system's analysis capabilities. In cartographic modelling, the problem (conceptual model) to be solved is decomposed into sub-models that can be solved separately. The model is composed in such a way that the solution of the individual sub-models can be combined to yield solutions to the original model (Bunn, 1984). Each sub-model can be further decomposed into smaller models that can be solved directly. The basic model of decision analysis is to break a decision problem into component parts, thereby allowing the decision maker to focus separately upon the critical issues. The pattern of relationships that allows the model to be decomposed is referred to as the decision model. Figure 6, shows the data flow in a GIS.

6.Data output

The output module (Figure 7), provides products required by the user and data analyst. These products include maps, graphs, tables, photographs, and magnetic devices. It should be noted that the quality of the displays is a function of the quality of the output devices. Output devices range in price-and hence in quality-from the low cost to the extremely expensive.

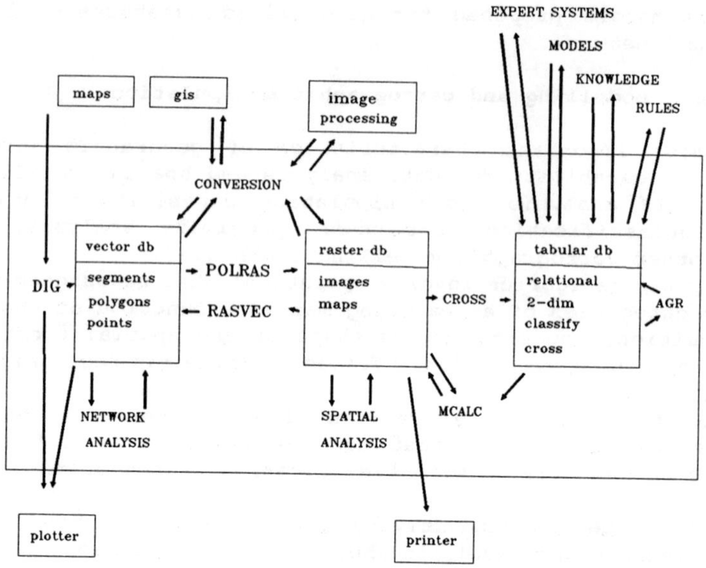

Figure 6. Data flow in a GIS (After Gorte, et al., 1988)

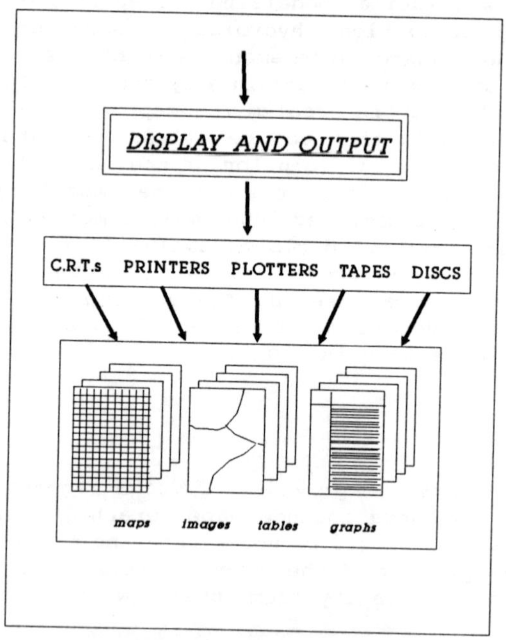

Figure 7. Data output

7.Available GIS

Geographic information systems are available both in PC/micro computers and mini and main frame computers. Tables 1 and 2 list a summary of some commercially available geographic information systems. It should be noted that these tables are by no means complete or exhaustive and do not necessarily reflect characteristics of current systems.

Table 1. GIS in mini and frame computers (From GIS World, 1989)

System Name	Hardware	Geometric storage	Attribute storage
ARC/INFO	VAX, PRIME IBM, DG	vector	relational
DELTAMAP	HP, SUN	vector	relational
INFOMAP	VAX	vector	relational
INTERGRAPH	VAX	vector	network
MAPS	VAX, PRIME	vector	relational
SICAD	SIEMENS	vector	relational
SYSSCAN	VAX	vector	relational
GEOBASED	VAX	vector	relational
SYSTEM 600	VAX	vector	relational

Table 2. GIS in PC/micro computers (From GIS World, 1989)

System Name	Hardware	Geometric storage	Attribute storage
ARC/INFO	IBM PC/AT SYSTEM 2	vector	relational
SPANS	IBM PC/AT SYSTEM 2	quadtree vector	
INFORMAP II	IBM/AT	vector	relational
ERDAS	IBM/AT	raster	
ILWIS	IBM/AT	raster vector	relational
PAMAP	IBM/AT	raster	relational
IDRISI	IBM/AT	raster	

References

Aronoff. S. 1989. Geographic information systems: A Management perception. Ottawa, WDL Publications.

Baxter, R.S. 1976. Some methodological issues in computer drawn maps. The cartographic journal, p.145-155.

Bryan, N.A. and A.L. Zobrist, 1976. Integration of socio-economic data and remotely sensed imagery for land use applications. Proceedings of the 2nd Annual Pecora symposium. Sponsored by the American Society of photogrammetry and the U.S. Geological survey. USA.

Bunn, D.W. 1984. Applied decision analysis. McGraw Hill Book Co, New York, USA.

Burrough, P.A. 1986. Principles of geographical information systems for land resource assessment. Clarendon Press, Oxford, UK.

Camarata, S.J. 1984. Digital image processing and geographic information systems: A mini/micro computer application interface. Technical papers of the 44th annual meeting of the American Congress on surveying and mapping, ASP/ACSM Convention, Washington, D.C., USA.

Dangermond, J. 1986. CAD vs. GIS. Computer graphics world. October.

Date, C. J. 1986. An introduction to database systems (4th ed). The systems programming series, Addison-Wesley.

Deucker, K.J. 1974. Urban geocoding. Annals, Association of American geographers, vol 64, p.318-325.

Deucker, C.J. 1979. Land resource information systems: A review of fifteen years experience. Geo-processing, 1, P.105-128.

Driel van, N. 1975. Geological information in a computer mapping system for land use planning. Proceedings of the international symposium on computer assisted cartography, Bureau of the Census and ACSM, Auto Carto II.

GIS World, 1989. Special reports on GIS technology '89: Results of the 1989 GIS World Geographic Information Systems Survey. Fort Collins, Colorado, GIS World Inc.

Gorte, B., R. Liem and J. Wind, 1988. The ILWIS software kernel. ITC Journal 1988-1, Enschede, The Netherlands.

Gribb, M.F. 1984. The combining of geographic and other information systems to create a new type of information system. Technical papers of the 44th annual meeting of the American Congress surveying and mapping, ASP/ACSM Convention, Washington, D.C.

Jerie, H.G., J. Kure, and H.K. Larsen, 1980. A systems approach to improving geo-information systems. ITC Journal 1980-4, Enschede, The Netherlands.

Kennedy, M. and C. Guin, 1975. Automated spatial data information systems: Avoiding failure. Urban study center, Louisville, USA.

Kennedy, M. and C.R. Meyers, 1977. Spatial information systems: An introduction. Urban study center, Louisville, USA.
Marble, D.F. and D.J. Peuquet (eds). 1983. Geographic information systems and remote sensing. Chapter 22, vol 1, Manual of remote sensing, American Society of Photogrammetry, USA.

Meijerink, A. M. J., C. R. Valenzuela and A. Stewart (eds). 1988. ILWIS, The integrated land and watershed management information system. ITC publication number 7. ITC, Enschede, The Netherlands.

Miller, W.R, 1975. A survey of geographically based information systems in California. Intergovernmental board on electronic data processing, State of california, May. USA.

Moberg, D. and J.M. Laefsky, 1982. Videodiscs and optical data storage. Byte, vol 7, no 6, p. 142-160

Moellering H. 1982. The challenge of developing a set of national digital cartographic data standards for the United States, Proceedings ACSM/ASP, Spring meeting, Denver. USA.

Nagy, G. and S.G. Wagle. 1978. Geographic data processing. Computing surveys, vol 11, No 2.

Orlof, R.W. 1985. Geo-graphics. P.C. World, January, 1985.

Peucker, T.K. and N. Chrisman, 1975. Cartographic data structures. The American cartographer, Vol 2.
Peuquet, D.J. 1977. Raster data handling in geographic information systems. Report R 77/7, Geographic information systems laboratory, State University of New York, Buffalo, New York, USA.

Phillips, R.L. 1974. Computer graphics in urban and environmental systems. Proceedings of IEEE, vol 62, no. 4, p.437-452.

Power, M. 1975, Computerized geographic information systems: An assessment of important factors in their design, operation and success. Washington University, Saint Louis, Missouri.

Star, J. E. and Estes, J. L. 1989. Geographic information systems: An introduction. Inglewood Cliffs, New Jersey. Prentice Hall.

Tomlinson, R. F. 1976. Computer handling of geographical data. UNESCO.

Valenzuela, C. R. 1988. ILWIS overview. ITC Journal, 1988-1., Enschede, The Netherlands.

Werner, P.A. 1974. National geocoding. Annals, Association of American geographers, vol 64, p.310-317.

COMPUTER SYSTEMS FOR GEOGRAPHIC INFORMATION SYSTEMS

Michael J.C. Weir
I.T.C.
P.O. Box 6, Boulevard 1945
7500 AA Enschede
The Netherlands

1. Introduction

A computer and associated peripheral devices are essential equipment for handling spatial data in a geographic information system. These devices are collectively known as hardware and, together with the system software to control their operation, form the basic elements of a computer system. Although it is not necessary to be a computer expert to work with GIS, an elementary knowledge of computer hardware and software is essential in order to understand how data are handled by the system. This chapter provides some brief notes on computer systems for GIS applications.

A basic computer system comprises the following hardware:
- the central processing unit
- storage devices (hard disk and diskette drive)
- keyboard
- monitor.

For GIS applications, this hardware alone is not normally sufficient. Additional devices will normally be required, namely:
- a digitizer and/or a scanner for inputing spatial data from existing maps;
- a high resolution colour monitor for displaying the results of GIS operations to the system user;
- a printer for printing raster maps, remote sensing images and textual information;
- a plotter for high-quality production of maps in vector form.

These peripheral devices for input and output are discussed separately in the chapters on these topics.

Digital computers range in size, capacity and capability from large mainframe computers, capable of carrying out a wide range of tasks for several users at high speeds, down to small personal computers (PCs). Until recently, most geographic information systems operated on multi-user minicomputers. With the increasingly favourable price performance ratio of computer systems, these machines are being replaced by smaller, less expensive workstations and powerful PCs.

A. S. Belward and C. R. Valenzuela (eds.),
Remote Sensing and Geographical Information Systems for Resource Management in Developing Countries, 297–300.

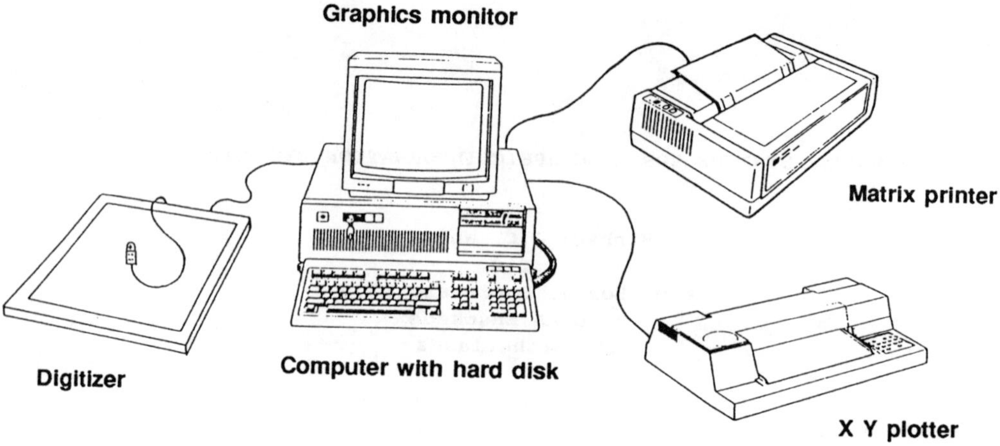

Figure 1. Hardware required for a simple PC-based GIS.

2.Central processing unit

The most important hardware component of a computer system is the Central Processing Unit (CPU). The CPU controls the transfer of data and performs basic arithmetic operations (addition and subtraction). Although these capabilities are excedingly limited, they can be performed at extremely high speeds, thus enabling complex manipulations to be performed on large amounts of data. The main CPUs found on personal computers are based on the Intel 80286, 80386 and 80486 processors.
The operations performed by the CPU are controlled by instructions in the form of a computer program. Computer programs are known as software. A distinction is made between the system software for controlling the various hardware devices and applications software, i.e. programs written to carry out specific tasks required by the user. All these instructions are normally written in a programming language such as BASIC, FORTRAN or Pascal and then translated into machine code. The machine code is a list of binary codes which are transmitted to the CPU as positive and negative voltages.

3.Internal memory

All computers have an internal memory which functions as a "working space" for holding programs and data. Every part of this internal memory is equally accessible to the CPU and for this reason it is known as

Random Access Memory (RAM). The capacity of the RAM is measured in bytes, one byte being composed of 8 bits of binary code. Depending on their power, PCs have a RAM capacity between 640 kilobytes (1Kb = 1024 bytes) and 4 megabytes (1Mb = 1024 x 1024 bytes). Internal memory space is not permanent and is cleared when the computer is switched off. RAM therefore cannot be used to permanently store data and programs. For this purpose, external (mass) storage devices are used.

4.External storage

Formerly, data and programs were stored on punched cards or paper tape. Nowdays the most common forms of storage are magnetic tape and magnetic disks, both of which can store data in a physically compact form.

Magnetic tape is held on either large reels or small cassettes.The advantage of magnetic tape is the possibility of storing very large amounts of data (even the most simple map will, in digital form, normally comprise several thousand coordinate pairs). Data is read from or written to the tape using a device known as a tape reader or streamer. Accessing data stored on tape is rather cumbersome. Data are stored sequentially and the tape may have to be repeatedly wound and rewound during complex manipulations.

Much faster access times are possible with magnetic disks. These may take the form of either:

- a fixed (hard) disk built into the computer, or
- flexible (floppy) diskettes which can be used on any machine provided it is fitted with the correct type of disk drive for reading from and writing to the diskette.

Hard disks have a relatively large capacity (PCs typically have hard disks with 20 or 40Mb capacity, but more is possible). Diskettes are have a much smaller capacity and are available in various sizes and capacities. The most common are 5.25 inch diskettes with 360Kb or 1.2Mb capacity and 3.5 inch diskettes with 720Kb or 1.4Mb capacity.

Although regularly used software and data may be held on a hard disk, all programs and data must be archived on tape or diskettes. It is important that data be archived as soon as possible after processing and that reserve (so-called "backup") copies of all software and important data be kept in safety, e.g. in a fireproof safe or at another location.

5.Operating systems

All programs and data are stored in the form of files. Normally each GIS operation will result in the creation of one or more new files. It is therefore essential to have some form of "housekeeping" facilities for handling these files. Such facilities are contained in special software known as the operating system which oversees the computer and all operations which it performs. Most personal computers operate under an operating system known as MS-DOS (Microsoft Disk Operating System). An elementary knowledge of the computer's operating system is essential for working with geographic information systems and other applications software.

6.Concluding remarks

There is a very wide range of computer hardware, software and supplies available on the market. A useful way to keep up-to date with developments in these areas is to refer to some of the many popular computer magazines available at most bookstores.
Because the computer hardware and software is such a rapidly changing field, no attempt has been made to provide details of the products of specific manufacturers in this short chapter.
Although most GIS vendors are happy to advise on the most appropriate hardware required to run their system, it is advisable to seek the advice of a competent consultant before making any major investment in either hardware or software.

DATA INPUT AND OUTPUT

Michael J.C. Weir
I.T.C.
P.O. Box 6, Boulevard 1945
7500 AA Enschede
The Netherlands

1.Data input

Before any spatial analysis or modelling operations can be carried out in a geographic information system, it is necessary to input the requisite data. The task of data input is generally the most time consuming-and therefore costly-aspects of GIS implementation and operation. Despite-or perhaps because of-this fact, the process of data input may not be given sufficient consideration. In particular, careful attention must be paid to quality control in order to ensure that the input data are of the highest quality. Without quality control, the time and expense of data collection and input may be wasted.

2.Data sources

Geographic information systems are essentially concerned with spatially referenced data. These data will normally be obtained from one or more of the following sources:

- existing maps
- aerial photographs
- satellite imagery
- data from airborne scanners
- field measurements
- other GIS databases.

Associated with the spatial data are the non-spatial (attribute) data obtained from field observations, point sampling, census figures, etc. A discussion of the techniques used to collect spatial data (e.g., land surveying, photogrammetry and satellite image processing) and of the field sampling techniques employed in resource surveys is beyond the scope of this chapter. Nevertheless, the GIS user should be aware of these techniques in order to assess the lineage and quality of the data which he or she is using. This question is becoming increasingly important as existing digital data sets become available to other users.

A. S. Belward and C. R. Valenzuela (eds.),
Remote Sensing and Geographical Information Systems for Resource Management in Developing Countries, 301–309.

3.Entering spatial data

The method of spatial data input depends primarily on the source of the data. The two principle approaches for inputting data from maps are:

- manual, by cartographic digitizing, and
- automatic, by scanning.

3.1.DATA INPUT BY CARTOGRAPHIC DIGITIZING

The most usual method of converting existing maps into digital form is by digitizing. Using a digitizer, all features are recorded as a series of X- and Y-coordinates. Data input from aerial photographs, using photogrammetric plotters fitted with encoders, is basically similar with the exception that the third dimension (terrain elevation) is also recorded as a Z-coordinate.

A digitizer comprises a tablet or table, on which the map is fastened, and a cursor by means of which lines and points on the map are located. In most digitizers the coordinates are measured by means of a fine grid of wire loops embedded into the table (figure 1). An wire coil in the cursor emits an electromagnetic pulse which is detected by the grid of wires and transmitted to the computer as a coordinate pair (expressed in millimetres in the digitizer XY coordinate system). Most cursors are fitted with four or more buttons for transmitting special signals for program control, e.g. to indicate the end of a line.

Digitizers are available in a range of formats, ranging from small 27cm x 27cm tablets up to large (1m x 1.5m) tables. Some models are back-lit to facilitate digitizing from drawing film or other translucent material.

The resolution, or precision, with which coordinates are recorded varies from device to device. In general, expensive, large format digitizers have a higher resolution - around 0.025mm. The absolute accuracy, or repeatability, is generally about three times the resolution.

The process of map digitizing involves:

- preparation of the map for digitizing
- the recording of the map coordinates (the actual digitizing)
- editing and correcting the data before storing in the map data base.
- entering the associated attribute codes.

3.1.1.Map preparation

Before starting digitizing it is necessary to ensure that the source map is complete (including a grid or graticule), correct, and free of physical damage (folds, tears, etc.). In some cases it may be advisable to redraft the map, retaining only those features which have to be recorded, thus giving the operator a clean, tidy document from which to work. If feature labels are to be added during digitizing, the operator must also be provided with a list of the relevant codes.

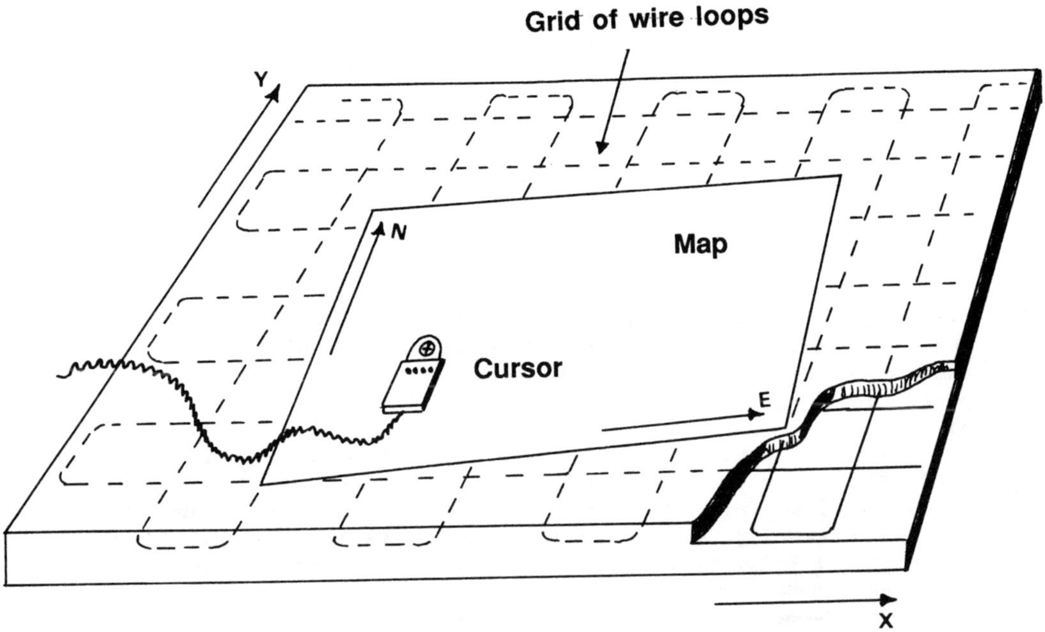

Figure 1. Digitizer

3.1.2.Recording the map coordinates

A digitizer records coordinates in units of millimetres (most digitizers can also record in inches if required). These coordinates are referenced to the X- and Y-axes of the table. Since terrain coordinates are normally required, the digitizer coordinates must be transformed to the terrain or map coordinate system. This transformation involves a rotation (to align the map with the table axes) and a scale change (from millimetres on the table to metres in the terrain). The relevant transformation parameters are computed by digitizing at least three control points. Usually additional known points are digitized to allow for correction due to map shrinkage, etc. Since all coordinates are referenced to the terrain coordinate system, the map may be removed from the table and then put back again later without affecting the results.

During digitizing, the progress of the work will be displayed on the monitor of the computer. This involves computing another transformation, namely from the digitizer to the screen coordinate system.

Point features are recorded as simple X,Y coordinate pairs. Linear features may be recorded in pointwise mode, i.e. at intervals selected by the operator, or continuously (in so-called "stream" mode). In the latter case, coordinates are recorded at either fixed distances or fixed time intervals. In both cases, it will normally be necessary to perform a data reduction procedure to remove superfluous data (figure 2).

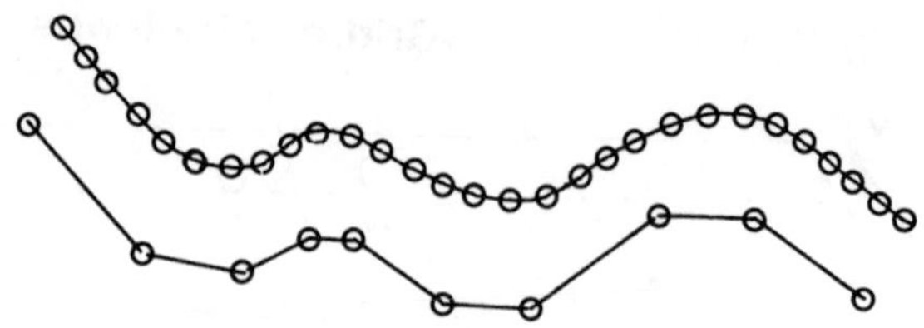

Figure 2. Data reduction

The digitizing of point and individual linear features such as roads does not cause any serious difficulties. Area features, such as land use units or administrativel/istricts are, however, a different matter. In manual map production it is more or less obvious that a given area is bounded by certain lines.

This can also be achieved simply by digitizing each area as a separate polygon. This approach has a serious disadvantage however - nearly all boundaries must be followed twice. Not only is this time consuming, but also it is virtually impossible to avoid creating gaps and overlaps between adjacent polygons which will, at best, make the plotted map untidy and at worst create many small spurious polygons during map overlay. Furthermore, there may be difficulties in dealing with "islands", i.e. polygons nested within other polygons.

To avoid these problems, it is preferable to digitize common boundaries once only. This means however, that the computer must be instructed how to form polygons from the individual line segments. In order to do this it is necessary to exactly define the topology of the map, i.e. the connectivity and relationships between lines, nodes and polygons. It is clear that this task, done more or less automatically in manual cartography, is not easy to define in a simple, logical way for the computer. Indeed, many early digitizing programs required a considerable amount of effort to build the topology by recording nodes and polygon centroids. In most modern digitizing software these tasks are handled automatically.

3.1.3.Automatic digitizing - line followers

Since the digitizing process is very time consuming-all lines on the map have to be followed manually-agencies with large workloads are showing an increasing interest in the use of devices for automating the process.

A line follower can be thought of as an "automatic" digitizer. The device comprises a detector which "locks" onto a line, follows it and records the coordinates. Despite the high degree of automation provided by line followers, some operator intervention is necessary to solve confusing situations, especially at line junctions. The equipment is therefore particularly useful for following continuous lines such as contours.

3.1.4.Editing

Since digitizing is a tedious operation, mistakes will inevitably occur. Editing facilities are therefore necessary. Common errors occurring during the process of map digitizing include:

- failure to snap lines together at nodes
- over- and undershooting of lines
- omission of lines or points
- double recording of lines or points
- incorrect feature coding
- incorrect location of (parts of) features.

Typical editing facilities required to correct such errors include:

- a zoom function to enlarge or reduce the area of attention currently displayed
- deletion of the last point(s) digitized
- deletion of complete lines (this facility is also required for updating existing digital data)
- a snap function to close a line onto a given point or node
- a function to move individual points, or even complete lines, to a new location or alignment
- special geometric functions, e.g. to square buildings.

3.1.5.Attribute entry

Feature labels, e.g. "road" or "land use boundary" are normally entered via the keyboard, cursor button or by digitizing within a special menu area reserved on the digitizer surface immediately before or after a line is digitized. Attribute labels for area features (polygons) are normally entered only after the topology of the digitized data has been checked and if need be corrected.

3.2.DATA INPUT BY CARTOGRAPHIC SCANNING

Since data input by manual digitizing is very time consuming, many large agencies make use of scanners. Although the actual scanning process is fast, it is still necessary to prepare a clean, unambiguous source document (often by redrafting the original map. The map is placed on a drum which rotates in front of a detector which records whether an image is present or not (figure 3.1). Alternatively, the map can be fixed to a table over which the scanner head moves in a series of parallel lines (figure 3.2). The lines of the original map are recorded as a series of small pixels forming a binary image (image, no image). This raster data must then thinned to reduce all lines to a single pixel width from which vector coordinates can be computed. Once this vectorizing process has been completed it will normally be necessary to edit and correct the data using the various functions listed above.

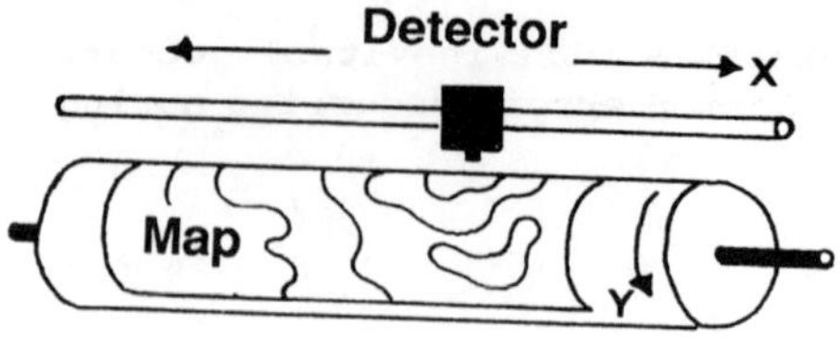

Figure 3.1 Drum scanner

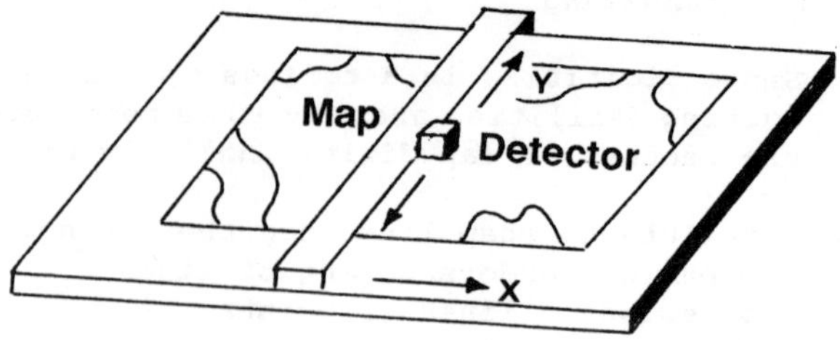

Figure 3.2 Flat bed scanner

Although there are a number of low cost instruments available[1], scanners are generally expensive and are mainly used by large mapping agencies. In general, the map to be scanned will be monochrome (it may be necessary to redraft the map or make copies from each printing plate separately), although some scanners use filters for colour separation. In all cases, the map to be scanned must be as clean as possible.

4. Conversion of vector data to raster format

Since many geographic information systems use raster format, the original digitized data must be converted (rasterized) before it can be manipulated by the system. The process of polygon to raster conversion involves the following steps:

- from the defined cell size, determine the number of rows and columns covering the map
- by finding the maximum and minimum y-coordinates in the list of digitized polygon boundary coordinates, determine the first and last row numbers in which the polygon falls
- find the points with y-coordinates corresponding to the centre of the first row and, from the corresponding x-coordinates, determine maximum and minimum column values
- assign the appropriate code to all cells of the row between the minimum and maximum columns
- repeat the above two steps row by row until the whole polygon has been rasterized.

[1] See: M. Bosma and J. Drummond (1989), Low cost scanner survey, ITC, Enschede:135pp.

5. Input of non-spatial data

Input of non-spatial data to the GIS database is generally performed by typing at the keyboard. Many systems also accept data files imported from commercial database and spreadsheet software packages. Certain types of non-spatial data may be recorded by automatic loggers (e.g. climatic or hydrologic data) or on field computers (e.g. forest inventory data) and read into the database.

6. Data output

The purpose of a geographic information system is to provide spatially-referenced information for use by planners, resource managers and other decision makers. This information will normally be presented in the form of maps and other graphical documents. The process of map output is therefore of great importance.

7. Output devices

Some means of output is required in order to present the results of GIS analysis to the system user. An important distinction is made between:

i) hard copy output, i.e. a permanent copy, usually in the form of a printed map, and
ii) soft copy in the form of an image on the computer monitor.

7.1. MONITORS

Not all computer monitors are suitable for displaying maps, satellite images and other graphical information. Even if a graphics monitor is available, it can only be used if the computer is fitted with a graphics card or board which controls the display. The data to form the image is stored in a memory space called a frame buffer. This physical display is in the form of a raster of pixels activated by a beam of electrons hitting a phosphor- coated surface. In a monochrome monitor each pixel is either "on" or "off". A colour monitor has three phosphor layers and three electron guns for displaying in red, green and blue. In addition, the frame buffer contains a number of so-called bit planes which allow the image to be displayed in a wide range of intensities. Most systems used for GIS and digital image processing have at least eight bit planes, enabling 28 or 256 colours to be displayed. The resolution of some common graphics cards is shown in table 1.

The colours assigned to different features are stored in the computer memory in a so-called Look-Up Table (LUT). In its most simple form a LUT is simply a list of input values (e.g. feature codes) and their corresponding colour. Such a LUT can be used with any colour output device, e.g. to select different pens for a plotter. LUTs for display on a colour monitor are slightly more complex since they contain three output values (red, green and blue) for each input value or code.

Table 1. Some common graphics display cards

Name	Resolution
Hercules	720 x 348 monochrome
IBM Colour Graphic Adaptor (CGA)	320 x 200 colour
IBM Enhanced Graphics Adaptor (EGA)	640 x 350 colour
Techmar	640 x 480 colour
Matrox	640 x 480 colour
Super VGA (Video Graphics Array)	800 x 600 colour
Number 9 Revolution	1024 x 768 colour

A high resolution monochrome monitor may be sufficient for handling simple vector data, but colour is obligatory in a GIS which incorporates some form of satellite image processing. Although a single monitor may suffice, many systems employ both a high resolution colour monitor for graphics display and a monochrome monitor for program commands and other text information.

7.2.HARD COPY DEVICES

The most common-and cheapest-hard copy output device is the dot matrix printer which can print alphanumeric characters at very high speed. Each letter or number is formed by a series of tiny dots produced by needles on the printing head striking an inked ribbon (figure 4). Cheaper printers have 9 needles, but better quality is obtained with 24 needle printers. Most matrix printers can be programmed to produce lines and other symbols but the cartographic quality is generally poor. Some printers have three ribbons (red, yellow and blue) for colour output. By overprinting, six different colours can be obtained.

Ink jet plotters (figure 5) give better quality colour output, because both the range of colours and the resolution is higher. As the name suggests, ink jet plotters work by squirting tiny drops of coloured ink onto the plotting paper.

Even higher resolutions can be obtained by means of an optical film writer. Film writers use a beam of light to transfer the image to black and white or colour film. The light intensity can be varied, e.g. through a brightness range of 256 levels. Optical film writers are relatively expensive and are particularly used for high quality output of satellite imagery.For high quality output, especially of vector data, a pen plotter is required. This device comprises a pen which is driven over the plotting surface by X- and Y-step motors.

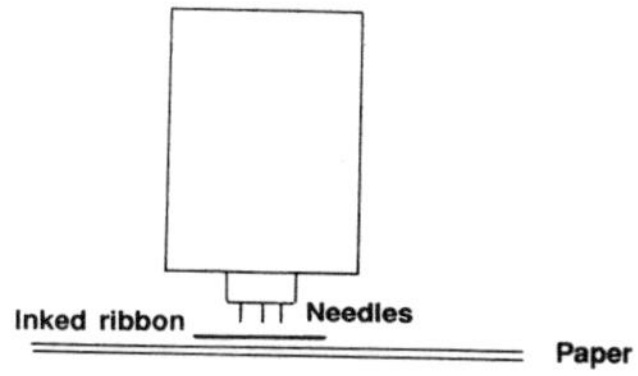

Figure 4. Dot matrix printer

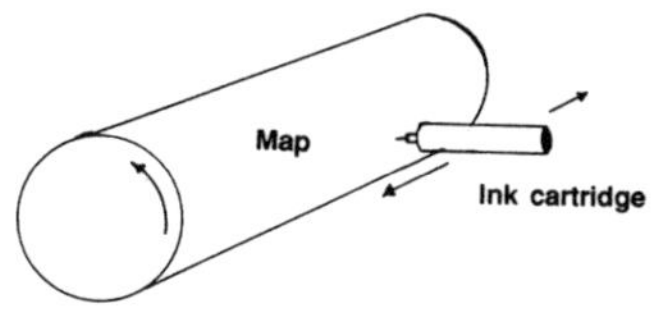

Figure 5. Ink jet plotter

Four basic instructions are therefore necessary - select pen, pen up, pen down, and move to Xi,Yi. Depending on cost, a wide range of maximum plotting formats (up to A0) are available. There are two main types of pen plotter - flatbed plotters and drum plotters (figure 6). For high quality cartographic output, a plotter with a motor step size of about 0.02 to 0.03mm is required. Instead of a pen, it is sometimes possible to use other means of producing an image, e.g. by scribing or by means of a light source projected onto photographic film.

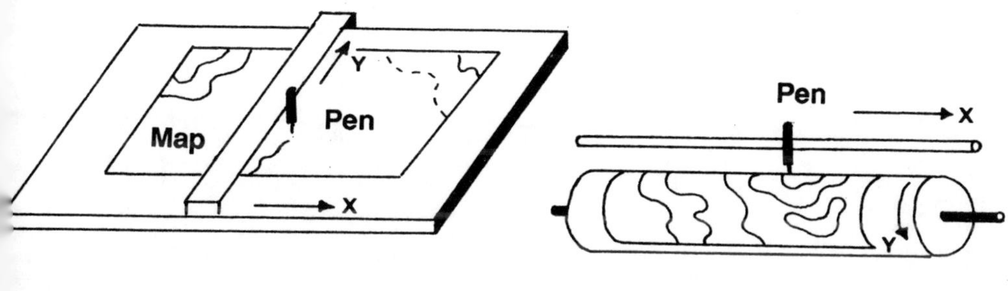

Figure 6. Pen plotters

SPATIAL DATABASES

Carlos R. Valenzuela
I.T.C
P.O. Box 6, Boulevard 1945
7500 AA Enschede
The Netherlands

1. Introduction

A geographic database is a collection of spatially referenced data that acts as a model of reality. Spatial or geographic data have a set of characteristics that make them different from data used in the information systems developed for medical, business or other applications.

Geographic data has tree major components: its **geographic position, attributes or properties,** and **time or dynamics.** In other words, where it is, what it is, and when did it exist.

Position. Position refers to the fact that each feature has a location that must be specified in a unique way. The location can be :

(a). **Absolute** such as coordinate position using a coordinate system like cartesian (X,Y) or global geographic (latitude, longitude); row/column position describing a position on a grid; position within some abstract framework such as street addresses, or

(b). **Relative** which is location with reference to some other object, eg. topologic position which is described through relationships to other objects-"adjacent to", "intersects with", "is contained in".

Geographic data may be stored at different levels of positional accuracy. To some extent the locational data are always imprecise at some level of detail. Data may be accurate to few centimetres, while other only to tens of meters.

Attributes. Attributes are often termed non-spatial data since they do not in themselves represent locational information. They are not recognized as having permanent locations with respect to other entities and they re invariant to changes in scale and projection. An attribute is a characteristic of an entity. Attribute value is the actual measurement that is stored in the database. These values have a degree of inaccuracy, which should be considered when using the data in analysis or modelling. A land use unit may not be 100 % pure or a soil unit has certain variability. Often this type of inaccuracy is not addressed by GIS users, but it is critical to recognize and take into account this deficiency.

Attribute data can be categorized as follows:

- **Nominal.** Values that establish identity. Attributes described by name with no specific order. (maize, wheat,...). Operations on this kind of data are frequencies and aggregate totals, no mathematical operations are

A. S. Belward and C. R. Valenzuela (eds.),
Remote Sensing and Geographical Information Systems for Resource Management in Developing Countries, 311–333.

allowed.
- **Ordinal.** This values establish an inherent order or ranking. (First, second,...). Operations include calculations of median and percentiles but no mathematical operations.
- **Interval.** Attribute with a natural sequence and equal intervals on a scale with an arbitrary zero point, where division does nor make any sense, eg. it makes sense to indicate that 30° C is 15° C warmer than 15° C, however it would not make sense to say that 30° C is twice as warm as 15° C. Operations include correlation and regression analysis, standard deviation, arithmetic mean.
- **Ratio.** Attributes with same characteristics as **interval** but with a natural zero or starting point (100, 200,... tons/ha). The zero point of weight is absolute but the zero point of the Celsius scale is not. All mathematical operations for real numbers are allowed, statistics include geometric mean and coefficient of variations.

Time. Time is important and critical since geographic information is referenced to a point in time or a period of time. Knowing the time when geographic data were collected is of utmost importance to using these data appropriately.

Temporal data can be described in terms of:
- **Time duration.** Time span of the current database
- **Temporal resolution.** The discrete time interval for which data sets are collected and aggregated per unit of time. For example, daily or monthly data on water availability for selected crops and yield predictions
- **Temporal frequency.** The frequency over time (rate) at which observations are collected, eg. monitoring temperature and other climatic factors to determine the frequency of frost conditions for selected time intervals.

2. Database concepts

Geographic information systems are increasingly being built around existing database management systems (DBMS) (Aronoff, 1989). They handle many functions, which otherwise would have to be programmed into the GIS. There are two approaches using DBMS: (a) they handle both, the spatial and non-spatial data, and (b) non-spatial data is accessed through the DBMS, while spatial data is handled directly by the GIS. Several GIS that have built-up their own DBMS also follow both approaches to access geographic data.

Several applications involve the use of only the non-spatial data before any spatial analysis is considered. For example, a land evaluation can be performed without having any geographic access (spatial data), provided a good attribute database for soil characteristics, crop requirements, and social-economic characteristics is available. Geographic access and spatial analysis at a later stage in the process will add a new dimension to the land evaluation model.

Some definitions of databases are given below (Hubber, 1981, Date, 1986):

" Database is a collection of **non redundant data shareable** between different application systems ".

" Collection of inter-related data stored together with as **little**

redundancy as possible to serve **one or more applications** in an optimal fashion ". Minimum data redundancy provides higher degree of data consistency.
" Combination of software and hardware that makes it possible and convenient to perform one or more tasks that involve handling large amounts of data ".
" Essentially nothing more than a computerized record-keeping system: a system whose overall purpose is to record and maintain information ".
" Collection of stored **operational data** used by the application systems of some particular **enterprise** ".

The database approach increases the benefits obtained from a GIS. Among the advantages are:

- **Reduction in data redundancy.** Sharing data by different applications instead of each having independent databases results in reduction of data redundancy. They act as a big file, different users may have different needs, but may use the same data. Individual portions of data may be shared by several users and each may use them for different purposes. Sharing is a consequence of the fact that the database is integrated.
- **Maintenance of data integrity and quality.** Databases should have as minimum redundancy as possible. Redundancy makes updating more difficult and may cause problems with the integrity of the database. Controlling the redundancy and updating procedures can be implemented more efficiently using a database. This will also help avoiding inconsistencies, since a proper design will establish and prescribe models, rules and standards which will have to be followed.
- **Security restrictions.** Databases include security tools to control access to the data. This is important particularly for inserting and deleting data. Only authorized users can have access (different levels of accessibility) to the database.

2.1 VIEWS OR SCHEMES

There are three ways in which the real world (enterprise) can be viewed: (a) as it really is [reality], (b) as it is perceived by people [descriptive representation], and (c) as it is described by symbols [data] (Hubbard, 1981). The data are stored in the database and maintained and manipulated by the application programs.

The data used to characterized the database can be perceived from three views: **Conceptual, external and internal** views as illustrated in Figure 1.

The **Conceptual view or conceptual scheme,** represents all the data requirements which are to be stored in the database. It is the primary means by which the integrated database is designed, specifying the total collection of data elements, their relationships, and constrains. For example;

- The data held on the entity HORIZON are (Horizon_no, Profile_no, Symbol, Depth,...,)
- Depth is expressed in cm
- Horizon_no and Profile_no are the identifiers (Primary key)

The conceptual view constitutes a relatively solid, stable view of the data. This view should be able to evolve with the enterprise as it evolves.

The **external view or external scheme,** describes the data requirements of an application program. The perceived external view represents what the users expect from the application program. The local external view, represents the actual data and the relationships required by the database to support the application program.

The **internal view or internal scheme,** represents the database itself. It is normally not seen by the user or application programmer. In the logical design, it is represented by a logical model that illustrates the data organized into the structures of a given DBMS. In the physical design, the model illustrates the actual implementation of these structures onto the physical storage devices.

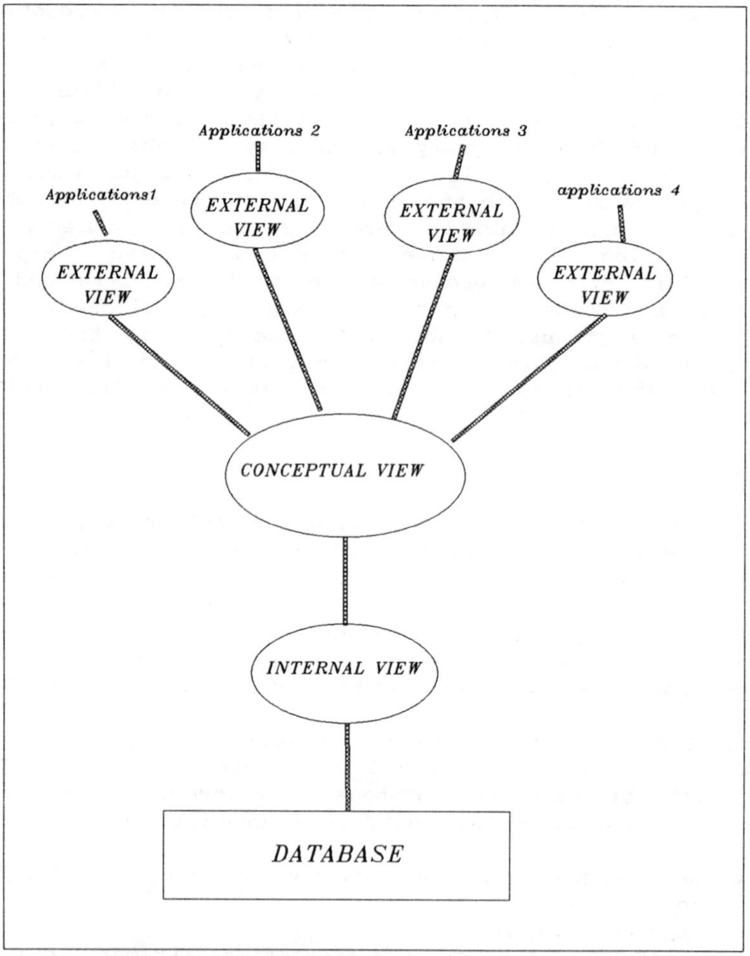

Figure 1. Three level DBMS architecture

2.2 DATABASE MANAGEMENT SYSTEMS

A database management system (DBMS) is comprised of a set of programs that provide facilities to manipulate and maintain the data in the database. They were developed to manage the sharing of data in an orderly manner and to ensure that the integrity of the database is maintained.
Some of the main characteristics of a DBMS are:

Data independence. Application programs are independent of access strategy, physical structure and characteristics of the physical storage devices where the application data is stored. Databases have **data independence**, i.e. insulation between how users perceive the data and how the data is physically stored. Users see data as a whole independently of the physical structure of the data. Individual portions of data may be shared by several users.

Data dictionary. The structure of the database is stored in the **data dictionary.** A data dictionary is a system specifically used for storing and retrieving information about the structures defined for a database system.
A data dictionary contains information about the data in the database.

Data structuring. A database containing large amounts of data items will be of limited use unless the data is structured in a meaningful way. The database management system should provide facilities to structure the data and be able to express the often complex relationships which may exist between data items.

Validation and Recovery. Sharing of data improves the consistency of data, but makes the enterprise potentially more vulnerable to errors. Erroneous data inserted into the database by one application will affect not only that application but any others which share that data. Consequently, the database management system should validate data before permitting it to be stored in the database, and should provide comprehensive recovery procedures.

Redundancy can be controlled. In a file processing environment,separate data files are used for each application and considerable data redundancy may result. These are sound reasons for maintaining multiple copies of some data. However, excessive data redundancy is expensive. In addition to the added data storage, an effective strategy must be provided to update the multiple copies of the data. a DBMS can be used to monitor and reduce the level of redundancy and where multiple copies of data are retained the DBMS can manage the updating procedures.

User views. A DBMS can provide a convenient user interface to create and maintain multiple user views.

2.3 DATABASE STRUCTURES

Geographic information systems have adapted the three main data models proposed for generalized databases. These models are: the hierarchical, network and relational.

The data model, on which a database system is based, represents the organization, description and manipulation of the database. It includes a set of primitive operations (commonly referred to as data language, data sub-language or data manipulation sub-language) that can be used to manipulate the data stored in the database.

In a **hierarchical structure**, the data are organized in a hierarchy or tree structure. This model is formed by a set of links connecting all record types in the tree. The **root** is the top of the hierarchy. The root has one or more lower elements related to it, and these, in turn, may also have one, or more lower level elements subordinated to them. As a consequence, each element is connected with only one higher element, termed **parent**, and with one or more lower elements, termed **children**. Figure 2 shows schematically the hierarchical model.

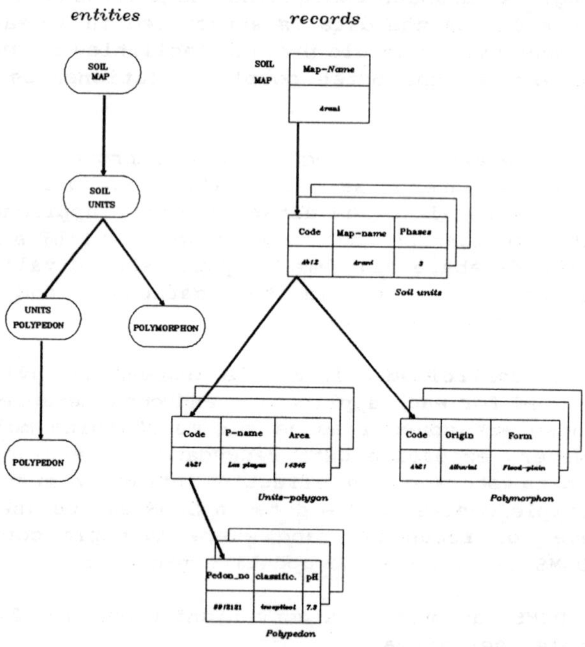

Figure 2. Hierarchical model

Since, there is only one link between elements, links need not to be named. They exist only between parent and children, hence, there is no connection between elements at the same level, which constitute one of the main disadvantages of this model. Hierarchical models are inflexible, new links between elements, once the tree is established, are not possible. linkages are only possible vertically and not horizontally or diagonally. Search and data access are only possible through the identifier or key attribute, but difficult through other attributes.

Hierarchical structures provide high speed of access to large datasets and are easy to update. Some data that possess an inherent hierarchy, such as census data organized by state, within state by city, within city by census track, can be adequately represented by a hierarchical model.

Network structures overcome some of the problems encountered in the hierarchical model. They allow entities to have multiple parents as well as any number of children, and no root is requires (Figure 3). As a result, network models have greater flexibility than hierarchical models for handling complex spatial relationships.

Figure 3. Network model

Links between entities of the same type are not allowed and while a record can have only one owner within a set, a record can be defined in any number of sets and thus, can have any number of owners.

They have less redundancy than hierarchical models, provide high speed retrievals and include methods for building and re-establishing new links. A drawback in this model is that for its implementation, all relations should be known beforehand and it is not as flexible as the relational model.

Hierarchical and network data structures are also called **navigational structures**, because the relations are constructed by links such as pointers, and application programmers can navigate through the database using these pointers.

A **relational structure** can be thought of as a collection of normalized relations (entities) defined in a collection of domains (Figure 4). A normalized relation may be viewed by the user as a two dimensional table, where each row of the table, known as **tuple**, corresponds to one element of the relation. Each column of the table, known as an attribute, contains entries that belong to the set of values constituting the underlying domain of the column. A domain is an abstract set of single data values where the domain underlying a column consists of precisely those data values that appear as entries in that column. Figure 5, illustrates a relational table.

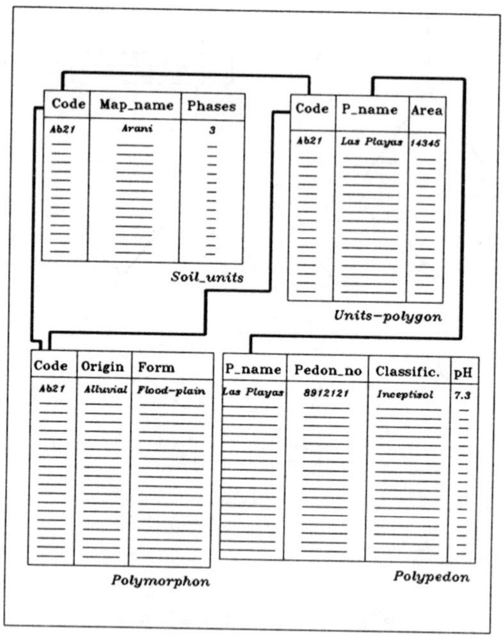

Figure 4. Relational model

An important feature of a relational data structure is that associations between tuples are represented by data values in columns drawn from a common domain. Of all attributes in a particular entity table, there is usually one attribute with values that uniquely identify the entries (tuples) in that entity table. This attribute is said to be the **primary key.** A primary key is not restricted to a certain single attribute; it can also be a combination of several attributes which together have the unique identification property. One entity table may sometimes contain more than one attribute combination providing the unique identification property.

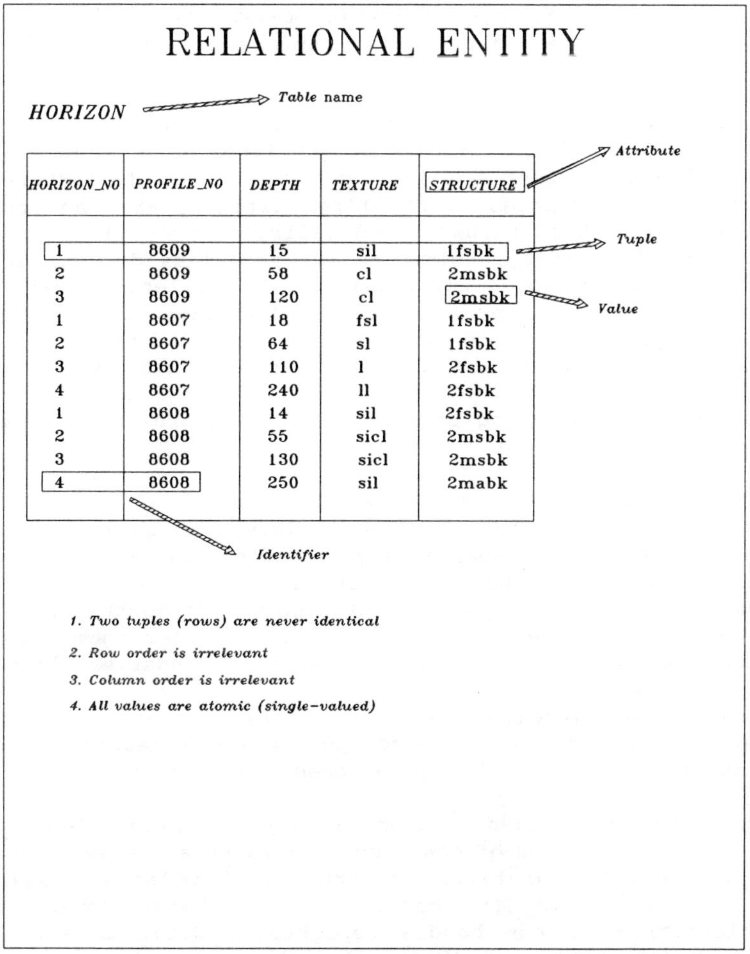

HORIZON_NO	PROFILE_NO	DEPTH	TEXTURE	STRUCTURE
1	8609	15	sil	1fsbk
2	8609	58	cl	2msbk
3	8609	120	cl	2msbk
1	8607	18	fsl	1fsbk
2	8607	64	sl	1fsbk
3	8607	110	l	2fsbk
4	8607	240	ll	2fsbk
1	8608	14	sil	2fsbk
2	8608	55	sicl	2msbk
3	8608	130	sicl	2msbk
4	8608	250	sil	2mabk

1. *Two tuples (rows) are never identical*
2. *Row order is irrelevant*
3. *Column order is irrelevant*
4. *All values are atomic (single-valued)*

Figure 5. Relational table

3. Geographic data types

There are four (4) primitive types of geometric entities to encode spatial data: points, lines, polygons and continuous surfaces.

Point data. Points (0-dimension -an object that has position in space but not length-) are the simplest type of spatial data. They can be considered to embrace all spatial entities that are positioned or described by a single X-Y coordinate pair.

Points can be of two kinds, observations relating to discreetly distributed phenomena such as oil and water wells or surface mining operations, and observations relating to contiguous distributions such as weather station reporting precipitation measurements of soil temperature.

The choice of entities which will be represented as points depends on the scale of map/study.

Line data. Line entities (1-dimension -an object having a length-) are all linear features built up of straight line segments made up of two or more coordinates. Line entities can be static (structural type) or dynamic (flow). Examples of the first category are the transportation infrastructure such as highways and railroads or utility infrastructure such as gas, electricity, telephone, water. They do not carry inherent spatial information about connectivity. A collective network of line segments defining existing traffic flows or capacities and stream flow data are examples of dynamic line data (Star and Estes, 1989). Lineaments and faults are examples of linear features that may or may not be connected.

An **arc, chain, or a string** is a set of n X-Y coordinate pairs describing a continuous complex line.

Polygons or area data. Polygons (2-dimension -an object having a length and width-) constitute the most common data type used in GIS. They are bounded regions. The boundaries may be defined by natural phenomena such as landforms or by man such as forest stands or land use units.

The most common types of polygons are the environmental/natural resource zones such as land cover, geologic, soils; social-economic zones such as census tracts; land records like land parcel boundaries, land ownership, tax information.

Polygons can be **adjacent** where the boundary of a polygon is shared by at least another polygon and **nested (holes and islands)** where areas of different attributes are wholly enclosed within another area.

Continuous surfaces. (3-dimension -an object having a length, width and height/depth-). Examples of continuous surfaces are elevation (as part of topographic data), rainfall, temperature, population density. Most commercially available GIS products can not handle true 3-dimensional data, although they can handle topographic data, usually as Digital Elevation Model (DEM), and display isometric views, contour maps and so on. Most DEM's use either gridded elevation matrices or triangular meshes (TIN) to allow for these terrain representations. In these cases, the elevation, or z-coordinate, is treated as a dependent variable. Some systems allow the draping of another mapped feature, such as soils, land cover data or satellite images (Landsat MSS or TM, SPOT) onto an isometric view of a topographic elevation surface, thereby creating an illusion of a 3-dimensional scene.

Common GIS applications involve the mapping of essentially 2-dimensional land surface phenomena such as land use, forestry or soils data. Some applications can be accomplished by reducing the 3-dimensional representation to a quasi 2-dimensional one through the use of surfaces. The surfaces which can represent bedding planes for example, can then be contoured or displayed as isometric views. However, in these cases, the elevation of the surface is not truly independent variable, and so these systems are best defined as quasi 3-dimensional or pseudo 3-dimensional.

4. Data structure

The user's perception of a phenomenon that actually exists or the human conceptualization of reality has to be represented and stored in a complete database.

There are several ways of organizing these geographic data in a computer. The data model of the earth or the description of entity and their relationships are represented in terms of basic entities such as points, lines, areas and surfaces. The non spatial data or the sets of attributes describing the scales of properties that apply to the entity are stored usually in relational database management information systems.

The spatial distribution of points, lines, areas and surfaces are represented in digital form in two basic types of spatial models; **tessellation or raster and vector** models. Figure 6 illustrates both models.

4.1 VECTOR MODELS

Vector is defined as a quantity with a starting coordinate and an associated displacement and direction. The approach used in the vector model is to precisely specify the position of the points, lines and polygons used to represent the conceptualization of reality (data model). The fundamental primitive, in the vector model, is the point. Objects are created by connecting points with straight lines, areas are defined by set of lines. The stretch of common boundary between two junctions (nodes) has various names, **edge** is favoured by graph theorists, **vertex** for the junctions, **chain** is officially sanctioned by the US National Standards, **arc** and **segments** are used by several systems.

In the vector model objects are represented by the points and lines that define their boundaries. The position of each object is defined by its placement in a map space that is organized by a coordinate reference system. Vector models have the line as the basic logical unit in a geographical context. A series of X-Y pointers location along the line are recorded as the components of a single data record. Points are recorded as lines of zero length, areas or polygons constitute lines with common beginning and ending points.

The most common vector models are the whole polygon structure and the topologic model.

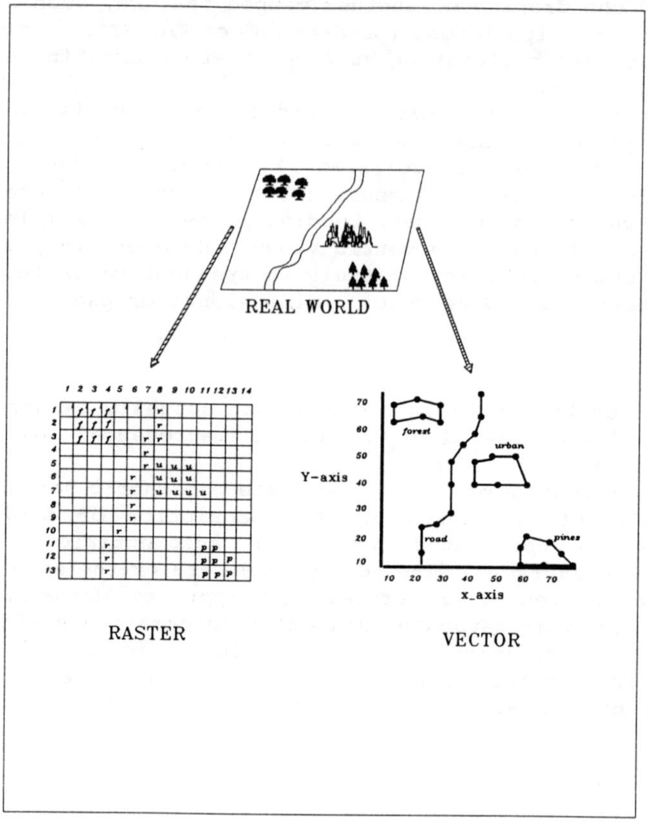

Figure 6. Representation of data in vector and raster models (After Aronoff, 1988)

The whole polygon structure. In the whole polygon structure or spaghetti model each polygon is encoded in the database as one logical record and it is defined by a string of X-Y coordinates representing a closed area. The structure of this model is very simple and easy to understand. Figure 7, illustrates this structure. The main disadvantages of this model are:

- lines between adjacent polygons must be digitized and stored twice
- no spatial relationships are retained, thus, existing spatial relationships between polygons must be derived through computation relationships between polygons
- islands are only graphical constructions
- editing and updating the database are carried out comparing it visually with a graphic output

This model is often used for applications that are limited to the simple forms of computer assisted cartographic production. The lack of information extraneous to the plotting process, such as spatial relationships, makes this model efficient for reproducing the original map (Peuquet, 1984).

The topologic model. In the topologic model the spatial relationship among entities are explicitly recorded in what is known as topologic model.

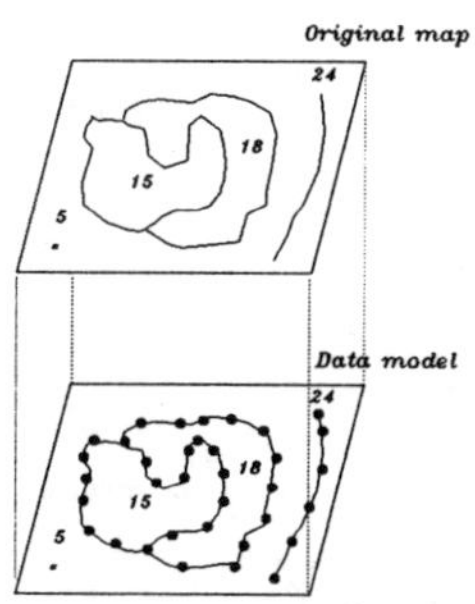

Data Structure

FEATURE	NUMBER	LOCATION
Point	5	$X\ Y$
Line	24	$X_1,Y_1,X_2,Y_2,X_3,Y_3,\ldots,X_n,Y_n$
Polygon	15	$X_1,Y_1,X_2,Y_2,X_3,Y_3,\ldots,X_1,Y_1$
	18	$X_1,Y_1,X_2,Y_2,X_3,Y_3,\ldots,X_1,Y_1$

Figure 7. Whole polygon structure (After Dangermond, 1982)

Topological data structures are considered to be essential for automated geographic and cartographic analysis where topological routines are the most efficient for processing operations such as automatic error detection and updating, and other polygonal analysis procedures. Topology is defined as the mathematical study of the relationships and transformations of geometric configurations. As defined for cartography and automated spatial analysis requires that all lines be explicitly and unambiguously linked into a completely digital line graph that can be interpreted in a computer environment.

The basic logical entity is a straight line segment that begins or ends at the intersection with another line or at a bend in the line. Each line segment is recorded with the coordinates of its two end points. In addition, the identifiers, or names of the polygons on either side of the line are recorded, thus the more elementary spatial relationship are explicitly retained and can be used for analysis and moreover, the basic data are stored in a non-redundant manner. The data structure of this model is depicted in Figure 8.

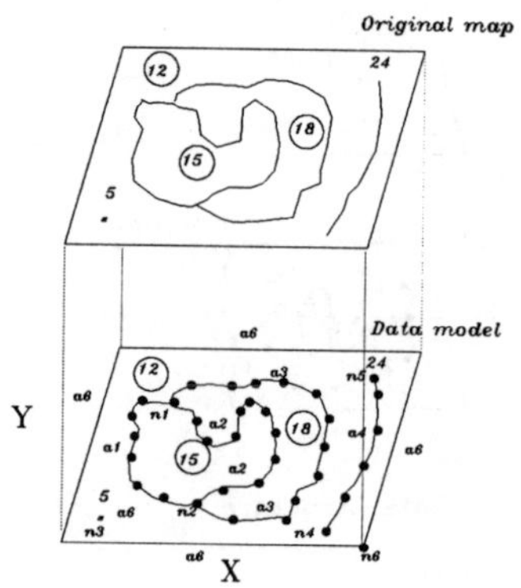

DATA ENCODING

Node	Arcs
n1	a1,a2,a3
n2	a1,a2,a3
n3	a6
n4	a4
n5	a4
n6	a6

Arcs	From Node	To Node	Right Polygon	Left Polygon
a1	n1	n2	12	15
a2	n2	n1	18	15
a3	n1	n2	18	12
a4	n4	n5	12	12
a5	n3	n3	12	12
a6	n6	n6	0	12

Polygon	Arcs
12	a1,a3,a4,a5,a6
15	a1,a2
18	a2,a3

Arc	Start	Coordinates	End
a1	X Y	X Y,X Y,X Y	X Y
a2	X Y	X Y,X Y,X Y	X Y
a3	X Y	X Y,X Y,X Y	X Y
a4	X Y	X Y,X Y,X Y	X Y
a5	X Y	X Y,X Y,X Y	X Y
a6	X Y	X Y,X Y,X Y	X Y

Figure 8. The topologic data model

Among the topologic models we find:

Dual Independent Encoding (DIME) model, is the best known model using the topologic concept. It was devised by the US General Census Bureau to store digitally urban areas with topologic formations for demographic analysis (U.S. Census, 1969). In the DIME file, each segment is spatially defined according to the definitions of the model, using both street addresses and UTM coordinates. The main disadvantage of this model is that since segments do not occur in any sequence, for the retrieval of any particular segment, a sequential, exhaustive search must be performed on the entire file. **Arc-node Structure or POLYVERT model,** implemented at the Harvard Laboratory for Computer Graphics (Peucker and Chrisman, 1975). This model overcomes the major retrieval difficultities by storing each type of data entity separately in a hierarchical data structure. The arc is defined as the basic line entity. It is a sequence of straight lines which begin and end as a node. A node is the intersection point between two arcs. Polygons are areas that are completely bounded by a set of arcs. The hierarchical structure allows selective retrieval of only specific classes of data at a time. Various attributes can be easily included and linked to the geometry.

Relational structure, where the topologic data is organized similarly as the arc-node model, but the attribute values are stored in relational tables using relational database management information systems. The attribute data are stored in tables where the columns are the different fields on attributes, one of these tables includes the points or nodes of the spatial database. Several commercial general purposes relational database management information systems are available in the market.

4.2 RASTER MODEL

One of the simplest data structures is a raster or cellular organization of spatial data. Raster (tessellation) models have as the basic data unit, a unit of space for which entity information is explicitly recorded. In a raster structure, a value for the parameter of interest -land use class, soil type, and so on- is developed for every cell in an array over space. Each cell in a raster file is assigned only one value, different attributes are stored as separate data files, Figure 9, illustrates the structure of the raster model.

Operations on multiple raster files involve the retrieval and processing of the data from corresponding cell positions in the different data files or layers. Basically, this overlay concept is like stacking the layers (two-dimensional arrays) and then analyze each cell location. The raster model is easily interfaces to the hardware devices commonly used for the input and output of spatial data.

The main raster models are the grid and other regular tessellation, nested tessellation, and irregular tessellation. Figure 10, shows the rasterization process.

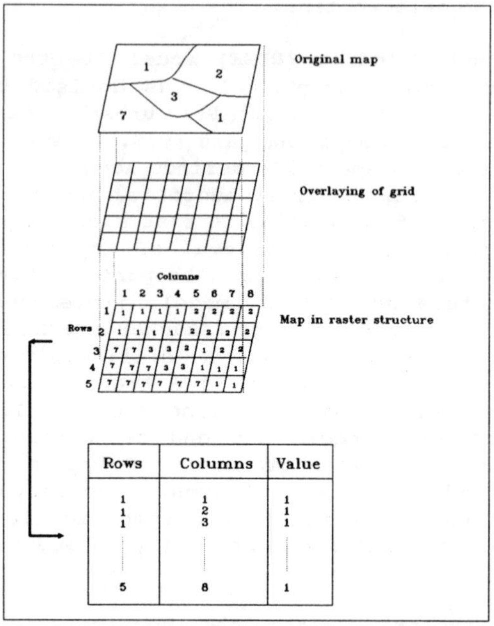

Figure 9. Structure of the raster model (After Dangermond, 1982)

Grid and other regular tessellation. In this model square, triangular, and hexagonal meshes have been used. The most widely used is the square grid. In this raster array, each cell is referred by a row and a column number and a number representing the type of value of the attribute being mapped. The origin of the raster is usually the upper left corner, this referencing system is different from the other, georeferencing systems like geographic coordinates where the origin is one specific point on the earth surface.

The size of the rater cell is sometimes confused with the **minimum mapping unit**, i.e., the smallest element we can uniquely represent in our data. Raster cell and minimum mapping unit are not the same. Each cell represents an area of the land surface, a minimum mapping unit is formed by one or more cells.

Since the attribute of each cell is stored as a unique value, the total number of values to be stored is the product of the number of lines times the number of columns. The smaller the area of land that each cell represents, the higher the resolution of the data. The size of the file increases rapidly with resolution, since the file size is related to the area of coverage, it increases by the square of the increase in resolution.

One of the limitations of square grids is that the diagonal neighbours are not the same distance away as neighbours in the four cardinal directions from a central point. In a hexagonal mesh all neighbouring cells of a given cell are equidistant from that cell's centerpoint. Radial symmetry makes this model advantageous for radial search and retrieval functions (Peuquet, 1984).

Triangular tessellation has the disadvantage that all triangles do not have the same orientation, making several procedures involving single cell comparisons very complex. However, this characteristic gives triangular

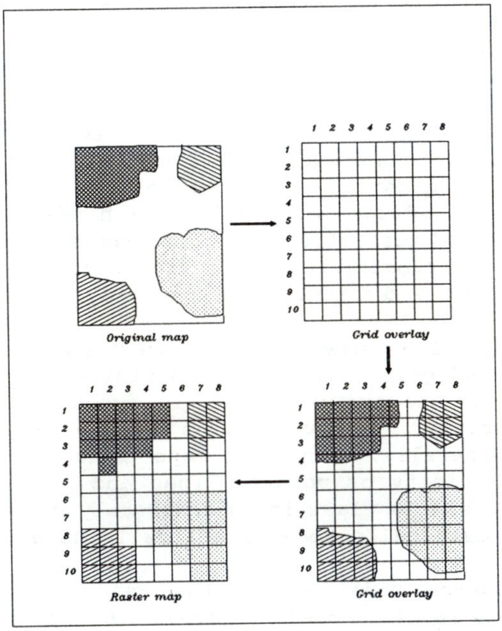

Figure 10. Rasterization process

structures a big advantage in representing terrain and other typical surface data. Hexagonal and triangular tessellations have a disadvantage that they can not be recursively subdivided into smaller cells of the same shape as the original cells, and a numbering system is more complex than that of the square model.

In terms of processing efficiency on general procedures to compute spatial properties such as area and centroid calculations, or to perform spatial manipulations such as overlays and windows, the algorithms initially devised for squared grids can be easily modified to work in the case of a triangular or hexagonal system. These in fact have the same order of computational complexity (Ahuja, 1983).

There are two major methods to convert the original map into a raster structure (rasterization).

- **Point sampling,** in which the value of the cell is of that attribute which occurs at a given point (usually the center) in the cell
- **Modal sampling,** in which the value of the cell is that of the predominant attribute contained within the boundaries of the cell.

When raster data is converted into vector data, usually, the boundary between two homogeneous units defines a line element. Because the basic unit for raster data is a cell, lines created by the conversion process may appear as "staircase" lines, instead of a smooth straight line. Another problem with raster to vector conversion is that a line defined by a series of cells in the raster domain may be converted to an elongated polygon, known as "fat" line. This may have to be processed to form a single line (Peuquet, 1981, Fabbri, et al., 1988).

Nested tessellation model. In this model cells are recursively subdivided into smaller cells with the same shape and orientation. The most studied model is the **quadtree** based on the recursive decomposition of a grid (Peuquet, 1984, Samet, 1989, Samet, 1990). A quadtree first encloses the area in consideration within a square, and subdivides the square into four subquadrants. Each quadrant is then recursively subdivided into subquadrants until all of them are uniform with respect to image value, or until a predetermined resolution level is reached. With their variable resolution and natural subdivisions into hierarchical patches, quadtrees are ideal for storing and handling large geographic areas (Mark and Lauzon, 1985).

The quadtree data model provides a more compact raster representation by using a variable-sized grid. Instead of dividing an area into one size cells, finer subdivisions are performed in areas with finer detail as illustrated in Figure 11. This provides a greater resolution only where it is needed. Generally, thematic maps require higher levels of resolution in the vicinity of points, lines and polygon boundaries. A large polygon with a single class is as accurately encoded with one large cell as with many small cells because all cells have the same attribute value.

The use of a quadtree model allows the use of large cells (coarse resolution) to encode large homogeneous areas. Higher level of resolution is used for areas that have higher spatial variability.

A disadvantage of quadtree models is that they appear to be not invariant to translation, rotation or scaling. Shape analysis and pattern recognition are difficult and need intermediate procedures (Burrough, 1986).

Irregular tessellation. Here an irregular net can be adjusted to reflect the density of data occurrences within an area. In dense areas the cells are small while in sparse areas cell are large. Irregular tessellations are the **Triangulated Irregular Network (TIN) or Delaunay triangles**, where each vertex of the mesh has an elevation value. The TIN model is an alternative to the regular raster of a Digital Elevation Model (DEM). It was developed in the early 1970's, as a simple way to build a surface from a set of irregularly spaced points (Peucker, et al., 1978). They are used to represent terrain data for digital elevation models, hill shading and other land surface representations. TIN is based on irregularly distributed points that are first triangulated into a series of connected facets (Poiker and Griswold, 1985).

Thiessen polygons or Voronoi diagrams also called **Dirichlet** tessellations constitute other irregular model where the polygons are convex and have a variable number of sides. They are used and are efficient in analysis of adjacency and proximity.

Irregular tessellations are difficult to generate and not well suited for a number of spatial analyses, such as overlaying procedures (Peuquet, 1984). The majority of the current applications in both image processing and geographic information systems can be handled by using raster data.

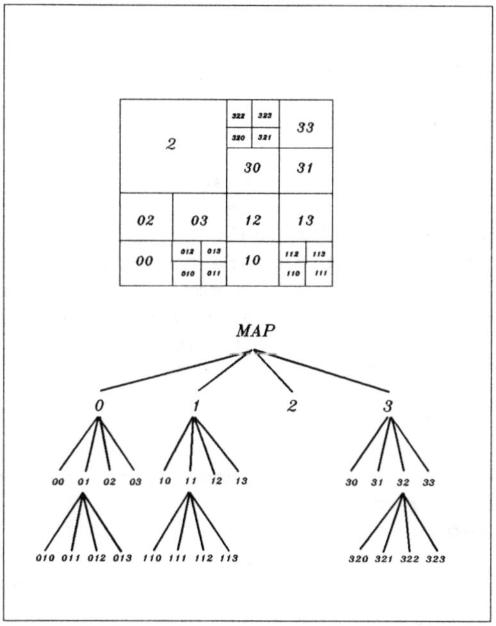

Figure 11. Quadtree model

However, the complete functions of a spatial data processor require both, raster and vector data types. Ideally a system should have data structure independence, i.e., the user should be insulated from the data structure when performing any given application, the selection of the appropriate data structure should be done by the system, selecting either raster or vector according to a decision based on both speed and accuracy of operation.

4.3. COMPARISON BETWEEN DATA STRUCTURES

Vector and raster models have both, advantages and disadvantages. Each approach tends to work best in situations where the spatial information is to be treated in a manner that closely resembles the data model. Raster models are generally well suited when the geographic information of interest is the spatial variability of a phenomenon, network analysis is best suited for vector models.

ADVANTAGES RASTER MODEL

- Simple data structure
- Overlay and combination of mapped data with remotely sensed data is easy
- Various kinds of spatial analysis are easy
- Simulation is easy because each spatial unit has the same shape

and size
- Technology is cheap and it is being energetically developed
- Same set of grid cells are used for several variables
- Simpler when doing your own programming

DISADVANTAGES RASTER MODEL

- Wasteful use of computer storage for spatially sparse data
- Errors in estimating perimeter and shape
- Network linkages are difficult to establish
- Projection transformations are time consuming
- Use of large cells to reduce data volumes means that phenomenologically recognizable structures can be lost and there can be serious loss of information
- Crude raster maps are usually less beautiful and accurate

ADVANTAGES VECTOR MODEL

- Good representation of phenomenological data structure
- Compact data structure
- Accurate graphics
- Topography described with network linkages
- Retrieval, updating and generalization of graphics and attributes are possible
- Widely used to describe administrative zones

DISADVANTAGES VECTOR MODEL

- Complex data structure
- Simulation is difficult because each unit has different topological form
- Overlaying of several polygon maps or polygon and raster maps is difficult
- Display and plotting can be expensive
- Technology is expensive
- Spatial variability is not implicitly represented

4.4.DATA COMPRESSION

Since many of the cells contain the same value as neighbouring cells, there is considerable redundancy. Significant reductions in size of the data files can be achieved by using data compression techniques.

There are several method for storing data more efficiently. Some of these methods are completely reversible, that is, the original data sets may be recovered exactly, other minimize the volume of the stored data by losing small and controlled amount of the original data.

Run length encoding. Run length encoding is a technique that allows the recovery of the original data. It makes use of the fact that many data sets have large homogeneous regions. In this method, the values of the attribute, number of cells in the run, and the row number are recorded (Figure 12).

The effectiveness of this method varies with the data set. In the worst case, when there are no repeating sequences at all along the rows of the array, the algorithm will actually double the size of the original data set.

An example of a data compression method that is not reversible, in other words that the original data is not recovered completely, is the quadtree model.

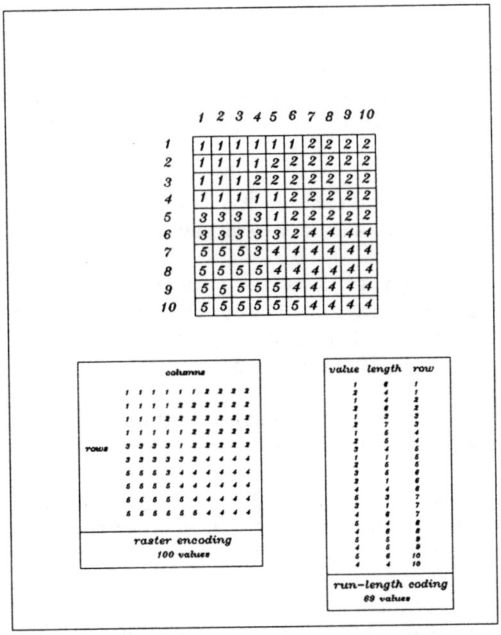

Figure 12. Run length encoding

References

Ahuja, N. 1983. On approaches to polygonal decomposition for hierarchical image decomposition. Computer vision, Graphics and Image processing, Vol 4.

Aronoff. S. 1989. Geographic information systems: A Management perception. Ottawa, WDL Publications.

Burrough, P. A. 1986. Principles of geographical information systems for land resource assessment. Clarendon Press, Oxford, UK.

Dangermond, J. 1982. A classification of software components commonly used if Geographic information systems. Proceedings U.S.-Australia workshop on the design and implementation of computer-based geographic information systems, p 70-91, Honolulu, Hawaii.

Date, C. J. 1986. An introduction to database systems (4th ed). The systems programming series, Addison-Wesley.

Fabbri, A.G., K.B. Fung and T. Kasvand, 1988. Issues in the integration of spatially-distributed data ancillary to remotely sensed images. Image analysis and processing II. Edited by V. Cantoni, V. Di Gesu and S. Levialdi. Plenum Publishing Corporation.

Hubber, G.U., 1981. Computer assisted database design. Van Nostrand Reinhold data processing series. New York.

Mark, D. M. and J. P. Lauzon, 1985. Approaches for quadtree-based geographic information systems at continental or global scales. Proceedings of Autocarto 7, Digital representations of spatial knowledge, Washington D.C.

Nagy, G. and S. G. Wagle. 1978. Geographic data processing. Computing surveys, vol 11, No 2.

Peucker, T. K. and N. Chrisman, 1975. Cartographic data structures. The American cartographer, Vol 2.

Peuquet, D. J. 1981. An examination of techniques for reformatting digital cartographic data/part 1: The raster-to-vector process: Cartographica, Vol 18, No. 1.

Peuquet, D. J. 1981. An examination of techniques for reformatting digital cartographic data/part 2: The vector-to-raster process: Cartographica, Vol 18, No. 1.

Peuquet, D. J. 1984. A conceptual framework and comparison of spatial data models. Cartographica 21(4), p. 66-113.

Poiker, T . K. and L. A. Griswold, 1985. A step towards displays and digital elevation models. Proceedings of Autocarto 7, Digital representations of spatial knowledge. Washington D.C.

Samet, H. 1989. The design and analysis of spatial data structures. Reading, Addison-wesley.

Samet, H. 1990. Applications of spatial data structures: Computer graphics, image processing and GIS. Reading, Addison-wesley.

Star, J. E. and Estes, J. L. 1989. Geographic information systems: An introduction. Inglewood Cliffs, New Jersey. Prentice Hall.

U. S. Department of Commerce, Bureau of the Census, 1969. The DIME geocoding system, in Report No. 4, Census Use Study.

DATA ANALYSIS AND MODELLING

Carlos R. Valenzuela
I.T.C.
P.O. Box 6, Boulevard 1945
7500 AA Enschede
The Netherlands

1. Introduction

The most important characteristic of geographic information systems is the provision of capabilities for spatial analysis functions. These functions use the spatial and non-spatial attributes in the database to answer questions about the real world.

The database in a geographic information system is a model of the real world that can be used to simulate certain aspects of reality. A model may be represented in words, in mathematical equations or as a set of spatial relationships displayed as a map.

2. Analysis functions

The objective of data analysis is to extract or query useful information to satisfy the requirements or objectives of decision makers at all levels of detail. An important use of the analysis is the possibility of predicting what will happen in another location or at another point in time. This ability provides the possibility to select the best possible alternative.

The range of analysis procedures have been subdivided into four categories; retrieval/reclassification/measurement, overlay, distance and connectivity and neighbourhood (Aronoff, 1989).

2.1. RETRIEVAL, CLASSIFICATION AND MEASUREMENT FUNCTIONS

In these functions spatial analysis and attribute data are retrieved, but only the attribute data are modified. No changes are made to the location of spatial elements and no new spatial elements are created.

2.1.1. Retrieval operations. Retrieval operations on the spatial and attribute data involve the selective search and manipulation, and output of data without the need to modify the geographic location of features or to create new spatial entities.

Retrieval operations include:

- Retrieval of data using geometric classifications. Specifying the spatial domain of a point, line or area, retrieve all spatial entities and non-spatial attributes contained in the entire or in

A. S. Belward and C. R. Valenzuela (eds.),
Remote Sensing and Geographical Information Systems for Resource Management in Developing Countries, 335–348.

portion of that spatial domain.

- Retrieval of data using symbolic specifications. Retrieve data using a name or code of an attribute. Example, retrieve effective depth and dominant texture of a given soil type.
- Retrieval of data using conditional and logical statements. Retrieve data that satisfies alphanumeric conditions using logical expressions. Example, retrieve all soil series with a pH range of 6.0 to 7.5 and silty clay textures.

Typical examples of queries on spatial and non-spatial data are:

- Where are the parcels with maize
- What is the land use in location X,Y
- What is the shortest distance between A and B through C and has optimum recreational conditions

2.1.2.Reclassification procedures. Reclassifying procedures involve operations that reassign thematic values to the categories of an existing map as a function of the initial value, the position, size or shape of the spatial configuration associated with each category, for instance a soil map, reclassified into a permeability map, or a terrain mapping unit (TMU) reclassified into a K-factor map, as illustrated in Figure 1. The process involves looking at the attribute for a single data layer and assigning an additional attribute, the new class name.

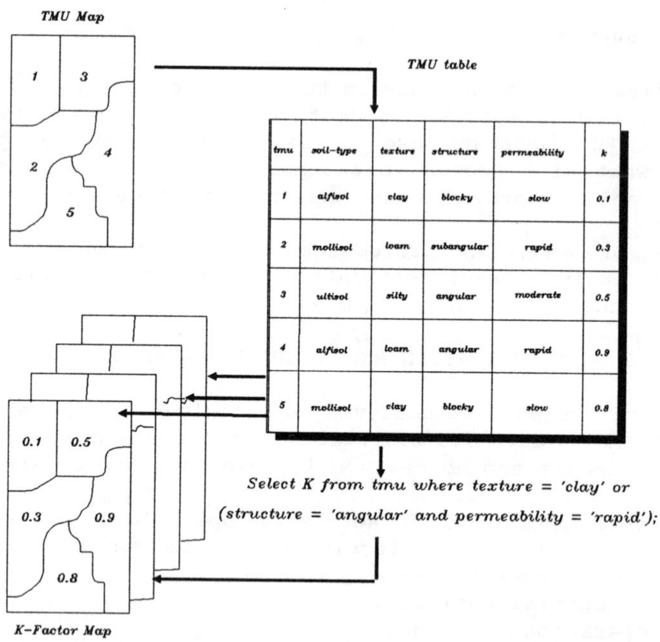

Figure 1. Reclassification using attribute data

Reclassification can also be performed in multiple data layers as part of an overlay operation. For example, a desirable area for recreational purposes might be an undulating forested area with well drained soils and non agricultural zone. Each of these factors may be presented in different layers. An overlay analysis could be used to identify the areas satisfying these criteria and then reclassify them as good recreational areas.

Reclassification procedures sometimes combine classes and make the classification less detailed, this process is called generalization or in vector systems using the topologic model, the process involves the elimination of boundaries and it is called map dissolve as illustrated in figure 2.

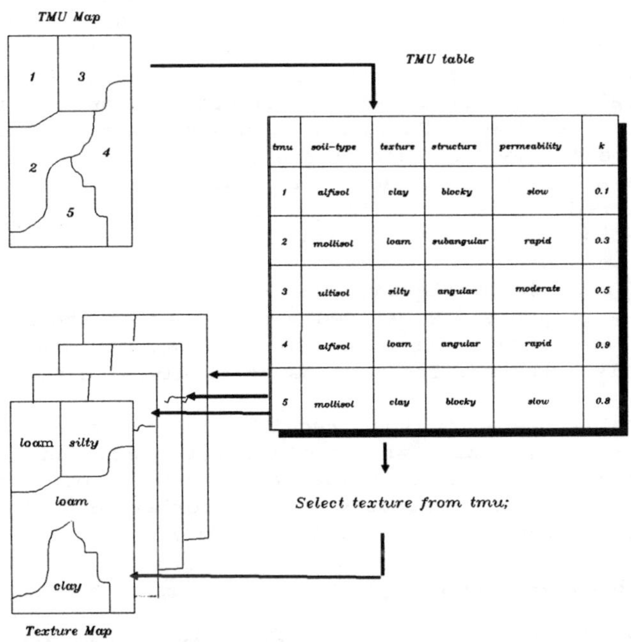

Figure 2.- Map generalization or map dissolve

2.1.3.Measurement functions. Measurement of spatial data involve the calculation of distances between points, lengths of lines, area and perimeter of polygons, and volumes. Measurements involving points include distances from a point to a point, a line or a polygon; enumeration of the total number of points as well as the enumeration of points falling within polygons (eg number of water wells within an agricultural area).

Volumetric measurements are used, very often, in engineering applications such as calculating the amount of cut and fill material necessary for road constructions.

2.2.OVERLAY OPERATIONS

Overlaying of maps result in the creation of a new map where the values assigned to every location on that map are computed as a function of independent values associated with that location on two or more existing maps. Overlaying operations create a new data set containing new polygons formed from the intersection of the boundaries of the two or more sets of separate polygon layers. In addition of creating new polygons based on the overlay of the multiple layers, these polygons have multiple attributes, i.e., the attributes which were given to each separate overlay before the composition occurred.

Arithmetical and logical overlay operations are common in all GIS software packages. Arithmetical overlay includes operations such as addition, substraction, division and multiplication of each value in a data layer by the value in the corresponding location in the second data layer. Logical overlay involves the selection of an area where a set of conditions are satisfied. For example, desirable areas for recreational purposes may be defined as forested areas with undulated topography, well drained soils and non-agricultural land use. If land use, landforms and soils are stored in separate layers in the database, then a logical overlay operation may be used to identify the areas where these conditions are satisfied.

Figure 3, depicts the overlaying concept in a vector structure (topologic overlay). New polygons are defined where areas of the first and second map overlap. An output attribute table containing data about the new polygons is also created. The process of subdividing polygons is called **clipping**.

It is a complex operation that requires intensive and time consuming processing time, especially when working with large and intricate data sets. Figures 4 and 5 depict the overlaying process of points in polygon and lines in polygon respectively. Overlaying using conditional statements are illustrated in figures 6, and 7.

2.3.NEIGHBOURHOOD OPERATIONS

Neighbourhood operations involve the creation of new data based on the consideration of "roving window" of neighbouring points about selected target locations. They evaluate characteristics of an area surrounding a specified target location. In all neighbourhood operations it is necessary to indicate one or more target locations, the neighbourhood considered around each target and the type of function to be executed on the attributes within the neighbourhood.

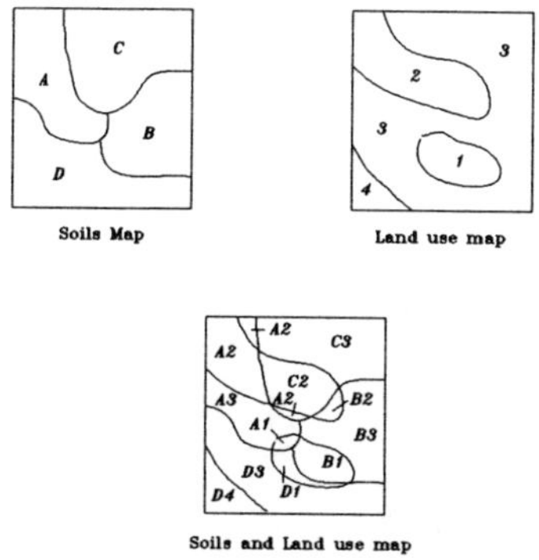

LAND USE

USE_CO	USE_NA
1	FOREST
2	AGRICULTURE
3	PASTURE
4	SHRUBS

SOILS

SOILS_CO	DEPTH
A	130
B	200
C	180
D	190

LAND USE / SOILS

SOIL-USE	USE_NA	DEPTH
A1	FOREST	130
A2	AGRICULTURE	130
A3	PASTURE	130
B1	FOREST	200
B2	AGRICULTURE	200
B3	PASTURE	200
C2	AGRICULTURE	180
C3	PASTURE	180
D1	FOREST	190
D3	PASTURE	190
D4	SHRUBS	190

Figure 3. Overlay procedure in a vector system

The typical neighbourhood operations in most GIS are search functions, topographic functions and interpolation.

2.3.1.Search functions. Search functions constitute one of the most commonly used neighbourhood function. They determine the value of each target feature according to some characteristic of its neighbourhood. The region of interest (search area) is usually square, rectangular or circular, the size is determined by the analyst.

These functions are of two kinds, those that operate with numerical (interval and ratio) data and the ones that are used with nominal and ordinal data. Typical functions on numerical data are the total, average, maximum, minimum and statistics such standard deviation or variance.

Functions on nominal and ordinal data are majority, counting operations such as number of different classes and frequencies. An example of a search function is the identification and enumeration of all farms with electricity in a given political region, or the identification of the land use in an area of 10 km along a principal road.

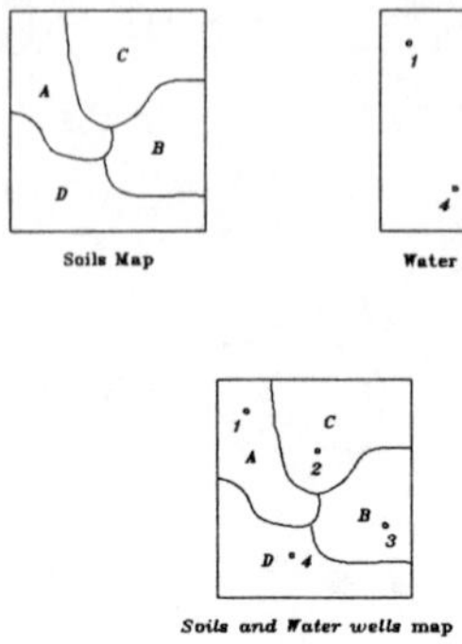

WATER WELLS

WELL_NO	QUANTITY
1	500
2	300
3	350
4	400

SOILS

SOILS_CO	DEPTH
A	130
B	200
C	180
D	190

WELLS / SOILS

SOIL-WELL	QUANTITY	DEPTH
A1	500	130
B3	350	180
C2	300	180
D4	400	190

Figure 4. Point in polygon overlaying

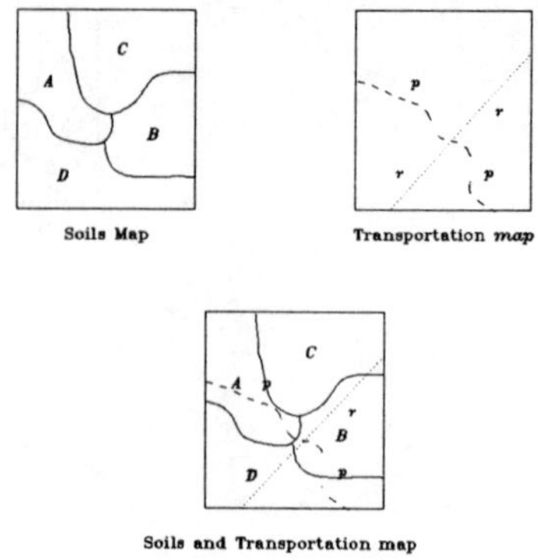

TRANSPORTATION

CODE_NO	DESCRIPTION
P	PRINCIPAL ROAD
R	RAILROAD

SOILS

SOILS_CO	DEPTH
A	130
B	200
C	180
D	190

TRANSPORTATION / SOILS

SOIL-TRANPORTATION	DESCRIPTION	DEPTH
AP	PRINCIPAL ROAD	130
BP	PRINCIPAL ROAD	200
BR	RAILROAD	200
CR	RAILROAD	180
DP	PRINCIPAL ROAD	190
DR	RAILROAD	190

Figure 5. Line in polygon overlaying

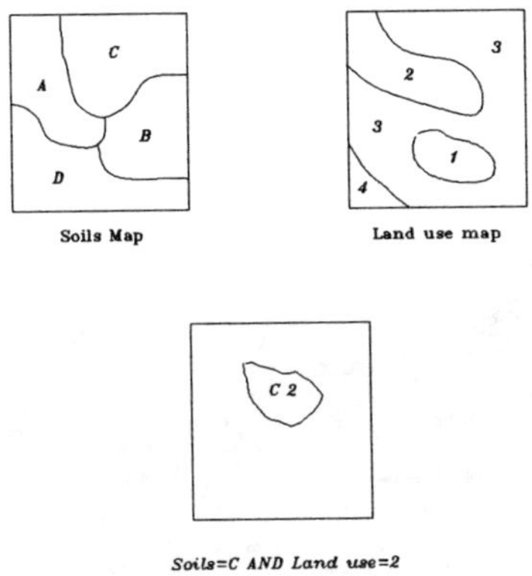

Figure 6. Map overlaying using **and** statement (intersection)

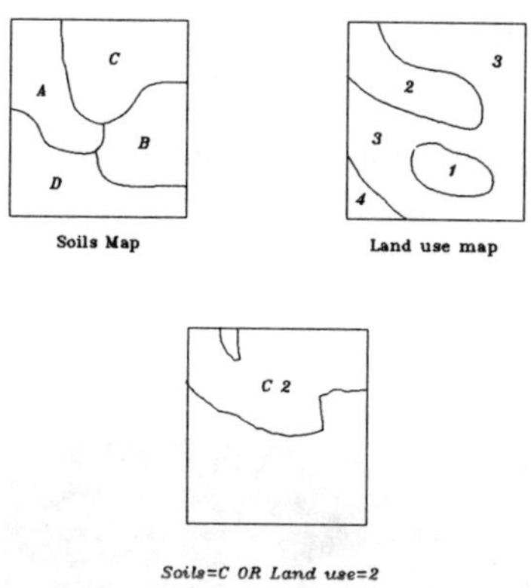

Figure 7. Map overlaying using **or** statement (union)

2.3.2.Topographic functions. The surface characteristics such as the slope, relief and form of an area are referred to as topography. The topography of a surface can be represented in a digital elevation model (DEM). A DEM represents a topographic surface in terms of a set of elevation values measured at a finite number of points (Tempfli, 1982), and contains terrain features of geomorphological importance such as valleys and ridges, peaks and pits.

Figures 8 and 9 illustrate an isometric view in a grid and "hillshading' representations, respectively. The area is located in El Oro, nearby Mexico City.

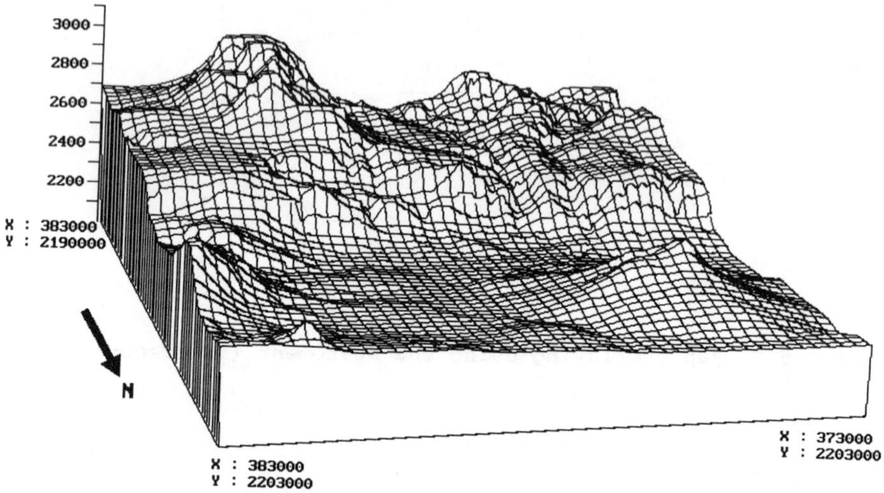

Figure 8. Isometric view of El Oro, grid representation.

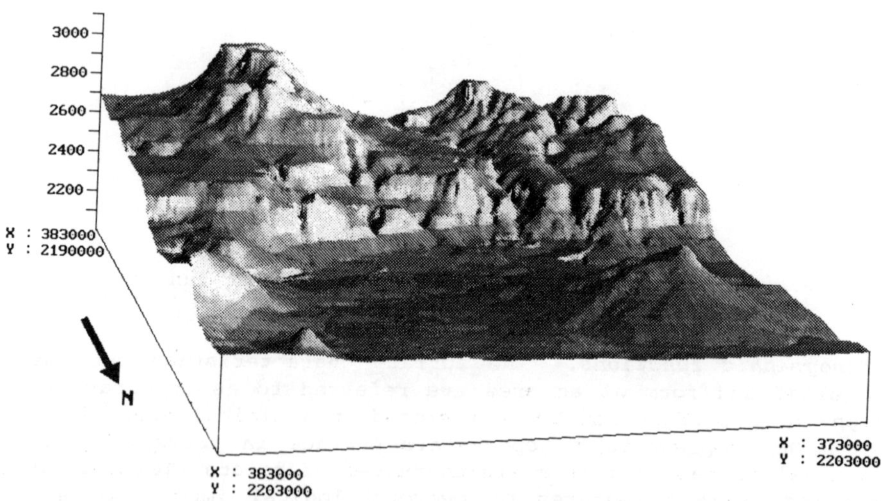

Figure 9. Isometric view of El Oro, hillshading representation.

DEM are commonly organized in regular grid formats because of their ease for direct computer manipulation. Due to their fixed spatial resolution, regular grids, are not adapted to changes in relief roughness. The use of smaller cell sizes to represent small changes, leads to large storage requirements and redundancy in less rugged terrains.

The triangulated irregular network (TIN) structure, organizes data in irregularly spaced triangular facets. This allows for additional information in areas of rough relief without redundancy in smoother areas.

Topographic functions are used to calculate values that describe the topography of an area. The most common transformations working with elevation data are the **slope**, which is defined as the rate of change of elevation and the **aspect** which is the direction that a slope faces. The maximum slope is called the **gradient**.

2.3.3.Interpolation. Interpolation is the procedure of estimating unknown values at unsampled sites using the known values of existing observations at neighbouring locations. It is derived from the latin **inter** (between) and the polish **polire** (refine). The term means refining by computing values in between. Extrapolation is the procedure of estimating values outside the area covered by available observations.

Point based interpolation is used to estimate values at predetermined locations using points of known locations and values, eg. meteorological stations, spot heights, porosity measurements. The output can be an isopleth map. This type is one of the most commonly used spatial interpolator in a GIS.

Aerial interpolation estimates the values in target zones using the known values from a source zone, eg. given population counts for census tracts, estimate the population for electoral zones.

Point and aerial interpolation include a variety of methods to predict values. Among them are, kriging, b-splines, trend surface analysis, fourier series and moving averages (Sampsom, 1978; Lam, 1983; Davis, 1986; Burrough, 1986).

The quality of the interpolation results is a function of the precision, accuracy, number and distribution of the known points used in the calculation and the manner in which the mathematical function models reality. The unknown values are then calculated according to this function. The choice of an appropriate model is, therefore, crucial in order to secure reasonable results.

The interpolation assumes that observations close together in space are more likely to have similar values than observations far apart. The values of the attributes to be interpolated in most cases must be interval or ratio variables.

2.4.CONNECTIVITY FUNCTIONS

Connectivity operations are those that estimate values by accumulating them over the area that is being traversed. The accumulated values can be qualitative or quantitative.

In order to calculate a connectivity function , it is required to specify the manner in which the spatial elements are interconnected, the rules (possibilities, constraints) that control the movements allowed along the spatial elements and a unit of measurement.

Connectivity functions are grouped into contiguity, proximity, network and spread operations.

2.4.1.Contiguity. Contiguity measures characterizes spatial units that are connected. A contiguous area is formed by a group of spatial units that have one or more common characteristics and constitute a unit.

Common measures of contiguity are the size of the contiguous area and the shortest and longest straight-line distance across the area. For example, search an undulated to rolling landscape under forest covering an area of minimum 1000 square km and no larger than 2000 square km with water bodies of at least 10 square km. The area under forest should not have sections narrower than 10 km and the water bodies wider than 10 km.

2.4.2.Proximity. Proximity involves the measurement of the distance between features. The measurement unit can be distance in length, travel distance in time or other units. In order to calculate proximity, the features or objects (roads, wells, houses, thematic units); the units of measure (meters, seconds); a function to calculate proximity (euclidian distance); and the area of interest should be specified. A buffer zone may be the result of a proximity analysis. Figure 10, illustrates a distance map in a watershed in El Oro, Mexico. The distances are calculated from roads and settlements.

Examples of proximity analysis are the determination of a buffer zone along an area selected as a forest reserve. This zone serves as a barrier protecting the reserve area from any type of exploitation activity. Other example is the identification of areas within 400 km of potential sites for office building and more than 1500 m from proposed housing and conservation locations.

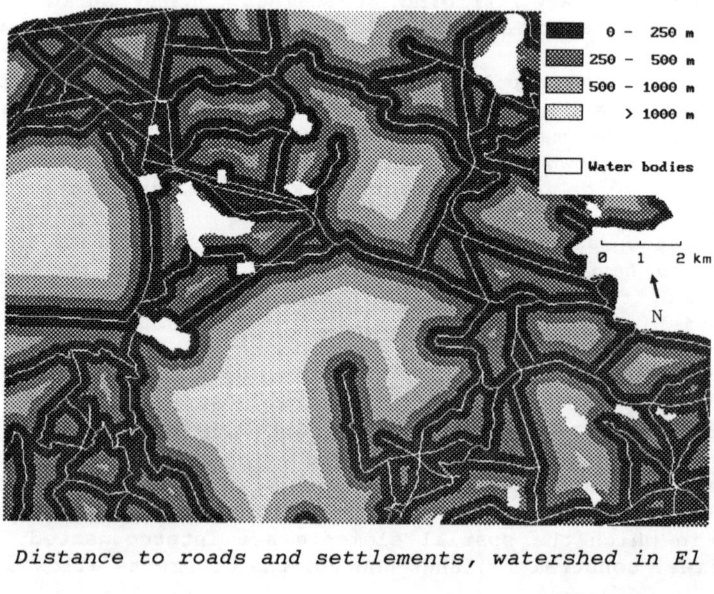

Figure 10. Distance to roads and settlements, watershed in El Oro, Mexico

2.4.3.Network. Network functions are commonly used in analysis that require moving resources from one location to another.

GIS is used to perform network analysis such as prediction of network loading like for instance transport of water and sediments in a fluvial system; route optimization such as airline scheduling, urban transportation services or municipal garbage collection; and resource allocation, eg. subdivision of municipal district into zones that can be efficiently serviced by hospitals and schools. In network analysis four components are usually considered: a set of resources (eg. sediments transported by water); one or more locations where the resources are located (eg. a fluvial system); a destination (eg. outlet of the watershed); and a set of constraints (eg. only permanent streams of higher order).

Examples of network analysis include scheduling the urban transportation service. The analysis include optimization of the route to service as many persons as possible in a determined zone. Other example of network analysis is the subdivision of a forested area into zones that can be efficiently covered by fire brigades in case of emergency. The analysis involves the identification of the most effective routing, so that fire brigades can arrive to any fire as rapidly as possible.

2.4.4.Spread. Spread functions have characteristics of the proximity and network functions. A spread function evaluates phenomena that accumulate with distance. They are used to calculate transportation time or cost over a complex surface. Examples of spread functions are the calculation of accumulated costs such as the estimation of the cost of a telephone line construction from an existing line trunk to all other locations in the area of interest; or the estimation of the time taken to walk in a relatively rugged terrain. The more you walk, the longer it takes to walk over similar terrains. This is a consequence of the cumulative effort over time. A topographic incline impedes movements differently in uphill, downhill or across slopes. Movements may be constrained by partial barriers that reduce the rate of movement or by absolute barriers that stop the movement completely.

3.Modelling

A model is a simplified representation of reality which presents significant features or relationships in a generalized form, i.e., it is a selective approximation of reality.

There are three main categories of models:

Descriptive models, which characterize or describe the real world. A map is an example of a descriptive model.

Predictive models are concerned with estimating what might occur under certain conditions. A well-known predictive model is the Universal Soil Loss Equation (USLE) soil erosion model.

Decision models are used to suggest certain courses of action to be followed in response to certain circumstances. Decision models may be considered as structured recommendations used in conjunction with descriptive and predictive models.

A GIS may be used in conjunction with a series of rules to model spatial processes. A good example is the forecasting of fire danger using parameters such as slope, elevation (derived from digital elevation models), vegetation type and biomass (interpreted and estimated from the analysis of satellite images), and meteorological data (also from satellite data).

A characteristic of modelling is the use of the attribute data, ie, each map has one or several tables that include a specific single datum (attribute) of the pertinent map. Use of the attribute data in combination with the graphical information (map) was illustrated in Figure 1. The TMU map (Meijerink, et al., 1988), is reclassified with the K-factor (soil erodability) obtained from the attribute database. Figure 11, depicts soil losses estimation using the universal soil loss equation (USLE)[Wishmeier and Smith, 1978]. The TMU map (homogeneous units) are reclassified with the KLS-factor (erosion susceptibility) and multiplied with the R-factor (rain erosivity) to obtain the potential soil erosion (soil erosion losses in ton/ha/year under standard conditions; slope length 22 m, slope gradient 9 percent and fallow bare soil). This map is then multiplied by the vegetation map reclassified with the C-factor (vegetative cover) to generate the soil erosion hazard map (soil losses in ton/ha/year) under the current conditions.

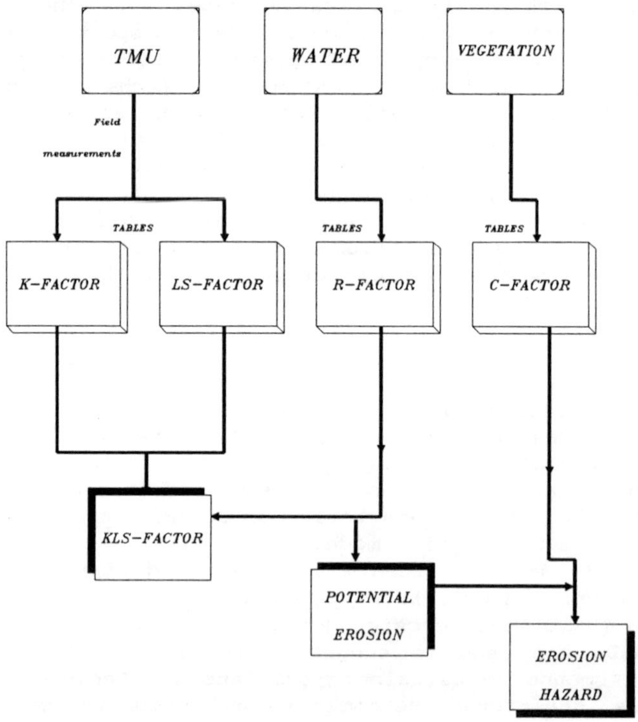

Figure 11. Schematic representation of an erosion model

In cartographic modelling, the problem (conceptual model) to be solved is broken down into sub-models that can be solved separately. The model is composed in such a way that the solution of the individual sub-models can be combined to yield solutions to the original model. Each model can be further broken down into even smaller models for solving directly (Nou, 1983). The basic model of decision analysis is to break a decision problem into component parts, thereby allowing the decision maker to focus separately upon the critical issues. The pattern that allows the model to be broken down is referred to as decision model (Bunn, 1984).

Another important characteristic of cartographic modelling in GIS is the fast overlays in windows (portions) of the database. This is particularly important when using complex interactive analysis procedures, because the analyst has the opportunity to analyze and evaluate-almost in real time-the result of the cartographic modelling procedure. If satisfied with the result, the entire area can be analyzed; if not, variables or procedures can be changed and a new trial made .

A major advantage of cartographic modelling is the need to define the problem clearly and to decide on the data required to solve it. The need to develop a clear, logical conceptual model with well defined spatial operations that can be linked together forces the user to think clearly about the steps needed to solve the problem and to make his/her methodology open to examination (Burrough, 1986). Several spatial decisions have to be taken during a cartographic modelling. There are some general steps involved during the decision making process. These steps are:

- identify the objective through a thorough analysis of needs and requirements. Collect the necessary data that is available and define the data gathering procedure for non-available data
- define the problem rigorously and in detail. State clearly the objectives, assumptions, limitations and constraints
- if there is more than one objective
    - define the relations between objectives in commensurate terms (usually express in costs), e.g. express in money the environmental impact of a road construction; the objectives are to minimize costs of construction and its impact on the environment
    - collapse the objectives into one objective, e.g. minimize sum of construction and environmental costs
- find the appropriate solution procedure and method
- find the solution to obtain the optimal answer to the problem

Most GIS's include spatial analysis capabilities. The range and ease to perform spatial analysis is ample. Complex operations can be performed, however, it is important that the analysis or the user of a system is critical and includes validation and data control quality assessments throughout the entire implementation process.

References

Aronoff. S. 1989. Geographic information systems: A Management perception. Ottawa, WDL Publications.

Bunn, D. W. 1984. Applied decision analysis. McGraw Hill Book Co, New York, USA.

Burrough, P. A. 1986. Principles of geographical information systems for land resource assessment. Clarendon Press, Oxford, UK.

Davis, J. C. 1986. Statistics and data analysis in Geology. John Wiley and sons.

Lam, N. 1983. Spatial interpolation methods: A review, The american cartographer, 10(2) p.129-149.

Meijerink, A. M. J., C. R. Valenzuela and A. Stewart (eds). 1988. ILWIS, The integrated land and watershed management information system. ITC publication number 7. ITC, Enschede, The Netherlands.

Nou, D. S. 1983. Expert computer systems. Computer, vol 16.

Sampsom, R.J. 1978. Surface II, revised edition, Kansas Geological survey, Lawrence, Kansas.

Wischmeier, W. H. and D. D. Smith. 1978. Predicting rainfall erosion losses: a guide to conservation and planning. USDA Agri Handbook No 537, Washington DC, USA.

ERRORS IN GEOGRAPHIC INFORMATION SYSTEMS

Michael J.C. Weir
I.T.C.
P.O. Box 6, Boulevard 1945
7500 AA Enschede
The Netherlands

1. Introduction

Geographic information systems are used to provide spatially referenced data and information to support decision making for resource management, land use planning, etc. It is therefore essential that products of GIS analysis and modelling operations be reliable and that decision makers can have confidence in using them. Maps and other data used in a GIS database are obtained by measurement (e.g. photogrammetric plotting) and therefore inevitably contain errors, no matter how small. In addition to these inherent errors, most GIS operations will themselves introduce new errors. Those who use GIS must therefore understand the nature of these inherent and operational errors and the way in which they are propagated through the system. Unfortunately, however, error are perhaps one of the least understood aspects of geographic information systems. Although it is no possible to remove errors from the system, they can at least be managed and kept to an acceptable minimum.

2. Precision and accuracy

All instruments for collecting the primary data used in a GIS, for example the photogrammetric plotters used to make maps or guages to measure stream run-off, can only measure to a certain level of precision. Precision is a statement of the smallest unit of measurement to which data can be recorded. Making measurements with a high degree of precision does not mean that the results are necessarily correct.

Precision should not be confused with accuracy. Accuracy is a statistical concept which states the liklihood or probability that a particular set of measurements is within a certain range of the true value, for example that there is a 95% probability that all features on a map are within 5 metres of there true position. Typical measures of accuracy include the root mean square error and the standard error of a mean value estimated from a sample. Classification accuracy (of land use for example) can be expressed in a similar way, e.g. there is a probability of X% that a certain class label is correct.

Confusing precision with accuracy can result in so-called "spurious accuracy". Spurious accuracy frequently occurs in computer calculation, for example, displaying areas calculated from coordinates obtained by

A. S. Belward and C. R. Valenzuela (eds.),
Remote Sensing and Geographical Information Systems for Resource Management in Developing Countries, 349–355.

digitizing a 1:50 000 map to the nearest square metre.

3.Sources of error in geographic information systems

Errors can arise at every stage of using a geographic information system, from the collection of the original data to the output and use of the resulting information (table 1)

Table 1. Sources of error in a geographic information systems

1. Errors in the source data

- geometric (positional) and semantic (classification) errors in the compilation of source maps
- geometric and classification errors in remotely sensed data
- errors in other source data, e.g. from field sampling
- innacuracies due to the vague ("fuzzy") character of natural boundaries, e.g. of vegetation or soil types
- errors due to the source data being out of date

2. Errors occurring during data input

- digitizing errors due to operator mistakes and limited precision of the digitizer
- errors in attribute data entry (typing errors)

3. Errors in data storage

- errors due to the limited precision with which coordinates and other numerical data are stored
- errors arising from vector to raster conversion

4. Errors in data analysis and manipulation

- propagation of errors during map overlay
- errors due to incorect use of formula (misuse of logic, etc.)
- errors arising from interpolation, e.g. for the determination of terrain slope from digitized contours

5. Errors in data output and application

- cartographic errors due to the limitations of output devices

- incorrect or inappropriate application of GIS products

A full treatment of all these sources of error is beyond the scope of this chapter. A comprehensive discussion of GIS errors can be found in chapter 6 of Burrough (1986). The examples described below illustrate three of the principal, but least obvious sources of error in geographic information systems.

4.Errors in source maps

Figure 1 shows five land cover maps prepared by five forest technicians trained in air photointerpreation and mapping. From an examination of figure 1 a number of geometric errors (due to the use of a simple photogrammetric plotter for transferring the delineated boundaries from photograph to the map control sheet) are apparent. More serious, however, are the obvious differences in the location of the boundaries due to differences in interpretation and classification of the various land cover types. Such differences are due to the limited resolution of the 1:40 000 aerial photographs and the fuzzy nature of the natural boundaries.

5.Errors due to rasterizing polygon data

Polygon to raster conversion can be considered a form of point sampling, in which the cell is assigned the value of that attribute which occurs at the centre of cell. Selecting appropriate dimensions of the grid cells used for data storage, analysis and map presentation is a fundamental problem. In adopting a given cell size, the user predefines the map resolution to be used in all subsequent operations. From a practical point of view, a fine grid (and thus a large number of cells) imposes limitations on processing time and on the physical size of the maps which can be output from the system. Although these problems are now becoming less serious with the availability of faster processors, inexpensive memory and high-resolution output devices, all raster based systems still require the user to make a concious choice of cell size.

The problems arising when a "true" polygon map is represented by an array of grid cells have long been recognized (Frolov and Maling, 1969), namely: i) errors in the position and form of the boundaries between map units; ii) errors in the calculated areas of individual map units. These errors occur in the cells which fall on the boundary between distinct map units. Studies by Frolov and Maling (op.cit.) and Switzer (1975) form the basis for a mathematical treatment of the errors resulting from gridding a map according to the attribute at the cell centre, as is usually employed in algorithms for polygon-raster conversion. For a review of these studies, see Burrough (1986). The size of the error depends not only on the cell size, but also on the complexity of the map, i.e. on the number of boundary cells.

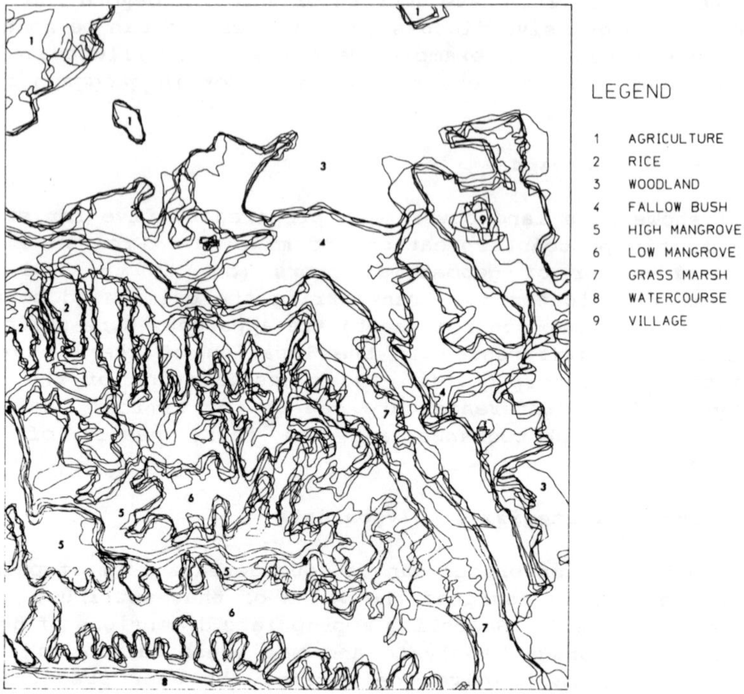

Figure 1. Superimposition of five land cover maps (see also Drummond, 1987)

Crapper (1980) discusses the similar problem of estimating areas by counting pixels in Landsat imagery in which perimeter cells with a mixed spectral signature occur.

Although procedures are available for assessing and predicting the accuracy of gridcell maps, there is little evidence that they are being routinely applied. An examination of published work suggests that when simple gridcell-based systems are used in practice, consideration is rarely given to the the selection of an appropriate grid cell size. In many cases little or no justification is given for the dimensions selected. In others, problems such as resolution are mentioned, but the final choice is apparently based on "educated guesswork". Since reliable area data and locational information are essential to many resource management applications, it is important that resource managers who use a GIS be aware of the positional and area errors which may be present in the products of the system.

Table 2 shows the areas of the different cover types as obtained from digitizing part of a 1:10 000 forest type map and by counting the cells in grid cell maps of various resolution created by rasterizing the original polygon data (Weir, 1988).

Table 2. Areas (ha) of forest types as computed from different grid cell maps

Type code	Source map	25m grid	50m grid	100m grid
10	21.45	21.50	20.50	14.00
11	17.00	17.38	16.00	10.00
12	12.25	12.56	13.25	6.00
21	32.29	32.25	32.50	33.00
22	7.80	7.50	7.25	8.00
23	0.99	1.13	1.00	1.00
31	86.46	86.25	89.25	111.00
32	4.08	4.06	4.25	3.00
33	0.98	0.94	0.75	---
40	9.41	9.25	7.75	10.00
41	0.66	0.63	0.75	1.00
45	4.47	4.63	4.50	2.00
Total	200.00	200.00	200.00	200.00

In order to assess the amount of information lost as a result of rasterizing, the number of discrete units and the length of the forest boundary in the test area were measured on each gridcell map and the results compared with the number of polygons and their perimeters as calculated from the original digitized boundary coordinates. The results (table 3) clearly reveal a reduction in the amount of map information as the cell dimensions increase. Particularly with a 25m grid, a number of units have been artificially created by fragmentation of elongated polygons during the rasterizing process. More than half of the total boundary length has been lost at 100 metre resolution.

Table 3. Loss of information due to polygon-raster conversion

	Source map	25m grid	50m grid	100m grid
Total number of map units	71	74	68	36
% reduction in perimeter	--	17.1	31.9	67.5

6.Error propagation in map overlay

One of the most commonly applied operations in geographic information systems is the overlaying of two or more maps. As the discussed above, every map in a GIS database will contain errors, either due to inaccuracies in the source data or due to errors arising from some form of computer processing, e.g. polygon-raster conversion. During the process of map overlay, all the errors in the individual maps contribute to the final error of the output map as shown in figure 2.

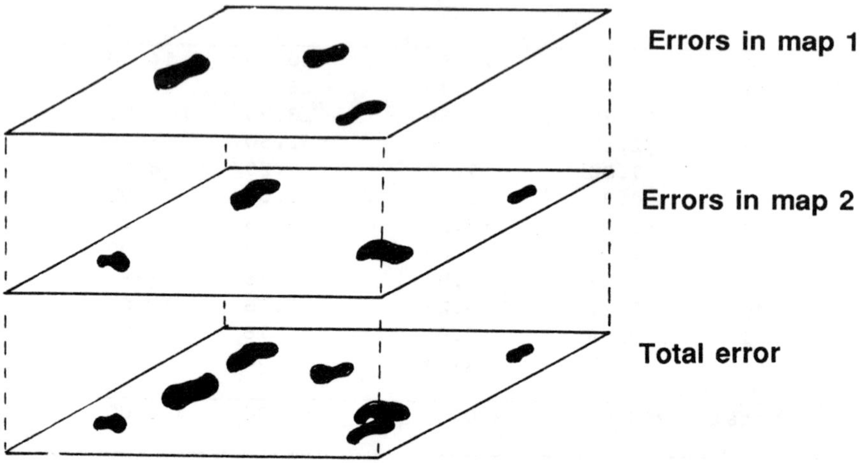

Figure 2. Error due to map overlay in a geographic information system

The implications of error propagation during map overlay can be serious (Bailey, 1988). MacDougall (1975) describes the result of overlaying six maps each of which on its own was considered to be of acceptable accuracy. The resulting overlay map was not significantly different from a random map. Although this may be an extreme case, the message is clear!

7.Concluding remarks

This chapter has discussed a number of sources of error in geographic information systems and given some indication of the magnitude which such errors may have. All persons using GIS technology - those involved in data collection, operators responsible for data input and processing and, most of all, decision makers who use the system output - must be aware of the errors which can occur and where possible take appropriate steps to keep such erors to an acceptable minimum.

References

Bailey, R.G. 1988. Problems with using overlay mapping for planning and their implications for geographic information systems. Environmental Management, 12(1):11-17.

Burrough, P.A. 1986. Principles of geographical information systems. Clarendon Press, Oxford: 193pp.

Crapper, P.F. 1980. Errors incurred in estimating an area of uniform land cover using Landsat, Photogrammetric Engineering and Remote Sensing, 46(10):1295-1301.

Drummond, J.E. 1987. A framework for handling error in geographic data manipulation, ITC Journal (1987-4):73-82.

Frolov, Y.S. and Maling, D.H. 1969. The accuracy of area measurement by point counting techniques, The Cartographic Journal,6(1): 21-35.

MacDougall, E.B. 1975. The accuracy of map overlays, Landscape Planning, 2:23-30.Switzer, P. 1975 Estimation of the accuracy of qualitative maps, in "Display and analysis of spatial data",(Davis and MacCullagh, eds), Wiley, New York:1-13.

Weir, M.J.C. 1985. Assessing the influence of cell size and sampling technique on forest resource data in a raster-based GIS, 9th Asian Conference on Remote Sensing, Bangkok.

SPATIAL DATA ANALYSIS IN RASTER-BASED GIS: AN INTRODUCTION TO GEOMETRIC CHARACTERIZATION

A. G. FABBRI
I.T.C.
P.O. Box 6, Boulevard 1945
7500 AA Enschede
The Netherlands

1. Introduction

A geographic information system, GIS, is a computer-based system for capturing and processing spatially distributed data of geographic nature. According to Aronoff (1989, p.1):

> *"A GIS is designed for collection, storage and analysis of objects and phenomena where geographic location is an important characteristic or critical to the analysis."*

The spatial distribution of objects or phenomena is a consequence of the processes that generated them and often it can be quantitatively characterized and used in a genetic interpretation. For example, the location of some landslides can be associated with active faulting in their vicinity, or the occurrence of tungsten ore deposits is associated with plutonic rocks. An analysis of the spatial distribution of all the features and phenomena which can be genetically related through a qualitative or a quantitative model can lead to assessing the strength of a spatial pattern and sometime it is possible to predict the location of more of such features. Spatial distribution deals with concepts such as: "coinciding in position with", "within a certain distance from", "adjacent to", "oriented in direction ... with strength ...." It deals with object shape as well, when this is characteristic of particular types of objects. In general, a model for spatially relating features of different types of geographically positioned datasets, requires a clear rationale, an absolute geographic positioning, and the registration of the overlapping features within the dataset. The accuracy of characteristics, including the position, is essential for the analysis of spatially distributed data.

A. S. Belward and C. R. Valenzuela (eds.),
Remote Sensing and Geographical Information Systems for Resource Management in Developing Countries, 357–388.

Digital images are processed so that new transformed images are obtained that are either more informative, or that isolate some desired features from the input data. A common digital form of image data, consists of a regular array of discrete values (grey level intensities, or symbolic labels) termed pixels, that are distributed at the intersections of a square grid or of a hexagonal grid, termed raster. In such images a pixel is surrounded by eight neighbours (4 at the corners of a square, and 4 on its sides) in a square tessellation or by six neighbours in the hexagonal one. Usually, to transform a digital image, the values of some of the neighbours of each pixel are analyzed to decide on the value to assign to the corresponding pixel in the transformed image. Such local procedures are termed local operators or neighbourhood operators. An approach that uses set theory for the analysis of images is mathematical morphology. Its theoretical background and its new potential for pattern recognition has been treated by Serra (1982). In mathematical morphology the local operators are termed structuring elements.

This lecture aims at providing the basic elements for generating spatial transformations of digital images. The assumption is made that much of the data in the earth sciences is generated by sensors in a rectangular array of picture elements or pixels, termed "raster". For this reason, a discussion of how to process image data in such arrays is sufficiently representative of most spatial manipulations in GIS. In addition, because image analysis by mathematical morphology (Serra, 1982; Fabbri, 1984) provides a generalization of techniques of spatial transformations and a statistical base for geometric characterization, elements of such morphology are used throughout this lecture.

Table 1: Comparison between raster and vector data models (modified after Aronoff, 1989, p. 166, Table 6.1).

RASTER	VECTOR
Advantages	
simple structure	more "compact" structure
easy overlaying	efficient topology encoding
efficient for high spatial variability	better suited for graphics (e.g., hand-drawn maps)
Disadvantages	
less compact (if not compressed)	more complex structure
topology more difficult	overlaying more difficult
jaggedness in output graphics is seen unless files are very large	inefficient for high spatial variability
	not effective for enhancement of digital images

2.Data Structures and Integrated Analysis Functions in a GIS.

In a computer there are two types of spatial data representations which are commonly used: the raster model and the vector model. The vector model consists of representations of objects by means of points, lines and polygons (bounded by points and lines!) as vectors in 2-D, or in 3-D, organized in chains or in sequences to form a data structure with explicit coordinates and pointers that connect objects in a hierarchy. The hierarchy must be known and has to be used to locate objects in a vector database. The raster model consists of a rectangular array of picture elements or pixels which are in one-to-one correspondence with small rectangular areas on the original picture material that they represent. The position of any feature in the original picture is only recorded to the nearest pixel. Pixel coordinates are implicit and absolute positioning is obtained by incrementing the X- and Y-coordinates of the pixel by multiples of the pixel size, from the absolute positioning of generally the upper left pixel in a column arrangement, in the direction of the first row.

In Table 1, a comparison is made between the two data models. Clearly, most computations can be performed on either data models, however, some computations are more convenient with one of the two models. The computations of interest in this lecture belong to the general analysis functions in a GIS. A classification of those functions is shown in Table 2.

We will describe few selected examples of operations of overlaying and neighbourhood and some connectivity functions. The purpose is not to be exhaustive, but to describe how such operations and functions can be generated in a raster data model. In any case, for visualization most plotting devices are based on a raster output format. Also, several GIS systems are now designed to be able to contemporaneously process vector data, raster data and tabular data in which more information is stored for complete data management. For this reason the distinction between the two data models is becoming less important. In the future, in the more advanced GIS the data models will be transparent to the user, i.e., the user will not be necessarily aware of the data model used for processing and/or the most computationally convenient model will be automatically selected by the system.

Table 2: A Classification of GIS Integrated Analysis Functions
(after Aronoff, 1989, p. 196, Fig. 7.5)

```
+----------------------------------------------------------------+
|        INTEGRATED ANALYSIS OF SPATIAL AND ATTRIBUTE DATA        |
|        |                                                        |
|        |                                                        |
|        +---Retrieval/Classification/---+---Retrieval            |
|        |   Measurement                 |                        |
|        |                               +---Classification       |
|        |                               |                        |
|        |                               +---Measurement          |
|        +---Overlay Operations                                   |
|        |                                                        |
|        +---Neighbourhood Operations-----+---Search              |
|        |                                |                       |
|        |                                +---Line-in-polygon and |
|        |                                |     point-in-polygon  |
|        |                                |                       |
|        |                                +---Topographic functs. |
|        |                                |                       |
|        |                                +---Thiessen polygons   |
|        |                                |                       |
|        |                                +---Interpolation       |
|        |                                |                       |
|        |                                +---Contour generation  |
|        |                                |                       |
|        +---Connectivity Functions-------+---Contiguity measures |
|        |                                |                       |
|        |                                +---Proximity           |
|        |                                |                       |
|        |                                +---Network             |
|        |                                |                       |
|        |                                +---Spread              |
|        |                                |                       |
|        |                                +---Seek                |
|        |                                |                       |
|        |                                +---Interdivisibility   |
|        |                                |                       |
|        |                                +---Illumination        |
|        |                                |                       |
|        |                                +---Perspective view    |
+----------------------------------------------------------------+
```

3.Operations between Images: Overlays[1]

Typically, "thematic mapping" consists of overlying maps of different types but corresponding to a same area on the ground. To support the overlying procedure, a rationale should be developed for spatially relating the mapping units, i.e., a model should be formulated that relates the processes (physical, chemical, human, etc.) represented in one map with those in other maps. Let us see how we can manage and analyze overlays in digital image processing.

Figure 1 exemplifies the labelling of overlaps of attributes. The attributes in a set of simple images are: lithology (A, B, and C), slope (S), and ground-water condition (I). A list of labels and of the corresponding frequencies is shown in Table 3.

If we recall the description of the "drilling-site selection" models in the expert system PROSPECTOR (see lecture on Expert System in Spatial Data Analysis), processing images of 38x34=1292 cells or pixels would have required 1292 runs of the system (i.e., through the inference network). However, as can be seen in Table 3, we have 12 combinations of labels in the overlay example shown in Figure 1a. For this reason, by using the "tabular approach" of Table 3, only 12 runs are required. This is feasible because of the contiguity and continuity of pixels with a same composite label. We term such property "connectivity" of the pixels belonging to the different images. Only 12 combinations exist in our example.

The statistics contained in Table 3, allows to take decisions on the usefulness of making up the various themes. The table itself can be augmented with more data about the single themes and then used in an inference network where selective retrievals are performed to derive more specialized themes. The various entries in Table 3 are the results of the spatial distribution of the pixel values in the three input maps shown in Figure 1.

---

[1] This section was taken from Fabbri and Kasvand (1988, p. 228-231)

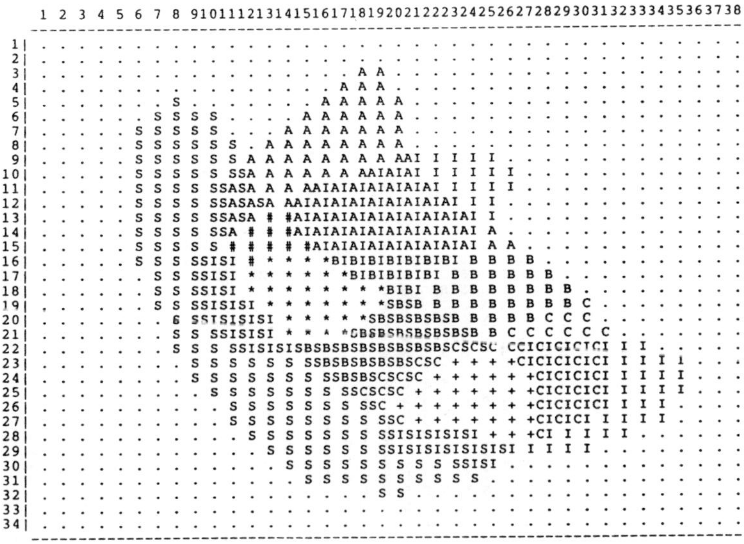

a

#=SAI, *=SBI, +=SCI

Figure 1. (a) Labelled image of intersections between images of soil class (S) in (b), ground-water conditions (I) in (c), and lithology (A, B, C) in (d), modified after Varnes (1974, p. 3, Fig. 1). Labels in (a) are: S, A, B, C, SA, SB, SC, SI, AI, CI, #=SAI, *=SBI, and +=SCI. Picture coordinates are added on top and to the left of the image in (a). The images are of sizes 38 pixels x 34 rows, i.e., 1292 pixels. Note that the expert system PROSPECTOR would have required 1292 runs. However, by using the "tabular approach" in Table 3, only 12 runs are required.

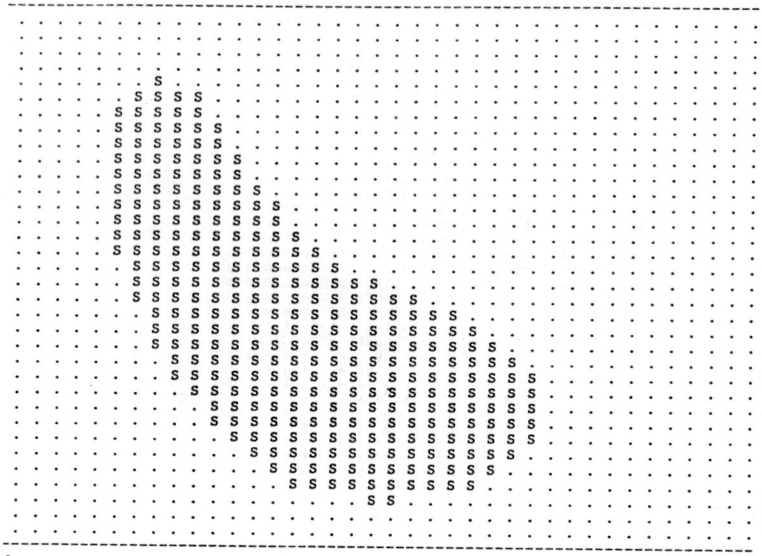

b

S = 316, . = 886

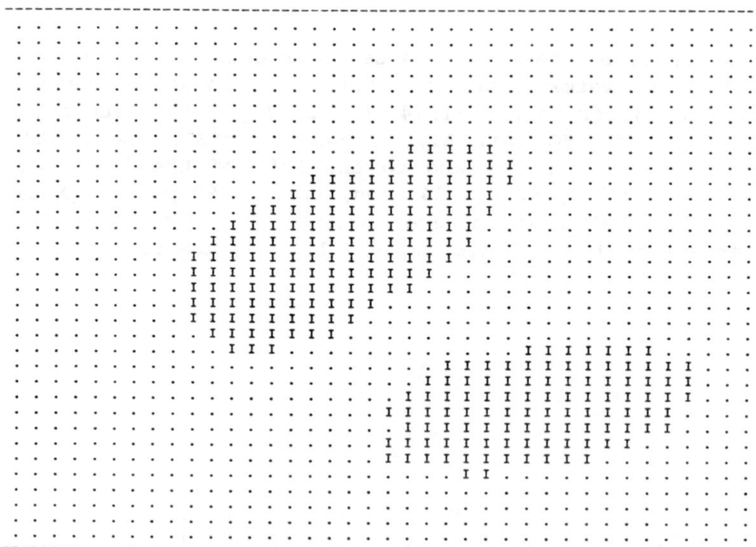

c

I = 245, . = 957

Figure 1: (continued).

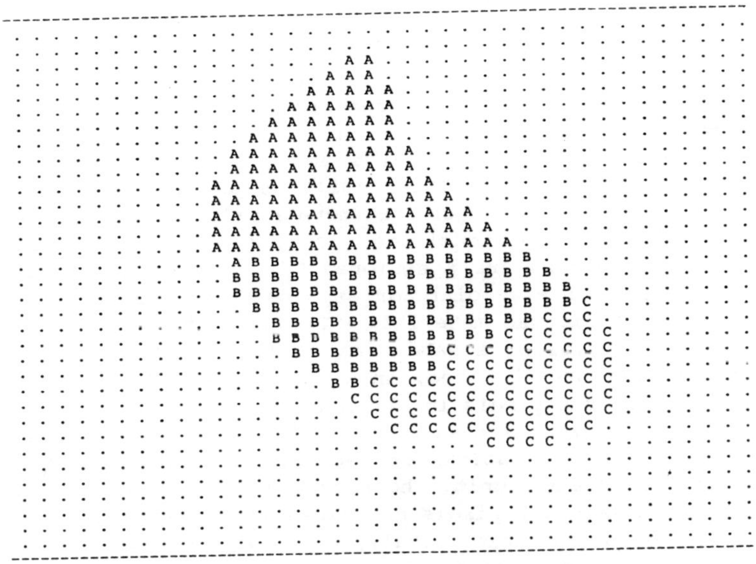

d

A = 122, B = 110, C = 82

Figure 1: (continued).

Table 3: Table of labels computed from the images in Figure 1. The row numbers in the left column represent the table address. From the pixel counts in the 7th column, a "contingency" table can be extracted.

	S	A	B	C	I	no.	.	.	etc. ...
1	S	.	.	.	.	150			
2	.	A	.	.	.	53			
3	.	.	B	.	.	31			
4	.	.	.	C	.	11			
5	.	.	.	.	I	29			
6	S	A	.	.	.	9			
7	S	.	B	.	.	29			
8	S	.	.	C	.	13			
9	S	.	.	.	I	33			
10	S	A	.	.	I	11			
11	S	.	B	.	I	34			
12	S	.	.	C	I	35			
13									
14									
15									

4. Spatial Transformations of Images[2]

In binary images there are only two grey levels (or color labels): black, the objects, and white, the background (or vice versa), usually indicated by the binary numbers 1 and 0, respectively. The description and quantitative characterization of binary images involve the geometrical properties of objects and relationships among objects (or shapes, or silhouettes) contained in them. This family of properties, however not necessarily limited to binary images, but more easily exemplified by them, including a periodicity of intensities, local properties or edges, is termed texture. The pixels are positioned in a two-dimensional image space, so as to be in point-to-point correspondence with small point-like areas in the original picture material. In applications where sets of maps and photographs covering a same area on the ground are digitized, sets of digital images are obtained, in which there is a point-to-point correspondence between the pixels in all images of the set. The analysis of these images can be organized as a hierarchically structured sequence of transformations and combinations of sets of many binary images, each corresponding to a single map unit. The result of the processing is a map pattern that contains all the pixels that identify areas associated with desirable properties such as statistically meaningful relationships to the occurrence of mineral or energy resources, or land stability for engineering geology purposes. To either aspects of image analysis, the set theory approach used in mathematical morphology (Matheron, 1972, 1975, Serra, 1982) offers powerful means for exact quantitative characterization.

Let us consider the simple artificial image shown in Figure 2, containing 12 black objects in a white background. The objects might represent a planar projection or a section containing some grains distributed in a finer matrix, or 12 holes in a porous material, obtained from a microscopic image, or the spatial distribution of a particular map unit. To extract information on geometrical characteristics of the objects, spatial transformations by structuring elements are computed. The digital image is considered as a set T of 36 x 36 = 1296 pixels that contains a subset A of the grains including all black pixels, and a subset Ac, or complement of A, containing all white pixels in the image. Figure 2 illustrates an image in which there are 278 black pixels, and 1296 - 278 = 1018 white pixels. Let us consider the transformations of the image set A, shown in Figure 3. There, a structuring element set B is used, that consists of a pair of black pixels that are horizontally adjacent. One of the two pixels is underlined to identify the origin of a short segment of length 2 pixels, the other indicates the end of the segment. A structuring element, B, is a small set of pixels that is swept across every pixel of a binary image set A, whose black-white value is changed to white-black according to the degree of coincidence (e.g., black with black) in its neighbourhood corresponding to B. This means that when the pixel at the origin of the structuring element happens to overlap a pixel

---

[2] This section was taken from Fabbri and Levialdi (1984, p. 589-601; Figs. 1-10).

in the image, the pixel itself and the surrounding pixels identify its neighbourhood for local computations. B in Figure 3 can be imagined as a set of 9 pixels in a 3 x 3 neighbourhood in which only the values of two black pixels are cared for in establishing a coincidence, and the values of the remaining 9 - 2 = 7 pixels are not cared for ("care" and "don't care" pixels).

```
   000000000111111111122222222223333333
   123456789012345678901234567890123456
  +------------------------------------+
01!....................................!
02!....................................!
03!....................................!
04!....................................!
05!....................................!
06!........**..........................!
07!......*****.....****................!
08!.....*******....****.....**...***...!
09!.....*******....*****..****..****...!
10!....********....***********..****...!
11!....********.....**********...***...!
12!....********......***...**..........!
13!....********........................!
14!....********.................**.....!
15!....****.**............****.****....!
16!.....***..**.....****..****.***.....!
17!......**...*.....****..***..****....!
18!...................**...**...***....!
19!.........**......****...............!
20!........****....****................!
21!........****......**................!
22!.........***........................!
23!........................****........!
24!........*.........**...****.........!
25!.......***........***..****.........!
26!.......***.......****...............!
27!.......***......*****....**.........!
28!.......***......****....***.........!
29!........**.......***...**...........!
30!.......***........*...**............!
31!........**...........**.............!
32!.........*..........***.............!
33!...........*******...*..............!
34!...........*****....................!
35!....................................!
36!....................................!
  +------------------------------------+

                  A               278
```

Figure 2: Artificial binary image set A of size 36 x 36 pixels and containing 12 black objects that occupy an area of 278 black pixels. Image coordinates identifying rows and columns are printed on the left and on the top of the illustration, respectively. Asterisks indicate the black pixels, dots indicate the white pixels.

```
   000000000111111111122222222223333333
   123456789012345678901234567890123456
  +------------------------------------+
01!....................................!
02!....................................!
03!....................................!
04!....................................!
05!....................................!
06!........*E..........................!
07!......****E.....***E................!
08!.....******E....***E.....*E...**E...!
09!.....******E....****E..***E..***E...!
10!....*******E....**********E..***E...!
11!....*******E.....*********E...**E...!
12!....*******E......**E...*E..........!
13!....*******E........................!
14!....*******E.................*E.....!
15!....***E.*E............***E.***E....!
16!.....**E..*E.....***E..***E.**E.....!
17!......*E...E.....***E..**E..***E....!
18!...................*E...*E...**E....!
19!.........*E......***E...............!
20!........***E....***E................!
21!........***E......*E................!
22!.........**E........................!
23!........................***E........!
24!........E.........*E...***E.........!
25!.......**E........**E..***E.........!
26!.......**E.......***E...............!
27!.......**E......****E....*E.........!
28!.......**E......***E....**E.........!
29!........*E.......**E...*E...........!
30!.......**E........E...*E............!
31!........*E...........*E.............!
32!.........E..........**E.............!
33!...........******E...E..............!
34!...........****E....................!
35!....................................!
36!....................................!
  +------------------------------------+
    v                     v         278-
    B = ** , B =** , C = A θ B       75=
                                   -----
                                    203
```

Figure 3: Erosion of binary image set A by a structuring element set B. The number of black pixels in the transformation is reduced from 278 to 203. The 75 black pixels that turned white are identified by E's. Both structuring element sets B and its reflection $\check{B}$ are shown at bottom. The pixel at the origin of the structuring element is underlined. Asterisks and dots indicate black and white pixels, respectively.

```
   000000000111111111122222222223333333
   123456789012345678901234567890123456
  +------------------------------------+
01!....................................!
02!....................................!
03!....................................!
04!....................................!
05!....................................!
06!......DD**..........................!
07!....DD*****...DD****................!
08!...DD*******..DD****...DD**.DD***...!
09!...DD*******..DD*****DD****DD****...!
10!..DD********..DD***********DD****...!
11!..DD********...DD**********.DD***...!
12!..DD********....DD***.DD**..........!
13!..DD********........................!
14!..DD********...............DD**.....!
15!..DD****D**..........DD****D****....!
16!...DD***DD**...DD****DD****D***.....!
17!....DD**.DD*...DD****DD***DD****....!
18!.................DD**.DD**.DD***....!
19!.......DD**....DD****...............!
20!......DD****..DD****................!
21!......DD****...DD**.................!
22!.......DD***........................!
23!......................DD****........!
24!......DD*.......DD**.DD****.........!
25!.....DD***......DD***DD****.........!
26!.....DD***.....DD****...............!
27!.....DD***....DD*****..DD**.........!
28!.....DD***....DD****..DD***.........!
29!......DD**.....DD***.DD**...........!
30!.....DD***......DD*.DD**............!
31!......DD**.........DD**.............!
32!.......DD*........DD***.............!
33!.........DD*******.DD*..............!
34!.........DD*****....................!
35!....................................!
36!....................................!
  +------------------------------------+
                                   278+
   v                       v       147=
   B = ***  ,  B = *** , D = A ⊕ B -----
                                   425
```

(In the figure, the first `*` of the first `***` and the last `*` of the second `***` are underlined.)

Figure 4: Dilatation if image set A by a structuring element set B. The transformation is increased from 278 to 425. The 147 white pixels that turned black are identified by D's.

```
   000000000111111111122222222223333333
   123456789012345678901234567890123456
  +------------------------------------+
01!....................................!
02!....................................!
03!....................................!
04!....................................!
05!....................................!
06!........**..........................!
07!......*****.....****................!
08!.....*******....****.....**...***...!
09!.....*******....*****..****..****...!
10!....********....***********..****...!
11!....********.....**********...***...!
12!....********......***...**..........!
13!....********........................!
14!....********.................**.....!
15!....****.**............****.****....!
16!.....***..**.....****..****.***.....!
17!......**...R.....****..***..****....!
18!...................**...**...***....!
19!.........**......****...............!
20!........****....****................!
21!........****.....**.................!
22!.........***........................!
23!........................****........!
24!........R.........**...****.........!
25!.......***........***..****.........!
26!.......***.......****...............!
27!.......***......*****....**.........!
28!.......***......****....***.........!
29!........**.......***...**...........!
30!.......***........R...**............!
31!........**...........**.............!
32!.........R..........***.............!
33!...........*******...R..............!
34!...........*****....................!
35!....................................!
36!....................................!
  +------------------------------------+
   v                                278-
   B = ** , B = ** ,                 75=
                                   -----
                  v
           E = (A ⊖ B) ⊕ B         203+
                                     70=
                                   -----
                                    273
```

(In the figure, the first `*` of the first `**` and the second `*` of the second `**` are underlined.)

Figure 5: Opening if image set A by a structuring elements set B: an erosion by B, followed by a Minkowski addition by B. The number of black pixels is first reduced from 278 to 203 by the erosion, and then it is increased to 276 by the Minkowski addition. The 5 black pixels that turned white by the opening are identified by R's.

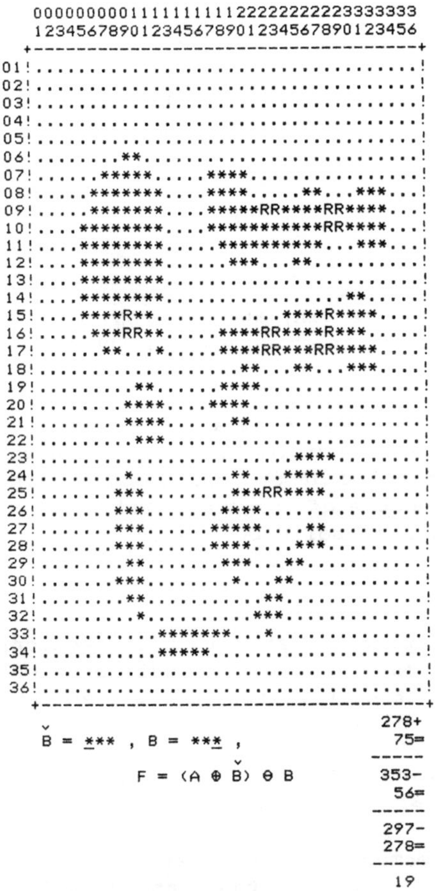

Figure 6: Closing of image set A by a structuring element set B: a dilatation followed by a Minkowski subtraction. The number of black pixels is first increased from 278 to 253 by the dilatation, and then reduced to 297 by the Minkowski subtraction. The 19 white pixels that turned black by the closing are identified by R's.

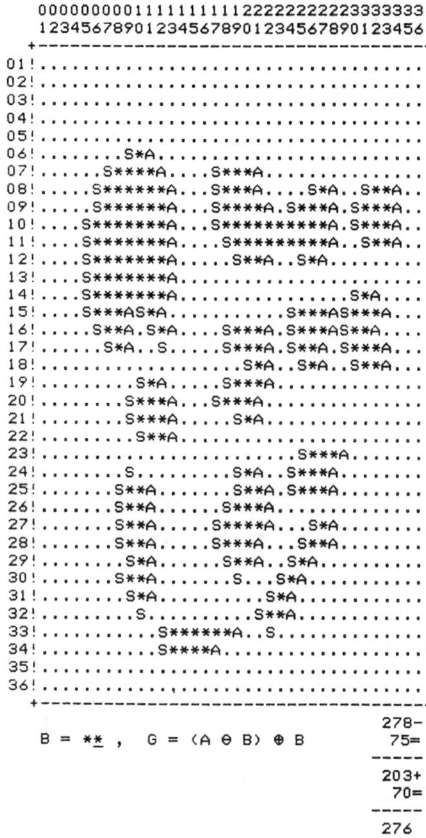

Figure 7: A minkowski subtraction of the image set A followed by a Minkowski addition by a structuring element set B. The transformed image G, is shifted one pixel to the right in respect to A, therefore, it is not entirely contained within A. The number of black pixels is first reduced from 278 to 203 by the Minkowski subtraction, and then it is increased to 276 by the Minkowski addition that follows. The black pixels that turned white in the subtraction are indicated by S's; the white pixels that turned black in the addition are indicated by A's.

Conversely, we may consider B as the set of all positions to which we translate A, and compute a logical operation between A and its translates (say the intersection between the two sets). The transformation of the image set A is computed as the logical union of all the intersections between A and its translates. Once B has been defined, an erosion of A by B generates a set C that can be indicated by the expression,

$$C = A \ominus \check{B}$$

where $\check{B}$ is the reflection of B respect to its origin, and the symbol $\ominus$ is the operator for the "shrinking" transformation in which some of the black pixels in A have become white in C.

The transformation shown in Figure 3, can be used to estimate the vertical elongation of the grains. The result of the transformation $A \ominus B$, which in this instance would affect the opposite edges of the black objects in A, has been termed Minkowski subtraction because of the published work of Minkowski (1911) on approximating the shape of solid bodies by integral geometry.

The reason for defining B and $\check{B}$ will be apparent later in this section when sequences of transformations will be exemplified. In this erosion transformation, a black pixel value in A that corresponds to the pixel at the origin of the reflection of B changes to white,

"if at least one black pixel in the neighbourhood identified by the reflection of B corresponds to a white pixel in the reflection of B."

An "expansion" type of transformation, termed dilatation and shown in Figure 4, is one in which the number of black pixels in the transformation is greater than that in the original untransformed image. The structuring element B is a longer segment than the one in Figure 3, a set of three black pixels. The expression,

$$D = A \oplus \check{B}$$

indicates a new image set D, $\check{B}$ is the reflection of B respect to its origin (underlined) and the symbol $\oplus$ is the operator of the expansion transformation. Clearly in this example, using the reflection of B generates changes in pixel values to the left of the black objects in A, while B would generate changes to the right. The latter transformation belongs to what has been termed a Minkowski addition. It can be generated also by shifting A one and two pixels to the right and computing a logical operation of union between the image sets A and its two translates.

In this dilation transformation a white pixel value in A that corresponds to the pixel at the origin of the reflection of B changes to black,

"if at least one black pixel in the neighbourhood identified by the reflection of B corresponds to a black pixel in the reflection of B."

While the erosion and dilatation shown so far represent one-step transformations, the transformation shown in Figure 5, consists of two steps, i.e., two successive transformations of the image set A by the set B (two black pixels), termed "opening", according to the expression,

$$E = (A \ominus \check{B}) \oplus B$$

To produce the opened image set E, the image set A is first eroded by B, and then a Minkowski addition by B is computed of the eroded image. In this instance the number of black pixels in E is less than the number of black pixels in A. All black pixels in A shorter than B in the horizontal direction have disappeared in the erosion and are not regenerated by the Minkowski addition that follows. This type of transformation "cleans" the black objects of small jagged edges. If performed with structuring elements of successively greater lengths 3,4 5 pixels, etc., it leads to granulometric characterization in the horizontal direction. The set of black pixels in the opened image set E in Figure 4, is entirely contained within the set of black pixels in A.

A "Closing" transformation of A by a set of three black pixels B, is shown in Figure 6. It provides the closed set F, according to the expression,

$$F = (A \oplus \check{B}) \ominus B$$

that indicates a dilatation of A by B followed by a Minkowski substraction by B. The number of black pixels in F is greater than the number of black pixels in A. The set of black pixels in A is entirely contained within the set of black pixels in F. Closing, in this instance, fills small holes or narrow gaps and bays in the black objects in the horizontal direction only. If performed with structuring elements of successively greater lengths, 4,5, 6 pixels, etc., it leads to characterizing interparticle distances.

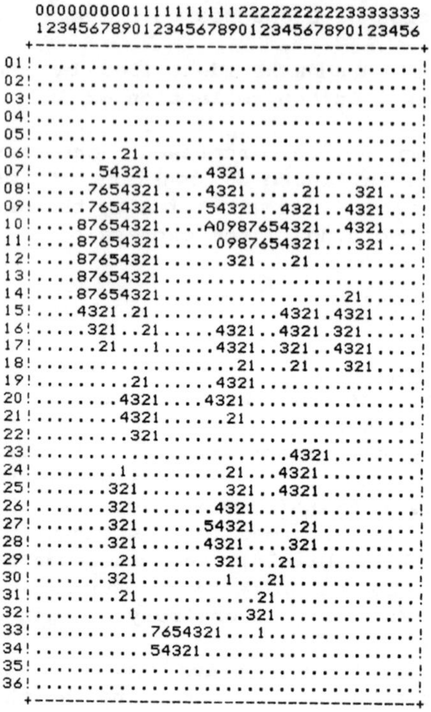

$\check{B}$ = <u>*</u>* , B = *<u>*</u>

$H(n) = A \ominus n\check{B}$, n=1, 2, ... , 11

Figure 8: Eleven successive erosions of the image set A by a structuring element set B. The black pixels that turned white after each erosion are indicated by successive erosion numbers 1,2,...9,0, and A. The number of black pixels remaining after each erosion are:
(0) = 278, (1) = 75, (2) = 70,
(3) = 52, (4) = 35, (5) = 14,
(6) = 10, (7) = 10, (8) = 7,
(9) = 2, (10) = 2, (11) = 1,
(12)= 0.

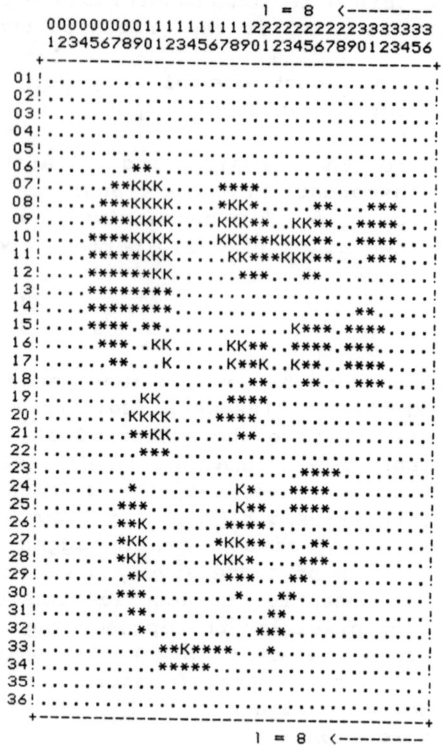

$\check{B}$ = <u>*</u> *

$K(1) = A \ominus \check{B}$

278/1296 = 0.2145
70/1008 = 0.0694

Figure 9: Geometrical covariance of the binary image set A produced by a structuring element set B consisting of two black pixels spaced L = 8 pixels apart. The pixel at the origin of the reflection of B is underlined. The transformation represents a shift of an identical copy of A, translated 8 pixels to the left respect to A. L is represented by the arrows at top and bottom of the illustration: the vertical strip of width L indicates the loss in overall overlap between A and its translate. The 70 black pixels in A that coincide with black pixels in the translate of A, are indicated by K's. The ratios of the number of black pixels in A and in K with the number of significant pixels in each are added below the illustration.

The reason why $\check{B}$, the reflection of B with respect to its origin, has been defined in mathematical morphology for erosions and dilatations, can be seen in Figure 7. There a transformation consisting of a Minkowski substraction of A by B, is followed by a Minkowski addition by B, according to the expression,

$$G = (A \ominus B) \oplus B$$

Where G, the image set resulting from the transformation is shifted one pixel to the right relatively to A. For this reason, the set of black pixels in G is not entirely contained within the set of black pixels in A. This transformation differs from the opening by B shown in Figure 5.

Characterization of shapes and textures is obtained by iterating transformations until a certain result is obtained, beyond which the transformation stops being informative. The function of the transformation, generally computed by counting the number of pixels after each iteration, is shown in Figure 8. There the image set A is eroded by the set B n times, for n = 1, 2, ..., 11 i.e., until all black pixels in A are changed to white pixels. This function produces a histogram of all horizontal segments within A of length 1 to 11. If performed in the different directions, this kind of transformation leads to granulometric characterization weighted in terms of either the number of black pixels that changed to white (in measure) or in terms of the number of chords (in number) detected after erosions of length nB. This topic has been studied by Matheron (1972) to provide the theoretical background to the quantitative morphological characterization of porous media.

The transformation of the image set A shown in Figure 9, is termed geometrical covariance. The set B is a structuring element consisting of two black pixels spaced 8 pixels apart in the horizontal direction. The transformation is obtained by shifting an identical copy of the image set A, 8 pixels to the left, and in that position computing the intersection between A and the shifted copy. In this instance, an 8 pixels wide vertical strip of pixels on the right side of A is left past the rightmost edge of the shifted copy. Therefore, the intersection is computed when the overlap is 36 x 28 pixels instead of the 36 x 36 in A. The strip of size 36 x 8 pixels represents the loss in significance. The transformation in Figure 10, is similar to the one in Figure 9, except that the spacing between the two black pixels in B is 16 pixels.

In this type of transformation $K(n) = A \ominus n\check{B}$ black pixels in A remain black only if there is another black pixel at a distance of L pixels (8 or 16 in Figures 9 and 10) to the right. The geometrical covariance function can be generated according to the expression,

$$K_{alpha}(l) = \text{mes}\,(A \ominus \check{B})$$

where A is the original image set, B is the structuring element used for the transformation (we use the reflection of B) that consists of two black pixels spaced l pixels apart in direction alpha (alpha = 0 for the horizontal direction), and mes indicates a measure of the transformed set, such as its area, given by the count of the black pixels in the transformed image set.

In the erosion function, exemplified in Figure 8, each transformation by the set B was in turn subjected to another transformation by the set B. This is not so for the geometrical covariance function where at each step it is always the set A that is transformed by a set B of successively greater length l. For small values of L, the geometrical covariance provides information about the shape of the objects in A; for larger values of l, it provides information on the spatial arrangements of the objects in A. Applications of this transformation are discussed by Serra (1982) and by Fabbri (1984).

The two transformations shown in Figure 11 are not restricted only to black pixels in the structuring element set coinciding with black and white pixels in the image set A, but also require that the white pixels in the structuring element coincide with white or black pixels in A. B1 represents a segment of length 4 black pixels with white pixels at both ends. In correspondence with its origin there is a black pixel. A shrinking type of transformation by B1 is termed "Hit-or-Miss" transformation. It is a generalization of the transformations described so far, according to the expression,

$$I = A \ominus \check{B1}$$

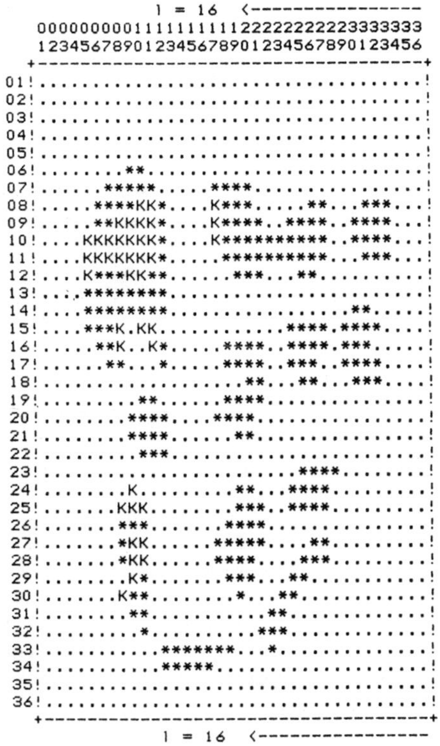

$\check{B}$ = *   *

$K(l) = A \ominus \check{B}$ 278/1296 = 0.2145   41/ 720 = 0.0569

Figure 10: Geometrical covariance of the binary image set A by a structuring element B consisting of two black pixels spaced L = 16 pixels apart. The 41 black pixels in A that coincide with black pixels in the translated copy of A are identified by K's. Additional information is given in Figure 9.

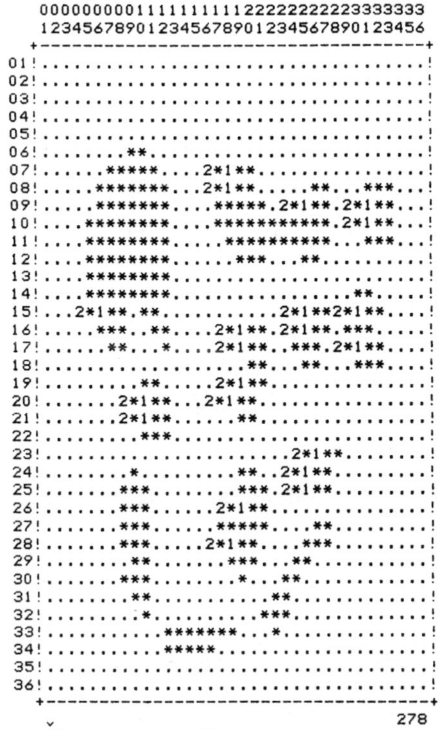

$\check{B}1$ = .*<u>*</u>**. , B1 = .**<u>*</u>*. ,

$\check{B}2$ = <u>.</u>****. , B2 = .****<u>.</u>

$I = A \ominus \check{B}1$ mes (I) = 21

$J = A \oplus \check{B}2$ mes (J) = 21

Figure 11: Hit-or-Miss transformation of the binary image set A by the structuring element sets B1 and B2, consisting of both black and white pixels. The pixels at the origin of the B1 and B2 are underlined. 1's indicate the 21 black pixels that turned white by a "shrinking" transformation by B1; 2's indicate the 21 white pixels that turned black by an "expanding" transformation by B2.

In this instance the transformation permits to identify and count all segments contained within the black objects in A that are exactly of length 4 pixels in the horizontal direction. Here, if all white pixels in the reflection of B1 coincide with white pixels in A, and all black pixels in the reflection of B1 coincide with black pixels in A, black pixels in A that correspond to the black pixel at the origin of the reflection of B1 are changed to white.

The identical count is obtained by an expanding type of Hit-or-Miss transformation by B2 of shape similar to B1, but with a white pixel at its origin, according to the expression,

$$J = A \oplus \check{B2}$$

In this instance, if the pixel coincidence is the same as in the previous one, it is the white pixels in A, in correspondence with the white pixel at the origin of the reflection of B2, that become black for each exact matching with the reflection of B2.

5.More Complex Spatial Transformations

In the previous section we have seen a few fundamental transformations of binary images. Here we will mention other spatial transformations which require longer and more complex sequences of elementary processing steps and overlay operations. Figures 12 and 13 show the results of eroding the image set A, in Figure 2, with black 8-connected and 4-connected structuring elements, respectively. This transformation identifies the internal circumference of the black objects: similarly an external circumference can be obtained by a dilatation. Such internal or external circumferences can be extracted by logical subtraction of the eroded from the original image or of the original from the dilatated image, respectively.

Erosion function and distance function are illustrated in Figures 14 and 15, respectively. Such transformations are quite common in GIS processing for corridor analysis. The object "skeletons", shown in Figure 16, can be used to facilitate the detection of object orientation patterns. The "skiz" or zone of influence transformation, shown in Figure 17, allows to generate the "tessellation" of the image plane assuming growth of the objects at a constant rate until the image plane is completely filled. It identifies all the pixels equally distant from the surrounding objects. Used with a random distribution of points, it generates Thiessen polygons or a Voronoi tessellation.

Object or component labelling, shown in Figure 18, is a transformation which uses connectivity (8- or 4-) to automatically identify all disconnected sets of pixels. Labels are assigned sequentially from top to bottom, comparing 2 adjacent rows at a time; all black objects in the image are 4-connected. Object 3 had to be relabelled because when computing reached row 08, this process did not know that it was labelling two extensions of a same object. Relabelling generated gaps in the sequence of labels, therefore when conflicts of labels were resolved (i.e., when one object had more than one label), overall relabelling had to be performed to obtain the uninterrupted sequence observed in this illustration. Labelling of objects with values 1 conventionally starts with label 2. In the illustration A, B, C and D, are one-digit symbols that correspond to 10, 11, 12, and 13. Note that 4-connectedness is assumed for the objects. Should, for instance, the pixel at row 30 and column 24 be originally white, i.e., 0 value, the object to the lower right, labelled C, would have been considered as two separate 4-connected objects, separated by an 8-connected background! Component labelling can be used to associate shape or distribution characteristics to individual objects and to selectively retrieve all objects with some common properties. For instance, all objects with a circumference greater than 1 and less than 3 in Figure 19, can be extracted from the labelled image in Figure 18. They correspond to the labels 4 to A, C, and D. Also, we can obtain the grain-size distribution by histogramming particle numbers and maximum circumference: i.e., 1 particle (D), 1 pixel; 9 particles (4-A, C, D), 2 pixels; 2 particles (3, B) 3 pixels; and 1 particle (2), 4 pixels.

Spatial characterization can be performed in many ways, particularly when multiple datasets are analyzed, and it can require long sequences of processing steps. It is important to develop precise strategies in spatial data analysis. In the earth sciences, "data-driven" procedures may be required to decide on a set of processing steps. This means that the steps to follow at a given point of an analysis are determined by the results of a previous transformation or operation, i.e., by the properties of the image. On the contrary, a "model-driven" procedure is one in which a particular goal drives the processing, independently of the properties of the image. It is the subject matter knowledge that allows to identify convenient strategies, either data- or model-driven. A simple example is discussed in the next section.

```
     000000000111111111122222222223333333
     123456789012345678901234567890123456
    +------------------------------------+
 01 |....................................|
 02 |....................................|
 03 |....................................|
 04 |....................................|
 05 |....................................|
 06 |.......EEE..........................|
 07 |......EE*EE.....EEEE................|
 08 |.....EE***EE....E**E....EEE...EEE...|
 09 |.....E*****E....E**EE..EE*E..EE*E...|
 10 |....EE*****E....EE**EEEE**E..EE*E...|
 11 |....E******E.....EE*EEEEEEE...EEE...|
 12 |....E******E......EEE...EE..........|
 13 |....E******E........................|
 14 |....E**EEEEE.................EE.....|
 15 |....EE*E.EE............EEEE.EEEE....|
 16 |.....EEE..EE.....EEEE..E*EE.E*E.....|
 17 |......EE...E.....EEEE..EEE..EEEE....|
 18 |...................EE...EE...EEE....|
 19 |.........EE......EEEE...............|
 20 |........EEEE....EEEE................|
 21 |........EE*E......EE................|
 22 |.........EEE........................|
 23 |........................EEEE........|
 24 |........E.........EE...EE*E.........|
 25 |.......EEE........EEE..EEEE.........|
 26 |.......E*E.......EE*E...............|
 27 |.......E*E......EE*EE....EE.........|
 28 |.......EEE......EE*E....EEE.........|
 29 |........EE.......EEE...EE...........|
 30 |.......EEE........E...EE............|
 31 |........EE...........EE.............|
 32 |.........E..........EEE.............|
 33 |...........EEEEEEE...E..............|
 34 |...........EEEEE....................|
 35 |....................................|
 36 |....................................|
    +------------------------------------+
                               ***
                Erosion by B = ***
                               ***
```

Figure 12: Erosion by a black 8-connected structuring element of size 3x3 pixels. Note the 4-connectedness of the set of pixels, marked with E's, whose value changed from black to white during the erosion.

```
     000000000111111111122222222223333333
     123456789012345678901234567890123456
    +------------------------------------+
 01 |....................................|
 02 |....................................|
 03 |....................................|
 04 |....................................|
 05 |....................................|
 06 |.......EEE..........................|
 07 |......E***E.....EEEE................|
 08 |.....E*****E....E**E....EEE...EEE...|
 09 |.....E*****E....E***E..E**E..E**E...|
 10 |....E******E....E****EE***E..E**E...|
 11 |....E******E.....E***EEE**E...EEE...|
 12 |....E******E......E*E...EE..........|
 13 |....E******E........................|
 14 |....E***E**E.................EE.....|
 15 |....E**E.EE............EEEE.E**E....|
 16 |.....E*E..EE.....EEEE..E**E.E*E.....|
 17 |......EE...E.....EE*E..E*E..E**E....|
 18 |...................EE...EE...EEE....|
 19 |.........EE......EE*E...............|
 20 |........E**E....EE*E................|
 21 |........E**E......EE................|
 22 |.........EEE........................|
 23 |........................EEEE........|
 24 |........E.........EE...E**E.........|
 25 |.......E*E........E*E..EEEE.........|
 26 |.......E*E.......E**E...............|
 27 |.......E*E......E***E....EE.........|
 28 |.......E*E......E**E....EEE.........|
 29 |........EE.......E*E...EE...........|
 30 |.......E*E........E...EE............|
 31 |........EE...........EE.............|
 32 |.........E..........E*E.............|
 33 |...........EEEEEEE...E..............|
 34 |...........EEEEE....................|
 35 |....................................|
 36 |....................................|
    +------------------------------------+
                                *
                Erosion by B = ***
                                *
```

Figure 13: Erosion by a black 4-connected structuring element of size 3x3 pixels. Note that the set of pixels marked with E's, whose value changed from black to white during the erosion, is in large part 8-connected. Compare with the image in Figure 12.

```
     000000000111111111122222222223333333
     123456789012345678901234567890123456
    +------------------------------------+
  01|....................................|
  02|....................................|
  03|....................................|
  04|....................................|
  05|....................................|
  06|........11..........................|
  07|......11221.....1111................|
  08|.....1223321....1221.....11...111...|
  09|.....1234321....12321111121..1221...|
  10|....12344321....12332222221..1221...|
  11|....12344321.....1232111221...111...|
  12|....12344321......111...11..........|
  13|....12332321........................|
  14|....12321221.................11.....|
  15|....1221.11............1111.1221....|
  16|.....121..11.....1111..1221.121.....|
  17|......11...1.....1121..121..1221....|
  18|...................11...11...111....|
  19|.........11......1121...............|
  20|........1221....1121................|
  21|........1221......11................|
  22|.........111........................|
  23|........................1111........|
  24|........1.........11...1221.........|
  25|.......121........121..1111.........|
  26|.......121.......1221...............|
  27|.......121......12321....11.........|
  28|.......121......1221....111.........|
  29|........11.......121....11..........|
  30|.......121........1...11............|
  31|........11...........11.............|
  32|.........1..........121.............|
  33|...........1111111...1..............|
  34|...........11111....................|
  35|....................................|
  36|....................................|
    +------------------------------------+

                                   *
          Erosion function: B =   ***
                                   *
```

Figure 14: Erosion function obtained by successively eroding the image set A in Figure 2, until no pixel value changed, and then by summing all eroded images with the original one.

```
     000000000111111111122222222223333333
     123456789012345678901234567890123456
    +------------------------------------+
  01|CBA99876556789876666789987789877789A|
  02|BA9887654456787655556788766787666789|
  03|A98765543345676544445677655676555678|
  04|987654332234565433334566544565444567|
  05|876543221123454322223455433454333456|
  06|76543211**12343211112344322343222345|
  07|654321*****12321****1233211232111234|
  08|54321*******1221****12221**121***123|
  09|54321*******1221*****11****11****123|
  10|4321********1221************11****123|
  11|4321********12321**********121***123|
  12|4321********123321***111**1232111234|
  13|4321********123432111222112321122345|
  14|4321********12343322232111121**12345|
  15|4321****1**123432111121****1****1234|
  16|54321***11**12321****11****1***12345|
  17|654321**121*12321****11***11****1234|
  18|765432112111234321 1**121**121***1234|
  19|876543221**123321****122112321112345|
  20|87654321****1221****1233223332223456|
  21|87654321****123321**1233222233334567|
  22|987654321***1234321123321111123445678|
  23|9876543211112344321 12321****12345678|
  24|87654321*122344321**121****123456789|
  25|7654321***12344321***11****123456789|
  26|7654321***1234321****1211112345678 9A|
  27|7654321***123321*****1221**123456789|
  28|7654321***123321****1221***123456789|
  29|87654321**1234321***121**1123456789A|
  30|7654321***12344321*121**1223456789AB|
  31|87654321**1222222121**1234456789ABC|
  32|987654321*1111111121***123456789ABCD|
  33|A9876543211*******121*123456789ABCDE|
  34|BA987654321*****11232123456789ABCDEF|
  35|CBA9876543211111223432 3456789ABCDEFG|
  36|DCBA9876543222223345434 56789ABCDEFGH|
    +------------------------------------+
                                   *
          Distance function: B =  ***
                                   *
```

Figure 15: Distance function obtained by successively dilatating the image set A in Figure 2, until no pixel value changed, and then summing all dilatated images. For facilitating visualization, the black pixels of the original image have been retained as asterisks, however, their value should now be 0, since the distance from the black pixels representing the objects is nil. The image can be generated also by eroding the complement of the image set A. Erosion from the edges of the image, however, has to be prevented by "padding" the outside of the image with 1s for all erosions.

```
   000000000111111111122222222223333333
   123456789012345678901234567890123456
  +------------------------------------+
01|....................................|
02|....................................|
03|....................................|
04|....................................|
05|....................................|
06|....................................|
07|........#...........................|
08|.........#......#...................|
09|........#.........#......#....#.....|
10|........#..........######......#....|
11|.......#...........#.....#..........|
12|........#...........................|
13|.......#.#..........................|
14|......#..#..........................|
15|.....#....#..................#......|
16|......#....#.............#...#......|
17|...................#....#.....#.....|
18|...................#................|
19|...................#................|
20|...........#.......#................|
21|.........##.........................|
22|....................................|
23|....................................|
24|........................##..........|
25|........#..........#................|
26|........#.........#.................|
27|........#.........#.......#.........|
28|........#........#.......#..........|
29|........#.........#....##...........|
30|........#.............#.............|
31|.........#............#.............|
32|.....................#..............|
33|............#.#.#...................|
34|...........#.#.#....................|
35|....................................|
36|....................................|
  +------------------------------------+
```

Possible skeleton of A

Figure 16: The 8-connected skeleton of the image set A, obtained by identifying the local maximum values in the erosion function shown in Figure 14, and by maintaining the 8-connectedness of the black pixels. Note the bifurcations obtained from the larger objects in the upper part of the image. Such segment crossings or junctions can be easily detected after skeletonization because they mark the position of black pixels surrounded by more than 2 pixels in a 3x3 neighbourhood. Skeletons, computed as "thin lines" (i.e., one pixel in width), are also used to better detect orientation patterns.

```
   000000000111111111122222222223333333
   123456789012345678901234567890123456
  +------------------------------------+
01|.............#........##....#.......|
02|.............#........##....#.......|
03|.............#........##....#.......|
04|.............#........##....#.......|
05|.............#........##....#.......|
06|........**...#........##....#.......|
07|......*****..#..****..##....#.......|
08|.....*******.##.****.###.**.#.***...|
09|.....*******.##.*****##****##****...|
10|....********.##.***********##****...|
11|....********..#..**********.#.***...|
12|....********..##..***###**..##......|
13|....********...#.....#######..######|
14|....********...##########..#.**....#|
15|....****#**...#......#.****#****....|
16|.....***##**..#..****##****#***.....|
17|......**.#.*..#..****##***##****....|
18|........#.###.#....**.#.**.#.***....|
19|.#######.**..##..****.##...#........|
20|........****.##.****..#######.......|
21|........****..##..**..####..##......|
22|########.***...########.......######|
23|###########...########..****........|
24|........*.#####...**.#.****.........|
25|.......***...##...***##****.........|
26|.......***...#...****.##############|
27|.......***..##..*****.##.**.........|
28|.......***..##..****.##.***.........|
29|........**...#...***.#.**...........|
30|.......***...##...*.#.**............|
31|........**.#######.#.**.............|
32|.........*#........#.***............|
33|..........#*******.#.*..............|
34|........##.*****...#................|
35|.#######...........#................|
36|#..................#................|
  +------------------------------------+
```

Possible skiz of A

Figure 17: A possible skiz of the image set A, obtained by identifying the local maximum values in the distance (or dilatation) function, shown in Figure 15, and by maintaining the pixel connectedness to identify all pixels at equal distance from the black objects. A skiz partitions the the image plane into the zones of influence of the black objects in it. Note the difficulty in deciding where to locate the local maximum distance when these are even multiple number of pixels. Compare with the image in Figure 15.

```
   000000000111111111122222222223333333
   123456789012345678901234567890123456
  +------------------------------------+
01|....................................|
02|....................................|
03|....................................|
04|....................................|
05|....................................|
06|........22..........................|
07|......22222.....3333................|
08|.....2222222....3333.....33...444...|
09|.....2222222....33333..3333..4444...|
10|....22222222....33333333333..4444...|
11|....22222222.....3333333333...444...|
12|....22222222......333...33..........|
13|....22222222........................|
14|....22222222.................55.....|
15|....2222.22............6666.5555....|
16|.....222..22.....7777..6666.555.....|
17|......22...2.....7777..666..5555....|
18|...................77...66...555....|
19|.........88......7777...............|
20|........8888....7777................|
21|........8888......77................|
22|.........888........................|
23|........................9999........|
24|........A.........BB...9999.........|
25|.......AAA........BBB..9999.........|
26|.......AAA.......BBBB...............|
27|.......AAA......BBBBB....CC.........|
28|.......AAA......BBBB....CCC.........|
29|........AA.......BBB...CC...........|
30|.......AAA........B...CC............|
31|........AA...........CC.............|
32|.........A..........CCC.............|
33|...........DDDDDDD...C..............|
34|...........DDDDD....................|
35|....................................|
36|....................................|
  +------------------------------------+
```

Component labelling of A

```
   000000000111111111122222222223333333
   123456789012345678901234567890123456
  +------------------------------------+
01|....................................|
02|....................................|
03|....................................|
04|....................................|
05|....................................|
06|........**..........................|
07|......*****.....****................|
08|.....*******....****.....**...***...|
09|.....***4***....**3********..*22*...|
10|....***44***....**33*******..*22*...|
11|....***44***.....**3*******...***...|
12|....***44***......***...**..........|
13|....********........................|
14|....********.................**.....|
15|....****.**............****.*22*....|
16|.....***..**.....****..*22*.*2*.....|
17|......**...*.....**2*..*2*..*22*....|
18|...................**...**...***....|
19|.........**......**2*...............|
20|........*22*....**2*................|
21|........*22*......**................|
22|.........***........................|
23|........................****........|
24|........*.........**...*22*.........|
25|.......*2*........***..****.........|
26|.......*2*.......****...............|
27|.......*2*......**3**....**.........|
28|.......*2*......****....***.........|
29|........**.......***...**...........|
30|.......*2*........*...**............|
31|........**...........**.............|
32|.........*..........*2*.............|
33|...........1111111...*..............|
34|...........11111....................|
35|....................................|
36|....................................|
  +------------------------------------+
```

Ultimate erosion of A

Figure 18: Component labelling of the image set A. Explanations are in the text.

Figure 19: Object "sieving" can be obtained by identifying the maxima in the erosion function, shown in Figure 14, and by overlying such maxima with the image of labelled objects (components) in Figure 18. Note that 2 objects, with labels 7 and A, respectively, contain two local maxima. This complex transformation is termed "ultimate erosion" in mathematical morphology: it retains the ultimate values, i.e., the pixel values just before the last erosion changes them to 0.

6.A Simple Example Using Operations and Transformations[3]

In the construction of thematic maps, selected coincidences between map features are extracted so that all areas on the ground associated with desirable properties for land use can be identified. Such an application has been discussed by Varnes (1974) in an analysis of the logic of geological maps and their use for engineering purposes. Figures 20 and 21, describe an example of feature extraction from two artificial images that correspond to a map illustrated by Varnes (1974, Fig. 19, p. 25).

In Figure 20, two digital images are shown in which the pixel values are labels that identify map unit membership of the pixels: each labelled pixel (labels A, C, 1, 2, and 3 in Fig. 20) indicates a black pixel in a black and white image representing that particular map unit. Two slope categories (A) and (C) in Figure 20a, correspond to three lithologic units, (1), (2), and (3) in Figure 20b. To extract the contact zones (D) between (A) and (C), and (E), between (2) and (3), identified by asterisks in Figure 21, the following transformations are performed. The sets (A), (C), (2), and (3) are dilatated by the structuring element set B, that consists of 5 black pixels, one at the origin and the other 4 above, below, to the right, and to the left. For the four units, the dilatation causes an outward expansion of one pixel. The intersection between the transformed images identify the contacts.

Furthermore, the intersection between the two contacts identify the coincident subset of both contacts. Such area that consist of six pixels, brings out the covariant changes that represent situations in which the classification criteria are not "genetically unrelated". This simple example illustrates that the reasoning behind the construction of thematic maps can be quantitatively portrayed with digital images by appropriate morphological transformations. The transformations exemplified belong to the very large family of local operators that require that for each pixel in an image to be transformed, the values of one or more pixels in its neighbourhood (including the pixel itself) are involved in a computation with the pixel values of the operator or structuring element.

In this example, a possible strategy for detecting boundary or contact spatial covariance was presented. Such covariance is of interest to engineering geologists. Other specialists might develop completely different strategies, as a function of their field of experience and of the characteristics of the dataset at hands. Uncertainties and other statistical attributes of the data can also be important parts of the spatial analysis.

---

[3] This section was extracted from Fabbri and Levialdi (1984, p. 601-603).

```
    0000000001111111111222
    1234567890123456789012
   +----------------------+
01!AAAAAAAAAAAAAAAACCCCCC!
02!AAAAAAAAAAAAAAAACCCCCC!
03!AAAAAAAAAAAAAAACCCCCCC!
04!AAAAAAAAAAAAAAACCCCCCC!
05!AAAAAAAAAAAAAACCCCCCCC!
06!AAAAAAAAAAAAAACCCCCCCC!
07!AAAAAAAAAAAAAACCCCCCCC!
08!AAAAAAAAAAAAACCCCCCCCC!
09!AAAAAAAAAAAAACCCCCCCCC!
10!AAAAAAAAAAAACCCCCCCCCC!
11!AAAAAAAAAAAACCCCCCCCCC!
12!AAAAAAAAAAAACCCCCCCCCC!
13!AAAAAAAAAAACCCCCCCCCCC!
14!AAAAAAAAAAACCCCCCCCCCC!
15!AAAAAAAAAAACCCCCCCCCCC!
16!AAAAAAAAAACCCCCCCCCCCC!
17!AAAAAAAAAACCCCCCCCCCCC!
18!AAAAAAAAACCCCCCCCCCCCC!
19!AAAAAAAAACCCCCCCCCCCCC!
20!AAAAAACCCCCCCCCCCCCCCC!
21!AAAAAACCCCCCCCCCCCCCCC!
22!AAAAAACCCCCCCCCCCCCCCC!
23!AAAAACCCCCCCCCCCCCCCCC!
24!AAAAACCCCCCCCCCCCCCCCC!
25!AAAAACCCCCCCCCCCCCCCCC!
26!AAAACCCCCCCCCCCCCCCCCC!
27!AAAACCCCCCCCCCCCCCCCCC!
28!AAACCCCCCCCCCCCCCCCCCC!
29!AAACCCCCCCCCCCCCCCCCCC!
30!AAACCCCCCCCCCCCCCCCCCC!
   +----------------------+
              mes (A) = 287
              mes (C) = 373
```

a

```
 0000000001111111111222
 1234567890123456789012
+----------------------+
!1111111111111111111111!
!1111111111111111111111!
!1111111111111111111111!
!1111111112222222111111!
!1111111122222222211111!
!1111111122222222211111!
!1111111222222222211111!
!1111111222222222211111!
!1111111222222222211111!
!1111111222222222111111!
!1111111122222221111111!
!1111111122222221111111!
!1111111122222221111111!
!1111111222222211111111!
!1111111222222211111111!
!1111111222222211111111!
!1111112222222211111111!
!1111122222222211111111!
!1111122222222211111111!
!1111333333333311111111!
!1111333333333311111111!
!1113333333333311111111!
!1113333333333111111111!
!1113333333333111111111!
!1113333333331111111111!
!1111333333311111111111!
!1111113333311111111111!
!1111111111111111111111!
!1111111111111111111111!
!1111111111111111111111!
+----------------------+
           mes (1) = 356
           mes (2) = 132
           mes (3) =  72
```

b

Figure 20. (a): artificial digital image of size 30x22 pixels whose labels (A) and (B) represent two different slope categories. (b): an image of the same size in registration with the line in (a), where labels (1), (2), and (3) represent three different lithologic units. The number of pixels with each label are listed below the illustrations.

```
   0000000001111111111222
   1234567890123456789012
  +----------------------+
01!AAAAAAAAAAAAAAA**CCCCC!
02!AAAAAAAAAAAAAAA**CCCCC!
03!AAAAAAAAAAAAAA**CCCCCC!
04!AAAAAAAAAAAAAA**CCCCCC!
05!AAAAAAAAAAAAA**CCCCCCC!
06!AAAAAAAAAAAAA**CCCCCCC!
07!AAAAAAAAAAAAA**CCCCCCC!
08!AAAAAAAAAAAA**CCCCCCCC!
09!AAAAAAAAAAAA**CCCCCCCC!
10!AAAAAAAAAAA**CCCCCCCCC!
11!AAAAAAAAAAA**CCCCCCCCC!
12!AAAAAAAAAAA**CCCCCCCCC!
13!AAAAAAAAAA**CCCCCCCCCC!
14!AAAAAAAAAA**CCCCCCCCCC!
15!AAAAAAAAAA**CCCCCCCCCC!
16!AAAAAAAAA**CCCCCCCCCCC!
17!AAAAAAAAA**CCCCCCCCCCC!
18!AAAAAAAA**CCCCCCCCCCCC!
19!AAAAAAFFF*CCCCCCCCCCCC!
20!AAAAA*FFFCCCCCCCCCCCCC!
21!AAAAA**CCCCCCCCCCCCCCC!
22!AAAAA**CCCCCCCCCCCCCCC!
23!AAAA**CCCCCCCCCCCCCCCC!
24!AAAA**CCCCCCCCCCCCCCCC!
25!AAAA**CCCCCCCCCCCCCCCC!
26!AAA**CCCCCCCCCCCCCCCCC!
27!AAA**CCCCCCCCCCCCCCCCC!
28!AA**CCCCCCCCCCCCCCCCCC!
29!AA**CCCCCCCCCCCCCCCCCC!
30!AA**CCCCCCCCCCCCCCCCCC!
  +----------------------+
```

```
 0000000001111111111222
 1234567890123456789012
+----------------------+
!1111111111111111111111!
!1111111111111111111111!
!1111111111111111111111!
!1111111112222222111111!
!1111111122222222211111!
!1111111122222222211111!
!1111111222222222211111!
!1111111222222222211111!
!1111111222222222211111!
!1111111222222222111111!
!1111111122222221111111!
!1111111122222221111111!
!1111111122222221111111!
!1111111222222211111111!
!1111111222222211111111!
!1111111222222211111111!
!1111112222222211111111!
!1111122222222211111111!
!111111FFF******1111111!
!11113FFF******11111111!
!1111333333333311111111!
!1113333333333311111111!
!1113333333333111111111!
!1113333333333111111111!
!1113333333331111111111!
!1111333333311111111111!
!1111133333111111111111!
!1111111111111111111111!
!1111111111111111111111!
!1111111111111111111111!
+----------------------+
```

```
            *
B̌ = B =   *_*_*
            *
```

$D = (A \oplus \check{B}) \cap (C \oplus \check{B})$ $\qquad$ $E = (2 \oplus \check{B}) \cap (3 \oplus \check{B})$

mes (D) = 60 $\qquad$ mes (E) = 18

$F = D \cap E$

a $\qquad$ b

Figure 21: Identification of a locality in which two sets of attributes, such as slope (a) and lithology (b), that are genetically independent, have contacts that coincide. The asterisks in (a) represent the black pixels in the image set (D), the intersection or overlap between the sets (A) and (C) dilatated by the structuring element the black pixels in the image set (E), the intersection of the sets (2) and (3) also dilatated by B. The coincidence between the contacts represented by the two sets (D) and (E) is the set (F) that contains 6 pixels. Asterisks indicate the sets (D), and (E); F's overprinted on asterisks indicate the set (F).

7. Statistical Pattern Integration

In a study on the integration of geological data with geochemical and geophysical data for gold exploration, GIS techniques were used (Bonham-Carter et al., 1988) in conjunction with methods of statistical regression. The study combined the digital data in weighted sums that predicted whether a particular drainage basin could contain gold mineralized areas. Corridors were computed around lineaments, the trace of anticlinal axes, and lithologic contacts, to derive thematic maps of the posterior probability to discover gold deposits in cells or subareas of one square kilometer. The a posteriori probability associated to cartographic units was used to identify small areas for which the a priori probability of discovering new mineralized areas was estimated by means of multivariate statistical analysis (Agterberg, 1989). Figure 22 shows how binary patterns, related to mineral deposits, can be combined for computing conditional probabilities, as discussed by Agterberg et al. (1990). For the deposits surrounded by a "small unit area", set D, unconditional probabilities, P(D) are estimated, i.e., the probability that any random unit area contains one deposit. The conditional probabilities, P(D¦B), P(D¦C) and P(D¦BC), are for deposits on rock unit B, in the corridors C, or in the intersection of both BC. The probabilities are estimated by counting the deposits that occur within the various map patterns. These probabilities are useful in performing the following tasks:

(1) the detection of optimum binary patterns by a measure of contrast, C, representing the correlation between a binary pattern and a point pattern, e.g., several corridors are compared;

(2) statistical integration by positive (W+) and negative (W-) weights representing the presence or absence of features based on the assumption of conditional independence tested comparing posterior probabilities with observed frequencies;

(3) assessment of uncertainty due to missing patterns that decreases when patterns are added.

This work discussed the theoretical background for the application of artificial intelligence to spatial data analysis and integration in the context of incomplete or/and uncertain information. The next section provides some concepts on the digital representation of spatial reasoning.

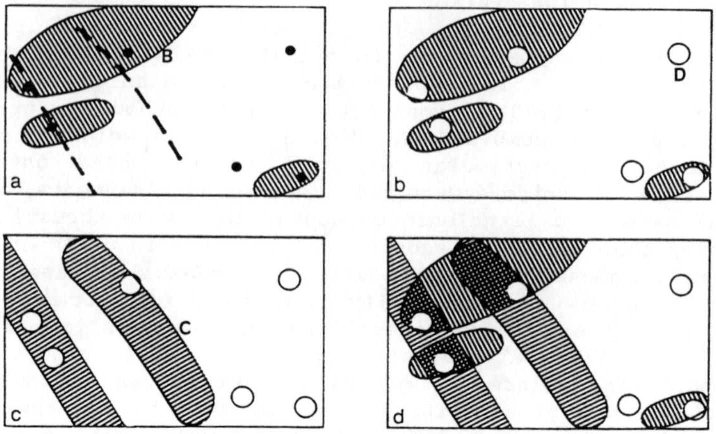

Figure 22: Illustration of how conditional probabilities can be computed by combining binary map patterns (redrafted after Agterberg et al., 1990, Fig. 1). (a) The distribution of mineral deposits (dots), the outcrops of rock map-unit B (NW dashed contoured area), and the traces of two faults (broken lines). (b) Spatial transformations of the mineral deposits into small circular areas, set D, and their relationships with the set B. (c) Spatial transformation of the fault traces into corridors, set C, and their relationships with the set D. (d) Overlay of the three images, sets B, C, and D.

8. Spatial Reasoning: Expert Systems and Digital Image Processing[4]

We have seen in the previous lectures that the central idea of expert systems is to combine evidence for and against a particular hypothesis. The combination may follow different kinds of rules, and no unanimity exists on how to combine uncertainties, belief or disbelief. It would require an entire course to deal with those topics!

In any case, given a particular measure of uncertainty, for digital image processing the uncertainties can be represented as 2-D mappings or images. The value of a pixel or its "grey level" may be interpreted as evidence for or against a particular hypothesis. The functions that weigh evidence

---

[4] This section was taken from Fabbri and Kasvand (1988, p. 224-230).

may be looked upon as special pixel-based operations between several registered images, which create a new image. Thus we will have an image of supporting evidence, an image of contrary evidence, an image of certainty factors for a particular hypothesis, etc. Pixel-to-pixel operations and Boolean operations (logical unions, intersections and complements) will be used. Conceptually, there is no difference between processing a single number of evidence and processing matrices of evidences (i.e., digital images). Weights for evidence can be represented as matrices of multipliers. The problem remains of how to determine the original probabilities and the weights, as it is in all expert systems.

In practice there are several problems to overcome in remote sensing image analysis and processing.

(1) Images in remote sensing are very large, i.e., millions of pixels and several bands;

(2) Many inputs to the inference network are images and their total number can be rather large if we consider both remote sensing and ancillary data;

(3) At each node of the inference network (hypothesis) an image of hypothesis is created;

and

(4) Large images take a long time to process.

However, images usually contain large areas where the contents or the information is the same or similar. Furthermore, the information in images is always "quantized" in amplitude as well as spatially. For example, the areas between two contour lines in a contour map of elevations can be considered to be of the same altitude, even though they seldom are in reality. To reduce storage requirements, the processing could be synchronized so that the images can be read sequentially from mass storage devices, e.g., one row of pixels at a time. To reduce computation time and to reduce the amount of data to be processed, the images should be quantized to acceptable levels for a given problem. Tables of non-spatial attributes will be used in conjunction with the images containing spatial attributes. Processing can also be refined in various ways, for example, by varying the spatial resolution of the images (multi-resolution), i.e., first a coarse analysis followed by a finer spatial resolution in areas of special interest.

Clearly, expert-system and image-processing methods can be merged in several ways to form optimal computational strategies for each problem area. Spatial data analysis and integration offers many possibilities: only a few of those have been discussed in this lecture.

References

Agterberg, F.P., 1989, Systematic approach to dealing with uncertainty of geoscience information in mineral exploration. Proc. 21st APCOM Symp., Application of Computers and Operations Research in the Mineral Industry, Las Vegas, Nevada, Feb. 27-March 2, 1989. Soc. of Mining Engineering Inc., Littleton, Colorado, p. 165-168.

Agterberg, F.P., Bonham-Carter, G.F., and Wright, D.F., 1990, Statistical pattern integration for mineral exploration. In, Gaal, G., and Merriam, D.F., Eds., Computer Applications in Resource Exploration. Oxford, Pergamon Press, in press.

Aronoff, S., 1989, Geographic Information Systems: A Management Perspective. Ottawa, Canada, WDL Publications, 294 p.

Bonham-Carter, G.F., Agterberg, F.P., and Wright, D.F., 1988, Integration of geological datasets for gold exploration in Nova Scotia. Photogrammetric Engineering and Remote Sensing, v. 54, n. 11, p. 1585-1592.

Fabbri, A.G., and Kasvand, T., 1988, Automated integration of mineral resource data by image processing and artificial intelligence. In, Chung, C.F., Fabbri, A.G., and Sinding-Larsen, R., Eds., Quantitative Analysis of Mineral and Energy Resources. Dordrecht, D. Reidel Publishing Co., p. 215-236.

Fabbri, A.G., and Levialdi, S., 1984, New computer architectures suitable for spatial analysis in the earth sciences. Sciences de la Terre, Serie Informatique Geologiques, n. 20, p. 587-612.

Fabbri, A.G., 1984, Image Processing of Geological Data. New York, Van Nostrand Reinhold, 244 p.

Matheron, G., 1972, Elements pour une Theorie des Milieux Poreux. Paris, Masson et Cie. Eds., 166 p.

Matheron, G., 1975, Random Sets and Integral Geometry. New York, John Wiley and Sons, 261 p.

Minkowski, H., 1911, Theorie der Konvexen Korper inbesondere Begrundung Oberflachen Begriffs. Gesammelte Abh., v. 2, p. 131-229.

Serra, J., 1982, Image Analysis and Mathematical Morphology. New York, Academic Press, 610 p.

Varnes, D.J., 1974, The logic of geological maps, with reference to their interpretation and use for engineering purposes. U.S. Geological Survey Prof. Paper 837, 48 p.

AN INTRODUCTION TO EXPERT SYSTEMS IN SPATIAL DATA ANALYSIS

A. G. FABBRI
I.T.C.
P.O. Box 6, Boulevard 1945
7500 AA Enschede
The Netherlands

1. Introduction

Artificial Intelligence (AI) involves the use of computer techniques to solve problems which traditionally have required human-like intelligence. A prerequisite to an AI program is the integration of numerous inputs to make a decision. The field includes the areas of natural language understanding, machine learning, computer vision and expert systems. Of these four areas, certainly the most rapidly growing is that of expert systems. Expert systems can be used as consultants or as learning tools. According to AI scientists, traditional algorithmic methods of data processing are not suited toward non-numeric reasoning; expert systems, however, solve problems through logical rules of inference. In this paper, an expert system is considered to be a package of special-purpose computer software which strives to emulate the competent skill and experience of an expert in a specific problem domain within the real world. The knowledge component of an expert system is generally separated from the reasoning mechanism. Ideally, it should have a reasoning capability that can be understood by the user.

2. Useful concepts and terminology

Before discussing expert systems and artificial intelligence approaches in general, it is necessary to introduce a few concepts, used in knowledge-base building and automated reasoning. The structure of an ideal expert system, represented in Figure 1, consists of the following parts: (1) a language processor for man-machine communication; (2) a "blackboard" for keeping track of the intermediate steps of the processing; (3) a knowledge base with facts (observations) and rules for problem solving; (4) an interpreter to apply the rules; (5) a scheduler to sequence the application of the rules; (6) a consistency enforcer to adjust the system behaviour when the knowledge is altered; (7) a justifier to explain the system's behaviour. Very few, if any, expert systems have all the components mentioned in Figure 1. However, in most expert systems, the knowledge base is kept separate from the reasoning mechanism.

There are several ways in which knowledge can be represented in order to solve a problem. Rules are the simplest form of knowledge relating facts

A. S. Belward and C. R. Valenzuela (eds.),
Remote Sensing and Geographical Information Systems for Resource Management in Developing Countries, 389–423.

or evidences together about a current problem: they are activated in order to generate processing action towards its solution. Examples of rules from different expert systems are shown in Table 1. Rules of this form are termed "production rules". As seen in the table, the IF part (the premise or condition) and the THEN part (the conclusion or action) may consist of more then one statement connected by "AND" or "OR".

Table 1 shows "production rules" of increasing complexity from I to V. In rule I, the premise consists of three parts, linked by "AND" relationships. In rule II, the conclusion consists of a list of values. Rule III has three parts in its premise, and the conclusion consists of four actions. In rule IV, the conclusion has a weight of 1.0 for the presence of suggestive evidence. Some rules provide numerical values of probability or confidence. An example is rule V where one weight (20) is associated with the presence and another (1) with the absence of evidence, and an a priori probability associated with the hypothesis confirmed by the conclusion. Different rules can be linked together into chains to form inference networks in rule-based expert systems. Figure 2 shows an example of an inference network from the expert system PROSPECTOR.

Another method of representing knowledge is based on "frames", in which features are associated with nodes that represent concepts or objects organized into a hierarchy. Each node has associated attributes and values which can be automatically inherited elsewhere in the hierarchy. To simplify the description of some facts, semantic networks and taxonomic networks (for storing an organized set of names) are frequently used.

While in conventional programs the flow of the computations is preset by the code itself, in AI programming the flow of computational steps may change and be driven by the data being supplied to the rules of frames, chained together in different structures. We term "forward-chaining" the procedure by which, given a number of facts, a search is performed for a consequent combination that complies with certain a priori specifications: i.e., we proceed from the facts to hypotheses. On the other hand, "backward-chaining" is based on a predefined set of hypotheses or models for which a search is made to confirm a hypothesis through the rules towards the relevant facts: i.e., we proceed from the hypothesis to the facts.

Table 1: Examples of rules from expert systems designed for geological applications.

Table 1: Examples of rules from expert systems designed for geological applications.
(I) Log analysis interpretation
"IF the water fit is normal and the Rw quality is excellent and there are more than 10 ft of hydrocarbons THEN the hydrocarbon presence is significant" (Apte' and Weiss, 1985, p. 586, ELAS).
(II) Volcanological prediction
"IF the preceding eruption dates from between 100 and 1000 years ago, THEN the probabilities that VEI (volcanic eruptivity index) are 0, 1, 2, 3, 4, 5 or 6 are 11, 5, 45, 26, 6, 4 and 1%, respectively" (Rosenthal-Sabroux et al., 1984, p. 107).
(III) Dipmeter test interpretation
"IF there is a delta-dominated, continental-shelf marine zone, and there is a sand zone intersecting the marine zone, and there is a blue pattern within the intersection, THEN assert a tributary fan zone: top = top of blue pattern; bottom = bottom of blue pattern; flow = azimuth of blue pattern" (Smith, 1984, p. 63, Dipmeter Advisor).
(IV) Seismic event identification
"IF (1) the focal depth of the event is deeper than 30 km, and (2) the location of the event is within a populated area THEN there is suggestive evidence (1.0) that the seismic event is an earthquake" (Liu, 1985, p. 461).
(V) Mineral exploration
"IF intrusive breccia, RCIB, is present THEN (20,1) suggestive morphology of igneous rocks, SMIR (0.03)," where RCIB is the evidence, SMIR is the hypothesis, (20 and 1) are the values of likelihood sufficiency (LS) and necessity (LN), respectively, associated with presence and absence of the evidence, and (0.03) is the prior probability associated with the hypothesis SMIR (Gaschnig, 1982, p. 51, Fig. 3.1, PROSPECTOR).

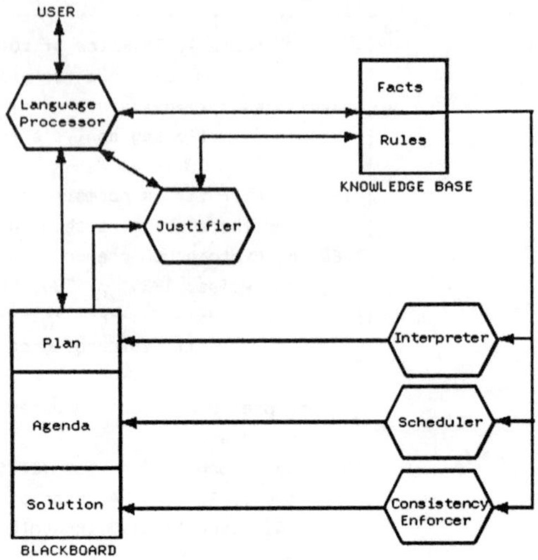

Figure 1. Ideal operational flowchart and structure of an expert system (after Hayes-Roth et al., 1983, p. 17, Fig. 1).

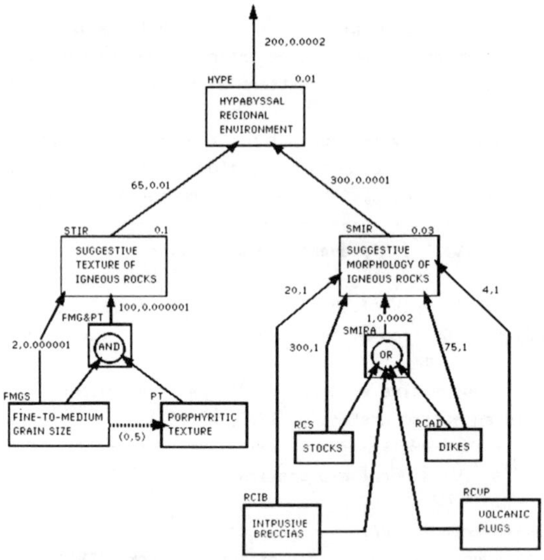

Figure 2. Part of an inference network for hypabbysal (HYPE) porphyry copper deposit model in PROSPECTOR (after Reboh, 1981, p. 15, Fig. 4). Logical (AND-OR), plausible (solid arrows), and contextual (dashed arrows) relationships or rules are represented. Inside the boxes are the statements of evidence or hypotheses; above to the left are the abbreviations, and to the right, the values of a priori probabilities. The values of likelihood sufficiency and necessity appear along the solid arrows; the context interval appears below the dashed arrows.

As we have seen in Table 1, rule (IV) deals with suggestive evidence, and rule (V) contains probable weights associated with the presence or absence of evidences and a priori probabilities associated with the hypotheses. An important criterion to distinguish expert systems, as we shell see, is the capability to handle uncertain evidence. Rule (V) in Table 1 uses two likelihood measures. The evidence can be modified during interaction by associating a certainty factor with it. In the case of PROSPECTOR, when evidence is volunteered, certainty is an integer value ranging between -5 and +5, to indicate absolute certainty of absence and of presence of the evidence, respectively. That interaction modifies the likelihood before the rule is activated towards the hypothesis connected to it in the inference network. The a priori probability for the hypotheses is in turn modified and so are other hypotheses in the network until a final goal is reached. An example of computations for uncertain evidence in PROSPECTOR can be described as follows (Duda et al., 1978). Interaction causes likelihood measures to be modified before the activation of rules towards the hypothesis. This modifies the a priori probability for the hypothesis. For example, as shown in Figure 2, the definite presence of the evidence RCIB multiples the "odds" of SMIR by factor 20, raising its probability from 0.03 to 0.382. The prior odds on SMIR are 0.03/(1-0.03)=0.030927, giving posterior odds on SMIR equal to 20*0.030927=0.61855 (1*0.030297=0.030297 for the observed absence), which corresponds to a probability 0.61855/(1+0.61855)=0.382 (0.03097/(1+0.039327)=0.030 for the observed absence). Here odds and probability are freely interchangeable through the simple relation O=P/(1-P), where O=odds, and P=probability. Hence, P=O/(1+O). Bayes' rule can be stated as O(H¦E)=LS*O(H), where E indicates the evidence, H the hypothesis, and LS is the likelihood sufficiency. Often a level of confidence can be associated with the volunteered piece of evidence E'. The level of confidence modifies the a priori probability of the hypothesis H as well as the other hypotheses in the network, until the goal state is reached. In our example the goal state is "a porphyry copper deposits is (is not) present." The modification can propagate throughout the inference network. Appendix C discusses in more detail PROSPECTOR's plausible inference scheme.

It should be remarked at this point that domain experts may find it difficult to provide figures for measures such as prior probability, and LS, because the applications themselves are imprecise. Also, experiments have demonstrated that the behaviour of such systems may be overly sensitive to perturbations, therefore good tools are necessary to properly balance the inference mechanism both while constructing it and during fine tuning of its operations.

3.Examples of expert systems derived from PROSPECTOR

This section discusses expert systems in geology which are representative of current research. Probably the most discussed expert system in geology is PROSPECTOR, a program developed to assist geologists in exploring for mineral deposits (Duda, 1980). Here, the reasoning process of the experienced geologist is emulated by evaluating the likelihood that a mineral prospect or region resembles one or more models built into a knowledge base. Several ore deposit models have been encoded. These reached different degrees of development in terms of their semantic networks, descriptive text and testing. The latter two tasks were the most labor-intensive and expensive.

Inference, semantic and taxonomic networks were used in PROSPECTOR. As shown in Figure 2, logical, plausible and context relations were distinguished in the reasoning processes represented in the networks (see Gaschnig, 1982). To facilitate the building of the models and the different networks, a knowledge acquisition system, KAS, containing a resident network editor, RENE, were programmed by Reboh (1981) as an empty shell for general-purpose knowledge engineering. In the interactive mode PROSPECTOR provided on-line consultation. In addition, two more modes of operation had been programmed: (1) a batch mode to process data from questionnaires and for model-performance testing, (2) a graphic input/output mode whereby maps were placed on a graphic tablet digitizer to obtain digital contour maps. Such maps provided spatially distributed evidence for "drilling-site-selection" models for which field and map evidence were combined (Duda et al., 1978; Duda, 1980). Inference networks were developed for these evidences. An image processing system produced and shared digital image files of up to 128 x 128 pixels or cells with PROSPECTOR. These files, evaluated by the inference network, produced a simplified digital map of a prospective area. Elementary spatial reasoning was built in the drilling-site-selection models by rating cells according to their linear distance from contours or traces of fractures (Duda et al., 1978). Fabbri and Kasvand (1988), and Fabbri et al. (1988) describe some inference nets in such models after Duda (1980) and Duda at al. (1978) in the context of computational vision for geology. Most systems based on PROSPECTOR or variants of it have retained the original inference network; however, PROSPECTOR II (McCammon, 1987) retained only the semantic network of PROSPECTOR. The remaining PROSPECTOR based systems (muPROSPECTOR and muPETROL, The Deciding Factor) were rewritten in a simplified form for microcomputers and did not retain PROSPECTOR's measures of uncertainty.

muProspector (McCammon, 1984 and 1986) is a simplified version of the PROSPECTOR expert system, which was developed for a microcomputer using the programming language muLISP (The Soft Warehouse, 1985). It was adapted so that others could tailor it to variety of problems such as: (1) mineral endowment estimation (Koch, 1986); (2) sedimentary basin analyses (Miller, 1986); (3) regional geological environments at the district and outcrop level (McCammon et al., 1984, McCammon, 1985).

muPetrol is an operational prototype expert system which uses muPROSPECTOR as a shell. It was constructed to provide the means to store the knowledge to classify sedimentary basins and to assist in estimating undiscovered petroleum resources. The knowledge base was described in detail (Miller, 1987a) and is available on floppy diskettes (Miller, 1987b). As was for muPROSPECTOR, muPetrol simply has the capability to use answers such as "yes", "no" or "I don't know" to the questions that it asks. It also can prompt the users to select one of a list of possible choices.

Also derived from the knowledge acquisition system KAS, developed from PROSPECTOR, is the microcomputer-based "The Deciding Factor" (Campbell and Fitzgerrel, 1985). It is an expert system shell, i.e., an inference engine devoid of knowledge base and rules. The Deciding Factor appears to be particularly easy to use due to the hierarchical and self contained nature of the interaction (help facilities). For this reason this shell will be described in the next section and will be used in the following two applications.

4.An expert shell: THE DECIDING FACTOR

This shell has a user-friendly interface for the interaction with two programs called EDITOR and CONSULTANT. The interactive EDITOR, a tool for building or modifying a knowledge base by formulating the various rules of a decision tree developed during the interaction, is shown in Figure 3. Evidences and hypotheses, termed factors and ideas, respectively, are built into hierarchically structured files termed "decision models". Logical relationships between ideas in a model can be specified using terms such as ALL, MOST, BEST, and WORST, that distinguish the factors that support an idea. Schematic examples of logical relationships are shown in Figure 4. Weights are associated with each particular idea or factor in a tree. The weights of factors or ideas supporting another idea can be automatically or manually normalized. Once a decision model has been constructed, the CONSULTANT is run to generate prompts for each factor and the user volunteers his/her confidence in it by entering integer values between -5 and +5 for absolute uncertainty and certainty of the evidence, respectively. The value of 0 indicates a "don't know" answer. This interaction is shown in Figure 5. Fine-tuning and testing of a model is performed by alternately using the two programs. Let us now consider two application examples of different complexity which lead to spatial reasoning.

VIEWING THE MODEL: COMPANY

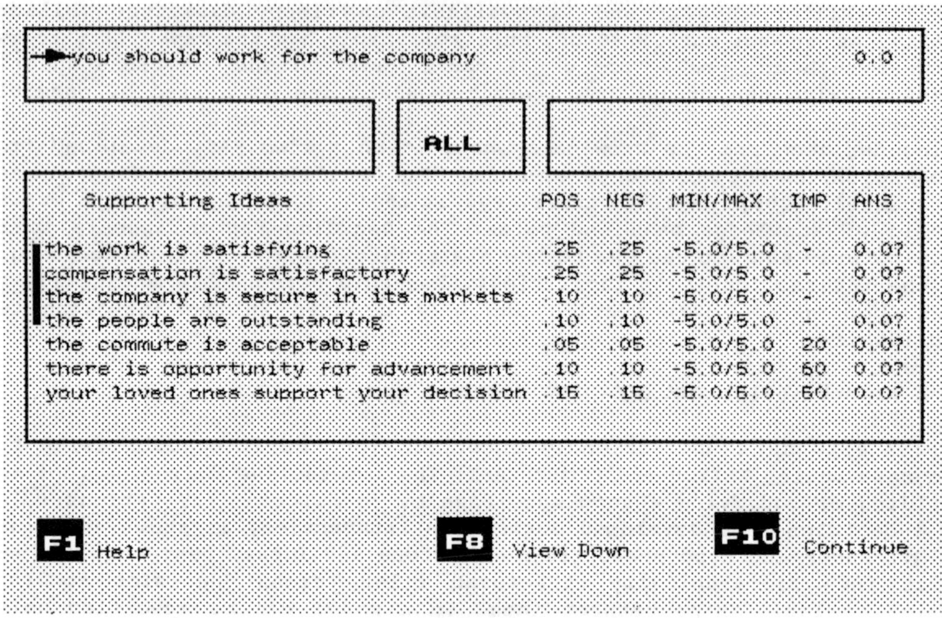

COMPANY

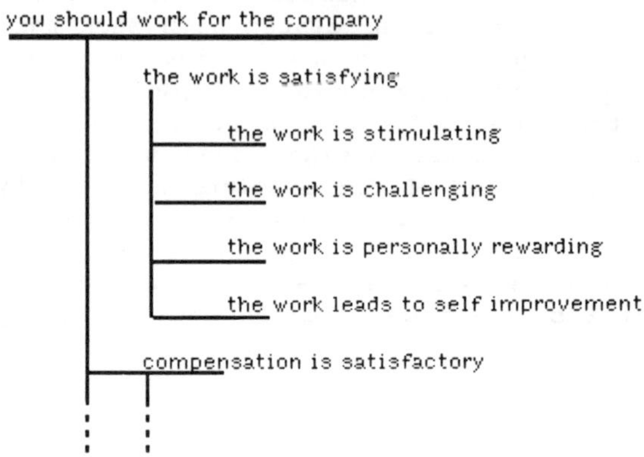

Figure 3. Building a decision model with The Deciding Factor. Above, the screen shows a main idea on the top, a decision logic ALL and a list of supporting ideas. Part of the corresponding ideas and factors are shown in the tree below. Also shown are positive and negative weights assigned to ideas, minimum and maximum answers, and the default importance assigned to them. The question marks to the right indicate the questions that require an answer. The black line to the left identify ideas that have further supporting factors appended to each (modified after Campbell and Fitzgerrel, 1985, pp. 9-10).

```
hypothesis (= main idea)
logical relationship (= ALL, MOST, BEST, WORST, C-ALL, ...
I  (i.e., support type; C- = conditional: if in range then)
I
I
I------factor (= supporting idea)
I
I
I------factor
:
:
I------factor
I
I
I------hypothesis (= supporting idea)
I      logical relationship
I      I
I      I
I      I------factor
I      I
I      I
I      I------factor
I      I
I      I
I      I------hypothesis
I      I      logical relationship
I      I      I
I      I      I
I      I      I------factor
I      I      I
I      I      I
I      I      I------factor
I      I      I
I      I      I
I      I      I------factor
I      I
I      I
I      I------factor
I
I
I------factor
```

```
+--------------------------------+
|ALL   All responses are combined |
|      to determine the support   |
|      for the hypothesis. Weights|
|      total 1.00. This is just   |
|      enough for a definite "yes"|
|      (5) or no (-5) judgement on|
|      the hypothesis.            |
|MOST  All responses are combined |
|      to determine the support   |
|      for the hypothesis. Weights|
|      may total more or less than|
|      1.00. This total may be    |
|      more or less than enough   |
|      for a definite judgement.  |
|BEST  Only the most strongly     |
|      supported factor supports  |
|      the hypothesis.            |
|WORST Only the least strongly    |
|      supported factor supports  |
|      the hypothesis.            |
|C-    Used as a prefix to the    |
|      logical relationships      |
|      listed above, e.g., C-BEST.|
|      It indicates conditional   |
|      logic: If the response of  |
|      the first supporting idea  |
|      is in the min. and max.    |
|      answer range, then the     |
|      remaining ideas are examin-|
|      ed and used to support the |
|      hypothesis. If the first   |
|      response is outside the    |
|      range, only this first     |
|      response is used for sup-  |
|      port, and the remaining    |
|      ideas are not examined.    |
+--------------------------------+
```

Note: (1) Only factors are questioned directly; (2) Positive or negative weights for factors (and hypotheses) range from +1.0 to -1.0; (3) Min. and max. answers for factors (and hypotheses) range between -5.0 and +5.0; (4) Importance for factors (only) ranges between 0% and 99%: it represents rate of belief and is used for assessing the % of reliability of the conclusion in the final report. Reliability is a function of the number entered for importance for a given factor and the strength of the answer (average imp. is 50).

Figure 4. Generalized inference network termed tree of logical decisions in The Deciding Factor. The logical relationships are described to the right; direct questioning, weighting, ranges for answers, importance for factors and reliability are described at the bottom.

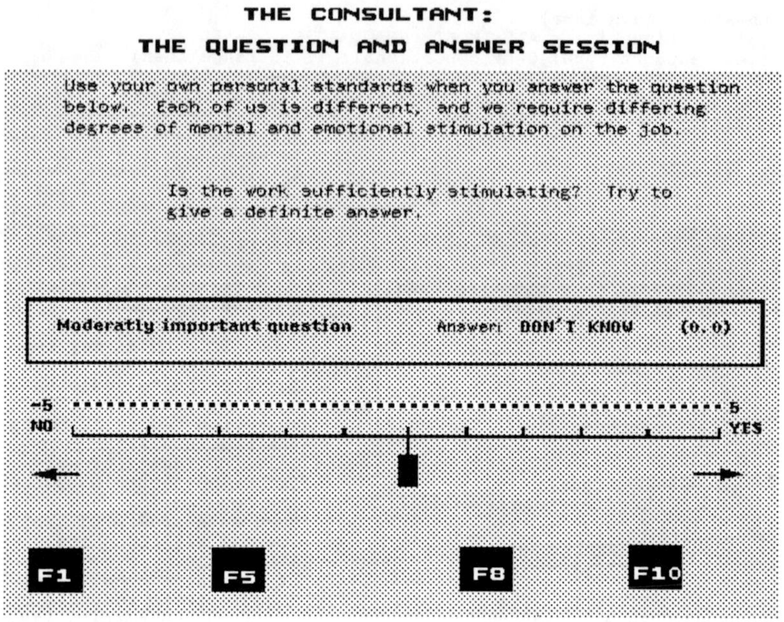

Figure 5. A prompt is answered quantitatively for a factor of moderate importance with values ranging between -5 and +5, in The Deciding Factor (after Campbell and Fitzgerrel, 1985, p. 14).

5. Example 1: Suitability for waste disposal

Typological regrouping refers to assembling geological map units into fewer units for particular purposes such as for assessing erosion risk using soil maps or conditions for engineering geology uses. Two maps of surficial deposits and of geologic conditions relating to waste disposal have been compiled by Hackett and McComas (1969). The ample explanations of the units in the legends of those maps enabled Varnes (1974) to do the following: (a) to estimate the distribution of 26 geologic units among 8 units of suitability for waste disposal; (b) to generate a data matrix for classifying the geologic units using essential, essential and unique, and inferential statements; and finally (c) to construct the tree of logical division shown in Figure 6. The tree was constructed so that the relative importance and effectiveness in discrimination were used to sequence the 9 questions necessary to a complete classification. For example, the presence of peat in a closed basin, observed on the two maps, seemed a critical feature because whenever peat occurred in it G3 was identified and conversely almost all G3 units corresponded to peat. In addition, the permeability of the surficial material clearly contributed to the separation of 7 out of 8 classes. For that reason the second question is about the permeability of the surficial material. Similar considerations were at the base of the remaining questions in Figure 6. Table 2 lists the description for the 8 map units of geologic conditions relating to waste disposal.

The knowledge contained in the two maps by Hackett and McComas (1969), and in the study by Varnes (1974) summarized in Figure 6 and Table 2, can be stored in an expert system (see later). Such knowledge has permitted the identification of the essential statements about the definition of all map units and has led to choosing the logical sequence for the operational questions for generating an efficient division. This was not a simple task and the chosen sequence presents some problems. For instance we may ask the following: "what is the meaning of the empty sets in Figure 6?" Such situations are common whenever overlays are used between different maps to generate new special maps by regrouping. Tables of combined map units and of their properties are generated which require expert manipulation and reinterpretation. The new units may be associated to different levels of uncertainty, frequency of occurrence and other logical and distributional properties. All the new units will have to be evaluated within the practical constraints of the theme being developed.

Let us now construct an expert system from the diagram in Figure 6, by using the expert shell "The Deciding Factor", described earlier (Campbell and Fitzgerrel, 1985). A decision model is shown in Figure 7 where the ends of the horizontal lines between yes and no (Y/N) answers, in the diagram of Figure 6, were made to correspond to complete certainties or complete uncertainties (values +5.0/-5.0) and the position of the vertical lines to the relative certainty on either positive or negative side of the don't know position at the center (value 0.0). To simplify the construction of the inference network, all 9 questions are asked first with nil positive or negative weights and then the hypothesis that there is evidence of suitability for waste disposal is assessed as the "BEST" one of 8 hypotheses that the evidence corresponds to any of the 8 units.

All evidences for each of the 8 hypotheses carry a weight share (positive or negative) proportional to the number of evidences having a weight in the identification of the particular map unit. Maximum and minimum threshold values for answers for evidences vary according to the position of the vertical lines in Figure 6. Appendix A shows the result of one consultation session with the expert system. Appendix B shows 4 screen dumps obtained during the interaction with this expert system.

6.Example 2: An idealized mineral deposit XYZ model

A more complex situation is represented in the inference network for a schematic mineral deposit model shown in Figure 8. Concepts of context rules, logical union (OR), intersection (AND) and complement (NOT) are used to express the relationships between stand alone ideas (evidences or factors) and hypotheses (ideas that receive support). Different values of pyrite concentration are used to compute the weights of its contribution to assessing the hypothesis of pyrite zone. The symbols and the values in Figure 8 are the same as those used in Figure 2. Note that in the decision model shown in Figure 9 (OR) relationships are resolved by BEST, and (AND) relationships by MOST of two factors! The relationships between pyrite concentration and no biotite, or those between the 3 hypotheses below the top one are resolved by ALL, which implies summing the contribution of their respective weights. Similarly to what was done in the diagram of Figure 8 for pyrite concentration, inference networks were built in PROSPECTOR for spatial relationships of cells digitized from maps or for observations in the field.

Table 2: Geologic conditions relating to waste disposal in McHenry county, Illinois (after Hachett and McComas, 1969, plate 2C, and Varnes, 1974, plate 1F). R indicates stop, major problems, impractical to overcome; Y indicates caution, major problems, controllable; and G indicates go, minor problems.

MAP UNIT	GEOLOGIC CONDITIONS
G-2	Area of thick clay-till overlaying dense bedrock. Only very small ground-water supplies available at depths of less than 500 feet. Potential for pollution of water supplies is low.
G-3	Area of ground-water discharge in small basins containing peat. Pollutants generally are confined in the area and unable to reach usable ground-water sources. Locally, use of these areas may be limited by periodic flooding. Pollution potential is low.
Y-1	Deposits of pebbly clay and silt of varying thickness and local deposits of sand, which might overlie potential ground-water sources. Potential for pollution is low to moderate.
Y-2	Thick deposits of dry permeable materials more than 30 feet above ground-water zone. Possibly a good area for sanitary landfills and a poor area for lagoons. Pollution potential is moderately low for landfills and moderate to high for lagoons.
Y-3	Areas of mixed drift with extreme range in character of materials. Materials range from clay to gravel. Individual site evaluation is essential, as pollution potential ranges from low to high.
R-1	Areas where depth to ground-water saturation is shallow and where permeable materials are present; widespread movement of pollution of land surface into shallow water sources and surface water bodies creates a moderately high potential for pollution.
R-2	Area of thick sand and gravel aquifers, at or very close to the surface. Ground-water levels are high so that waste would be disposed in the saturated zone. Locally, this area is subject to periodic flooding. Potential for pollution on site is high, but pollutants might be rapidly diluted.
R-3	Area of thin drift over highly fractured bedrock, or area of bedrock outcrops. Potential for pollution of bedrock aquifers is high.

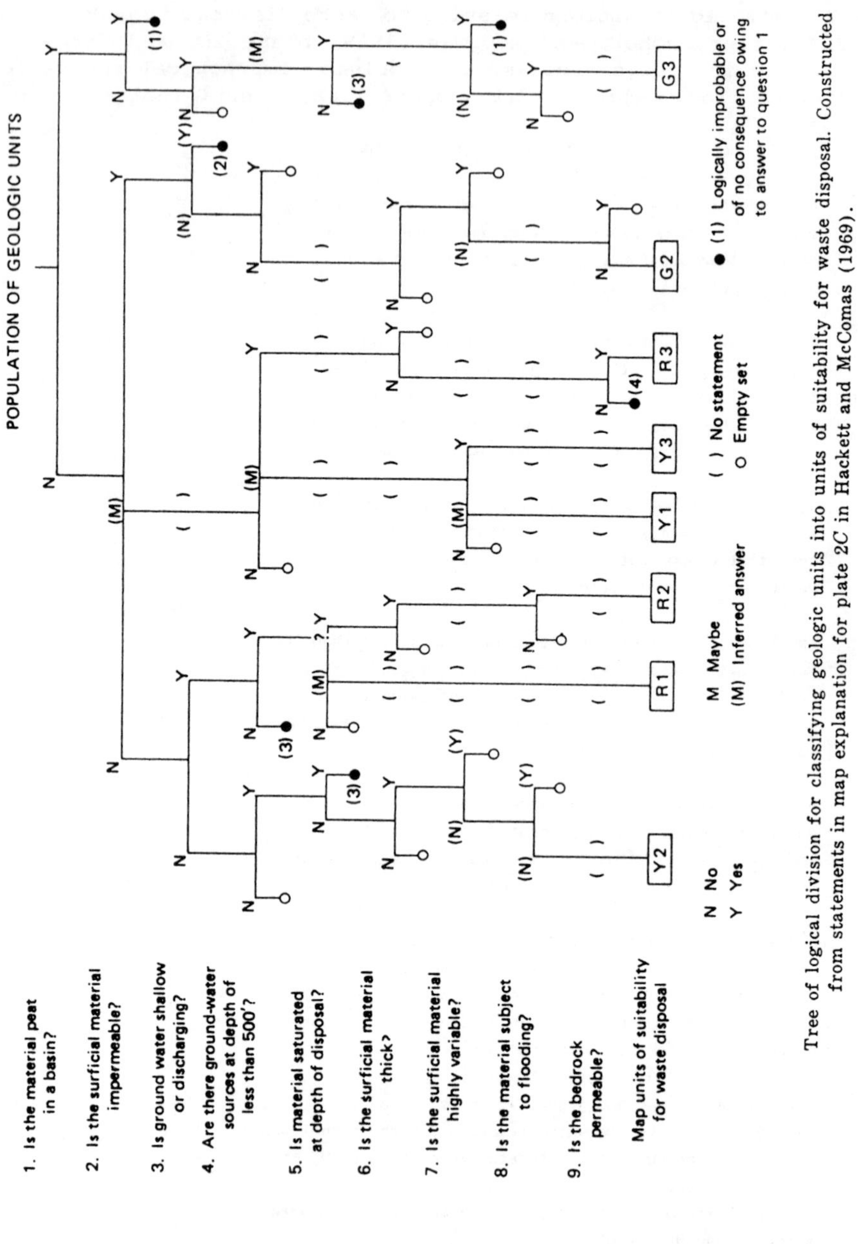

Figure 6. The tree of logical division after Varnes (1974, p. 24, Fig. 18).

```
                DECISION MODEL TO EVALUATE WHETHER
            THE TERRAIN IS SUITABLE FOR WASTE DISPOSAL


the terrain is suitable for waste disposal
ALL
I
I
I      POS 0.00 NEG 0.00 MIN -5.0 MAX 5.0 IMP 50
I------the material peat is in a basin
I
I
I      POS 0.00 NEG 0.00 MIN -5.0 MAX 5.0 IMP 50
I------the surficial material is impermeable
I
I
I      POS 0.00 NEG 0.00 MIN -5.0 MAX 5.0 IMP 50
I------ground water is shallow or discharging
I
I
I      POS 0.00 NEG 0.00 MIN -5.0 MAX 5.0 IMP 50
I------there are ground water sources at depth of
I      less than 500 feet
I
I
I      POS 0.00 NEG 0.00 MIN -5.0 MAX 5.0 IMP 50
I------material at depth of disposal is saturated
I
I
I      POS 0.00 NEG 0.00 MIN -5.0 MAX 5.0 IMP 50
I------the surficial material is thick
I
I
I      POS 0.00 NEG 0.00 MIN -5.0 MAX 5.0 IMP 50
I------the surficial material is highly variable
I
I
I      POS 0.00 NEG 0.00 MIN -5.0 MAX 5.0 IMP 50
I------the material is subject to flooding
I
I
I      POS 0.00 NEG 0.00 MIN -5.0 MAX 5.0 IMP 50
I------the bedrock is permeable
I
I
I      POS 1.00 NEG 1.00 MIN -5.0 MAX 5.0
I------there is evidence of suitability for waste
       disposal
```

```
BEST
I
I
I      POS 1.00 NEG 0.00 MIN -5.0 MAX 5.0
I------map unit may be G3
I      ALL
I      I
I      I
I      I      POS .17 NEG .17 MIN .5 MAX 5.0 IMP 50
I      I------=the material peat is in a basin
I      I
I      I
I      I      POS -.17 NEG -.17 MIN -5.0 MAX -.5 IMP 50
I      I------=the surficial material is impermeable
I      I
I      I
I      I      POS .17 NEG .17 MIN .5 MAX 5.0 IMP 50
I      I------=ground water is shallow or discharging
I      I
I      I
I      I      POS .17 NEG .17 MIN 1.0 MAX 5.0 IMP 50
I      I------=material at depth of disposal is saturated
I      I
I      I
I      I      POS -.17 NEG -.17 MIN -5.0 MAX -2.0 IMP 50
I      I------=the surficial material is highly variable
I      I
I      I
I      I      POS .17 NEG .17 MIN 0.0 MAX 5.0 IMP 50
I      I------=the material is subject to flooding
I
I
I      POS 1.00 NEG 0.00 MIN -5.0 MAX 5.0
I------map unit may be G2
I      ALL
I      I
I      I
I      I      POS -.13 NEG -.17 MIN -5.0 MAX 0.0 IMP 50
I      I------=the material peat is in a basin
I      I
I      I
I      I      POS .13 NEG .17 MIN 2.5 MAX 5.0 IMP 50
I      I------=the surficial material is impermeable
I      I
I      I
I      I      POS -.13 NEG -.17 MIN -5.0 MAX 0.0 IMP 50
I      I------=ground water is shallow or discharging
I      I
I      I
I      I      POS -.13 NEG -.17 MIN -5.0 MAX 0.0 IMP 50
I      I------=there are ground water sources at depth of
I      I      less than 500 feet
I      I
I      I      POS .13 NEG -.17 MIN -1.0 MAX 5.0 IMP 50
I      I------=the surficial material is thick
```

Figure 7. The decision model constructed from the tree in Figure 6 using The Deciding Factor.

```
I
I
I       POS 1.00 NEG 0.00 MIN -5.0 MAX 5.0
I------map unit may be Y1
I       ALL
I       I
I       I
I       I       POS 0.00 NEG -1.00 MIN -5.0 MAX 0.0 IMP 50
I       I------=the material peat is in a basin
I       I
I       I
I       I       POS .14 NEG -1.00 MIN -2.0 MAX 2.0 IMP 50
I       I------=the surficial material is impermeable
I       I
I       I
I       I       POS .14 NEG -1.00 MIN -2.0 MAX 1.5 IMP 50
I       I------=there are ground water sources at depth of
I       I       less than 500 feet
I       I
I       I
I       I       POS .14 NEG -1.00 MIN -2.0 MAX 2.0 IMP 50
I       I------=the surficial material is highly variable
I
I
I       POS 1.00 NEG 0.00 MIN -5.0 MAX 5.0
I------map unit may be R2
I       ALL
I       I
I       I
I       I       POS -.13 NEG -.17 MIN -5.0 MAX 0.0 IMP 50
I       I------=the material peat is in a basin
I       I
I       I
I       I       POS -.13 NEG -.17 MIN -5.0 MAX -1.5 IMP 50
I       I------=the surficial material is impermeable
I       I
I       I
I       I       POS .13 NEG -.17 MIN -.5 MAX 5.0 IMP 50
I       I------=ground water is shallow or discharging
I       I
I       I
I       I       POS .13 NEG .17 MIN .5 MAX 5.0 IMP 50
I       I------=there are ground water sources at depth of
I       I       less than 500 feet
I       I
I       I
I       I       POS .13 NEG .17 MIN 1.5 MAX 5.0 IMP 50
I       I------=material at depth of disposal is saturated
I       I
I       I
I       I       POS .13 NEG .17 MIN .5 MAX 5.0 IMP 50
I       I------=the surficial material is thick
I       I
I       I
I       I
```

```
I       I
I       I
I       I       POS -.13 NEG -.17 MIN -5.0 MAX -.5 IMP 50
I       I------=the surficial material is highly variable
I       I
I       I
I       I       POS -.13 NEG -.17 MIN -5.0 MAX -1.5 IMP 50
I       I------=the bedrock is permeable
I
I
I       POS 1.00 NEG 0.00 MIN -5.0 MAX 5.0
I------map unit may be R3
I       ALL
I       I
I       I
I       I       POS -.17 NEG -.25 MIN -5.0 MAX 0.0 IMP 50
I       I------=the material peat is in a basin
I       I
I       I
I       I       POS .17 NEG -.25 MIN -2.0 MAX 2.0 IMP 50
I       I------=the surficial material is impermeable
I       I
I       I
I       I       POS .17 NEG .25 MIN 2.0 MAX 5.0 IMP 50
I       I------=there are ground water sources at depth of
I       I       less than 500 feet
I       I
I       I
I       I       POS -.17 NEG -.25 MIN -5.0 MAX -.5 IMP 50
I       I------=the surficial material is thick
I       I
I       I
I       I       POS .17 NEG .25 MIN .5 MAX 5.0 IMP 50
I       I------=the bedrock is permeable
I
I
I       POS 1.00 NEG 0.00 MIN -5.0 MAX 5.0
I------map unit may be Y3
I       ALL
I       I
I       I
I       I       POS 0.00 NEG -1.00 MIN -5.0 MAX 0.0 IMP 50
I       I------=the material peat is in a basin
I       I
I       I
I       I       POS .17 NEG -1.00 MIN -2.0 MAX 2.0 IMP 50
I       I------=the surficial material is impermeable
I       I
I       I
I       I       POS .17 NEG -1.00 MIN -2.0 MAX 1.5 IMP 50
I       I------=there are ground water sources at depth of
I       I       less than 500 feet
I       I
I       I       POS .17 NEG 0.00 MIN 2.5 MAX 5.0 IMP 50
I       I------=the surficial material is highly variable
```

Figure 7. (Continued)

```
I      I      POS .13 NEG .17 MIN 2.5 MAX 5.0 IMP 50
I      I------=the material is subject to flooding
I
I
I      POS 1.00 NEG 0.00 MIN -5.0 MAX 5.0
I------map unit may be R1
I      ALL
I      I
I      I
I      I      POS -.17 NEG -.25 MIN -5.0 MAX 0.0 IMP 50
I      I------=the material peat is in a basin
I      I
I      I
I      I      POS -.17 NEG -.25 MIN -5.0 MAX -1.5 IMP 50
I      I------=the surficial material is impermeable
I      I
I      I
I      I      POS .17 NEG .25 MIN .5 MAX 5.0 IMP 50
I      I------=ground water is shallow or discharging
I      I
I      I
I      I      POS .17 NEG .25 MIN .5 MAX 5.0 IMP 50
I      I------=there are ground water sources at depth of
I      I      less than 500 feet
I      I
I      I
I      I      POS .17 NEG -.25 MIN -1.5 MAX 1.0 IMP 50
I      I------=material at depth of disposal is saturated
I
I
I      POS 1.00 NEG 0.00 MIN -5.0 MAX 5.0
I------map unit may be Y2
       ALL
       I
       I
       I      POS -.11 NEG -.14 MIN -5.0 MAX 0.0 IMP 50
       I------=the material peat is in a basin
       I
       I
       I      POS -.11 NEG -.14 MIN -5.0 MAX -1.5 IMP 50
       I------=the surficial material is impermeable
       I
       I
       I      POS -.11 NEG -.14 MIN -5.0 MAX 0.0 IMP 50
       I------=ground water is shallow or discharging
       I
       I
       I      POS .11 NEG -.14 MIN -1.0 MAX 5.0 IMP 50
       I------=there are ground water sources at depth of
       I      less than 500 feet
       I
       I
       I      POS -.11 NEG -.14 MIN -5.0 MAX -.5 IMP 50
       I------=material at depth of disposal is saturated
       I

I
I      POS .11 NEG .14 MIN .5 MAX 5.0 IMP 50
I------=the surficial material is thick
I
I
I      POS -.11 NEG -.14 MIN -5.0 MAX -.5 IMP 50
I------=the surficial material is highly variable
I
I
I      POS -.11 NEG -.14 MIN -5.0 MAX -.5 IMP 50
I------=the material is subject to flooding
```

Figure 7. (Continued)

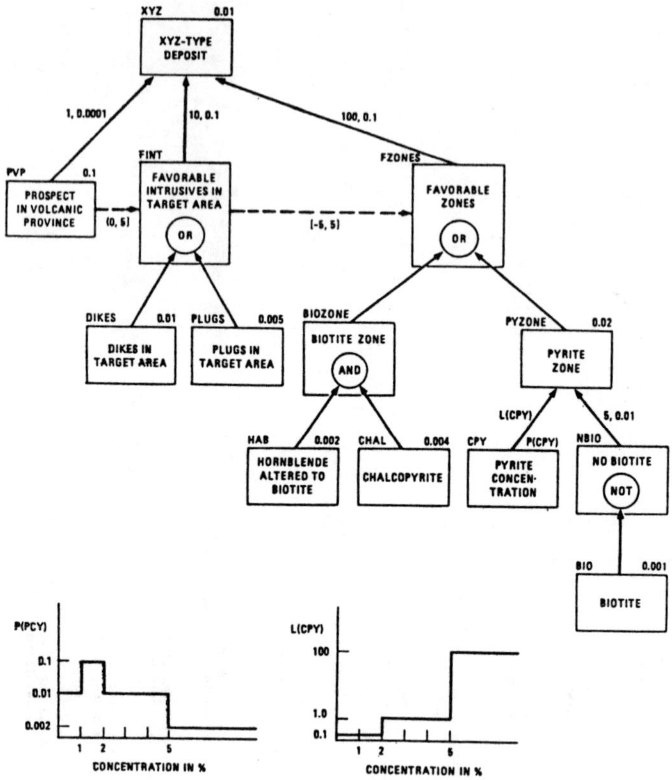

The XYZ-model

Figure 8. Inference network for the idealized XYZ deposit model in PROSPECTOR (after Reboh, 1981, p. 98, Fig. B-1).

```
              DECISION MODEL TO EVALUATE WHETHER
                  THE DEPOSIT IS OF XYZ TYPE

The Deposit is of XYZ type
ALL
I
I
I   POS .01 NEG .83 MIN -5.0 MAX 5.0
I---the prospect is in volcanic province
I   C-ALL
I   I
I   I
I   I   POS 1.00 NEG 1.00 MIN 0.0 MAX 5.0 IMP 50
I   I-C-you have proof of volcanism
I   I
I   I
I   I  If response in range, then
I   I
I   I
I   I   POS 1.00 NEG 1.00 MIN -5.0 MAX 5.0
I   I---there are favorable intrusives
I       C-ALL
I       I
I       I
I       I   POS 1.00 NEG 1.00 MIN -5.0 MAX 5.0
I       I-C-you have intrusives in target area
I       I   ALL
I       I   I
I       I   I
I       I   I   POS .50 NEG .50 MIN -5.0 MAX 5.0 IMP 50
I       I   I---there are dikes in target area
I       I   I
I       I   I
I       I   I   POS .50 NEG .50 MIN -5.0 MAX 5.0 IMP 50
I       I   I---there are plugs in target area
I       I
I       I
I       I  If response in range, then
I       I
I       I
I       I   POS 1.00 NEG 1.00 MIN -5.0 MAX 5.0
I       I---there are favorable zones
I           ALL
I           I
I           I
I           I
I           I
I           I   POS .50 NEG .50 MIN -5.0 MAX 5.0
I           I---it is a biotite zone
I           I   C-MOST
I           I   I
I           I   I
I           I   I   POS 1.00 NEG 1.00 MIN .5 MAX 5.0 IMP 50
I           I   I-C-hornblende altered to biotite is present
I           I   I
I           I   I
I           I   I  If response in range, then
I           I   I
I           I   I
I           I   I   POS 1.00 NEG 1.00 MIN .5 MAX 5.0 IMP 50
I           I   I---chalcopyrite is present
I           I
I           I
I           I   POS .50 NEG .50 MIN -5.0 MAX 5.0
I           I---it is a pyrite zone
I               ALL
I               I
I               I
I               I   POS .50 NEG .50 MIN -5.0 MAX 5.0
I               I---pyrite concentration is sizeable
I               I   BEST
I               I   I
I               I   I
I               I   I   POS 1.00 NEG 1.00 MIN -5.0 MAX 5.0 IMP 50
I               I   I---pyrite percentage is between 0 and <1
I               I   I
I               I   I
I               I   I   POS 1.00 NEG 1.00 MIN -5.0 MAX 5.0 IMP 50
I               I   I---pyrite percentage is between 1 and <2
I               I   I
I               I   I
I               I   I   POS 1.00 NEG 1.00 MIN -5.0 MAX 5.0 IMP 50
I               I   I---pyrite percentage is between 2 and <5
I               I   I
I               I   I
I               I   I   POS 1.00 NEG 1.00 MIN -5.0 MAX 5.0 IMP 50
I               I   I---pyrite percentage is                 >5
I               I
I               I
I               I   POS .50 NEG .50 MIN -5.0 MAX 5.0 IMP 50
I               I---there is no biotite
I
I
I   POS .09 NEG .08 MIN -5.0 MAX 5.0
I---=there are favorable intrusives
I
I
I   POS .90 NEG .08 MIN -5.0 MAX 5.0
I---=there are favorable zones
```

Figure 9. The decision model constructed from the inference network in Figure 8 using The Deciding Factor.

7.Spatial reasoning in PROSPECTOR

Besides the basic function of matching field observations to ore-deposit models, facilities for using graphical input in the selection of drilling sites for porphyry copper deposits were added to PROSPECTOR (Duda et al., 1978; Duda, 1980) for inference networks that integrated geochemical, geophysical and structural geologic evidence from maps. Figure 10 shows a portion of an inference network for "favourable alteration" for drilling site selection. According to those authors, basic situations from digitized map-samples or cells are: (i) being within some region, or (ii) near some contour or line. For transition regions, a piecewise "favourability function" was used to gradually reduce the probability from 1 to 0 according to distances within a predefined range. Figures 11a and 11b show the digitized data for structures and for alteration for the Island Copper deposit in British Columbia, Canada (resolution is 128 cells by 128 cells). Figure 11c shows that the favourable structures, i.e., the zones of influence from fractures and from their intersections, have been identified by computing simple spatial transformations. Figure 11d shows a final favourability digital map of the area. Figure 12 shows a portion of an inference network for favourable structures in a drilling site selection model in PROSPECTOR. It is similar to what was constructed for the spatial information in Figure 11c. F(d, a, b) is the distance function used to assign weights to map evidence (spatial evidence) XNCM1 to XNIO1 where a=100m and b=500m, and to XNFI1 where both a and b are equal to 0m (i.e., spatial coincidence). For example, F for evidence XNMCM1 is 1 if d varies from 0m to 100m. Then it decreases linearly to 0 for d equal to or greater than 500m.

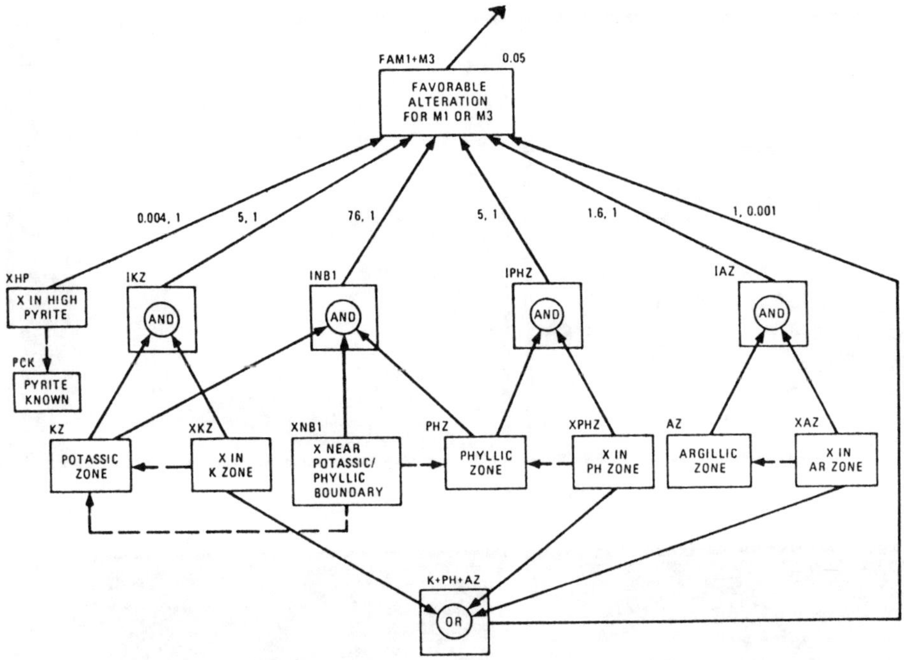

Figure 10. Inference network for drilling site selection (simplified after Duda et al., 1978, p. 109). The hypothesis of favourable alteration for deposit types M1 or M2 is represented. In the boxes (termed spaces) are evidences and hypotheses. Arrows represent rules, broken arrows represent context-constraint rules (i.e., one evidence or hypothesis must be assessed before another one can be assessed), and values beside the arrows represent the weights for LS and LN, respectively. The value of a priori probability assigned to the hypothesis FAM1+M3 is 0.05. Intermediate spaces are for logical implications (i.e., relationships).

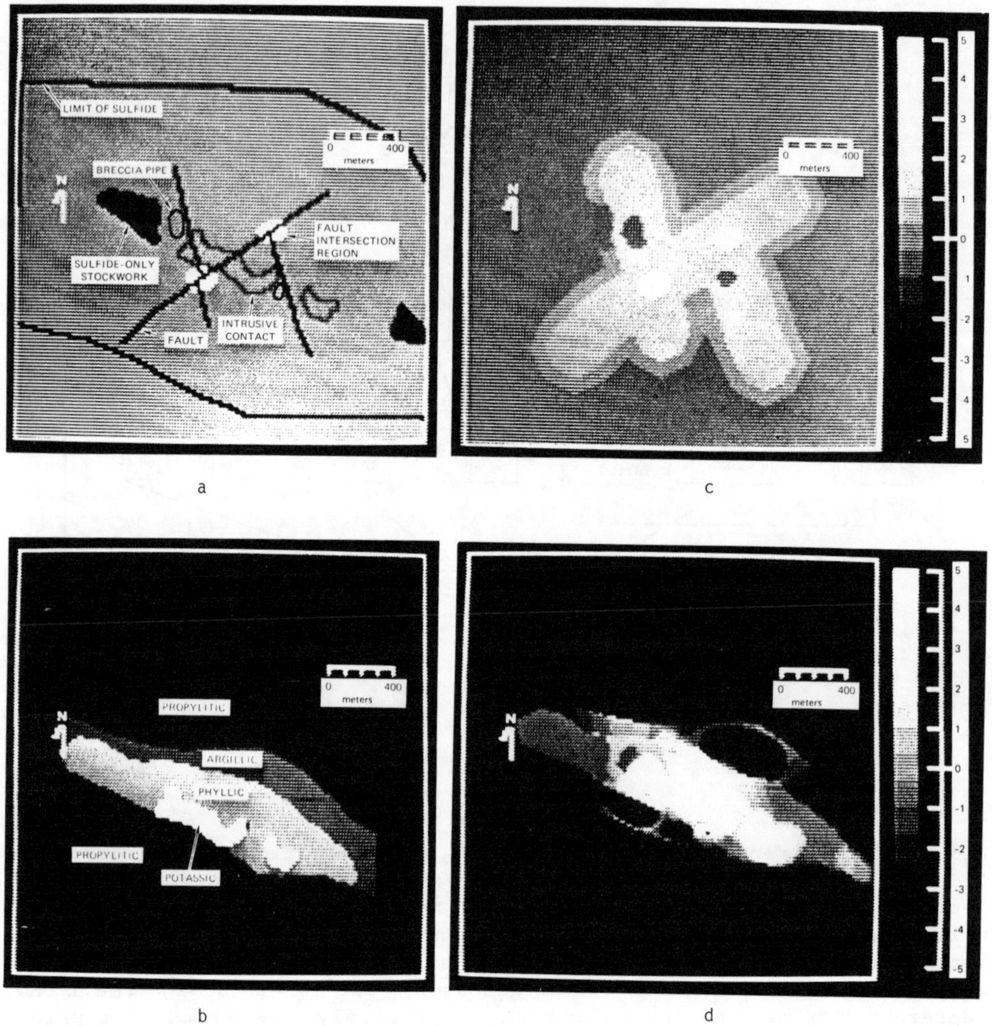

Figure 11. Graphic displays for Island Copper study area, British Columbia, Canada. (a) the structures, (b) the alterations, (c) the favourable structures, and (d) the final favourability map (after Duda et al., 1978, p. 119-122). The values between -5 and +5 represent the certainties as grey levels in the illustration of monitor displays. Resolution is 128x128 cells or pixels.

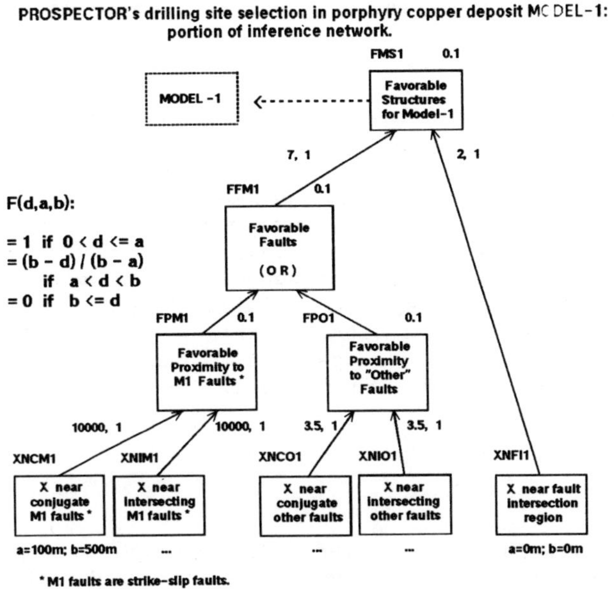

Figure 12. An inference network for favourable structures in PROSPECTOR's drilling site selection model. Explanation is in text.

8. Concluding remarks

We have seen, in simple terms, the basic principles of expert systems, of production rules, and of inference networks. Expert knowledge can be stored in the form of rules, as was done in several geological expert systems (Yatabe and Fabbri, 1988, 1989) some of which were derived from PROSPECTOR. Using the expert shell "The Deciding Factor", two applications were introduced from engineering geology map analysis and for ore deposit identification in mineral exploration. Many other applications can be made in other fields of the geosciences either with similar systems or with much more powerful ones. In all such applications, one of the most difficult tasks is to translate expert knowledge into a viable computational form.

Particular attention is now being given to computer reasoning in the analysis of spatially distributed data. The research areas of geographic information systems and digital image processing are now more and more making use of artificial intelligence in what is termed computational vision. The drilling site selection models in PROSPECTOR represent an early stage of development of those methods. Many of the data processing tasks in a GIS can be guided and their results evaluated using expert rules, operations between images (or maps) and spatial transformations to characterize both geometrical attributes and the distribution of map features (including their spatial relationships, i.e., connectivity, adjacency, distance, orientation, overlap, etc.). Uncertainty and confidence in either map form or table form can also be associated with such map features for automatic theme derivation by spatial data analysis.

References

Apte', C. V. and Weiss, S. M., 1985, An approach to expert control of interactive software systems. Trans. Patt. Anal. and Machine Intell., v. 7, n. 6., p. 586-591.

Campbell, A. and Fitgerrell, S., 1985, The Deciding Factor user's manual. Software Publishing Corporation, Mountain View, CA, 52 p.

Duda, R. O., 1980, The PROSPECTOR system for mineral exploration. SRI International, Stanford, CA, Final Report, April 1980, 120 p.

Duda, R. O., Hart, P. E., Barret, B., Gaschnig, J. G., Konolige, K., Reboh, R. and Slocum, J., 1978, Development of the PROSPECTOR consultation system for mineral exploration. SRI International, Stanford, CA, Final Report covering Oct. 1, 1976 to Sept. 30, 1978, 193 p.

Gaschnig, J., 1982, PROSPECTOR: An expert system for mineral exploration. In, D. Michie, ed., Introductory Readings in Expert Systems, Gordon and Breach Science Publishers, New York, p. 47-64.

Hackett, J. E. and McComas, M. R., 1969, Geology for planning in McHenry County. Illinois Geol. Survey Circ. 438, 31 p.

Hayes-Roth, F., Waterman, D. and Lenat, D., eds., 1983, Building Expert Systems. Addison-Wesley, Reading, MA, 444 p.

Koch, G. S., Jr., and Papacharalampos,D., 1988, GEOEVALUATOR, An expert system for resource appraisal: a demostration prototype for kaolin in Georgia, U.S.A. In, Chung, C. F., Fabbri, A. G. and Sinding-Larsen, R., 1988, Quantitative Analysis of Mineral and Energy Resources. D. Reidel Publishing Co., Dordrecht, p. 513-527.

Liu, H. H., 1985, A rule-based system for automatic seismic discrimination. Patt. Recog., v. 18, n. 6, p. 459-464. McCammon, R. B., 1984, Recent developments in PROSPECTOR and future expert systems in regional resource evaluation. Proc. 9th Pecora Symp. on Spatial Information Technologies for Remote Sensing Today and Tomorrow, Oct. 2-4, 1984, Sioux Falls, SD, p. 241-248.

McCammon, R. B., 1986, The muPROSPECTOR mineral consultant system: A microcomputer-based expert system for regional resource evaluation. U.S. Geological Survey Bulletin 1967, 35 p.

McCammon, R. B., 1987, PROSPECTOR II - and experimental computer- based diagnostic consultant for mineral resource assessment. Proc. MGUS '87 Conf., Redwood City, CA, April 13-15, 1986, p. 49 (abstract).

Miller, B. M., 1986, Building an expert system helps classify sedimentary basins and assess petroleum resources. GEOBYTE, v. 1, n. 2, p. 44-48, 83-84. Miller, B. M., 1987a, The muPetrol expert system for classifying world sedimentary basins. U. S. Geol. Surv. Bull. 1810, 87 p.

Miller, B. M., 1988b, The muPetrol expert system for classifying world sedimentary basins. U. S. Geol. Surv. Open File Report 87-404, 5 p., (with floppy disk).

Reboh, R., 1981, Knowledge engineering techniques and tools in the PROSPECTOR environment. SRI International, Stanford, CA, Technical Note 243, june 1981, 149 p.

Rosenthal-Sabroux, C., Schlienger, D. and Sabroux, J.-C., 1984, Vers un systeme expert volcanologue: Sciences de la Terre, Serie Informatique Geologique, n. 20, p. 99-112.

Smith, R. G., 1984, On the development of commercial expert systems. AI Magazine, v. 5, n. 3, p. 61-77.

The Soft Warehouse, 1985, muLISP-85 reference manual. Honolulu, Hawaii, 221 p.

Varnes, D., 1974, The logic of geological maps with reference to their interpretation and use for engineering purposes. U. S. Geol. Surv. Prof. Paper 837, 48 p.

Yatabe, S. M. and Fabbri, A. G., 1988, Artificial intelligence in the geosciences: a review. Sciences de la Terre, Serie Informatique Geologique, n. 27, v. 1, p. 37-68.

Yatabe, S. M. and Fabbri, A. G., 1989, Putting AI to work in geoscience. Episodes, v. 12, n. 1, p. 10-17.

Appendix A.

A consultation with The Deciding Factor using the decision model in Figure 7.

INTRODUCTION TO WASTE8

This consultation will look at whether the terrain is suitable for waste disposal.

The conclusions will be based upon all of the following.

Whether the material peat is in a basin.

Whether the surficial material is impermeable.

Whether ground water is shallow or discharging.

Whether there are ground water sources at depth of less than 500 feet.

Whether material at depth of disposal is saturated.

Whether the surficial material is thick.

Whether the surficial material is highly variable.

Whether the material is subject to flooding.

Whether the bedrock is permeable.

Whether there is evidence of suitability for waste disposal.

BEGIN DISCUSSION OF WASTE8

This consultation will look at whether the terrain is suitable for waste disposal.

To what degree do you believe that the material peat is in a basin?

MAYBE SO (2.0)

To what degree do you believe that the surficial material is impermeable?

THINK NOT (-3.0)

To what degree do you believe that ground water is shallow or discharging?

PERHAPS (1.0)

To what degree do you believe that there are ground water sources at depth of less than 500 feet?

DON'T KNOW (0.0)

To what degree do you believe that material at depth of disposal is saturated?

MAYBE SO (2.5)

To what degree do you believe that the surficial material is thick?

DON'T KNOW (0.0)

To what degree do you believe that the surficial material is highly variable?

THINK NOT (-3.0)

To what degree do you believe that the material is subject to flooding?

PERHAPS (1.5)

To what degree do you believe that the bedrock is permeable?

DON'T KNOW (0.0)

Here we will look at whether there is evidence of suitability for waste disposal.

This is one of the key ideas in this consultation.

Let's look at whether map unit may be G3.

Earlier you provided information leading to the moderately positive (2.0) evaluation of whether the material peat is in a basin.

That information is being used here.

Earlier you provided information leading to the moderately negative (-3.0) evaluation of whether the surficial material is impermeable.

That information is being used here.

Earlier you provided information leading to the slightly positive (1.0) evaluation of whether ground water is shallow or discharging.

That information is being used here.

Earlier you provided information leading to the moderately positive (2.5) evaluation of whether material at depth of disposal is saturated.

That information is being used here.

Earlier you provided information leading to the moderately negative (-3.0) evaluation of whether the surficial material is highly variable.

That information is being used here.

Earlier you provided information leading to the slightly positive (1.5) evaluation of whether the material is subject to flooding.

That information is being used here.

Your information has led to a moderately positive (2.2) evaluation of whether map unit may be G3.

Here we will look at whether map unit may be G2.

Earlier you provided information leading to the moderately positive (2.0) evaluation of whether the material peat is in a basin.

That information is being used here.

The most recent information has eliminated any possibility that map unit may be G2.

Here we will look at whether map unit may be R3.

Earlier you provided information leading to the moderately positive (2.0) evaluation of whether the material peat is in a basin.

That information is being used here.

The most recent information has eliminated any possibility that map unit may be R3.

Here we will look at whether map unit may be Y3.

Earlier you provided information leading to the moderately positive (2.0) evaluation of whether the material peat is in a basin.

That information is being used here.

The most recent information has eliminated any possibility that map unit may be Y3.

Here we will look at whether map unit may be Y1.

Earlier you provided information leading to the moderately positive (2.0) evaluation of whether the material peat is in a basin.

That information is being used here.

The most recent information has eliminated any possibility that map unit may be Y1.

Here we will look at whether map unit may be R2.

Earlier you provided information leading to the moderately positive (2.0) evaluation of whether the material peat is in a basin.

That information is being used here.

The most recent information has eliminated any possibility that map unit may be R2.

Here we will look at whether map unit may be R1.

Earlier you provided information leading to the moderately positive (2.0) evaluation of whether the material peat is in a basin.

That information is being used here.

The most recent information has eliminated any possibility that map unit may be R1.

Here we will look at whether map unit may be Y2.

Earlier you provided information leading to the moderately positive (2.0) evaluation of whether the material peat is in a basin.

That information is being used here.

The most recent information has eliminated any possibility that map unit may be Y2.

This concludes questioning on one of the key ideas in this consultation.

Your information has led to a moderately positive (2.2) evaluation of whether there is evidence of suitability for waste disposal.

At this point, enough information has been gathered to come to a final conclusion.

Your information has led to a moderately positive (2.2) evaluation of whether the terrain is suitable for waste disposal.

This evaluation is very slightly (29%) reliable because you tended to answer the important questions with little certainty.

KEY IDEAS IN WASTE8

There is a moderately positive (2.2) evaluation that the terrain is suitable for waste disposal.

This is based on all of the following.

Idea	Response	Influence
The material peat is in a basin	2.0	0.0
The surficial material is impermeable	-3.0	0.0
Ground water is shallow or discharging	1.0	0.0
There are ground water sources at depth of less than 500 feet	0.0	0.0
Material at depth of disposal is saturated	2.5	0.0
The surficial material is thick	0.0	0.0
The surficial material is highly variable	-3.0	0.0
The material is subject to flooding	1.5	0.0
The bedrock is permeable	0.0	0.0
There is evidence of suitability for waste disposal	2.2	2.2

INCREASING RELIABILITY IN WASTE8

Stronger responses to these ideas could increase reliability from 29% to 100%

Idea	Your Current Response	Potential Increase in Reliability
There are ground water sources at depth of less than 500 feet	0.0	11%
The surficial material is thick	0.0	11%
The bedrock is permeable	0.0	11%
Ground water is shallow or discharging	1.0	9%
The material is subject to flooding	1.5	8%
The material peat is in a basin	2.0	7%
Material at depth of disposal is saturated	2.5	6%
The surficial material is impermeable	-3.0	4%
The surficial material is highly variable	-3.0	4%

SUMMARY OF WASTE8

There is a moderately positive (2.2) evaluation that the terrain is suitable for waste disposal.

These are the ideas that led to this conclusion.

Idea	Response
The material peat is in a basin	2.0
The surficial material is impermeable	-3.0
Ground water is shallow or discharging	1.0
There are ground water sources at depth of less than 500 feet	0.0
Material at depth of disposal is saturated	2.5
The surficial material is thick	0.0
The surficial material is highly variable	-3.0
The material is subject to flooding	1.5
The bedrock is permeable	0.0

Appendix B.

The Deciding Factor screen dumps obtained while visualizing the decision model in Figure 7.

--▸ the terrain is suitable for waste disposal 0.0?

ALL

Supporting Ideas	POS	NEG	MIN/MAX	IMP	ANS
the material peat is in a basin	0.00	0.00	-5.0/ 5.0	50	0.0?
the surficial material is impermeable	0.00	0.00	-5.0/ 5.0	50	0.0?
ground water is shallow or dischargin	0.00	0.00	-5.0/ 5.0	50	0.0?
there are ground water sources at dep	0.00	0.00	-5.0/ 5.0	50	0.0?
material at depth of disposal is satu	0.00	0.00	-5.0/ 5.0	50	0.0?
the surficial material is thick	0.00	0.00	-5.0/ 5.0	50	0.0?
the surficial material is highly vari	0.00	0.00	-5.0/ 5.0	50	0.0?
the material is subject to flooding	0.00	0.00	-5.0/ 5.0	50	0.0?
the bedrock is permeable	0.00	0.00	-5.0/ 5.0	50	0.0?
↓ there is evidence of suitability for	1.00	1.00	-5.0/ 5.0	-	0.0?

F1 Help F8 View Down F10 Continue

--▸ there is evidence of suitability for waste disposal 0.0?

BEST

Supporting Ideas	POS	NEG	MIN/MAX	IMP	ANS
↓ map unit may be G3	1.00	0.00	-5.0/ 5.0	-	-5.0?
↓ map unit may be G2	1.00	0.00	-5.0/ 5.0	-	-5.0?
↓ map unit may be R3	1.00	0.00	-5.0/ 5.0	-	-5.0?
↓ map unit may be Y3	1.00	0.00	-5.0/ 5.0	-	-5.0?
↓ map unit may be Y1	1.00	0.00	-5.0/ 5.0	-	0.0?
↓ map unit may be R2	1.00	0.00	-5.0/ 5.0	-	-5.0?
↓ map unit may be R1	1.00	0.00	-5.0/ 5.0	-	-5.0?
↓ map unit may be Y2	1.00	0.00	-5.0/ 5.0	-	-5.0?

F1 Help F7 View Up F8 View Down F10 Continue

--▶ map unit may be G3 -5.0?

ALL

Supporting Ideas	POS	NEG	MIN/MAX	IMP	ANS
the material peat is in a basin	.17	.17	.5/ 5.0	50	0.0?
the surficial material is impermeable	-.17	-.17	-5.0/ -.5	50	0.0?
ground water is shallow or dischargin	.17	.17	.5/ 5.0	50	0.0?
material at depth of disposal is satu	.17	.17	1.0/ 5.0	50	0.0?
the surficial material is highly vari	-.17	-.17	-5.0/-2.0	50	0.0?
the material is subject to flooding	.17	.17	0.0/ 5.0	50	0.0?

F1 Help F7 View Up F10 Continue

INTERPRETING THE DECISION MODEL SCREEN

The main idea appears here. Evaluated response appears here.

Supporting ideas appear here.

ALL

Supporting Ideas POS NEG MIN/MAX IMP ANS

The type of logic in use appears here.

Positive and negative weights, acceptable response ranges,
and importance of each idea appear here.

Weights of each answer given appear here.

Use the ↑ ↓ keys to position the pointer opposite a supporting idea.
To view the idea, press F8. Press F7 to return.

PgDn More F10 Continue

Appendix C.

PROSPECTOR'S PLAUSIBLE INFERENCE SCHEME[1]

Likelihood sufficiency and necessity

Figure 13 shows part of a PROSPECTOR's model for porphyry copper deposits, PCDA. The two numbers associated with each inference rule are its LS and LN values (likelihood sufficiency and likelihood necessity values), respectively. The number appearing above each box, representing a "nonterminal node" is the prior probability of that assertion's being true.

Example: The existence of stocks (RCS) is a more highly sufficient indicator of "suggestive morphology of igneous rocks" (SMIR), i.e., LS=300, than is the existence of either dikes (RCAD), intrusive breccias (RCIB) or volcanic plugs (RCVP), i.e., LS=75, 20, and 4, respectively. Similarly, "favourable level of erosion" (FLE) is highly sufficient and highly necessary factor for establishing "favourable regional environment" (FRE), i.e., LS=5700 and LN=0.0001).

On the other hand, the existence of a "preintrusive through going fault system" (OTFSYS) is only mildly sufficient and mildly necessary for establishing "favourable regional environment" (FRE). Hence, the positive LS or negative LN votes of (FLE) are weighted much more heavily than those of (OTFSYS).

Let us analyze how logical and plausible relations can be combined as building blocks to accomplish the intent of the economic geologist designing the model. Let us consider the section of our (PCDA) model concerned with establishing (SMIR), "suggestive morphology of igneous rocks." There are four positive indicators for establishing a suggestive morphology of igneous rocks (SMIR), namely (RCIB), "intrusive breccias", (RCS) "stocks," (RCAD) "dikes," and (RCVP) "volcanic plugs." Each of these factors contributes independently to establishing (SMIR), although to different degrees. The absence of any one of these four factors individually is unimportant (i.e., LN=1 for those rules).

However, if it is known that none of these factors is present [implying that the disjunctive node (SMIRA) is false], then the probability of a "suggestive morphology of igneous rocks" is essentially zero [LN=0.0002 for (SMIRA)]. Note that in defining an inference network for a model, the object is to induce the model designer to articulate such statements, and then to translate the statement into network constructions.

---

[1]Note: This text was extracted from Reboh (1981, pages 14-20).

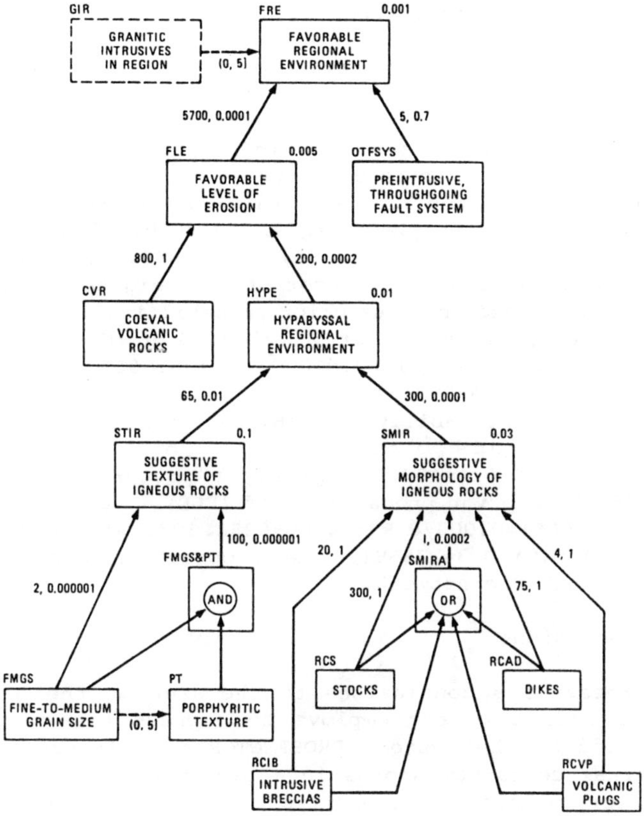

Sample Inference Network from a Prospector model

Figure 13: Sample inference network for PCDA, a model for porphyry copper deposit in PROSPECTOR (after Reboh, 1981, p. 15, Fig. 4).

Propagation of the effect of a piece of evidence.

How does the effect of a piece of evidence propagate upward through the model?

Example: Suppose that the user has indicated only that "intrusive breccias" are present but this is definite (i.e., +5). This fact multiply the odds of (SMIR) by a factor 20, hence raising its probability from 0.03 to 0.382 [the prior odds on (SMIR) equal to 20*0.030927=0.61855, which corresponds to a probability of 0.61855/(1+0.61855)=0.382]. This in turn increases the odds of (HYPE) by a factor of 300 weighted by the degree to which (SMIR) has increased from its prior probability, i.e., by the factor 300*(0.382-0.03)=108.866. [The prior probability of (HYPE) is 0.01 and the prior odds are 0.01/(1-0.01)=0.010101. 108.866*0.010101=1.0996554; 1.0996554/(1+10996554)=0.5237313].

Hence, the posterior probability of (HYPE) is 0.52373, which in turn increases the odds of (FLE) by a factor of 200*(0.52373-0.01)/(1-0.01)=103.78, giving a posterior probability for (FLE) of 0.34276. [The prior probability of (FLE) is 0.005, and the prior odds are 0.005/(1-0.005)=0.005/0.995=0.0050251. 103.78*0.005251=0.5215048. 0.5215048/(1+0.5215048)=0.3427559]. The propagation continues in this manner upward through the network.

Certainty of conclusions.

PROSPECTOR expresses its conclusions to the user on the same -5 to +5 certainty scale that the user employs to express his certainty about evidence requested by the system. PROSPECTOR maps internal probability values to external certainty scores in a piecewise linear fashion, such that the posterior certainty is proportional to the difference between the posterior probability and the prior probability.

For instance, if the user wants to express that evidence E is present to some degree, he must specify a probability P(E¦E') that is greater than the prior probability P(E), and thus must know P(E) so that he can choose a value of P(E¦E') relative to the reference point P(E). Instead of asking for P(E¦E'), PROSPECTOR asks for the certainty measure C(E¦E'). This can be viewed as the posterior probability normalized with respect to the prior probability. It is arbitrarily scaled from -5 to +5, and corresponds to the piecewise linear function of P(E¦E'), shown in Figure 14, chosen so that P=0 corresponds to C=-5, P=P(E) corresponds to C=0, and P=1 corresponds to C=+5.

$$C(E|E') = \begin{cases} 5 * \dfrac{P(E|E') - P(E)}{1 - P(E)} & \text{if } P(E|E') > P(E) \\[2ex] 5 * \dfrac{P(E|E') - P(E)}{P(E)} & \text{otherwise} \end{cases}$$

This function is also used by PROSPECTOR when communicating to the user the uncertainties for conclusions.

Figure 14: Relation between certainty and probability in PROSPECTOR (after Reboh, 1981, p. 20, Fig. 6).

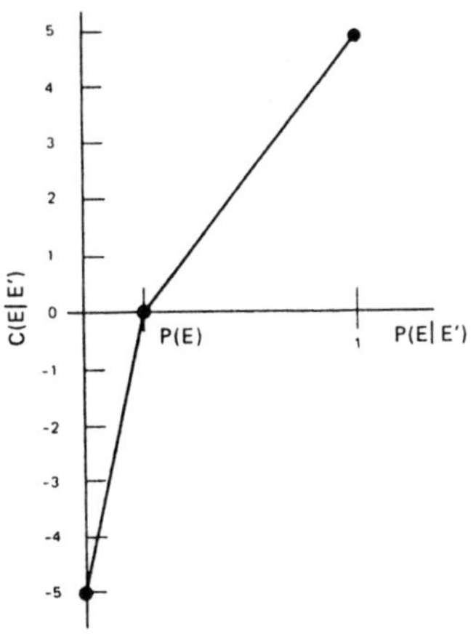

Relation between Certainty and Probability

Example: Since the prior probability of (FLE) is 0.005, a posterior probability of 0.34276 corresponds to a posterior certainty of, 5*(0.34276-0.005)/(1-0.005)=1.697. Similarly, a posterior probability of 0.001 corresponds in this case to a posterior certainty of, 5*(0.001-0.005(/(0.005)=-4.

Note: See Shortliffe, E.H. and Buchanan, B.G., 1975, [A model of inexact reasoning in medicine. Math. Biosci., v. 23, p. 351-379] for a description of the subjective certainty scale used in the MYCIN medical diagnosis system. See also Shortliffe, E.H., 1976 [Computer-based medical consultations: MYCIN. American Elsevier Publ. Co., New York].

LAND USE MODEL USING A GEOGRAPHIC INFORMATION SYSTEM

Carlos R. Valenzuela
Hans de Brouwer
Allard Meijerink
I.T.C.
P.O. Box 6, Boulevard 1945
7500 AA Enschede
The Netherlands

1. Introduction

The need for improved methods for resource management and environmental hazard assessment is acute in many large watersheds in Asia. Thus, it would be useful to prepare a scenario of actual and estimated future land use, taking into consideration the potential suitability of the land for a range of crops, the existing farming systems, the erosion susceptibility, availability of land and other related data. The need for a versatile computerized system to handle such a vast amount of data has led to the development of the Integrated Land and Water Information System (ILWIS).

A leading principle for the design of the GIS was to enable interactions between biophysical simulations of crop production, on-site land degradation with related off-site effects, and social-economic considerations. The latter should, on the one hand, direct the choice of simulations in view of national policies, trends in population and agricultural production, and on the other hand, generate an input at a detailed level, such as land degradation due to changes in farming systems.

The identification of the user needs in Indonesia, in the broad field of watershed management and agricultural development, indicated that a multitude of tasks had to be performed. The first tasks were in the biophysical domain of monitoring land use, prediction of crop yield, estimation of erosion, of sediment yields of sub-catchments and the sediment transport in view of downstream problems.

At the beginning of the project, no clear concept could be formulated on the difficult interface between the social-economic and biophysical domains. What emerged was the necessity to design a flexible system with transparent modules and easy exchange of data and results. It was decided to develop and test a rule base, with social-economic criteria and transformation procedures, in an interactive, empirical way by working in the large test area of the Ogon Komering catchment in South Sumatra.

The structure to achieve the planning functions may be visualized in Figure 1. Database contents should be transformed with procedures and models in the rule base to produce scenarios such as future erosion hazards or other land degradation activity coupled to expected land

A. S. Belward and C. R. Valenzuela (eds.),
Remote Sensing and Geographical Information Systems for Resource Management in Developing Countries, 425–441.

pressure. The scenario consisting of maps, tables or graphs may be represented and steps to reduce undesired developments can be interactively evaluated with a GIS (Van der Putte, 1988).

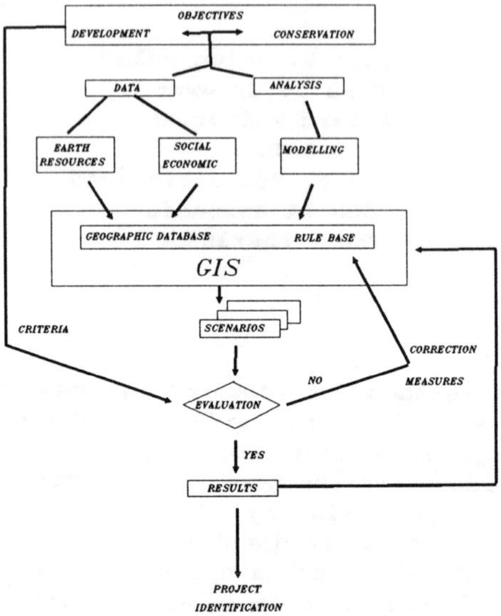

Figure 1. The role of geographic information systems in planning

Such a scenario for the Upper Komering catchment is shown in Figure 2. It depicts the predicted future land use based on suitability for coffee, the dominant crop, erosion hazard under coffee, availability of the land, accessibility and population pressure with the present types of farming systems (De Meijere, et al., 1988)

2.The ILWIS system

ILWIS integrates image processing capabilities, tabular databases and conventional geographic information system characteristics. The transformation of data into information requires more than the data storage, retrieval, manipulation and display. The conceptualization of the system takes into consideration that the end users of such a system usually do not have a thorough knowledge of computers. All operations are performed through a user friendly menu, which allows the user to concentrate on the applications rather than on learning the intricacies of the system. Experience users can, however, perform operations through commands and/or command files.

Data acquisition from aerospace images-for information extraction-is an integral part of the system. This is important in data-scarce regions and allows effective monitoring. Remote sensing data can be entered into the raster database from either computer compatible tapes (CCT's) or high density floppy disks. Conversion programs to read data directly from tape

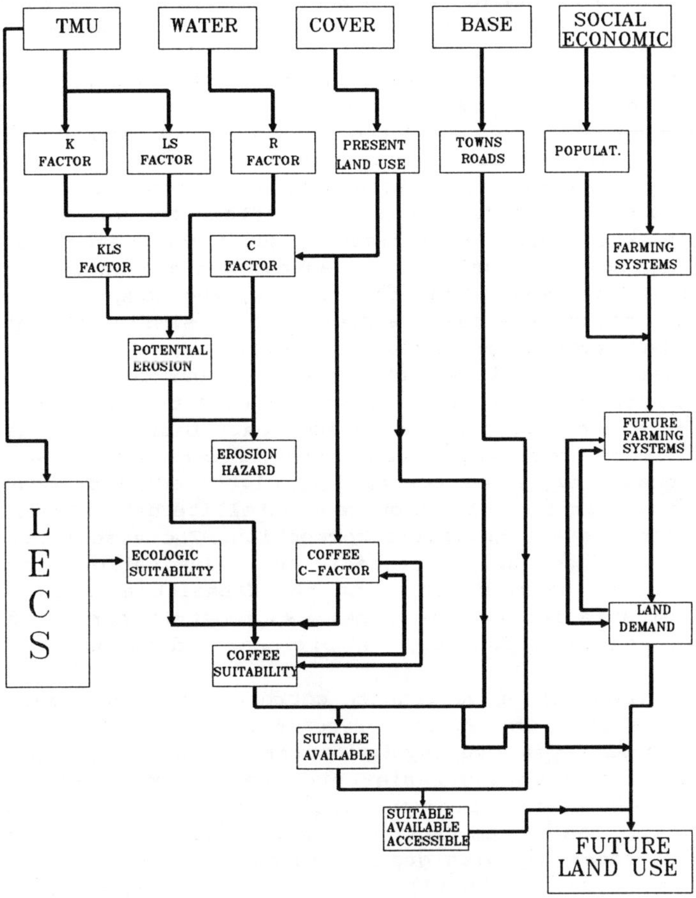

Figure 2. Flow chart of the scenario of future deforestation and land pressure

drives or other image processing systems are available. Analogue data are transformed into a vector format using a user friendly digitizing program.

Complex modelling procedures can be easily executed through the MAPCALCULATOR, which integrates tabular databases with spatial databases; it includes an easy-to-use modelling language and the possibilities of using functions and macros. Fast overlaying procedures constitute one of the main characteristics of the system. Complex procedures can be rapidly executed in portions of the study area on video memory. After an evaluation and assessment of the results of this modelling procedure, it can then be applied over the entire area.

Tabular and spatial databases can be used both independently and on an integrated basis. A versatile program TABLECALCULATOR, provides direct communication between the ILWIS kernel and outside models, statistical packages, raw data files, databases and other systems. Several knowledge driven queries can be performed in the relational database, the results of these queries are then integrated with spatial data. One example of the use of relational databases independently and yet integrated with spatial databases is in land evaluation studies, where several of the queries (crop requirements) are executed in tables without any spatial relationship, while several other queries involve the spatial domain (land qualities). The integration of both databases occurs in the matching of land qualities with crop requirements to attain optimal use of the land.

Computational procedures and efficient use of the system are improved by the appropriate use of modelling processes. Not all analysis involve the use of spatial databases; whenever possible, knowledge driven queries in the relational database should be preferred to similar operations in the spatial database (vector or raster domains).

Image processing capabilities integrated with cartographic modelling and relational databases constitute a powerful tool that allows the analysis of data which was not possible only few years back. ILWIS incorporates conventional image processing capabilities. Pre-processing and processing of data are possible. Pre-processing involves radiometric and geometric correction, although most of the current satellite data (SPOT, TM) already include radiometric and geometric correction. The system has facilities for corrections of sensor, atmospheric and illumination effects. These corrections can be carried out using the Mapcalculator or with special functions available for this purpose. Radiometric corrections includes the choice of nearest neighbour, and bilinear and bi-cubic interpolation routines.

Facilities to add UTM or geographic coordinates onto a map or an image are also available. Geometric corrections and addition of geographic coordinates facilitates overlaying operations between maps and remote sensing images, thus making easier the use of remote sensing data for monitoring activities that incorporate historical data (old maps and other ancillary data). The system provides the interactive selection of control points from a vector map with geographic coordinates.

Data processing several basic image analysis capabilities, such as histogram manipulation, automatic stretch display, user defined filters, transfer function manipulation and other standard functions. It includes calculation of covariance and correlation matrices, eigenvalues and eigenvectors and other statistics. A user friendly sampling program allows sampling by pixel, feature space plot analysis and sample and class statistics. Several classifier algorithms can be used. Before classifying an entire image, the behaviour of the different classifiers can be compared through an interactive pixel classification routine.

The system provides black and white or color hard copy output products in either vector or raster format. It supports standard peripherals such as EPSON, IBM, Tektronix and HP printers and plotters.

ILWIS is a system based on an IBM-AT (or compatible) personal computer with a high resolution monitor for colour display, a digitizer for data input and a colour printer or plotter for data output. A high resolution board (e.g., MATROX PG 640A or Extended VGA) is used for efficient and fast manipulation of vector and raster data.

Although ILWIS was designed primarily as a stand alone system, it can also be used as a working station to a larger national GIS system.

The program for analysis of raster files (maps or images), MCALC, is an important module of the system. Arithmetical (+, -, *, /), mathematical (sqrt, exp, sin, cos, atan), relational (>, >=, =, <, <=, <>), boolean (not, and, or, xor) and conditional (if ... then...) operations can be performed simultaneously on several maps or images and on attribute tables.

The image processing module contains all the standard programs for displaying, enhancing, classifying and transforming remote sensing data. The system includes the following specifications.

1. The system is capable of integrating raster, vector and tabular data. In this way, the advantages inherent to each domain can be fully utilized (Burrough, 1986): topology, geostatistical applications, digital terrain models, remote sensing processing, fast overlaying techniques, interfaces with outside models.

2. Acceptance of mapping units and remote sensing data. Scarcity of basic data makes imperative the need of appropriate methods for acquisition and structured storage of data concerning terrain (geology, geomorphology, soil geography), cover (land-use and land-cover type describing floristic composition, phenology and structure). This requires the integration of aerial photography interpretation and remote sensing image processing for feature extraction. For mapping units, aggregating natural associations (geomorphologically based terrain mapping units -TMUs-), an open-ended method for data detail (Meijerink, 1988). The coupling to the soil data base is described by Valenzuela (1988). The method is a compromise between rapidly acquired data (aerospace, field and existing data) and input variables required for simulation models. The well-known problem of data redundancy by systematic mapping and specific model input should be viewed in the perspective of conditional and relational modelling. The use of relational databases has reduced most of the problems for applications such as updating, estimation of non-direct input parameters by using simple or complex transfer functions.

3. Outside models. The GIS communicates with outside models in two ways: (a) assists in preparation of input data, which may consist of spatial analysis first (Bouma, 1988) and (b) preparation of spatial model results, for spatial combination with other maps.

4. Operators in GIS, interaction. Not all problems can be solved with existing simulation models. The system should be capable of translating expert knowledge. This can be achieved by: (a) using conditional and relational operators, and (b) the capability of these and other essential operators to address several data sources simultaneously and efficiently to allow near real-time interaction. It is important to realize that certain geographic entities, such as soil type associated to farming systems, may be of greater use to an expert than the attributes alone.

5. Operational efficiency, user friendliness. There will be a multitude of operations, which may be of different nature. A complex system such as ILWIS defeats its purpose if no care is taken to make modules and data exchange transparent, hide cumbersome operations from users, include adequate utilities, and make optimum use of the display capabilities.

3. Main modules

The main modules generated for this application were the base module, terrain-soil geography, land cover and land use, water and social-economic. These modules have both the spatial and non spatial components. The base module, includes data on administrative boundaries, transportation and communication facilities, settlements, sub-watershed and other related information.

3.1. TERRAIN-SOIL GEOGRAPHY MODULE

The terrain-soil geography (terrain mapping unit/TMU) module, describing and explaining the main land units and their corresponding soils (Table 1). The TMU is based on the widely-tested ITC system of geomorphic mapping. It has the geologic-geomorphologic foundation as the key factor and includes aspects of the physiographic approach; it also incorporates morphometric descriptions (Meijerink, 1988). A terrain mapping unit (TMU) groups the following aspects:

- geomorphologic origin and physiography
- lithology
- morphometry
- soil geography

A neutral term "terrain" is required to minimize confusion with such concepts as "land units" of the FAO land evaluation and land system classification described by Brink, et al. (1966). The term "geomorphologic units" is not fully appropriate because sub-differentiation may be based on soil distribution. The word "physiography" has been avoided because the meaning varies from the American Geologic Institute's "study of the genesis and evolution of land forms" to descriptions of land that include vegetation and land use. The term "terrain" also has its associations, particularly with parametric descriptions, related to trafficability and engineering geology, but it is perhaps the most neutral term at present.

A few definitions may help to understand the nature of terrain mapping units:

- a TMU describes a natural division of the terrain that can be distinguished on aerial photographs and can be verified on the ground. The units are differentiated on the basis of photomorphic properties in the stereo model
- a TMU is a unit which groups interrelationships of landform, lithology and soil, and it forms a first stratification for a sampling procedure
- a TMU differs from another-adjoining-unit because either the landforms are evidently different or the phenomena associated with the landform (nature of the weathered zone, lithology, types of soil) differ

- in terms of a GIS, a TMU may be described as the geographic location (polygons) of entities which relate to a unique set of attributes
- a TMU is a legend unit of a conventional thematic map

Table 1. Geomorphology: terrain mapping unit (TMU) classification criteria (From Meijerink, 1988)

PRIME DETERMINANTS

General sequence is: lithology or origin (interchangeable), specific origin, morphometry

ORIGIN
Of landforms or complexes (fluvial, volcanic, structural, etc)

SPECIFIC ORIGIN
Differentiation of landforms or associations of landforms within the main origin: physiographic criteria may be use in conjunction with analytical geomorphologic criteria, such as age, tectonic or climatic history, processes

LITHOLOGY
Lithologic or lithostratigraphic units or complexes, usually grouped in formations: mixtures of unconformable beds (overburden) and underlying lithology

MORPHOMETRY
Internal relief, valley density, slopes

SECONDARY DETERMINANTS

No general sequence

SUB-UNITS
Differentiation of important/relevant or large landform or association of landform elements; related to specific origin

SOIL UNITS
Soil units may in some cases form the basis of differentiation of sub-units; in exceptional cases even for main units (as a combination of lithology and specific origin). Soil units are generally associated with geomorphologic or physiographic main units, sub-units or catena elements

CATENAS AND SLOPE PROPERTIES
Characteristic sets of crests, slopes and valleys and their properties (cliffs, screes, etc)

TERTIARY DETERMINANTS

No sequence

PROCESS-RELATED ASPECTS
Landforms dominated by specific denudative or erosive processes, such as large landslides, lahards, badlands, etc

NATURAL HAZARDS
When desired, hazard criteria may be used for subdifferentiation, such as flood frequency, duration and depth, fault zones, etc

SELECT PROPERTIES
Generally environmental artifacts, such as mining dumps, refuse pits, sediment trapping areas, levelled ground, etc

3.2. THE LAND COVER AND LAND USE MODULE

The land cover and land use (CUMU) module, with the main characteristics of the vegetative cover and the dominant use type in the watershed. It includes a classification of cover, based on nature, density and height, in addition to generally accepted land use classifications.

The land cover and land use database is a separate module. This separation was considered as an expedient because:

- land cover and land use are sometimes mapped independently of soil and terrain, and
- land cover and land use change in time, in contrast to the terrain and soils, and thus this part of the database may need more frequent updates.

The land cover and land use database can be used for separate mapping of land cover and land use and also the latter maps can be integrated with terrain and soil mapping.

The database consists of two parts (Wijngaarden and Kooiman, 1988): one part organizes the data on the different land cover and land use types. This part does not include specific geographic locations. The second part describes the relationship between land cover and land use and the mapping units. This relationship can be with the land cover and land use mapping units or the terrain and the soil mapping units.

The first part has not been fully developed yet. Any existing description or classification method can be used, depending on the objectives of the project and/or the characteristics of the study area.

In the design of the second part, much attention was paid to the occurrence of complex mapping units, the need to link the land cover and land use database with the terrain and soil database, and the possibilities for monitoring and sequential mapping of land cover.

The CUMU database, like the TMU database, uses the ORACLE Relational database management information system. An overview of the land cover and land use database is given in Figure 3.

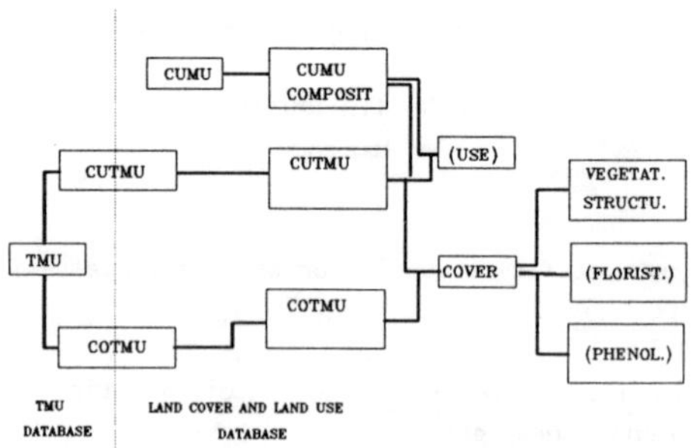

Figure 3. Overview of the land cover and land use database (After Wijngaarden and Kooiman)

3.3. THE WATER MODULE

The water module, presenting information useful to hydrologic models and water balance calculations. It includes station data (rainfall, evaporation factors, discharges and other measured variables). A large number of standard processing methods is included, which is used for hydrologic analysis and to supply input data to other operations and models, such as agro-climatology, crop yield modelling, erosion and others.

The model contains procedures for preprocessing raw hydrologic station data and applying general, simple models. A number of more complex hydrologic and erosion simulation models are also available, which were derived from outside sources and belong to "public domain software". Some of these models run on more powerful computers, but the input is prepared using ILWIS (Meijerink, et al., 1988).

Data preprocessing concerns mainly checking, correcting and estimating missing data. The raw data, as supplied by the data collection agencies (meteorologic services, water resource or irrigation agencies) are kept in separate source files. A selection is made of the required data for specific purposes. After preprocessing, the corrected and aggregated data can be transferred to the ILWIS non-spatial database or to a separate data file. For example, mean catchment rainfall can be derived from several stations (raw data) after using an estimation procedure for missing records (corrected data) and application of an area-weighted extrapolation method (aggregated data).

3.4. THE SOCIAL-ECONOMIC MODULE

The social-economic module, providing information concerning social and economic characteristics in the area. Farming systems are the basis of the economic evaluation and planning functions of the system. Data concerning the nature of the farming system are collected on the basis of the land cover/ land use map to include geographic variations.

For instance, in the study area land belongs to the village community, and allocation to local farmers is regulated by the village community council. Once land is occupied and/or improved, its boundaries are recognized within the village community. If the land is abandoned after a few seasons, it reverts to the community. If land is planted with tree crops, ownership of the trees remains vested in the planter.

Spontaneous immigrants can obtain land by expanding into forest lands free of claims, by marriage to locals, buying claims from locals or by inheritance. According to the agricultural census of 1983, most farmers (91 percent) cultivate their own land. Pure tenancy rarely occurs. Only three percent of the farmers cultivate under share cropping or fixed rents (Zabel. et al., 1988).

Local farmers are traditionally "shifting cultivators", but "pure" shifting cultivation systems are no longer found in the area. Population growth is the main dynamic factor that has change shifting cultivation into a more permanent system. Shifting and permanent cultivation tend to overlap to a large extend, and combinations of permanently cultivated wetland rice or coffee with shifting cultivation of upland rice or coffee are common. Part of the coffee land is often under-utilized because of lack of labour and/or low coffee prices. These lands are cleared and maintained again as soon as exploitation becomes profitable again.

4.The land-use model

The database was used in a land-use model developed for planning and resource management at a regional level. Social-economic and physical resource data are combined to predict future developments in land use. The planning should be done in such a way that sustainable production of crops is possible without exhausting the natural resources of the area, in particular, the problem of soil erosion has to be addressed.
The population of Indonesia of approximately 165 million (1981) grows at a rate of 2.34% per year. Most of the population is concentrated on the islands of Java and Bali. These islands comprise only 7% of the total land area.
The farm holdings in java and Bali are very small and the land is intensively cultivated. Little attention is paid to soil conservation and even the steep to very steep slopes are used for cultivating annual crops. On the other islands, shifting cultivation is still practiced while logging operations and transmigration settlements are responsible for extensive deforestation practices.
4.8 million ha have been identified as land in a critical to very critical state of erosion. This erosion process causes:

- shallowing of the soil depth,
- decrease in soil fertility,
- deterioration of the runoff conditions; peak flows increase and low flows decrease,
- increase in sediment load in the rivers causing flooding, sedimentation of reservoirs and shallowing of rivers.

Because of the importance of coffee production, both for the farmers and the Indonesian government, the land use model was directed towards the estimation of coffee producing areas. As a study area the Upper Komering watershed, South Sumatra, Indonesia was selected.
The Upper Komering watershed lies approximately between 4°12' and 5°00' Southern latitude and 103°35' and 104°27' Eastern longitude and covers an area of about 4.260 km^2. The altitude of the area ranges from 76 m above sea-level in Sukoraja to 2232 m at the summit of the Gunung Pesagi.
The area has a typical wet tropical (monsoonal) climate with an average yearly rainfall between 2500 and 3250 mm. During the North-East monsoon (November - April) heavy rainfall occurs; drier conditions prevail during the South-East monsoon (June - September). During the SE monsoon rainfall may be erratic. Sometimes these months receive no rainfall at all. In other years they may receive over 250 mm per months.
The minimum daily temperatures in the lowland and the low hills area of the watershed are around 22 - 23° C. The maximum daily temperatures vary from 30 -33° C. The temperature higher in the mountains may drop to an average of 8 - 12° C. The monthly variation is very small.

The oldest rocks in the area are mesozoic and occupy only a small portion of the area. They occur generally as complex folded strike ridges and consist of low grade metamorphics and hard sedimentary rocks, mixed with some volcanic material. The majority of the area consist of Tertiary rocks, mainly shales, tuffaceous shales and marls with at some places sandstone and hard limestone.

The morphology of the area is largely determined by Quaternary volcanism and tectonism (volcanic activities, pyroclastic activities, faulting and mass movements).

Eight general land use / land cover types can be distinguished in the area: rice/paddy fields, plantations (coffee, rubber, pepper), bushland, grassland, savanna and forest.

The prime occupation in the area is agriculture. For both, the Sumatranese farmers as well as the re-settled Javanese farmers, rice and especially coffee growing are the most important source of income.

The study area covers six (6) districts: Martapura, Simpang, Muaradua, Bandung Agung, Muaradua Kisam and Pulau Beringing. Only the first four have a good infrastructure of asphalted roads.

In order to simulate a scenario for near-future developments, and assuming that no restrictive measures are taken, the following procedure was adopted:

1. Based on interpretation of aerial photographs, satellite images and extensive field tests a soil erodibility map was generated. The map depicts the soil erodibility as a function of the soil properties, the slope gradient and the slope length.

2. A rainfall erosivity map was prepared using data obtained from meteorologic stations in the area. An empirical function was developed to estimate the erosivity of the rain. This was further refined including an interpolation procedure taking into consideration the orographic influence.

3. A combination (overlay) of the erodibility and the erosivity maps produced the potential erosion map. This map portrays the erosion in the area assuming that no vegetation is present.

4. The protective influence of the vegetation is incorporated using the land cover/land use map, after reclassification into a protective map. This map combined with the potential erosion map generated the erosion hazard map as illustrated in Figure 4a. This map shows the erosion under the present situation.

5. The TMU's ecologic suitability for coffee was assessed using the Land Evaluation Computer System (LECS). Figure 4b, shows the suitability map for coffee. LECS was developed by FAO (Wood and Dent, 1983) and calibrated for the area (Elbersen, et al., 1988).

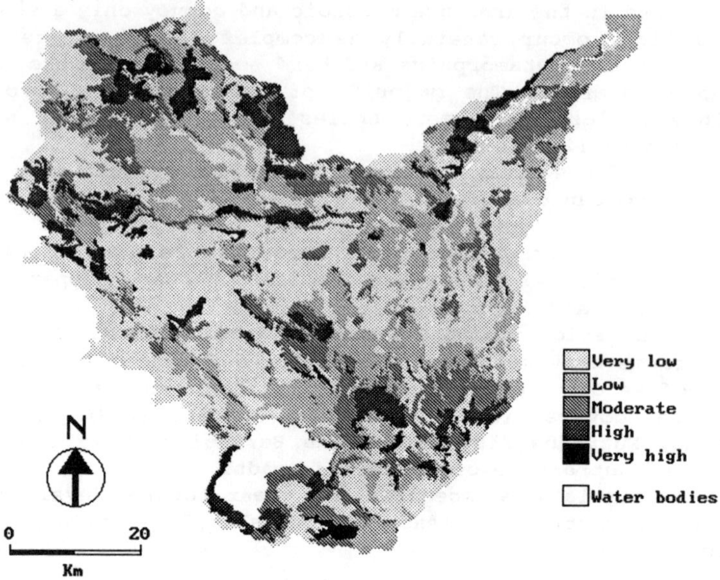

Figure 4a. Soil erosion hazard map of the Komering watershed

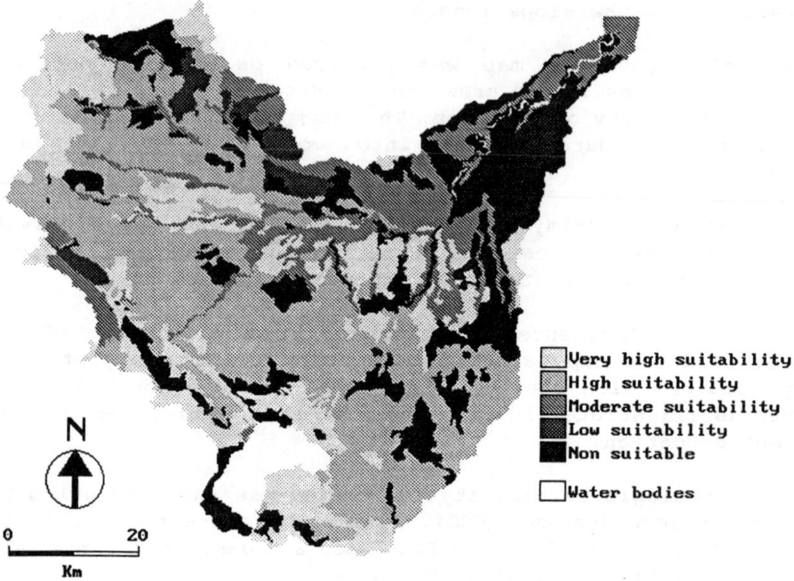

Figure 4b. Ecologic suitability for coffee, Komering watershed

6. In order to determine the erosion under coffee, a hypothetical scenario which assumed that the entire watershed was planted with coffee was generated. This is illustrated in figure 4c. The combination of the erosion map and the ecologic suitability for coffee generates the map depicting all the suitable areas for coffee in the watershed. This map is illustrated in Figure 4d.

7. A map showing all the areas available for future development was generated using aerial photographs, satellite images and field checks. Available areas are all areas that were not under agriculture yet as illustrated in Figure 4e.

8. The combination of areas suitable for coffee and land available for agriculture produces all areas that are suitable and available for coffee.

9. The social-economic study indicated that developments of new coffee producing areas would only take place in areas within 5 km from roads, villages, and/or agricultural parcels. Using connectivity analysis a distance maps was produced. Figure 4f, illustrates the distance map.

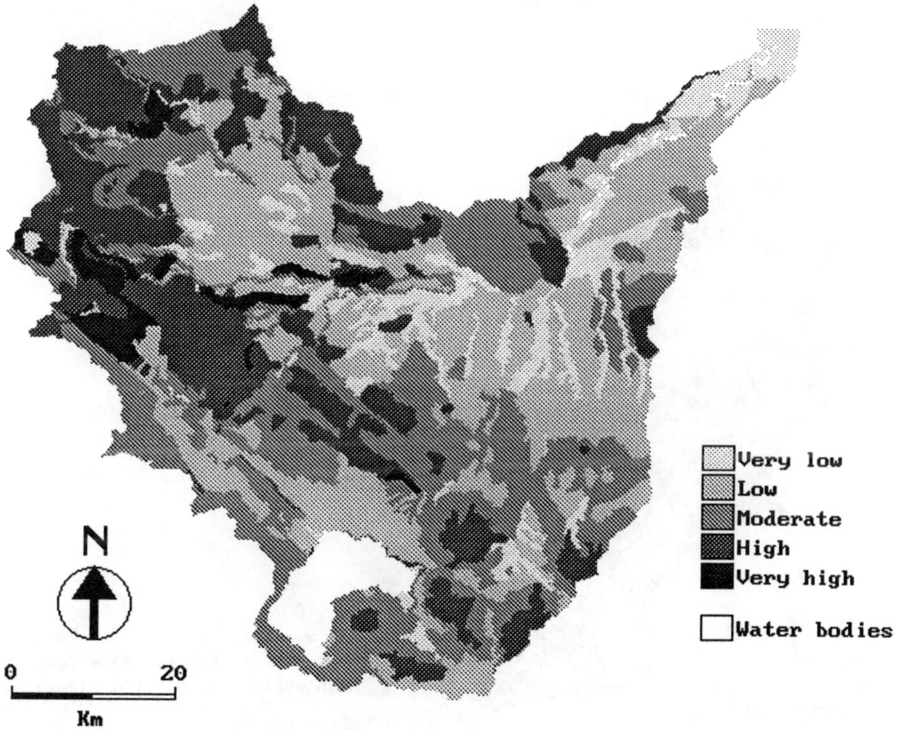

Figure 4c. Erosion hazard under coffee, Komering watershed

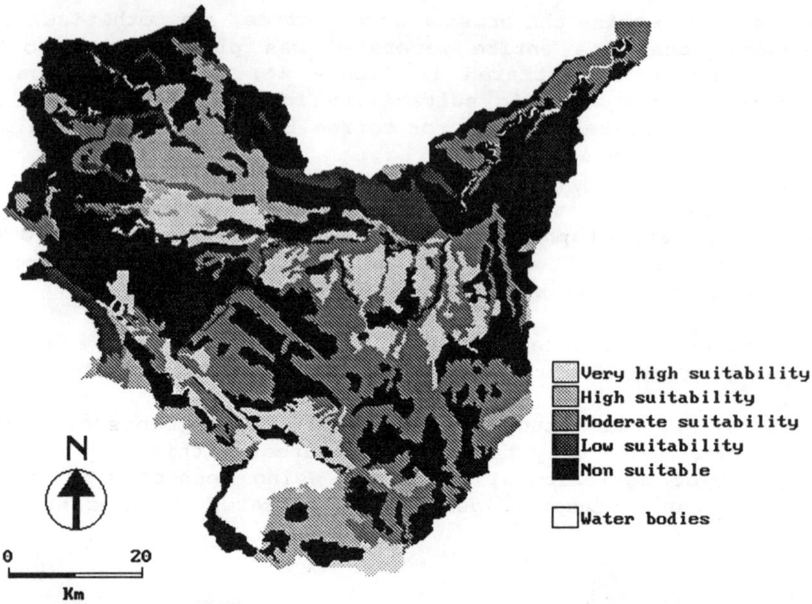

Figure 4d. Land suitability for coffee, Komering watershed

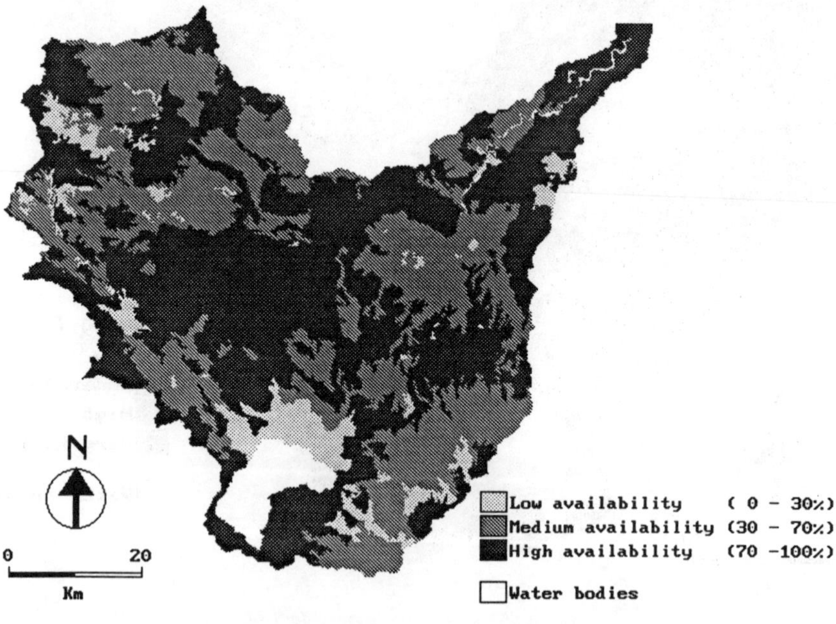

Figure 4e. Land availability for coffee, Komering watershed

10. Combining the suitable and available map and the accessibility map, all the areas where coffee can be grown are obtained and visualized.

11. The land-use situation for the year 2000 was simulated. Surveys and statistical analysis were used to assess trends in population growth, dynamic changes in farming systems, minimum farming size to predict future land demand. This was then compared to the availability of land suitable and accessible for coffee. Figure 4g, illustrates the areas available for coffee between 1986 and 2000.

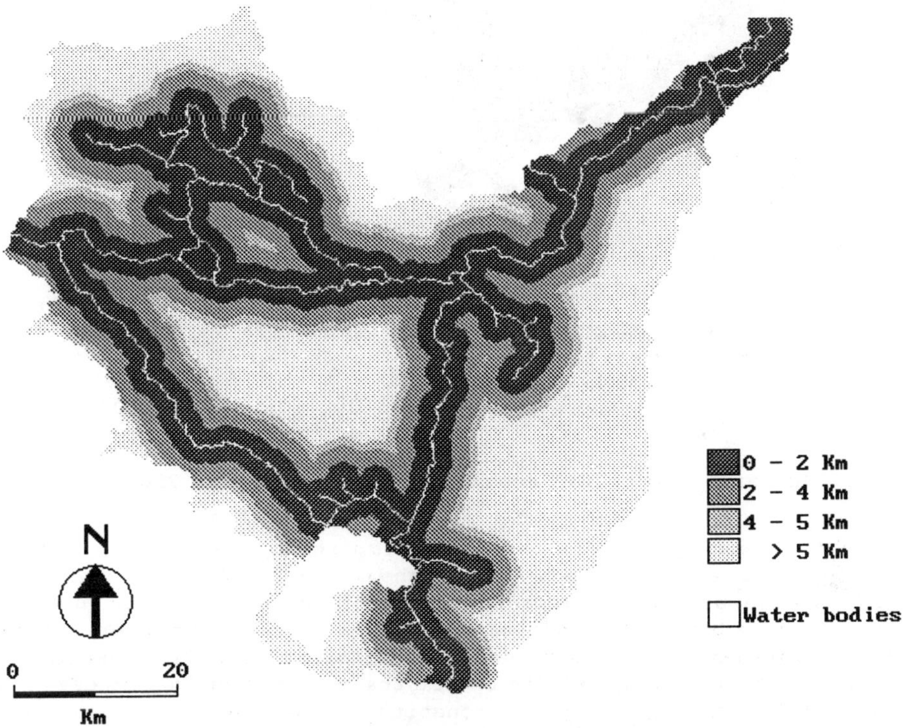

Figure 4f. Distance map, Komering watershed

5. Concluding remarks

The use of geographic information systems for planning natural resources is becoming increasingly more common. ILWIS constitutes an effective interactive system for spatial analysis. It provides enough analytical tools to effectively handle spatial and non-spatial databases, in addition to outside models.

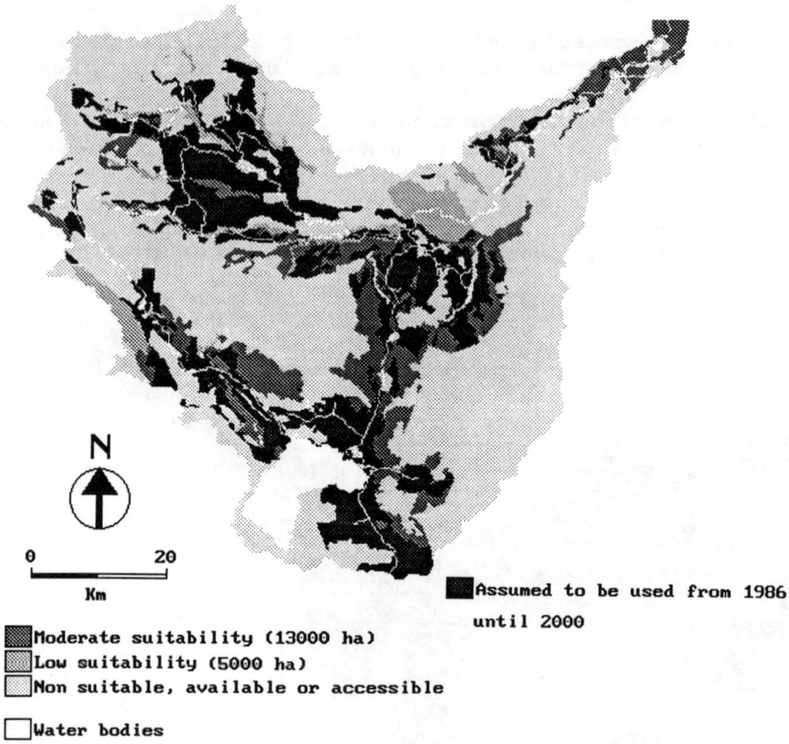

Figure 4g. Scenario for coffee expansion from 1986 until 2000

Building the geographic database is an important task that has to be properly planned and executed. Care should be taken in obtaining a database that is accurate, reliable, appropriate, complete and updated. Handling of these data with a GIS can be very efficient if properly done.

The use of simulation models to develop different scenarios and provide alternatives to decision makers is one of the main advantages of GIS. The land use model shown in this paper constitutes one example of the use of GIS for land use planning. Several other applications in this field and in other earth resources disciplines are also possible.

The nature of the system, PC-based, constitutes a limitation to handle large amounts of data, especially if working with large projects or at a national level. However, being an analytical tool, ILWIS benefits of a low cost platform which can be used in decentralized laboratories even in unfavoured situations such as those found in developing countries.

In spite of the accessibility to processing offered by the user-friendliness of the system described here, the user should be remained that the use of models or rule bases should always incorporate validation procedures and an assessment of the accuracy and reliability of the results. The next phase of the study incorporates the monitoring of the simulation model. Efforts are being made towards a cost-benefit study which will evaluate the financial benefits of using a low-cost PC-based GIS.

References

Bouma, J. (1988). Land qualities in space and time. Symposium Land Qualities in Space and Time. Wageningen.

Burrough, P.A. (1986). Principles of Geographic information systems for land resource assessment. Clarendon Press, Oxford, 194 p.

Brink, A.B., A. Mabut, R. Webster and P.H.T. Beckett. (1966). Report of the working group on land classification and data storage. MEXE Rep No 940, Christchurch.

Date C.J. (1986). An introduction to database systems (4th ed). The systems programming series. Addison-Wesley.

Elbersen, G.W.W, Ismangun, D.S. Sutaatmadja and A. A. Solihin (1988). Small-scale soil survey and automated land evaluation. ITC Journal 1988-1, p. 51-59.

Gorte, B., R. Liem and J. Wind. (1988). The ILWIS software kernel. ITC Journal 1988-1, p. 15-22.

Meijerink, A.M.J., (1988). Data acquisition and data capture through terrain mapping units. ITC Journal 1988-1, p. 23-44.

Meijerink, A.M.J., N.H.W. Donker, H. de Brouwer and T. Iswidi. (1988). The water module: a toolbox for processing hydrologic data. ITC Journal 1988-1, p. 76-82.

Meijere, de J.C., B. Mardanus and A. van de Kasteele. (1988). Land use modelling for the upper Komering watershed. ITC Journal 1988-1, p. 91-95.

Putte van der R. (1988). Farming systems and land use modelling for watershed management. ITC Journal 1988-1, p. 83-86.

Valenzuela, C.R. (1988). Soils geography. ITC Journal 1988-1, p. 45-50.

Wijngaarden, van W., A. Kooiman. (1988). CUMU: The land cover and land use database. ITC Journal 1988-1, p. 60-66.

Wood, S.R. and F.J. Dent. (1983). LECS methodology. Ministry of Agriculture, Government of Indonesia/FAO-AGOF/INS/78/006, manual 6.

Zabel, P., B. Mardanus, N. Surayana, T. H. Widiananto and J.C. de Meijere. (1988). Farming systems in the upper Komering watershed. ITC Journal 1988-1, p. 87-90.

Soil geographic database: structure and application examples

J Alfred Zinck* and Carlos R Valenzuela*

Reprinted from ITC Journal 1990-3, pp 270-293

ABSTRACT

The design of a soil database—an important component of a geographic information system—involves the description of both the spatial and non-spatial data. Individual soil map unit delineations are stored in the spatial database, while external and internal features of soil units are stored in the non-spatial database. Geomorphic criteria, implemented for delimiting the map unit boundaries, also play a fundamental role in the identification and characterization of soil distribution patterns and the understanding of soil variability. This article describes the conceptualization of a geoform-soil complex and its implementation for creating a soil database model. Database components, entities and analysis operations are outlined. Existing classification systems are used to identify the geomorphic units and their soil components. Conventional rules for determining soil map units are presented. Two application cases (in Venezuela) illustrate the use of the database for automated interpretation of soil data at different spatial levels and for different purposes. ILWIS was used to implement the database and process the application cases.

Through soil survey, large quantities of data are gathered from different sources, *ie*, photo and image interpretation, field observations and determinations, and laboratory analyses. Efficiency in storing and processing these primary data, and in displaying the derived soil information, is considerably increased when using computer-assisted procedures operating on appropriately structured databases.

Early soil databases aimed essentially at the storage, statistical treatment and retrieval of single or combined soil attributes. The advent of geographic information systems (GISs) led to the introduction of the spatial component in soil databases, improving the conventional approach to soil cartography by allowing the combination of ancillary data with the digital processing of remote sensing documents, and also stimulating new outlooks through, for example, computer-assisted and automated soil mapping.

Soils are recognized as three-dimensional bodies on the landscape (soilscape) whose spatial distribution and variability are strongly controlled by, among other features, the geomorphic factor. Modern soil surveys make intensive use of geomorphology for the delimitation of natural soil bodies and explanation of soil genesis. In fact, the identification of soil geographic distribution patterns can be reliably inferred from the structure of the geomorphic environments, depositional as well as erosional ones. This intimate association between geoforms and soils derives from the specific position of both at the Earth's surface, along the interface between lithosphere and atmosphere, as modified by energy and matter fluxes controlled by biotic and hydrologic factors. Approaches to soil cartography based on the unique geoform-soil relationships have blossomed under different labels, *eg*, morphopedology [9, 16] and geopedology [26]. This interdependence has also been recognized as a fundamental concept for structuring soil databases or soil-terrain databases (*eg*, ILWIS, SOTER), and for determining the basic spatial units (polygons as individual delineations of map units) to be introduced into the graphics modules of soil databases.

A soil (geographic) database is a tool for the capture, storage, processing and display of soil data and information generated by soil surveys from original sources. Displayed information carries an aggregated value in relation to primary data because of the capabilities offered by the system for analysis, modelling and graphic combinations, but the output information is intrinsically only as good as the input data. Quality control of primary data is therefore a major concern to

* Department of land resource surveys and rural development, ITC.

A. S. Belward and C. R. Valenzuela (eds.),
Remote Sensing and Geographical Information Systems for Resource Management in Developing Countries, 443–475.

guarantee a reliable implementation of interpreted soil information for land use planning. The use of a soil database in a GIS, however, considerably enhances the possibilities of satisfying users' needs through the capabilities of data processing, automated data interpretation, map overlaying and building decision alternatives (scenarios).

The present article aims to outline the structure of a soil database using geomorphic entries. The implementation of the database through ILWIS for automated interpretation of soil data at different spatial levels (*eg*, local, regional) and for different purposes (*eg*, agricultural, engineering) is demonstrated using two Venezuelan case studies.

SOIL DATABASE STRUCTURE

CRITERIA AND CONCEPTUAL MODEL

Typically, soil survey data are derived from three sources (see Figure 1): (1) visual interpretation and digital processing of remote sensing documents, including aerial photographs, radar and spectral images; (2) field observations and (instrumental) determinations of biophysical features, as well as observations of social and economic features; (3) laboratory analyses of mechanical, physical, chemical and mineralogic properties.

The relative importance of these three sources varies according to the scale and the purpose(s) of the survey. Generally speaking, the larger the scale of the final soil map, the more field observations and laboratory determinations are required to ensure an appropriate level of information.

Since soils as well as geoforms are three-dimensional bodies, external and internal features (in relation to the Earth's surface) are described, measured and classified with the purpose of establishing and delimiting soil map units.

By combining information from sources 1 and 2 (above), environmental site conditions and areal dynamics (erosion, flooding, land use changes, etc) are described, and map unit boundaries are delineated. Here, the implementation of geomorphic criteria through photo and image interpretation and field prospecting plays a fundamental role for the identification and characterization of soil distribution patterns and the comprehension of spatial soil variability.

By combining sources 2 and 3, properties of soil materials, geomorphic cover formations and geologic substrata are characterized and quantified. The horizon—a vertical subdivision of a soil profile with lateral extension—is the basic unit of data capture. Horizon and substratum information is aggregated into observation profiles, modal pedons and modal morphons. Variations among identification profiles, correlated with a given representative entity (morphon and pedon), are expressed in terms of ranges of characteristics for each taxon present in a map unit.

At this stage, the data available consist of (1) point observations with some dimensioning of the spatial variations of the soil characteristics and (2) a spatial frame delimited essentially on the basis of geomorphic surface criteria. Geomorphic soil map units (*ie*, geopedologic units) are obtained by matching the two.

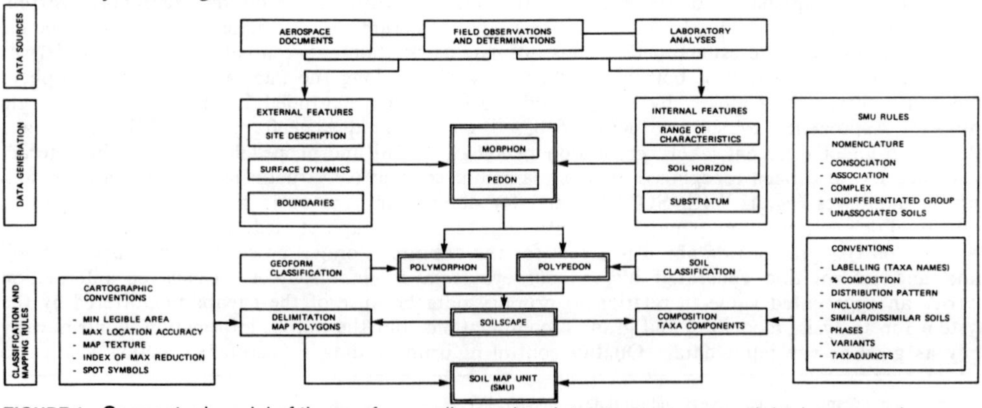

FIGURE 1 Conceptual model of the geoform-soil complex data structure, geopedologic approach

For mapping purposes, both objects—soils and geoforms—must be given identifying labels (taxonomic names) using existing classification systems. This conventional approach of classifying before mapping is adopted here [1]. Assemblages of contiguous pedons, forming polypedons, are classified by comparison with established taxonomic entities. A similar conceptual path was designed for the classification of geomorphic units, allowing progression from the description and sampling entity (morphon) to the classification entity (polymorphon).

A basic geomorphic unit (polymorphon) can host one or more polypedons. For example, Entisols and Inceptisols can occur intermingled in the same recent alluvial levee position. The combination on the landscape of such a polymorphon with its associated polypedons results in a soilscape unit.

Because of the inherent spatial anisotropy of soil material, more accentuated than that of geomorphic material, soil delineations are seldom totally homogeneous, requiring that the taxa components be identified and their proportions quantified using conventional soil mapping rules [17, 23]. Similarly, the delimitation of polygons obeys a set of cartographic conventions ensuring proper legibility of the soil map [6]. Thus cartographically and taxonomically controlled soilscape units, as unique combinations of geomorphic polygons with their corresponding soil content, result in soil map units (SMUs).

From this conceptual model of the geoform-soil complex (Figure 1), a database model can be created and a database structure can be designed.

STRUCTURAL COMPONENTS

A typical soil database (SDB) comprises a set of components, each represented by one or more files [2]. The structure of the database, as well as the nature and number of files, can vary according to specific purposes, but some main components are always found in any type of setup to cover the successive steps of data aggregation from horizon to map unit. The present database model comprises seven main structural components which are briefly introduced below (Figure 2).

Geoforms

Geoforms, considered as spatial and temporal soil formation frames and as soil cartographic "containers", are classified according to a taxonomic system with six hierarchic levels of ordination

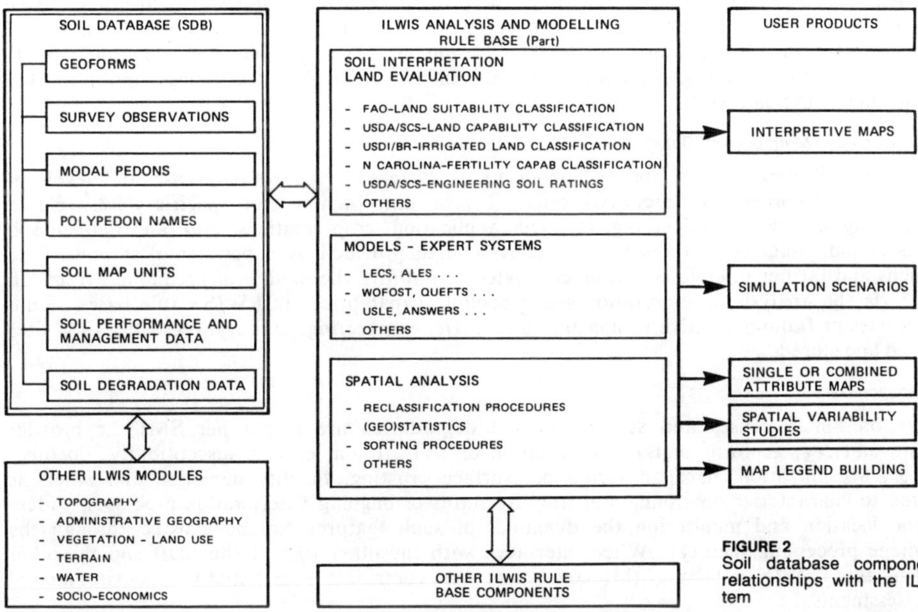

FIGURE 2
Soil database components and relationships with the ILWIS system

[26]. Improvements in the definition and hierarchy of diagnostic attributes, with differential weights according to geoform classes, will allow the geoform component to operate in the future as a sort of expert system. Splitting into several files or even into separate components is envisaged to accommodate the concepts of morphon and polymorphon, similar to the pedon and polypedon components.

Survey Observations

Routine point data gathered from auger holes, minipits and full-size pits during the soil identification and systematic mapping phases of a conventional soil survey are assembled in a permanent record of site observations. Horizon, whole profile and site data are registered in separate files. This information is used for selecting representative pedons and establishing the variation ranges of characteristics. Primary data capture using small field computers and standard description sheets [3] can be easily interfaced with the ILWIS data input facility. Coding of soil properties and horizon designations is according to Valenzuela [22].

Modal Pedons

This component concentrates the maximum range of detailed point data for each representative pedon. Complete profile and site descriptions and analytic data are recorded, covering altogether environmental features, morphologic, physical, chemical, mineralogic and biologic soil characteristics, as well as local soil uses and agronomic management histories. The USDA pedon coding system is used [11, 18].

Polypedon Names

Recognized and clearly established soil bodies (polypedons), named according to national or international taxonomic classification systems (*eg*, FAO [5] or USDA-SCS [19]), are recorded at any categoric level per project or study area. Regional and national soil correlation schemes can be developed from such partial soil catalogues.

Soil Map Units

The component hosting the soil map units forms the core of the database. Information on the kind of map units, taxa components, composition percentages, distribution patterns, ranges of characteristics, phases and inclusions are stored here for each SMU and SMU delineation. Two main files thus operate within this component: (1) an attribute file corresponding to the listing of all SMUs present in a project area or broader perimeter; (2) a graphics file containing SMU delineations as digitized polygons. Both files are interfaced among themselves and all other component files of the database.

Soil Performance and Management Data

Data on soil performance and behaviour under different types of land use and levels of management are recorded for benchmark soils. Typical data include crop specific yields, related to identified agricultural practices (*eg*, fertilizer application, crop rotations, etc), and defined soil map units or individual taxa components. Relevant data provided by experimental stations, field observations and farmer interviews can be correlated to improve the quality of predictions from soil properties via the analysis, interpretation and modelling capabilities of ILWIS's rule base. Similarly, successes or failures relative to non-agricultural (*ie*, engineering, sanitary, recreation) soil uses are collated and stored.

Soil Degradation Data

Data on kinds, causes and severity of soil degradation are stored per SMU or broader geographic unit (*eg*, geoform units). Information on soil erosion, erosion susceptibility, flooding, salinization, alkalinization, hardpan formation, surface crusting, fertility depletion and others is highlighted to characterize environmental fragility status or ongoing deterioration processes. Identifying the location and monitoring the dynamics of such features can be assisted through the digital image processing facility. When interfaced with the other files of the SDB and the other modules of the ILWIS database, this component can contribute substantially to environmental impact assessments.

The soil database acts as an independent module which can be easily interfaced with all other modular constituents of ILWIS, such as topography, administrative geography, vegetation and land use, terrain, water (hydrology-climate) and social and economic data. SDB benefits also from the data manipulation and transformation facilities of ILWIS for digital image processing, map overlays, statistical treatment, and modelling via the Oracle database management system (DBMS) [21].

Some of the components are described in the following sections, in particular as they concern fundamental aspects of database design, classifying geoforms, establishing soil map units and soil interpretation.

DATABASE DESIGN AND DATA ANALYSIS

The logical design of the database is illustrated in Figure 3, which shows the entities included and their relationships. Since many-to-many relationships are not allowed in a logical data structure of a relational database, one combination entity matching SMU and polypedon had to be created to guarantee one-to-one or one-to-many relationships.

The entities were created and stored in the Oracle relational DBMS [21]. The pedon and horizon data were coded using the USDA system [18] to allow data interchange with other users.

The soil data needed for spatial analysis, once stored in the database, can be retrieved as an ASCII file using the facilities provided by the relational DBMS and converted into tabular format using the ILWIS conversion facilities. Conversely, tabular data can be converted into the database format.

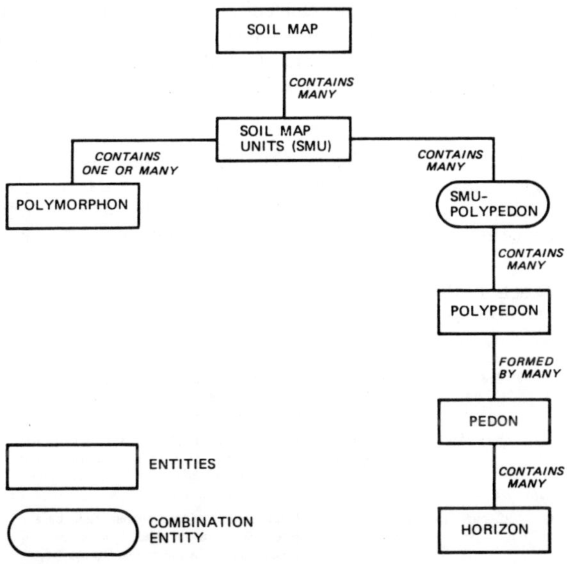

FIGURE 3 Logical design of the soil database

The "working dataset" can be analyzed, updated and deleted using the tabular database facilities. The tabular data are also used to create attribute maps using reclassification procedures which do not involve the generation of new polygons but rather an aggregation of existing polygons (map units, delineations), or interpretive maps using rules and/or existing criteria. Figure 4 shows the scheme for data integration and spatial analysis.

Since Oracle is no longer an integral part of ILWIS [21] and if no other relational database is available, the database can be designed and stored in the ILWIS non-spatial database. Although the latter is not a truly relational DMBS, it can perform most of the operations typical of a relational database. Some common characteristics of database management information systems,

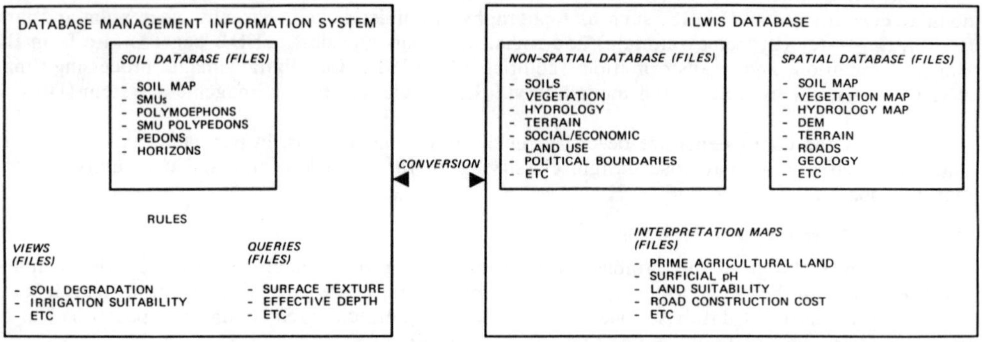

FIGURE 4 Data integration and spatial analysis

such as data independence, data dictionary, monitoring and tuning, are not yet present in the ILWIS non-spatial database.

GEOMORPHIC ENTRIES

If geomorphology is assumed to be one of the driving soil forming factors and soil mapping criteria, concepts provided by this discipline can conveniently be used for soil data and database structuring. The use of computer-assisted procedures and the requirements of data systems, however, impose on geomorphology new constraints for scientific systematization as to geoform classification, terminology and definition of diagnostic attributes.

System Structure

Using the principles of hierarchy theory [7], an attempt was made to build a hierarchic classification system of geoforms comprising six categoric levels, each defined by a specific generic concept, as shown in Table 1 [26]. Using the existing body of knowledge in geomorphology, identified taxa taken from the general literature were assigned to the various categories within the

TABLE 1 Synopsis of the geoform classification system (from [26])

Level	*Category*	*Generic concept*	*Short definition*
6	Order	Geostructure	Large continental portion characterized by a broad geologic structure (*eg*, cordillera, geosynclinal basin, shield)
5	Suborder	Morphogenetic environment	Broad type of biophysical medium originated and controlled by a style of internal and/or external geodynamics (*eg*, structural, depositional, erosional, etc)
4	Group	Landscape	Large portion of land characterized by a repetition of similar relief types or an association of dissimilar relief types (*eg*, valley, plateau, mountain, etc)
3	Subgroup	Relief/molding	• Relief as determined by a given combination of topography and geologic structure (*eg*, cuesta, horst, etc) • Molding as determined by specific morphoclimatic conditions or morphogenic processes (*eg*, glacis, terrace, delta, etc)
2	Family	Substratum	• Lithology of hard rocks (*eg*, gneiss, limestone, etc) • Facies of soft cover formations (*eg*, periglacial, lacustrine, alluvial, etc)
1	Subfamily	Landform	Conspicuous basic geoform type, characterized by an unique combination of geometry, dynamics and history (*eg*, levee, dune, solifluction lobe, backslope, etc)

system according to their compatibility with the required level of abstraction. Because of the lack of consensus in the reference sources used, some classes are still tentative as to their definition, naming and hierarchic position in the system.

Diagnostic Attributes

To make the system operational, one key task is the selection, definition and hierarchization of diagnostic attributes used for both the setup of taxonomic class limits and the practical identification and classification of geoforms by comparison with established taxa at any categoric level. To that end, four classes of attributes are used (Table 2):
(1) morphographic attributes (describing the geometry of geoforms)
(2) morphometric attributes (measuring the geoforms)
(3) morphogenic attributes (determinig the origin and evolution of geoforms)
(4) morphochronologic attributes (circumscribing the temporal context)

In terms of identification and classification of geoforms, attributes 1 and 2 have a rather low differentiating power. They are used preferentially at higher categoric levels and mainly for descriptive purposes. In contrast, attributes 3 and 4 are highly diagnostic but need more field and laboratory support. They are preferentially used at lower categoric levels. This attribute set can be implemented in an expert system for identification and classification of geoforms.

TABLE 2 Attributes used for identifying and classifying geoforms (from [26])

MORPHOGRAPHIC ATTRIBUTES – GEOMETRY OF GEOFORMS
Topography
Shape of the Topography
Profile of the Topography
Exposure (aspect)
Planimetry
Configuration of Geoforms
Contour Design of Geoforms
Drainage Pattern
Surrounding Conditions (connectivity)
Bordering Unit (adjacency)
MORPHOMETRIC ATTRIBUTES – DIMENSION OF GEOFORMS
Relative Altitude (relief amplitude, internal relief)
Valley Density (drainage density)
Slope Steepness
MORPHOGENIC ATTRIBUTES – DYNAMICS OF GEOFORMS
Particle Size Distribution
Structure (geogenetic and pedogenetic)
Consistence (Atterberg limits)
Mineralogy
Morphoscopy (exoscopy)
MORPHOCHRONOLOGIC ATTRIBUTES – HISTORY OF GEOFORMS
Absolute age
Radiometric Techniques, Dendrochronology, Tephrochronology, Varve Counting, Analysis of Historic and Prehistoric Events
Relative age
Pollenanalysis, Cultural Artefacts, Pedostratigraphy
Pedostratigraphy
Weathering Degree of Parent Material
Degree of Soil Morphologic Development
Leaching Indices
Status of Adsorption Complex
Clay Mineralogy

System Applications

The geoform classifcation system has been actively used for delimiting soil units. It allows also for systematic structuring of soil map legends, reflecting the strong relationship between soils' and geoforms' formation and geographic distribution (Figure 5). Moreover, soil interpretation maps can use the hierarchic geomorphic frame as legend entry. This allows easy generalization of information from lower to higher levels according to the application scales and purposes, *eg*, land use planning at local, regional or national levels (as will be shown through the two case studies).

Because it plays a leading role in soil survey, especially in providing the geographic reference units, *ie*, the cartographic frame for soil mapping in terms of relatively homogeneous delineations, it is judicious to use the geomorphic factor for determining the main entries to the data structure.

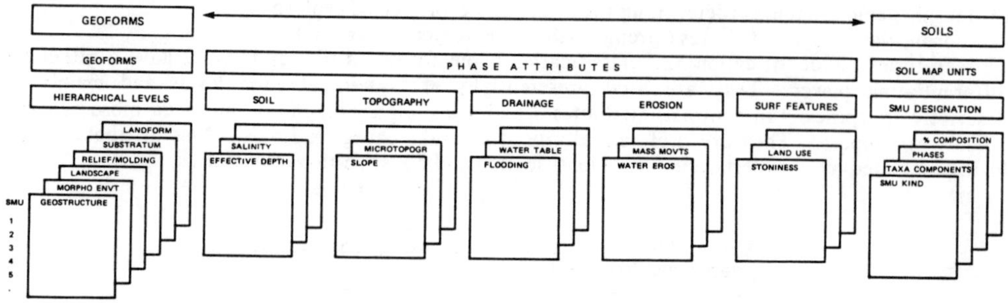

FIGURE 5 Soil map legend structure

SOIL MAP UNITS

From Taxonomic to Cartographic Units

Pedon and soil map unit are, respectively, the initial and final rings of a conceptual and operational chain linking soil taxonomy to soil cartography (Figure 6). The pedon (*ie*, soil profile) is the starting entity, in which description and sampling of soil material take place and whose position on the land surface depends on the surveying technique used for site selection (*eg*, random, systematic, oriented). A cluster of adjacent and similar pedons, fitting within the class limits of a taxonomic unit, is a polypedon. This is the classification entity for which the range of variation of the most important properties is established. An association of similar or distinct soil bodies (polypedons) in a given landscape unit (in the broad sense) forms a soilscape entity. Soilscape units are represented cartographically as soil map units. Each single occurrence of a given soil unit on a map is called a "soil delineation"—an individual polygon.

Soil map units are established on the basis of rules [6, 17, 23] which allow for selecting the appropriate kind of SMU (nomenclature) and provide the SMU identification criteria (conventions). Five SMU types are commonly used in soil mapping: consociation, association, complex, undifferentiated group and unassociated soils [17]. The degree of heterogeneity (impurity) of SMUs increases from consociation to unassociated soils. Accuracy of predictions in terms of crop

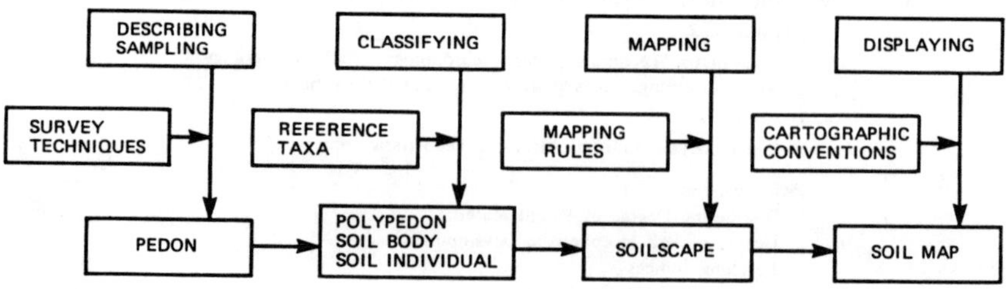

FIGURE 6
Soil mapping chain

selection, soil management and land use decreases in the same direction. The SMU nomenclature applies to any categoric level of the soil taxonomy system used.

Conventions for Establishing SMUs

A soil map unit is determined by means of three basic composition requirements: (1) the names of the taxa components; (2) the proportion of each taxum component in terms of surface area within the map unit; (3) the geographic distribution pattern of the taxa components (Figure 1).

Taxa sharing limits of diagnostic properties in the classification system used are considered as similar in terms of use, management and behaviour (*eg*, Mollic Ustifluvents and Fluventic Haplustolls). The presence of similar soils increases the degree of homogeneity of SMUs. Conversely, the occurrence of large amounts of dissimilar soils, having no common taxonomic boundaries, leads to an opposite effect, therefore differently affecting soil potentials.

Smaller quantities of similar or dissimilar soils are handled as cartographic inclusions or impurities. For dissimilar soil inclusions, the proportions allowed vary according to whether the inclusions are limiting or nonlimiting for soil use and management.

Properties considered as taxonomically non-diagnostic but relevant for soil use and management are expressed by means of phases. These are functional subdivisions of taxonomic units at any categoric level of the classification system, created for and adjusted to fit the practical objectives of each soil inventory.

Variants and taxadjuncts are cartographic concepts used only at series level for the provisional mapping of soil bodies which are either not extensive enough (*ie*, variants) or not different enough from recognized series (*ie*, taxadjuncts) to warrant the establishment of new taxa.

Nomenclature of SMU Classes

Matching these conventional criteria with the SMU compositional requirements is conducive to the selection of the best fitting SMU kind in each situation. The consociation, for example, is the most homogeneous SMU type, in which 75 percent or more of any given map delineation corresponds to a single taxon (or miscellaneous area) and similar soils. Nonlimiting dissimilar inclusions may not exceed 25 percent of the delineation and limiting ones not more than 15 percent.

Increasing amounts of dissimilar soils included in an SMU and increasing fuzziness or complexity of the soil distribution patterns result in SMU classes with more heterogeneity than the consociation. The main differences are reported in Table 3. The steps in establishing the map

TABLE 3 Kinds of map units and differentiating criteria (from [23])

Kinds of map units	Main taxa components			Dissimilar inclusions %			Distribution pattern criteria	Land use and management of main components	Mapping scale	Soils giving names
	Number	Degree of similarity	Dominant soil %	Non-limiting	Limiting	Limiting contrasting				
Consociation	one	similar	≥ 50	≤ 25	≤ 15	≤ 10	known and defined pattern main components present in each delineation separately mappable at map scale	similar	any	one soil
Association	two or more	dissimilar	≤ 50	≤ 25	≤ 15	-	same as consociation	dissimilar	any	two or more
Complex	two or more	dissimilar	≤ 50	≤ 25	≤ 15	-	same as association except not separately mappable at map scale	dissimilar	large	two or more
Undifferentiated group	two or more	dissimilar	variable	≤ 25	≤ 15	-	undefined pattern main components not present in each delineation	similarly restricted	smaller	two or more
Unassociated soils	two or more	dissimilar	variable	> 25	> 15	-	unknown pattern main components not present in each delineation	dissimilar	very small	two or more

units, either manually or through a computer- assisted procedure, are illustrated in Figure 7.

SOIL INTERPRETATION AND LAND EVALUATION

Multipurpose Use of Soil Survey Information

Any human activity having some physical expression on the landscape uses soil as a resource. Soil information may therefore be included in land use planning for many purposes [24]:

(1) agronomic purposes, considering the soil as a medium for cultivated plant growth, and thereby influencing the possibilities to increase the production of food and raw materials

(2) engineering purposes, considering the soil as a source of construction materials, or as a support for load- bearing engineering structures, and thereby influencing urban expansion, industrial site selection and transportation route development

(3) sanitary purposes, considering the soil as a medium for storage or filtration of urban, industrial or agricultural wastes, and thereby influencing the disposal or recycling of residues

(4) recreational and aesthetic purposes, considering the soil as a support for leisure areas or as a landmark, and thereby influencing the possibilities to improve the quality of human life

Soils thus effect many human activities. Some soils are simultaneously appropriate for more than one of these purposes. This inevitably leads to conflicts, in particular between agricultural and non- agricultural soil allocations. Choices therefore have to be made among various land use alternatives, frequently leading to the sacrifice of future agricultural land in favour of more lucrative engineering uses. The availability of soil information will not necessarily contribute by itself to avoiding such inappropriate practice. There must still be efficient transmission of the information to the users, who must be willing to implement it.

Even nowadays, soil information varies widely in quality and quantity from country to country. Especially in developing countries, the nature, location and value of the soil resource over large regions are still unknown. Available information is frequently inappropriate when compared with the objectives of land use planning, in particular at local and regional levels. Simultaneously, the development of soil databases, geographic information systems and modelling is creating an increased need for more and improved soil data.

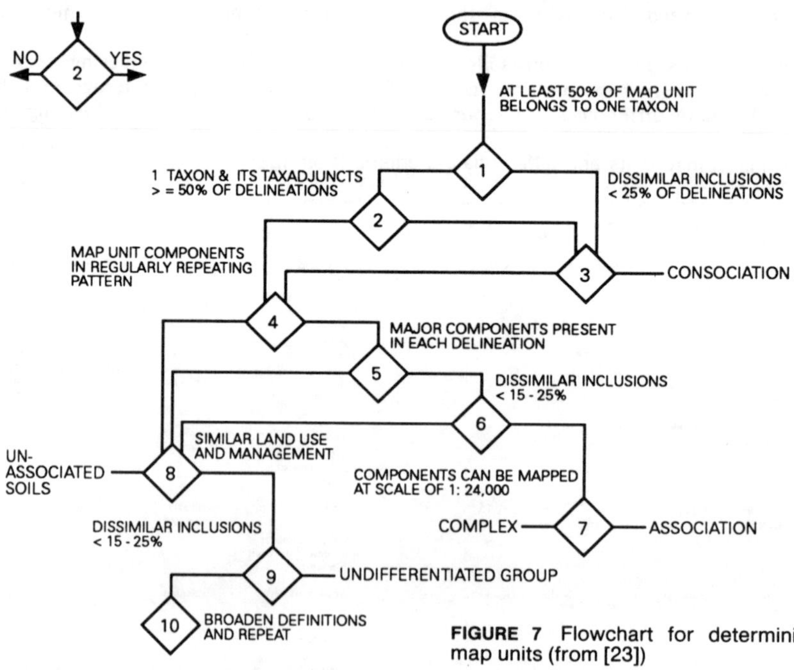

FIGURE 7 Flowchart for determining kinds of map units (from [23])

Reference Systems for Soil Survey Interpretation

Soil information is generally transferred to users via soil interpretation and land evaluation. Since land is a broader concept than soil, soil interpretation and land evaluation are not synonymous; they are, however, frequently used interchangeably. In practice, land evaluation is more commonly employed when dealing with agricultural purposes and soil interpretation when concerning engineering applications.

As opposed to natural systems for scientifically classifying objects occurring in nature (*eg*, plants, rocks, soils), land evaluation and soil interpretation use technical systems designed for practical application fields. Some of the latter are of general purpose (*eg*, FAO land suitability classification [4]; USDA-SCS land capability classification [10]). Others have a more specific orientation (*eg*, North Carolina's fertility capability soil classification [14]). A main difference is whether land is evaluated for specific land utilization types (*ie*, specific crops, as by the FAO system) or for more comprehensive land use classes (as by the USDA-SCS system). The concepts of land suitability and land capability are used to highlight this difference. Following the route traced by the California Storie index [15], parametric approaches have been developed to introduce more rigor in data manipulation and to quantify land evaluation [13]. Approaches based on productivity indices and crop yield prediction through modelling have a similar purpose [8, 12]. (Many of the current land classification and evaluation systems are now available in computerized versions.)

Soil Survey Interpretation Using ILWIS

In ILWIS, the soil database (as noted above) functions as a module coupled to other modules (topography, terrain, water, vegetation and cover, etc) and to the system's rule base. This allows processing and transformation of data into user-friendly information through interpretation schemes, modelling or simple reclassification (Figure 2).

For the application case studies treated below, two land classification systems were implemented for agricultural and engineering interpretation.

Irrigated Land Classification Using the specifications provided by the irrigated land classification manual [20] and adjustments tested in Venezuela in applying it for land classification in irrigation schemes, a rule-based system was built as shown in Table 4 [24, 31].

Six irrigated land classes were recognized. Arable land was classified in the first four classes in order of decreasing suitability. From the arable land, irrigable areas were selected on the basis of water availability and distribution facilities. Class 5 was tentatively not arable, pending technical and economic feasibility studies, *eg*, concerning water distribution from the source to the land. Unsuitable land was grouped into class 6. Three subclasses were established on the basis of soil, topography (including erosion) and drainage limitations (Figure 8). Management units

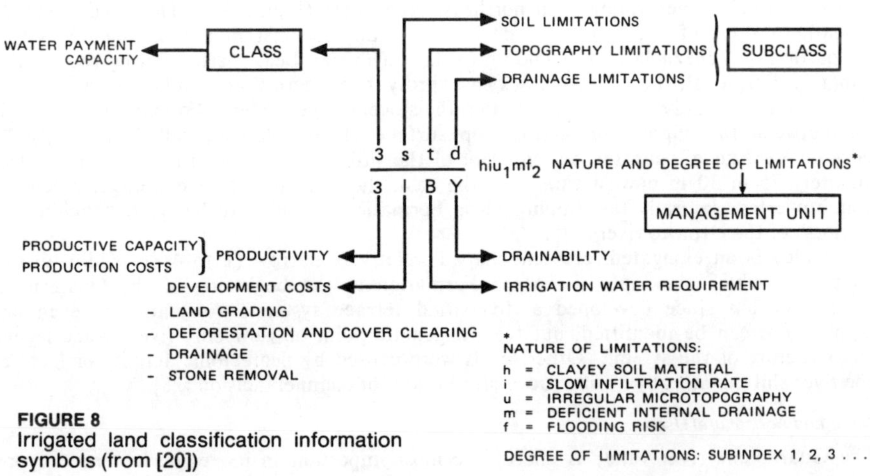

FIGURE 8
Irrigated land classification information symbols (from [20])

reflected practices required to overcome non-permanent recurrent limitations. Production costs incurred as a consequence of the former were combined with the natural soil productive capacity to make up the overall soil productivity. Permanent limitations were removed by terrain conditioning, *ie*, deforestation and clearing, grading, artificial drainage and stone removal, and expressed as development costs. Additionally, estimates of drainability and irrigation water requirements were specified for each class. The overall meaning of an irrigation class is a quantified expression of the water payment capacity of any classified tract of land.

Diagnostic criteria used for determining irrigated land suitability classes are listed in Table 4, together with quantified class limits corresponding to general irrigated crop requirements. This table was introduced in the ILWIS rule base to allow automated matching of irrigated crop requirements with the land qualities offered by each soil map unit.

Soil Interpretation for Engineering Land Uses Similarly, diagnostic criteria for determining land suitabilities for selected civil and sanitary engineering works were incorporated in the ILWIS rule base, using the USDA-SCS requirement tables designed for engineering soil interpretation [17]. Soil properties and environmental conditions involved in the land utilization types selected for the application case studies are listed in Table 5. An example of the engineering rating procedure is given in Table 6.

The Output Products The ILWIS analysis and modelling facilities applied to the primary and processed data stored in the soil database generate such user products as interpretation maps, land use simulation scenarios for decision making, and several spatial analysis products (*eg*, soil attribute maps, spatial variability of soil properties, automated legend building, etc). Examples are included below.

EXAMPLE 1: GUARAPICHE RIVER VALLEY, VENEZUELA

OBJECTIVE AND ASSUMPTIONS

The basic objective of the case study was to carry out agricultural and engineering soil interpretations for a simulated irrigated land development plan, using a computerized soil database within the context of a GIS. Fundamental assumptions regulating the stepwise development of the scenario were:

- application and effectiveness of river flood control measures aiming at the protection of fertile alluvial soils
- optimization of available irrigation water (water storage by dam-reservoir, selective water distribution, minimization of water distribution costs)
- preservation of prime agricultural land from engineering uses

REGIONAL SETTING

The Guarapiche river valley is in northeast Venezuela (Figure 9). The headwaters develop on the southern slope of the Cordillera de la Costa, an east-west mountain range bordering the northern portion of Venezuela and extending to the Caribbean Sea. After exiting the cordillera at approximately 300 m asl, the valley crosses eastwardly, over more than 120 km, a low-lying plateau formed by alluvial sandy sediments of Plio-Pleistocene age (Mesa Formation). The relative elevation between the slightly undulating top surface of the plateau, called regionally "Mesas Orientales de Venezuela", and the bottom floor of the valley varies from 140 to 150 m upstream to approximately 25 to 30 m downstream. To the east, the river has formed a large, recent alluvial plain which overtops laterally the dipping Mesa Formation and plateau before debouching into the Atlantic north of the Orinoco river delta.

The valley is an elongated, crescent-shaped, relatively narrow (1 to 4 km wide), dissymetric tectonic graben which sank into the Mesa Formation at the beginning of the Quaternary, and where the river has since developed a diversified terrace system. As many as seven or eight topographic steps can be identified, but from a genetic point of view only five terrace levels form the basic structure of the sytem. Other levels were caused by neotectonic activity or local erosion when the river shifted laterally during successive periods of channel incision [28].

Land, Water and Agricultural Uses

The Guarapiche river valley is one of the most important in its region in terms of presence

TABLE 4 Diagnostic criteria for determining irrigated land suitability classes (from [31])

Land qualities and characteristics	*CLASS 1 highly suitable*	*CLASS 2 moderately suitable*	*CLASS 3 marginally suitable*	*CLASS 4 suit for spec uses*	*CLASS 6 not suitable*
		TOPOGRAPHY - COVER			
GENERAL SLOPE %	0.1-3	0.1-5	0.1-8	< 16	> 16
MICROTOPOGRAPHY					
1 Degree of microrelief irregularity (rills, gullies, flood channels, etc)	Even to slightly irregular	Till moderately irregular	Till strongly irregular	Till very strongly irregular	Excessively irregular
2 Terrain grading requirements (2.5-3 Bs/m^3)*					
(a) Ground movement	< 500 m^3/ha	< 1000 m^3/ha	< 1500 m^3/ha	Grading not feasible, requiring sprinkler or drip irrigation	Excessive grading cost
(b) Average cut and fill magnitude	< 10 cm	< 20 cm	< 30 cm		
FIELD SIZE AND FORM (for irrigation pattern design)					
1 Smallest field size	4 ha	2 ha	1 ha	1 ha	< 1 ha
2 Shortest irrigation furrow length	150 m	100 m	50 m	50 m	< 50 m
SURFICIAL STONINESS					
1 Stoniness class	Till class 2 with mainly stones	Till class 2 with mainly stones and some boulders (< 30 cm)	Till class 2 with mainly stones and boulders (< 60 cm)	Till class 3 with mainly stones; till class 2 with mainly boulders	More stoniness than acceptable in irrigation class 4
2 Removal of stones and boulders (to depth of 25 cm) (20-30 Bs/m^3)*	⩽ 50 m^3/ha	⩽ 100 m^3/ha	⩽ 150 m^3/ha	⩽ 200 m^3/ha	> 200 m^3/ha
CLEARING OF VEGETATION COVER	Till heavy < 1200 Bs/ha*	Till very heavy < 2000 Bs/ha	Idem	Idem	-
		DRAINAGE			
SURFICIAL WATER EXCESS					
1 Flooding from streams	Without or exceptional (centennial)	Periodic (10 years)	Periodic (10 years)	Yearly (seasonal)	Permanent
2 Waterlogging from rainfalls	-	Seasonal spots	Seasonal generalized	Seasonal without outlet	Permanent
GROUNDWATER LEVEL (capillary fringe not included) Minimum depth from the soil surface during the rainy season	⩾ 120 cm	⩾ 90 cm	⩾ 75 cm	⩾ 45 cm	< 45 cm
DEFICIENT INTERNAL DRAINAGE (poor seasonal aeration)	No limitation	Slightly limited	Moderately limited	Strongly limited	Very strongly limited
DRAINAGE REQUIREMENT (ditches at parcel level: 2.5-3 Bs/m^3)*	Not required	⩽ 200 m^3/ha	⩽ 400 m^3/ha	⩽ 400 m^3/ha	> 400 m^3/ha
		EROSION			
PRESENT OR PAST EROSION	Slight sheet erosion, some shallow rills	Shallow (< 25 cm depth), dispersed rills	Moderately deep (< 50 cm), dispersed rills	Deep (< 100 cm), dispersed rills	Dense rills or gullies
EROSION SUSCEPTIBILITY	None or very slight	Slight	Moderate	Strong	Severe

TABLE 4 (cont)

Land qualities and characteristics	*CLASS 1 highly suitable*	*CLASS 2 moderately suitable*	*CLASS 3 marginally suitable*	*CLASS 4 suit for spec uses*	*CLASS 6 not suitable*
		GENERAL REQUIREMENTS			
PRODUCTIVITY	High	Moderate	Low	Limited to special crops under special management	Very low
1 Productive capacity	High	High to moderate	Moderate to low	Variable	Very low
2 Production costs	Low	Moderate to low	High to moderate	Variable	Very high
TERRAIN DEVELOPMENT COSTS* (deforestation, grading, drainage, stone removal)	Low < 1.500 Bs/ha	Moderate < 3.000 Bs/ha	High < 5.000 Bs/ha	Variable	Very high > 5.000 Bs/ha
WATER REQUIREMENTS (gross water volume per irrigation cycle)	Low to moderate < 70 cm to < 120 cm	Low till high < 70 cm to ⩾ 120 cm	Idem	Idem	-
DRAINABILITY	Good	Good to slightly restricted	Till restricted	Till poor	Not drainable
WATER CONDUCTION (technical feasibility)	Isolated position: 6i + (arable class) High position: 5h + (arable class); Very high position: 6h + (arable class)				
		SOIL			
TEXTURE					
1 Class	Sandy loam to friable clay loam	Loamy sand to very permeable clay	Loamy sand to permeable clay	Loamy sand to clay	Sand or very heavy clay
2 % clay	< 35%	< 50%	< 60%	< 80%	> 80%
EFFECTIVE DEPTH (DEPTH PHASE)	Deep and very deep	Moderately deep	Moderately surficial	Surficial	Very surficial
1 To permeable materials (sand, gravel, stone, boulder)					
(a) Fine sandy loam or finer textured, easily workable soil material	⩾ 90 cm	⩾ 60 cm	⩾ 45 cm	⩾ 15 cm	Soil shallower than depth requirements of class 4
(b) Medium or coarser sandy loam	⩾ 105 cm	⩾ 75 cm	⩾ 60 cm	⩾ 15 cm	"
(c) Loamy sand	-	⩾ 90 cm	⩾ 75 cm	⩾ 30 cm	"
2 To semi-permeable materials (non-indurated pans, penetrable and porous rocky substratum) As depth of soil material over depth of penetrable substratum	⩾ 45 cn over 150 cm	⩾ 35 cm over 120 cm	⩾ 25 cm over 90 cm	⩾ 20 cm over 60 cm	Soil shallower than depth requirements of class 4
3 To impermeable materials (hardpans, impenetrable and no-porous rocky substratum)					
(a) Full depth of soil material	⩾ 150 cm	⩾ 120 cm	⩾ 105 cm	⩾ 90 cm	Soil shallower than depth requirements of class 4
(b) Depth of soil material with at least 15 cm gravel over the impermeable substratum	⩾ 135 cm	⩾ 105 cm	⩾ 90 cm	⩾ 75 cm	"
(c) Depth of loamy sand or sandy loam	⩾ 135 cm	⩾ 105 cm	⩾ 90 cm	⩾ 75 cm	"

TABLE 4 (cont)

COARSE FRAGMENTS					
Randomly distributed or in lenticular concentrations (weighted average)					
1 Within the minimum effective depth allowed in each class	< 5 %	< 5 %	< 5 %	< 5 %	⩾ 5 %
2 Between 25 and 90 cm depth with fine sandy loam or finer textured matrix	< 5 %	< 20 %	< 35 %	< 35 %	⩾ 35 %
3 For 0 - 25 cm depth, stoniness classes are applicable	-	-	-	-	-
CONSISTENCE (dry)	⩽ hard	⩽ very hard	⩽ very hard	⩽ extremely hard	-
STRUCTURE					
Within the minimum effective depth allowed in each class					
1 Crumb, granular, blocky	Weak to strong	Very weak to strong	Very weak to strong	Very weak to strong	Single-grained, massive, columnar, generalized platy, polyhedrons with open cracks
2 Prismatic	weak breaking into blocky	Moderate	Strong	Strong	
3 Platy	-	-	-	In surficial horizon	
4 Polyhedrons	-	-	-	Filled cracks	
PERMEABILITY HYDRAULIC CONDUCTIVITY cm/hr (in most limiting horizon)	Moderate: 2.0-6.25	Mod slow: 0.5-2.0 Mod rapid: 6.25-12.5	Slow: 0.25-0.5 Rapid: 12.5-25.0	Very slow: 0.12-0.25 Very rapid: 25.0-40.0	< 0.12 > 40.0
INFILTRATION cm/hr (plowed soil)	2.0-6.5	0.8-6.5	0.4-6.5	0.2-6.5	< 0.2 or > 6.5
AVAILABLE WATER CAPACITY					
Water depth within the effective soil depth till a maximum 90 cm depth	⩾ 12 cm	⩾ 8 cm	⩾ 6 cm	⩾ 3 cm	< 3 cm
SALINITY - EC ds/m (saturation extract at 25°C) The values may be slightly higher in permeable soils	< 4	< 8	< 12	< 20	> 20
ALKALINITY					
1 pH (paste)	< 9 unless the soil is calcareous	Idem	Idem	Idem	> 9
2 Exchangeable sodium (ESP %)	⩽ 10	⩽ 15	⩽ 15	⩽ 20	> 20
ACIDITY (pH water 1:1)	> 6.0	> 5.0	> 4.5	⩽ 4.5	-

Note: class 5 tentatively not arable, pending technical and economic feasibility studies

* 1 US$ = 4.6 Bs (1978 value)

TABLE 5 Some properties used for engineering applications of soil information (from [17])

USDA TEXTURE
- Permafrost
- Too Clayey (thickest layer between 25 - 150 cm)
- Too Sandy (thickest layer between 25 - 150 cm)

UNIFIED CLASSIFICATION
- Thickest Layer Between 25 - 150 cm

AASTHO CLASSIFICATION
- Thickest Layer Between 25 - 100 cm

FRACTION > 7.5 cm (weight %)
- Weighted Average to 100 cm Depth
- Weighted Average to 150 cm Depth

SHRINK-SWELL POTENTIAL
- Thickest Layer Between 25 - 100 cm

DEPTH TO BEDROCK (cm)
- Hard
- Soft

DEPTH TO CEMENTED PAN (cm)
- Thick
- Thin

PERMEABILITY (cm/hr)
- Bottom Layer (± 2 m depth)
- Between 60 - 150 cm Depth (percolation rate)
- Between 60 - 100 cm Depth (filtering capacity)

DEPTH TO HIGH WATER TABLE
- Ground Water Table
- Apparent Water Table
- Perched Water Table
- Surface Ponding

FLOODING (PERIODICITY)

SODIUM ADSORPTION RATIO (0 - 100 cm)
- or Great Groups (Natric, Halic)
- or Alkali Phases

SOIL REACTION (pH) AT ANY DEPTH

SALINITY (dS/m) AT ANY DEPTH

SLOPE %

DOWNSLOPE MOVEMENT

DIFFERENTIAL SETTLING

TOTAL SUBSIDENCE

FORMATION OF PITS

POTENTIAL FROST ACTION

TABLE 6 Example of engineering rating table applied to soil and environmental limitations for septic tank absorption fields (from [17])

Property	Limits			Restrictive Feature
	Slight	Moderate	Severe	
USDA texture	-	-	Ice	Permafrost
Total subsidence (cm)	-	-	> 60	Subsides
Flooding	None	Rare	Common	Flooding
Depth to bedrock (cm)	> 180	100-180	< 100	Depth ro rock
Depth to cemented pan (cm)	> 180	100-180	< 100	Cemented pan
Depth to high water table (cm)	- > 200	- 130-200	Surface < 130	Ponding Wetness
Permeability (cm/hr): 60 - 150 cm depth 60 - 100 cm depth	 5-15 -	 1.5-5 -	 < 1.5 > 15	 Percs slowly Poor filter
Slope (%)	< 8	8-15	> 15	Slope
Fraction > 7.5 cm (1) (weight %)	< 25	25-50	> 50	Large stones
Downslope movement	-	-	(2)	Slippage
Formation of pits	-	-	(3)	Pitting

(1) Weighted average to 100 cm
(2) If the soil is susceptible to movement downslope when loaded, excavated, or wet, rate "severe slippage"
(3) If soil is susceptible to the formation of pits caused by the melting of ground ice when the ground cover is removed, rate "severe pitting"

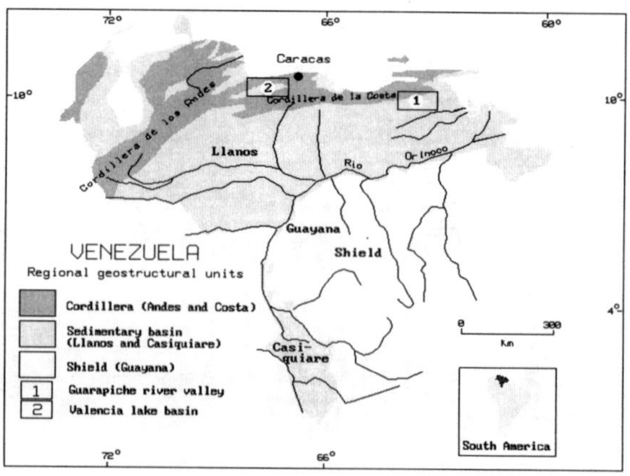

FIGURE 9 Venezuela regional structural units

and extent of fertile soils. There is a long tradition of growing irrigated crops (tobacco, vegetables, orchard trees) on the lower terraces by pumping or diverting water directly from the river channel. Higher terraces are devoted to traditional rainfed agriculture (maize, cotton, beans and vegetables) and the sandy mesa soils to ranching or groundnut plantations. The scarcity of river channel water during the dry season (November-April) does not allow irrigation of more than 1500 to 2000 ha. Expansion of the irrigable surface area requires additional water supply from either groundwater sources or an eventual river dam-reservoir.

Floodplain and lower terraces comprise the most fertile soils and host the most intensive agriculture, but are also exposed to flooding by the overflowing river. The active floodplain of the Upper Holocene is subjected to yearly inundations during the rainy season. The Lower Holocene terrace or inactive floodplain is covered by river water once or twice every 10 years. The lower terrace belonging to the Upper Pleistocene can be ravaged by catastrophic floods occurring once or twice per century. For both reasons—for flood control to protect the younger terraces and their high soil potentials and for increasing the quantity of water available for irrigation—a dam (El Guamo) is planned upstream at the outer fringe of the cordillera. The irrigable surface area can thus be expanded to 5000 ha and even 6500 ha, considering a water recuperation factor of 30 percent through recycling of excess irrigation water along the valley [30].

DATASET

The dataset used here corresponds to a small area of approximately 1430 ha (Figure 10) extracted from a second-order (semi-detailed) soil survey map covering 132000 ha at the scale of 1:25000. Aerial photographs (scales 1:20000 to 1:25000) were interpreted for preliminary identification and delineation of geoforms (to the landform level) and selection of transects for field surveys. Approximately seven observations were recorded per 100 ha, but this density was increased on the lower and middle terraces where the higher-potential soils are located. Soils were identified, classified and mapped at the series level according to the USDA soil taxonomy [19].

The map "window" comprised 12 soils series and three miscellaneous areas. The former were mapped as consociations. More heterogeneous terrain tracts were delineated as complexes of miscellaneous areas and higher level taxa (Figure 10). The soil map legend of Table 7 illustrates the strong relationship between geoforms and soil units. Figure 11, together with Table 8, displays the whole set of 39 map units as consociations of soil series phases and complexes of more comprehensive taxa [29].

Representative pedons of each soil series and additional supporting profiles were described, sampled and analyzed. Ranges of characteristics were established using the whole set of observations. Relevant environmental, morphologic, physical and chemical data were introduced into the soil database for processing and analysis.

DATA PROCESSING, ANALYSIS AND MODELLING

After digitizing the geomorphic soil map and inputting the basic pedon data and ranges of characteristics, three kinds of products were generated by matching map polygons with selected attribute data: (1) a digital elevation model; (2) several single soil and environmental attribute maps; (3) a set of interpretive maps for agricultural and engineering land use purposes. Figure 12 illustrates the sequential approach to data transformation and information generation.

Digital Elevation Model (DEM)

The available topographic map of the area (scale 1:100000, 20 m contour intervals) was insufficient to build a DEM compatible with the detail level of the soil map. Direct field estimates of the heights of the fluvial terraces and other topographic steps in relation to the local base level of the Guarapiche river (as extracted from the geomorphic soil map) provided much better reference data for displaying graphically the staircase relief configuration of the area (Figure 13) and the associated soil units (Figure 14). Relative elevations in relation to the river bed level were as follows:

- annual floodplain (Qo^a): 2 m
- periodic floodplain (Qo^b): 4 - 5 m
- low terrace (Q1): 8 - 10 m
- middle terrace (Q2): 15 - 20 m (several differentiated levels)

FIGURE 10 Guarapiche valley: soilscape units (from [29])

- high terrace (Q3): 38 - 40 m
- mesa (very high terrace) (Q4): 75 - 80 m

In developing the case study, the DEM was used effectively for:

- displaying the spatial distribution pattern of soil map units on the landscape
- visualizing areas effected by floods of various return periods (seasonal, 10 years, 100 years)
- determining the feasibility of pumping irrigation water from the river channel to higher terraces
- highlighting microclimatic conditions in relation to aspect, dominant winds and panoramic views for appropriate human settlement site selection

TABLE 7 Legend of soilscape units, Guarapiche valley (from [29])

GEOFORMS					SOILS		
Landscape	*Relief/molding*	*Substratum Depositional facies*	*Landform Position*	*Relative age (1)*	*Soil series and miscellaneous areas (2)*	*Taxonomic classification (3) (above series level)*	*Common cartographic inclusions*
Valley	Annual floodplain (seasonal flooding)	Coarse alluvial active flood sediments	Point bars and levees	Qo^a	TAM	Usthorthents–Ustipsamments–riverwash complex	-
	Periodic floodplain or very low terrace (1–2 floods/10 year period)	Sandy and loamy alluvial load excess deposits	Levee	Qo^b	Guatatal	Typic Ustifluvents, coarse-loamy (calcareous)	Mollic Ustifluvents and Fluventic Ustropepts, coarse-loamy
		Silty alluvial overflow deposits	Overflow mantle	Qo^b	Guarapiche	Fluventic Ustropepts, fine-silty	Fluventic Haplustolls, fine-silty
	Vales, fans and glacis	Coarse colluvio-alluvial deposits	-	Qo - Q1	TCAM	Ustipsamments–Ustropepts–Dystropepts complex	-
	Low terrace	Sandy and loamy alluvial load excess deposits	Levee	Q1	San Félix	Cumulic Haplustolls, fine-loamy	Cumulic Haplustolls, coarse-loamy
		Silty alluvial overflow deposits	Overflow mantle	Q1	Santo Domingo	Cumulic Haplustolls, fine-silty	Cumulic Haplustolls, coarse-silty and fine-loamy
		Clayey alluvial overflow deposits	Overflow basin	Q1	Canaguaima	Cumulic Haplustolls, fine	-
	Middle terrace	Sandy and loamy alluvial load excess deposits	Levee and crevasse splay	Q2	Cardones	Typic Haplustalfs, fine-loamy	Arenic Haplustalfs, fine-loamy
		Clayey alluvial overflow deposits	Overflow basin	Q2	Bejucal	Entic Chromusterts, very-fine	Typic Chromusterts, very-fine
	Glacis and middle glacis-terrace	Coarse colluvio-alluvial deposits	-	Qo - Q2	Caicara	Alfic Ustipsamments	Psammentic Haplustalfs, sandy, and Arenic Haplustalfs, fine-loamy
	Glacis and high glacis-terrace	Coarse colluvio-alluvial deposits	-	Qo - Q3	Comunidad	Typic Usthorthents, coarse-loamy, nonacid	Typic Ustipsamments and buried Ustults
	Terrace (all levels)	Variable	Riser (scarp)	Var	TS	Terrace scarp-plateau escarpment complex	-
Plateau	Slightly undulating mesa	Coarse alluvial splay deposits	Splay mantle	Q4	Maturin	Paleustults, fine	Paleustults, fine-loamy and Arenic subgroups
	Glacis and swale filling	Coarse colluvial deposits	-	Qo - Q3-4	Potrerito	Typic Ustorthents, sandy, siliceous	Typic Ustipsamments and buried (Plinthic) Ustults
	Escarpment	Variable	-	post-Q4	TS	Terrace scarp- plateau escarpment complex	-

(1) Relative age determined mainly on the basis of pedostratigraphic criteria
Qo^a - Upper Holocene Q1 - Upper Pleistocene Q3 - Older Middle Pleistocene
Qo^b - Lower Holocene Q2 - Younger Middle Pleistocene Q4 - Lower Pleistocene

(2) Unless otherwise indicated, SMU are consociations of soil series. See in Table 8 the detailed legend of soil map units (with phases)

(3) Unless otherwise indicated, the mineralogy and soil temperature regime families are respectively mixed and isohyperthermic

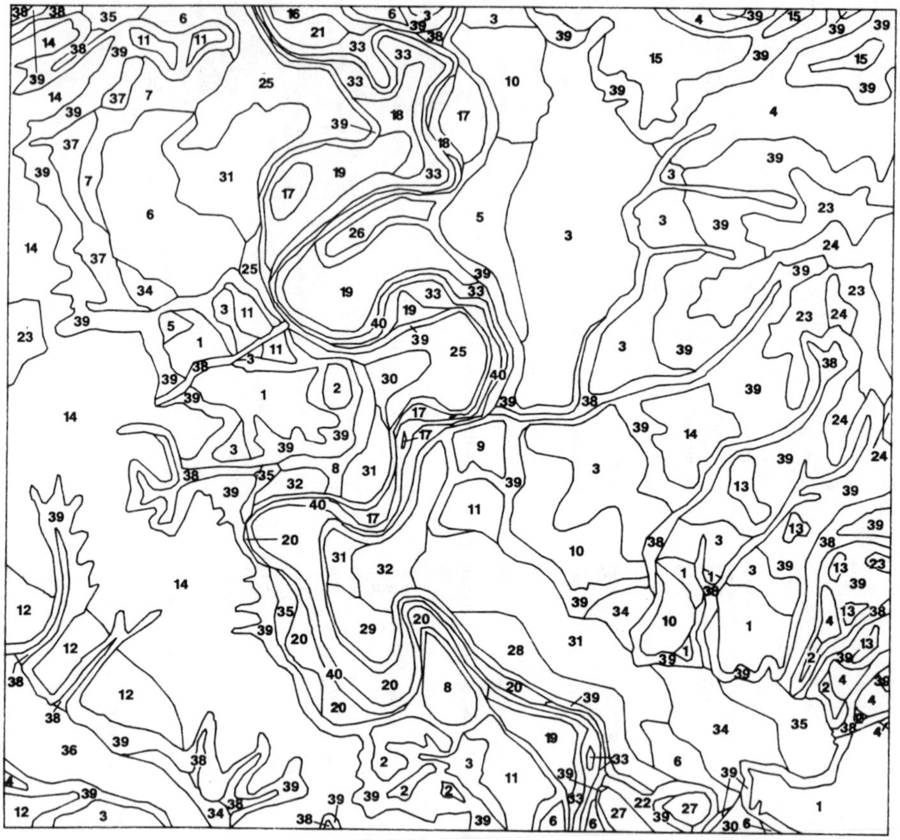

FIGURE 11
Guarapiche valley: soil map units and delineations (polygons; see Table 8), (from [29])

Single Attribute Maps

For illustrative purposes, single attribute maps were obtained by selective extraction of attribute (soil properties or environmental features) and graphics (SMU) data. Examples of surface texture and soil reaction classes are shown in Figures 15 and 16, respectively. Extrapolation of recorded river floods over their corresponding geomorphic surfaces (terrace levels) as shown on the DEM led to the elaboration of a flood hazard map (Figure 17).

Agricultural Soil Potentials

General crop requirements for irrigated agriculture (as expressed in Table 4) were included in the ILWIS rule base. Similarly, land characteristics and qualities of the 39 map units of the study area were recorded in the soil database. By matching supply and demand of land qualities, an irrigated land suitability map was obtained in two versions: one without flood control (map not displayed) and one assuming that river inundations will be mitigated by the proposed upstream El Guamo dam (Figure 18). Two additional maps were derived from the latter: the first (Figure 19) shows the distribution of arable land (irrigated land classes 1 to 4), and the second (Figure 20) shows the location of prime agricultural land (irrigated land classes 1 and 2). Although most of

TABLE 8 Legend of soil map units, Guarapiche valley (from [29]), (see Figure 11)

CONSOCIATIONS OF SOIL SERIES PHASES AND COMPLEXES OF MISCELLANEOUS AREAS, WITH HIGHER LEVEL TAXA (slope percentage in parentheses)
1 Bejucal silty clay (0.5 - 1)
2 Bejucal silty clay (0.5 - 1), isolated position
3 Caicara loamy sand (1 - 3)
4 Caicara loamy sand (1 - 3), isolated position
5 Caicara sandy loam (1 - 3)
6 Canaguaima silty clay loam (0.1 - 0.5)
7 Canaguaima silty clay loam (0.1 - 1), moderately irregular microtopography
8 Canaguaima silty clay (0.1 - 0.5)
9 Cardones loamy sand (1 - 3), moderately irregular microtopography
10 Cardones sandy loam (1 - 3)
11 Cardones sandy loam (1 - 3), moderately irregular microtopography
12 Comunidad loamy sand (1 - 3)
13 Comunidad loamy sand (2 - 5), isolated position
14 Comunidad sandy loam (1 - 3)
15 Comunidad sandy loam, moderately surficial (2 - 4)
16 Guarapiche silt loam, moderately deep (0.3 - 1), moderately irregular microtopography
17 Guarapiche silty clay loam (0.3 - 1), moderately irregular microtopography
18 Guatatal loamy sand (0.5 - 1), moderately irregular microtopography
19 Guatatal sandy loam (0.5 - 1), moderately irregular microtopography
20 Guatatal sandy loam, moderately deep (0.5 - 1), moderately irregular microtopography
21 Guatatal sandy loam, surficial (0.5 - 1), moderately irregular microtopography
22 Guatatal sandy clay loam, moderately deep (0.5 - 1), moderately irregular microtopography
23 Maturin loamy sand, colluvial cover (1 - 3)
24 Potrerito loamy sand (1 - 3)
25 San Felix sandy loam (0.5 - 1)
26 San Felix sandy loam, deep (0.5 - 1)
27 San Felix sandy loam (0.5 - 1), moderately irregular microtopography
28 San Felix clay loam (0.5 - 1)
29 San Felix clay loam (0.5 - 1), moderately irregular microtopography
30 Santo Domingo silt loam (0.3 - 0.6)
31 Santo Domingo silty clay loam (0.3 - 0.6)
32 Santo Domingo silty clay loam (0.3 - 0.6), moderately irregular microtopography
33 TAM - Mixed alluvium, variable surface textures (0.5 - 1)
34 TCAM - Mixed colluvio-alluvium, variable surface textures 2 - 4)
35 TCAM - Mixed colluvio-alluvium, variable surface textures, moderately deep (3 - 6)
36 TCAM - Mixed colluvio-alluvium, variable surface textures, moderately deep (0.1 - 1), seasonally flooded
37 TCAM - Mixed colluvio-alluvium, variable surface textures (16 - 24)
38 TCAM - Mixed colluvio-alluvium, variable surface textures (5 - 7), isolated position
39 TS - Terrace scarps and plateau escarpment, steep
40 River

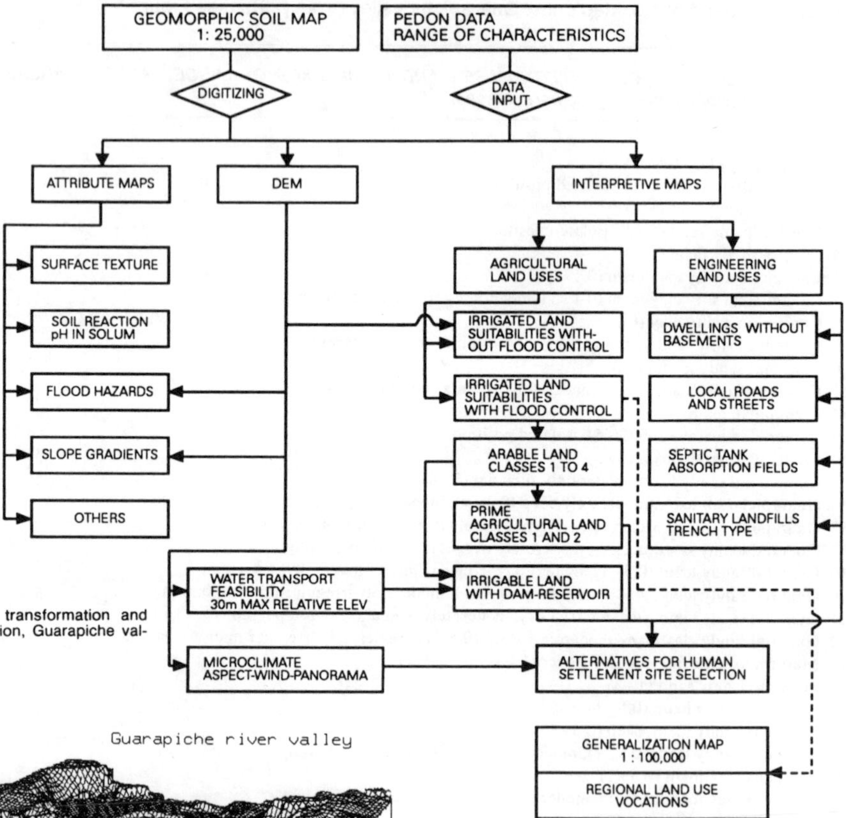

FIGURE 12
Flowchart of data transformation and information generation, Guarapiche valley

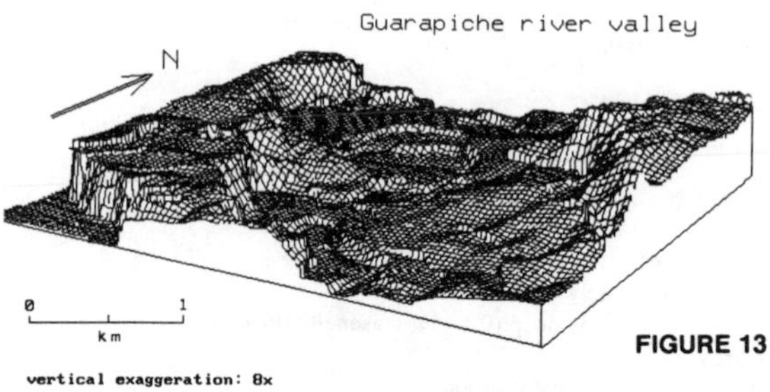

FIGURE 13 Guarapiche valley DEM

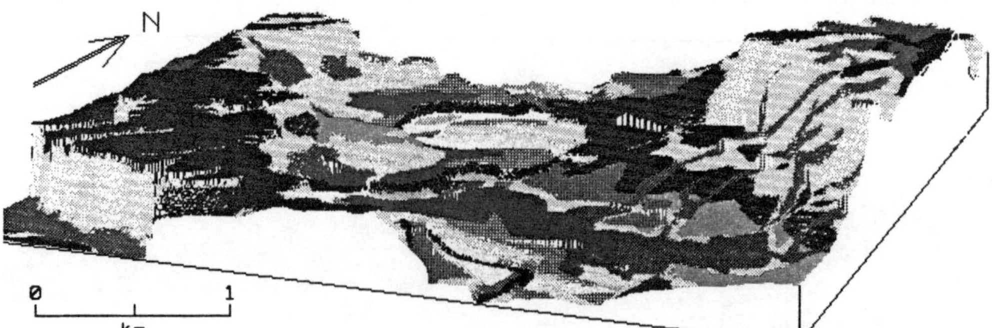

FIGURE 14 Guarapiche valley DEM and soilscape units, vertical exaggeration 8X (see Figure 10)

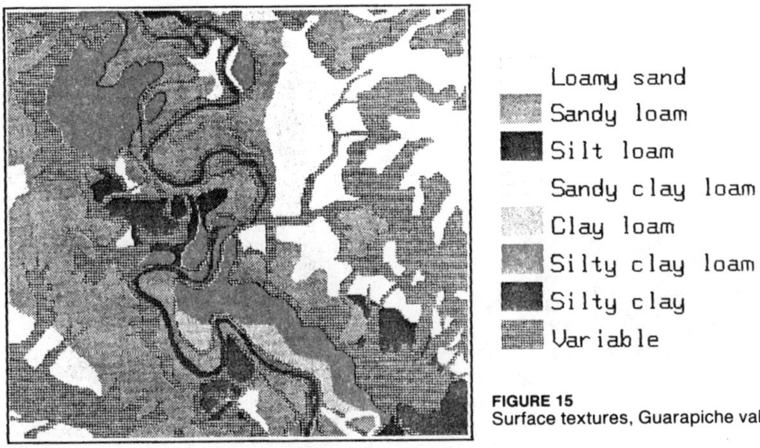

FIGURE 15
Surface textures, Guarapiche valley

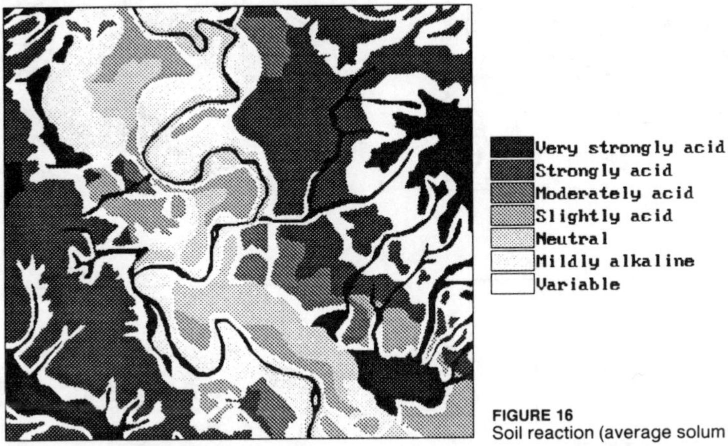

FIGURE 16
Soil reaction (average solum pH)

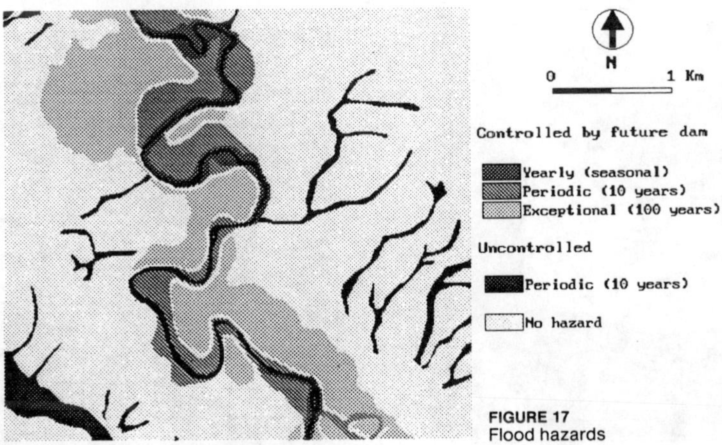

FIGURE 17
Flood hazards

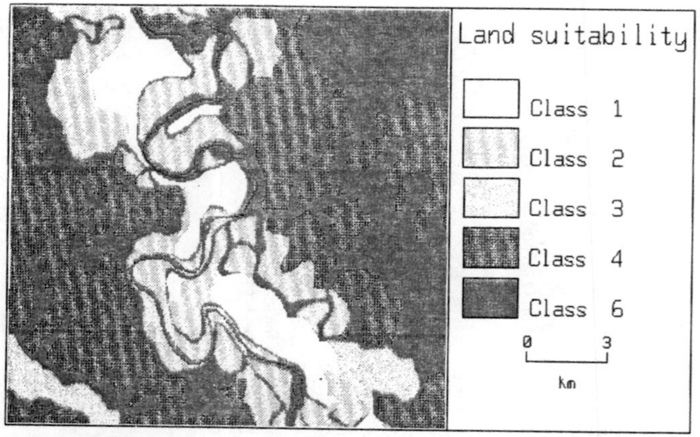

FIGURE 18 Guarapiche valley, irrigated land suitability classes (assuming flood control)

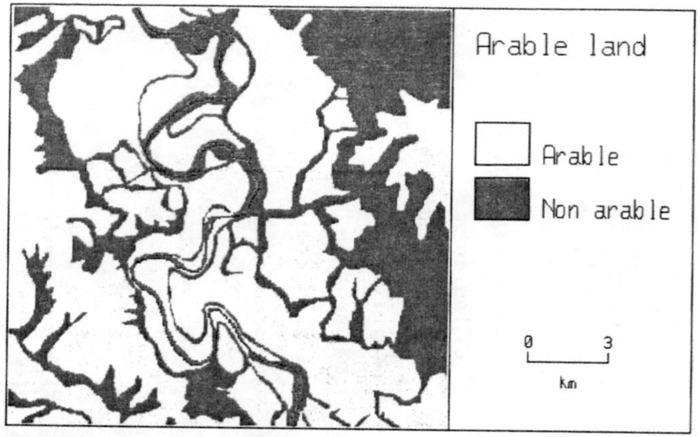

FIGURE 19 Arable land for irrigation purposes

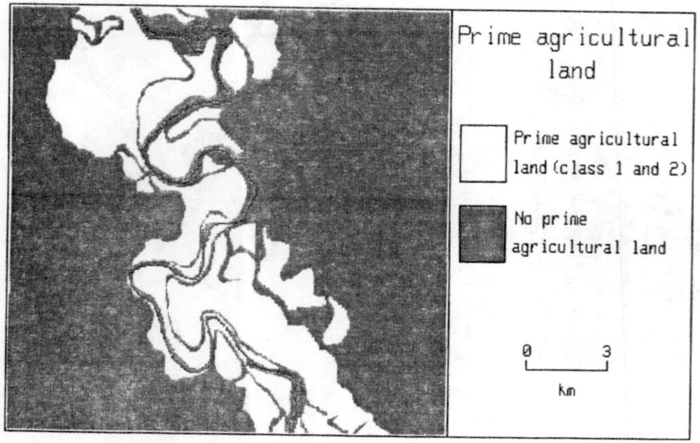

FIGURE 20 Prime agricultural land for irrigation purposes

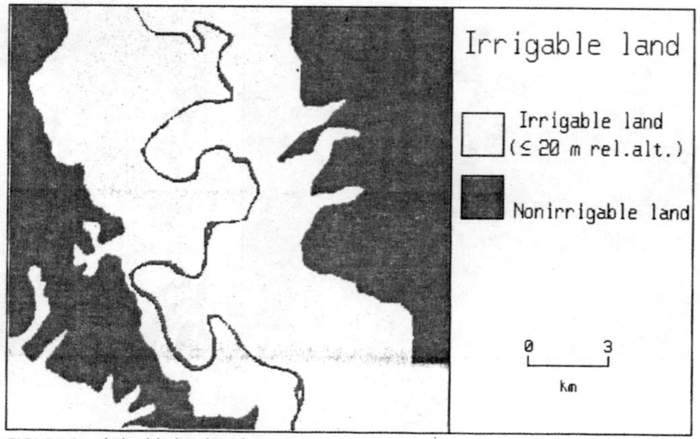

FIGURE 21 Irrigable land perimeter

the area—with the exception of the terrace risers, plateau escarpment and narrow vales incising the stepped terraces—is arable, only a small part of it is actually prime agricultural land. Under present conditions, these high value areas are exposed to flood hazards, but would be protected by the dam.

Arable land was then compared with the surface area that could be irrigated with the expanded water source of the dam reservoir, *ie*, 6500 ha. The water budget shows a positive balance: there is more water available than needed. The entire arable area may not merit irrigation, however, because of the high costs of moving the water from the main water channel (the river) to the higher terraces. For scenario development, a maximum water distribution height of 30 m was assumed to determine the irrigable perimeter (Figure 21). This restriction also allowed water to be reserved for irrigation of downstream prime agricultural land.

Engineering Land Use Limitations

The development of the irrigation plan assumed that a new human settlement would be established near the irrigation perimeter. Sets of relevant soil properties (Table 5) were interpreted against USDA-SCS rating tables [17] to determine the limiting factors and severity of limitations affecting the various SMUs for four selected civil and sanitary engineering land uses: dwellings without basements, local roads and streets, septic tank absorption fields and trench-type sanitary landfills (Figures 22 to 25).

Alternative Human Settlement Sites

By superimposing limitation maps for dwellings, streets and septic tanks, several alternative settlement sites were obtained. Sanitary landfills must be located certain distances from habitation sites.

Selection of the first choice site was based on two strongly exclusionary scenario assumptions. First, prime agricultural land would be preserved by law from non-agricultural uses. As a consequence, site alternatives overlapping with class 1 and 2 land areas were eliminated, and potential settlement sites became restricted to the higher valley terraces and plateau mesas. Coincidentally, such landscape positions also benefitted from better environmental conditions in terms of panoramic exposure and good air circulation. As a second scenario hypothesis, it was assumed that all four engineering land use limitations would rate "slight". In such highly selective conditions, only two small areas in the northeastern corner were rated suitable for a settlement (Figure 26).

For a second choice, the range of possible sites was broadened by reducing the severity of restrictive conditions. For example, preservation of prime agricultural land was no longer an imperative. In turn, limitations on engineering land uses should not rate worse than "moderate", especially for environmentally sensitive uses such as sanitary landfills and septic tank absorption fields. This scenario would accept some lowland areas as moderately appropriate human settlement sites (Figure 26).

Other alternatives are possible when considering only site conditions for dwellings and local roads, or if more flexibility is introduced into the modelling scheme by accepting higher severity ratings and therefore also higher conditioning and/or management costs.

Regional Land Use "Vocations"

By generalizing the irrigated land suitability map of the whole survey project at scale 1:25000, a regional land use "vocation" map at scale 1:100000 was created (Figure 27). (Land use "vocation" is defined here as an extension of the familiar land use "capability" to generalized potentials for broader use types such as irrigated agriculture, rainfed agriculture, rangeland, forest cover, etc.) The generalization process used the hierarchic structure of the geoform classification system for aggregating information from landform level to relief level to landscape level. Simultaneously, land suitability classes were clustered into vocation classes (see Table 9).

EXAMPLE 2: VALENCIA LAKE BASIN (TURMERO VALLEY), VENEZUELA

OBJECTIVE AND ASSUMPTIONS

Urban fringes are highly dynamic areas where differential market values of land are

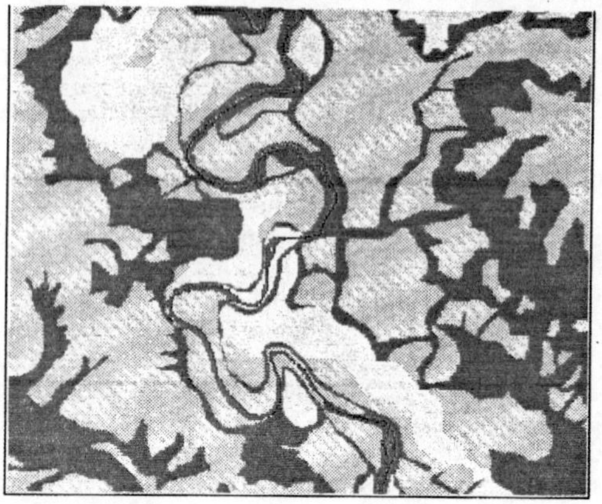

FIGURE 22 Soil limitations for dwellings without basements

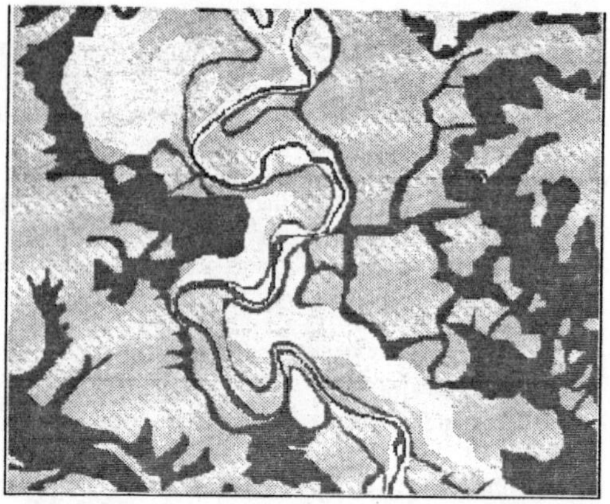

FIGURE 23 Soil limitations for local roads and streets

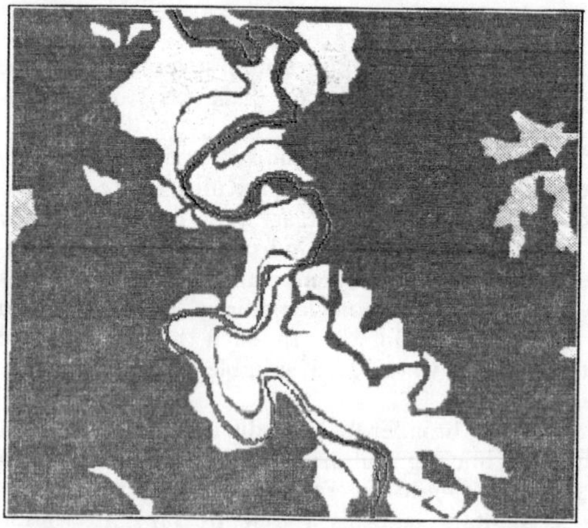

FIGURE 24 Soil limitations for septic tank absorption fields

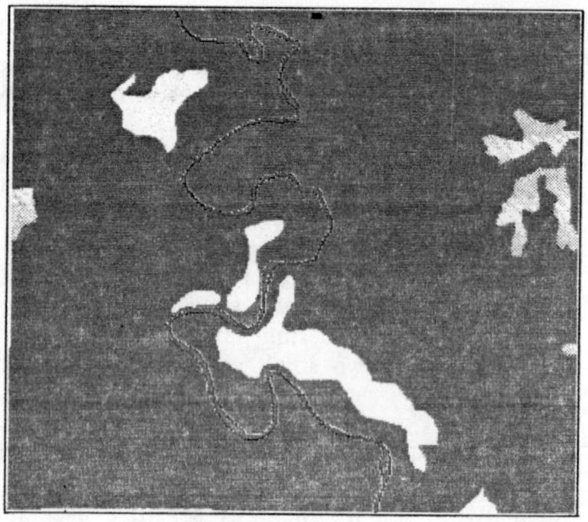

FIGURE 25 Soil limitations for trench-type sanitary landfills

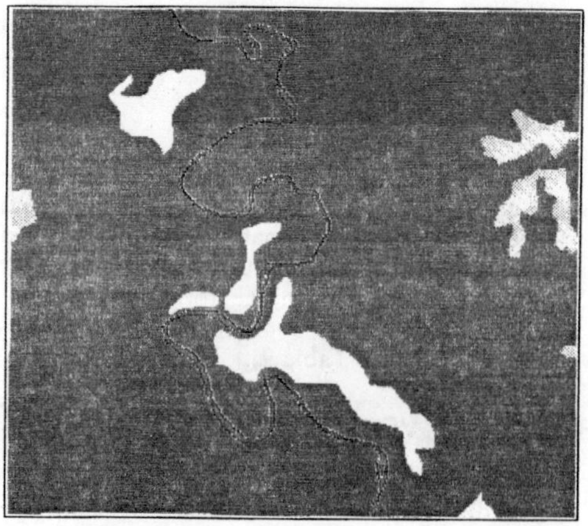

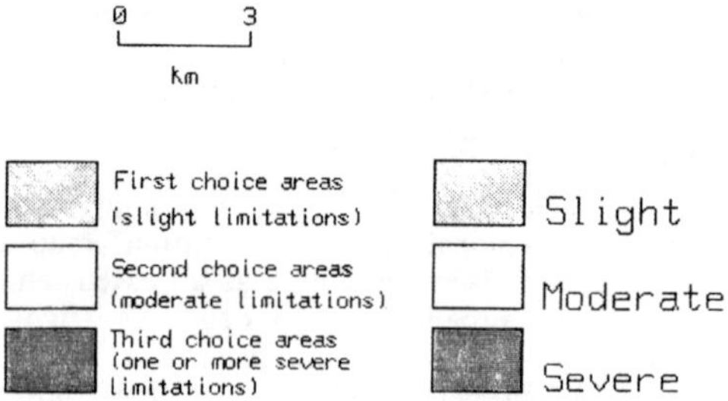

FIGURE 26 Soil and environmental limitations affecting the selection of human settlement sites

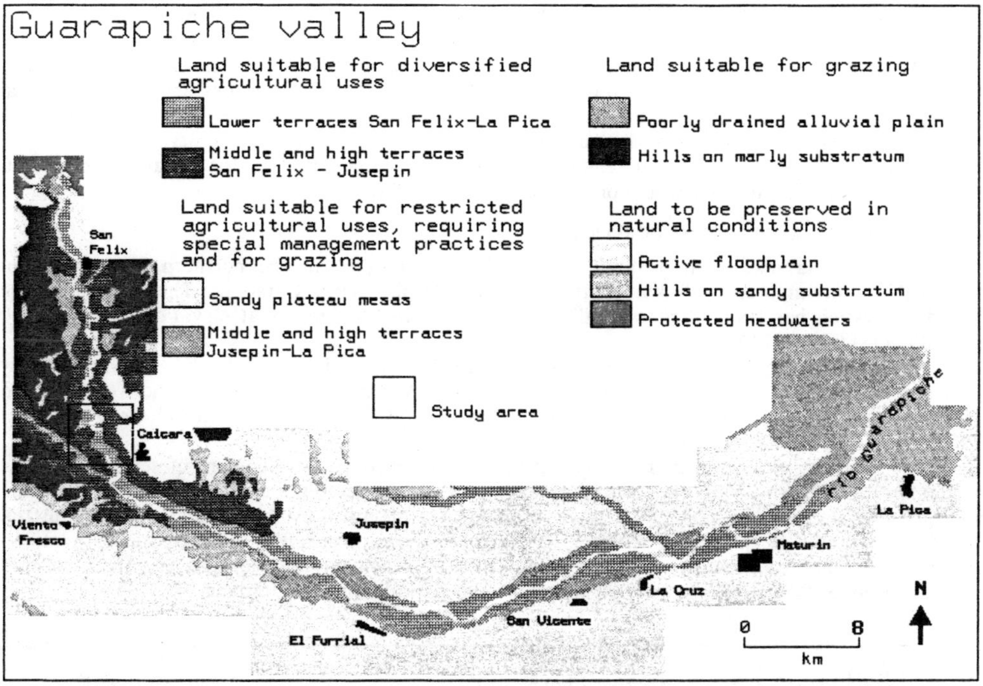

FIGURE 27 Regional vocations for broad agricultural land uses, Guarapiche valley (from [30])

TABLE 9 Guarapiche valley - Agricultural land use limitations and vocations (from [30])

GEOMORPHIC SETTING			*LIMITATIONS*									*SOIL POTENTIALS*	
Landscape	*Relief/ molding*	*Sector*	*Water* 1	2	3	*Land* 4	5	6	7	8	*Improvements needed*	*Irrigated land suitability classes*	*Land use vocations*
Valley	Active floodplain		N	E	N	N	M	E	M	M	Flood control	6	River channel protection
	Lower terraces	Upstream	M	S+M	S	N	N+M	N+S	N	N	Flood control Land grading Irrigation	1, 2, 3	Diversified agricultural uses
		Downstream	S	M	S	N	S+M	N+S	N	N	Land grading Irrigation Drainage at parcel level Fertilization	2, 3, 4	Diversified agricultural uses
	Middle and high terraces	Upstream	M	N+S	E	S	M	S+M	S	S+M	Land grading Irrigation Fertilization Conservation	4	Restricted agricultural uses with special management practices
		Downstream	S+M	N+M	E	S	E	M	S	S+M	Irrigation Drainage at parcel level Fertilization Conservation	4	Pastures and restricted agricultural uses
Plain	Active floodplain and low terrace		N	E	S	N	M	N	N	N	Fertilization Generalized drainage	3, 4	Restricted agricultural uses and pastures
Plateau	Sandy mesas		E	N	E	M	E	M	S	M	Irrigation Fertilization Conservation	4	Pastures and restricted agricultural uses
Hill land	Hills on marly substratum		M	N	E	E	E	M	S	M	Conservation	6	Pastures
	Hills on sandy substratum		E	N	E	M	E	E	S	M	Conservation	6	Nature protection
Mountain	Ridges, hogbacks		S	N	E	E	E	E	S+M	E	Conservation	6	Headwaters protection

LIMITATION KINDS 1 - Moisture deficiency; 2 - Moisture excess; 3 - Surface water; 4 - Ground water; 5 - Soil; 6 - Topography; 7 - Present erosion; 8 - Potential erosion

LIMITATION DEGREES N - None; S - Slight; M - Moderate; E - Severe (extreme)

determined not by natural soil variability, but by spatial connectivity, neighbourhood conditions, particular interests of pressure groups and political decisions on land allocation. Rational periurban land use planning could be improved, however, by using soil information [27]. In the following case study, small-city location alternatives are proposed as substitutes for sprawling urban fringe expansion. The basic planning policy accepts, among other things, such fundamental assumptions as:

- the preservation of prime agricultural land to minimize long-term social costs
- the selection of appropriate engineering and environmental sites for urban development, avoiding competition with agriculture on highly productive soils

REGIONAL SETTING AND LAND USE CONFLICTS

The Valencia lake basin is an endorreic graben thrusted in the northern coast range of Venezuela, approximately 150 km southwest of the capital city, Caracas. The bottom of the closed depression has been filled with alluvial and lacustrine sediments at elevations between 400 and 500 m asl (Figure 28).

Approximately half of the depositional area is prime agricultural land. Similar high-quality land is relatively scarce (2 percent) in Venezuela. The Valencia lake region is also well endowed with water resources. As a consequence, it has always been an agricultural area of great importance for the country's economy.

Over the last years, however, the explosive and uncontrolled development of several cities, in particular Valencia and Maracay, caused a set of problems reflecting a strong competition between non-agricultural land uses—including physical expansion of cities, establishment of new industrial facilities, extension of the traffic network, demand for recreational areas and disposal of wastes—and highly intensive agriculture. In the early 1980s, the non-agricultural land use areas covered approximately 36000 ha of the 166000 ha of the flat depositional bottom land of the basin. They

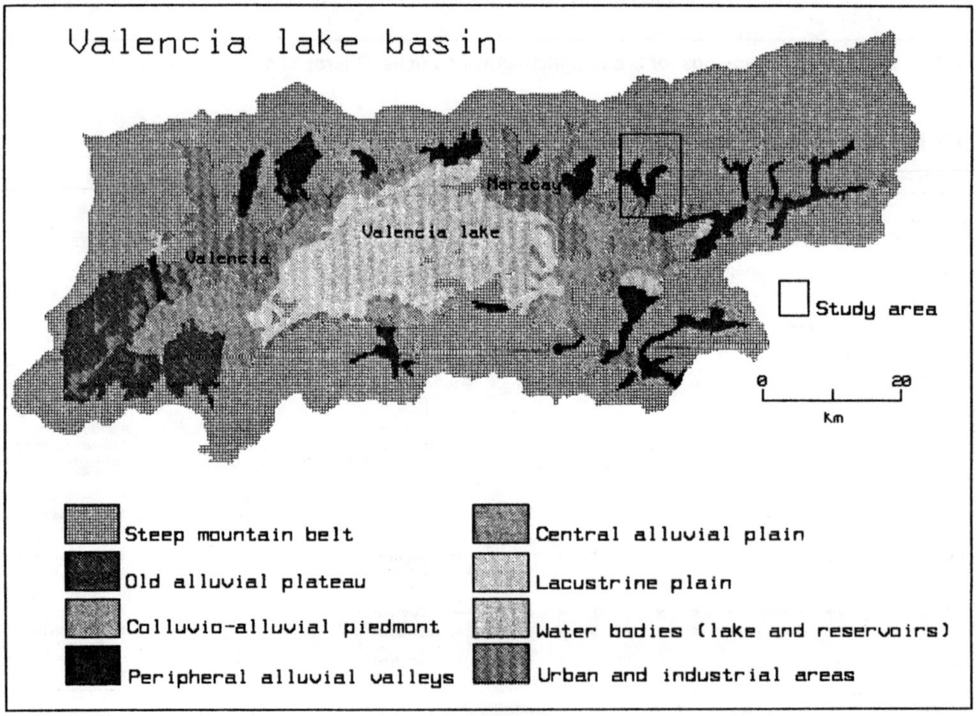

FIGURE 28 Landscape units, Valencia lake basin (from [25])

are expanding now at a rate of 2000 ha per year. As a result, the increasing urban consumption of prime agricultural land is generating serious land use conflicts at the periphery of the expanding cities.

In a pessimistic perspective, explosive urban expansion would be accepted as unavoidable, and the whole (physically closed) basin would be considered as "urbanizable", regardless of the high agricultural productivity of the land. This scenario would simply project the present development trends into the future, assuming that no attempt is made to coherently organize the geographic space. Such a spontaneous land use evolution would result in forming a disordered conurbation of some four million inhabitants, from which any type of agricultural activity would probably be excluded [25].

A rational planning scenario, on the contrary, would aim at limiting the expansion of the non-agricultural land uses by implementing a strong prescriptive scheme. The highly suitable land could be preserved for agriculture in the central plain and peripheral valleys of the basin (Figure 28). Urban expansion, in turn, would be shifted towards other types of landscape (piedmont and plateau), where soils are marginally suitable for agriculture but appropriate for engineering and other non-agricultural uses [25].

Using a small sample area in one of the eastern tributary valleys (Turmero valley), the implications of such a planning scenario on the spatial distribution of urban and agricultural land uses were analyzed and visualized.

DATASET

The flat depositional areas of the valley floor and the slightly inclined glacis and fans of the peripheral piedmont were covered by a second-order soil survey at scale 1:20000 and an observation density of approximately one point per 10 ha. The surrounding mountainous areas were inventoried using a low observation density, third-order survey at scale 1:100000. The depositional heterogeneity of the valley and piedmont landscapes led to establishing 32 map units, mainly as associations and some consociations of phases of soil families, in a surface area no larger than 3000 ha [31].

DATA PROCESSING, ANALYSIS AND MODELLING

The chain of operations from primary cartographic documents—*ie*, 1:100000 scale topographic map and 1:20000 scale geomorphic soil map with the supporting attribute data (pedons and ranges of characteristics)—to the determination of land use options is illustrated in Figure 29.

Interpretation Maps and DEM

A digital elevation model (DEM) was constructed from a topographic map (scale 1:100000). On the basis of a derived slope gradient map, all (mountainous) areas having slopes greater than 20 percent were discarded from further scenario development because of severe topographic restrictions effecting irrigation as well as urbanization (Figure 30).

Irrigated land suitabilities were successively generalized from management to subclass to class units to generate levels of information compatible with different land use planning objectives, some of which are indicated on Figure 29. Through the last generalization stage (*ie*, irrigated land classes), broad land use vocations were determined, allowing for spatial zoning of the area (Figure 31).

Simultaneously, a set of soil interpretation maps for engineering land uses was generated covering some basic technical requirements (*ie*, dwellings, streets, sanitary landfills, septic tanks, shallow excavations, sources of construction and fill materials, playgrounds) applied to the selection of suitable areas for urban development (Figure 29).

Planning Alternatives

In the context of the land use conflicts in the area, priority was given to the preservation of prime agricultural land. As a consequence, land suitable for diversified agriculture, located mainly along the valley floor (floodplain and terraces with less than 5 percent slope), was set aside as unavailable for urban expansion.

In turn, future urban development would be channeled towards the piedmont areas with slopes between 5 and 20 percent. Soils there are only marginally suitable for agriculture because

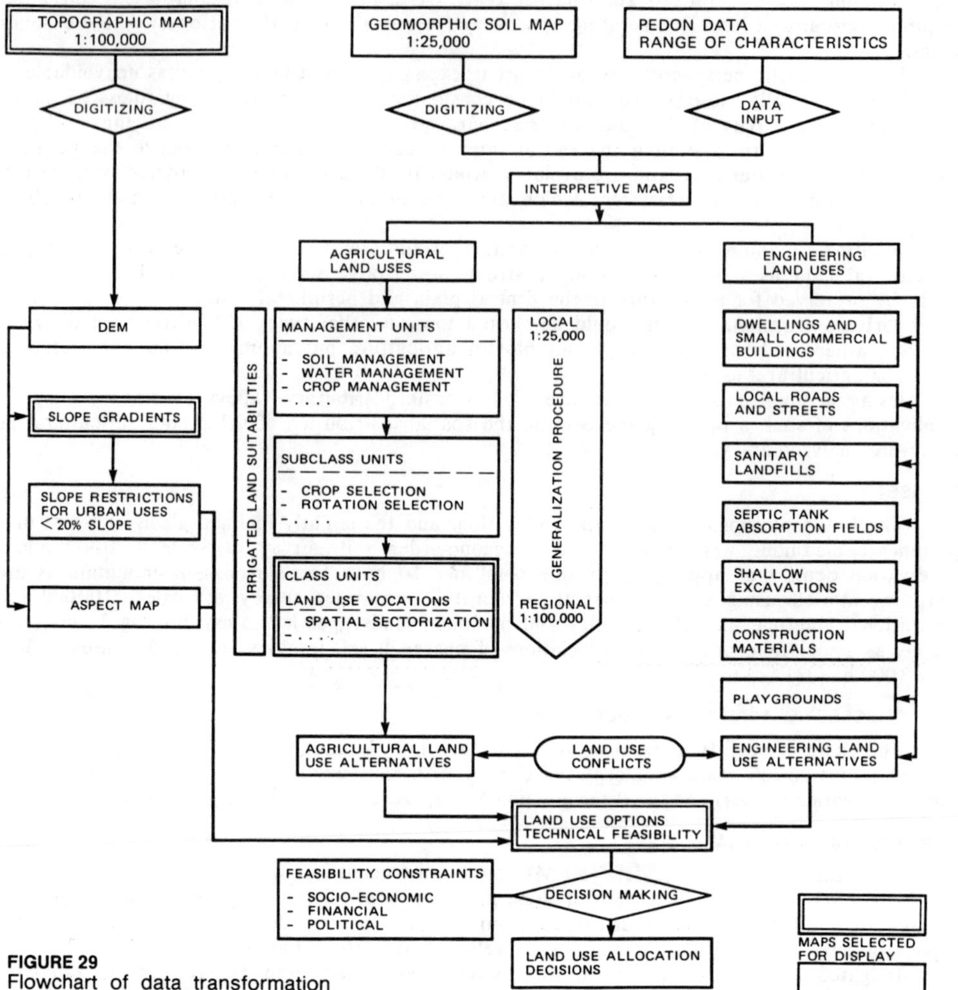

FIGURE 29
Flowchart of data transformation and information generation, Turmero valley

of coarse textures and low fertility; good drainage, high bearing capacity and low shrink-swell potential, however, make them appropriate for engineering uses. The resulting agricultural and urban land use options, based solely on technical feasibility criteria, are shown in Figure 32. Final land use allocation would be controlled by social, economic, financial and political constraints not taken into consideration here.

The desirable outcomes of such a rational planning scenario would be the diametric opposite of the unacceptable results of uncontrolled urbanization.

CONCLUSIONS

A conceptual model of the geoform-soil complex is used for proper soil cartography and provides the basis for the design of a soil database. The soil database can be used to implement

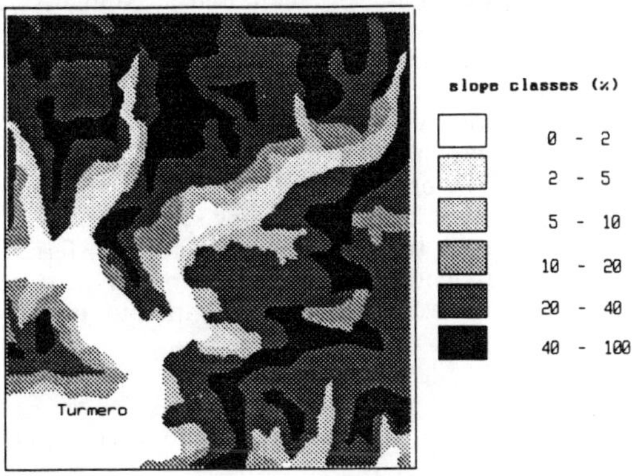

FIGURE 30 Turmero valley slope gradients

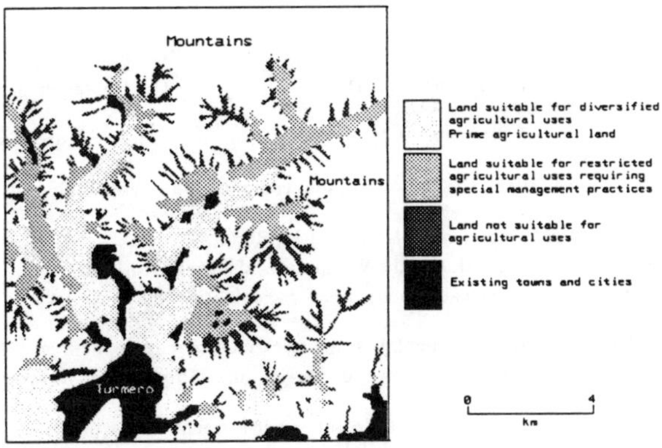

FIGURE 31 Turmero valley agricultural land use vocations

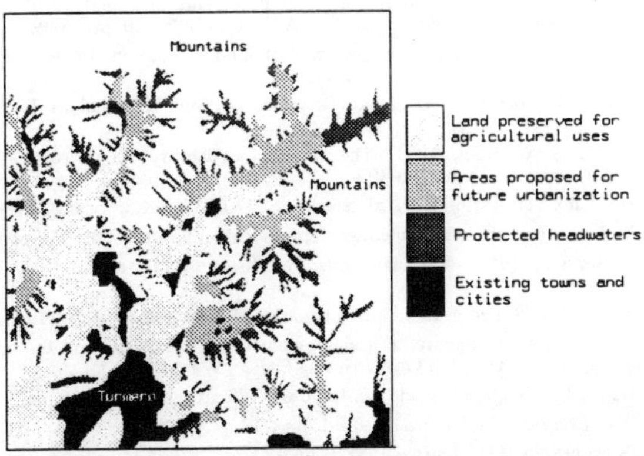

FIGURE 32 Turmero valley land use planning options

knowledge-based approaches for establishing soil map units. Similarly, general rule bases for testing and improving the reliability of soil maps can be handled through the soil database.

A proper integration of soil spatial and non-spatial databases allows the creation of attribute and interpretive maps. The creation of attribute maps involves the use of simple reclassification procedures, while interpretive maps are the result of interpretation of the soil attributes. General rule bases can be established for automatic data interpretation. The interpretive maps thus generated can be used in turn with GIS operations (which involve the use of other databases) for cartographic modelling and development of land use planning scenarios.

Soil data are usually transformed into information and transferred to end users as soil interpretation and land evaluation products. Geographic information systems constitute an efficient and versatile tool to automate the transformation of soil data into soil information.

ACKNOWLEDGEMENT

We wish to thank Jose Luis Palacio and Hans de Brouwer for their technical assistance.

REFERENCES

1 Bie, S W. 1980. Computer-assisted soil mapping. In: D R F Taylor (ed). The Computer in Contemporary Cartography. John Wiley and Sons Ltd, New York, pp 123-149.

2 Dumanski, J, B Kloosterman and S E Brandon. 1975. Concepts, objectives and structure of the Canada soil information system. Can J Soil Sci 55, pp 181-187.

3 Elbersen, G W W and R Catalan. 1986. The use of portable computers in physiographic soil surveys. Proc Internat Soil Sci Cong, Hamburg.

4 FAO. 1976. A Framework for Land Evaluation. Soils Bulletin 32, Rome.

5 FAO-Unesco. 1986. The FAO-Unesco Soil Map of the World, revised legend. World soil resources report 60, Rome.

6 Forbes, T, D Rossiter and A van Wambeke. Guidelines for Evaluating the Adequacy of Soil Resource Inventories. SMSS Tech Monograph No 4.

7 Haigh, M J. 1987. The holon: hierarchy theory and landscape research. Catena Suppl 10, pp 181-192.

8 Keulen, H van and J Wolf (eds). 1985. Modelling of Agricultural Production: Weather, Soils and Crops. Pudoc, Wageningen.

9 Kilian, J. 1974. Etude du milieu physique en vue de son aménagement. Conceptions de travail. Méthodes cartographiques. L'Agronomie Tropicale, Vol XXIX, 2-3. pp 141-153.

10 Klingebiel, A and P Montgomery. 1961. Land Capability Classification. Agri Handbook 210, USDA Soil Conservation Service, Washington DC.

11 Mausbach, M J and G T Stubbendieck. 1987. Microcomputer processing and analysis of pedon descriptions. In: W U Reybold and G W Petersen (eds). Soil Survey Techniques. SSSA Spec Publ No 20, pp 33-39.

12 McCormack, D E. 1987. Soil potential ratings, a special case of land evaluation. In: K J Beek, *et al* (eds). Quantified Land Evaluation Procedures. ITC Publ 6, Enschede, pp 81-84.

13 Riquier, J, D L Bramao and J P Cornet. 1970. A New System of Soil Appraisal in Terms of Actual and Potential Productivity. FAO, Rome.

14 Sanchez, P A, W Couto and S W Buol. 1982. The fertility capability soil classification system: interpretation, applicability and modification. Geoderma 27, pp 283-309.

15 Storie, R E. 1978. The Storie Index Soil Rating Revised. Spec Publ 3203, Univ California, Berkeley.

16 Tricart, J and J Kilian. 1979. L'écogéographie et l'aménagement du milieu natural. Edit Maspéro, Paris.

17 USDA, Soil Conservation Service. 1983. National Soils Handbook. Superint of Doc, US Govt Printing Office, Washington DC.

18 USDA, Soil Conservation Service. 1988. Pedon Description Program. Beta version 1.0, Internal publ, Washington DC.

19 USDA, Soil Survey Staff. 1975. Soil Taxonomy: a basic system of soil classification for making and interpreting soil surveys. Agr Handbook 436. Superint of Doc, US Govt Printing Office, Washington DC.

20 USDI, Bureau of Reclamation. 1953. Irrigated Land Use, Land Classification Manual. Denver.

21 Valenzuela, C R. 1988. ILWIS overview. ITC Journal 1988-1, pp 4-14.

22 Valenzuela, C R. 1988. Soils geography. ITC Journal 1988-1, pp 45-50.

23 Wambeke, A van and T Forbes (eds). 1986. Guidelines for Using Soil Taxonomy in the Names of Soil Map Units. SMSS Tech Monograph No 10.

24 Zinck, J A., 1977. Ensayo sistemico de organizacion del levantamiento de suelos. Ministerio de Obras Publicas, Venezuela.

25 Zinck, J A. 1981. Conflits d'utilisation et possible destins des terres dans la dépression du Lac de Valencia, Venezuela. Thèse Doct. Univ Bordeaux, France.

26 Zinck, J A. 1988. Geomorphology and Soils. Internal publ, ITC, Enschede.

27 Zinck, J A. 1990. Soil information and land use conflicts in urban fringes. A Venezuelan land use planning scenario case study. Proc 14th internat cong of soil science, Kyoto.

28 Zinck, J A and P L Urriola. 1970. Origen y evolucion de la Formacion Mesa. Un enfoque edafologico. Ministerio de Obras Publicas, Venezuela.

29 Zinck, J A and P L Urriola. 1971. Estudio edafologico del Valle Guarapiche, Estado Monagas. Ministerio de Obras Publicas, Venezuela.

30 Zinck, J A and P L Urriola. 1973. Lineamientos para el ordenamiento espacial del valle Guarapiche. Revista Agronomia, Venezuela, No 13, pp 14-24.

31 Zinck, J A, M Garcia and J Perez. P. 1975. Estudio de suelos semi-detallado de la depresion del Lago de Valencia, Estados Aragua y Carabobo. Bloque de levantamiento No 18, Valles de Turmero. Ministerio de Obras Publicas, Venezuela.

BIBLIOGRAPHY

Bie, S W (ed). 1975. Soil information systems. Proc 1st meeting ISSS WG-SIS, Wageningen 1975. Pudoc, Wageningen.

Bliss, N B and W U Reybold. 1989. Small scale digital soil maps for interpreting natural resources. J Soil and Water Cons, pp 30-34.

Burrough, P A. 1986. Principles of Geographical Information Systems for Land Resources Assessment. Clarendon Press, Oxford.

Burrough, P A. 1987. Natural resources databases: conceptual units, data structure and natural variation. In: K J Beek, *et al* (eds). Quantified Land Evaluation Procedures. ITC Publ 6, Enschede, pp 60-65.

Burrough, P A and S W Bie (eds). 1984. Soil Information Systems Technology. Proc 6th meeting ISSS WG-SIS, Norway 1983. Pudoc, Wageningen.

CNRS. 1972. Cartographie géomorphologique, Travaux de la RCP77. Edit Centre National de la Recherche Scientifique, Mémoires et Documents 12, Paris.

Cunningham, R L, G W Petersen and C J Sacksteder. 1984. Microcomputer delivery of soil survey information. J Soil and Water Cons, pp 241-243.

Giltrap, D J. 1983. Computer production of soil maps: I, Production of grid maps by interpolation; II, Interactive map display. Geoderma 29, pp 295-325.

Girard, M C (ed). 1981/82. Traitement informatique des données du sol. Proc 3rd meeting ISSS WG-SIS, Paris 1981. Inst Nat Agron, Paris-Grignon.

Guth, P L, E K Ressler and T S Bacastow. 1987. Microcomputer program for manipulating large digital terrain models. Computers and Geosciences 13, No 3, pp 209-213.

Huddleston, J H. 1984. Development and use of soil productivity ratings in the United States. Geoderma 32, pp 297-317.

Informatique et Biosphère. 1971. Glossaire de pédologie. Description de l'environnement en vue du traitement informatique. Paris.

Kukachka, F R. 1987. The microcomputer: a valuable tool for field soil survey. In: W U Reybold and G W Petersen (eds). Soil Survey Techniques. SSSA spec publ No 20, pp 49-55.

Lamp, J. 1987. Prospects of soil information systems. In: K J Beek, *et al* (eds). Quantified Land Evaluation Procedures. ITC Publ 6, Enschede, pp 57-59.

Mausbach, M J and W U Reybold. 1987. In support of GIS in the SCS: SIS. In: K J Beek *et al* (eds). Quantified Land Evaluation Procedures. ITC Publ 6, Enschede, pp 77-80.

Meijerink, A M J. 1988. Data acquisition and data capture through terrain mapping units. ITC Journal 1988-1, pp 23-44.

Meijerink, A M J and C R Valenzuela. 1986. Terrain-soil module. ILWIS publ series M2, ITC, Enschede.

Moore, A W and S W Bie (eds). 1977. Uses of soil information systems. Proc 1st Australian meeting ISSS WG-SIS, Canberra 1976.

Moore, A W, B G Cook and L G Lynch (eds). 1981. Information systems for soil and related data. Proc 2nd Australian meeting ISSS WG-SIS, Canberra 1981. Pudoc, Wageningen.

Reybold, W U and G W TeSelle. 1989. Soil geographic databases. J Soil and Water Cons, pp 28-29.

Rudeforth, C C. 1975. Storing and processing data for soil and land use capability surveys. J Soil Sci 26, No 2, pp 155-168.

Sadovski, A N and S W Bie (eds). 1977. Developments in soil information systems. Proc 2nd meeting ISSS WG-SIS, Bulgaria 1977. Pudoc, Wageningen.

Shields, J A and D R Coote. 1987. SOTER Procedures Manual for Small-scale Map and Database Compilation. ISRIC, Wageningen.

USDA, Soil Conservation Service. 1971. Guide for Interpreting Engineering Uses of Soils. Superint of Doc, US Govt Printing Office, Washinton DC.

Wilding, L P. 1988. Improving our understanding of the composition of the soil-landscape. In: Proc internatl interactive workshop on soil resources: their inventory, analysis and interpretation for use in the 1990s. Minneapolis, pp 13-39.

Zinck, J A. 1980. Valles de Venezuela. Cuadernos Lagoven.

INTEGRATION OF GIS AND REMOTE SENSING IN LAND USE AND EROSION STUDIES

Gerardo Bocco
C.R. Valenzuela
I.T.C.
P.O. Box 6, Boulevard 1945
7500 AA Enschede
The Netherlands

1. Introduction

The effective utilization of the large amounts of spatial data produced by remote sensing systems is dependent upon the existence of an efficient, geographic data handling and processing system that will transform the data into usable information (Marble and Peuquet, 1983). A GIS represents the most effective mechanism for making use of the data captured by remote sensing systems and also enhances the effectiveness of this data capture operation through correlation of remote sensing input with data already stored in a GIS (Marble, 1984).

Remote sensing classification systems could, in many instances, benefit from access to highly accurate, ancillary ground data which could significantly improve (spectral) classification procedures (Marble and Peuquet, 1983).

Different classification accuracies have been reported in the literature. Thompson and Henderson (1984a) found that Thematic Mapper (TM) separated soil moisture regimes with about 60 % accuracy throughout the growing season of corn and soybeans. The same authors (1984b) found 56 % agreement between TM results and the United States Department of Agriculture (USDA) soil survey results for upland clay soils. Anuta et al. (1984) found accuracies higher than 75 % for several cover types. In a study of inherent error in the overlaying of several thematic maps, Walsh et al. (1987) calculated that the accuracy of land cover determined by a multispectral classification of Landscat MSS data was 57 %. Jadkowski and Ehlers (1989) used a combination of unsupervised classification and image interpretation techniques for land use/cover analysis with SPOT data (10 m spatial resolution) and obtained accuracies as high as 93 %.

One of the most common and frustrating problems in multispectral classifications using statistical decision rules is the inconsistency in the identification of information classes, i.e., the improper transformation of spectral classes in information classes.

This inconsistency occurs because of similarities in spectral responses: different information classes may have the same or similar spectral responses in all spectral bands. When the classifier is run and a decision

A. S. Belward and C. R. Valenzuela (eds.),
Remote Sensing and Geographical Information Systems for Resource Management in Developing Countries, 477–490.

rule is applied to the data set, the classifier cannot differentiate among them. Several of these classes, however, can be visually separated by a trained and experienced interpreter.

The aim of this paper is to describe for natural resource scientists a classification method that interactively integrates conventional satellite image processing techniques, GIS procedures and the interpreter's expertise (as formalized in maps and point data). Specifically, our objective was to create a GIS-assisted map of land use using a multispectral cover classification upgraded by other digital image processing techniques and geomorphic/ landscape field information (see Bocco and Valenzuela, 1988).

The approach was tested in part of a drainage basin (approximately 250 km^2) in central Mexico (Figure 1). The area is located in the Mexican Volcanic Belt, a large physiographic province of Quaternary volcanic origin where more than 50 % of the Mexican population resides. Rain-fed agriculture is the main economic activity in the area, and accelerated erosion processes are a widespread phenomemon.

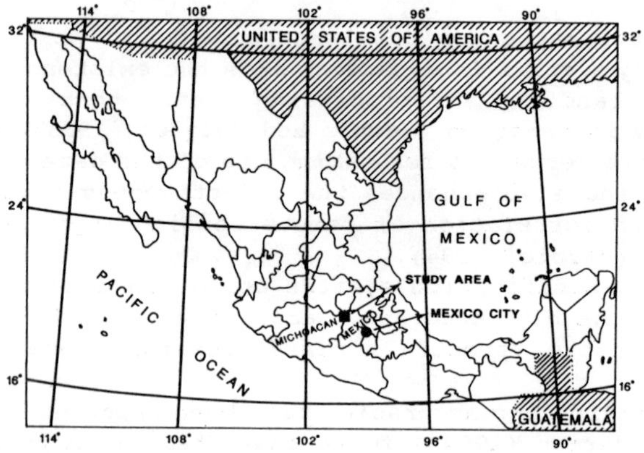

Figure 1. Location of the study area

2.Processing

The use of remote sensing products in a GIS is illustrated in Figure 2. The processing followed during the analysis is shown schematically in Figure 3.

Landsat Thematic Mapper (TM) data were geo-referenced to a Universal Transverse Mercator (UTM) projection using 30 control points and a second-order algebraic transformation equation. The accuracy of the registration was 0.5 pixel (rms error).

2.1.IMAGE ENHANCEMENT

Single bands and histograms were carefully analysed and different colour composites were created using different combinations of enhanced bands. Enhancement techniques improve both visual landscape interpretation and the sampling of spectral classes for a (supervised) classification.

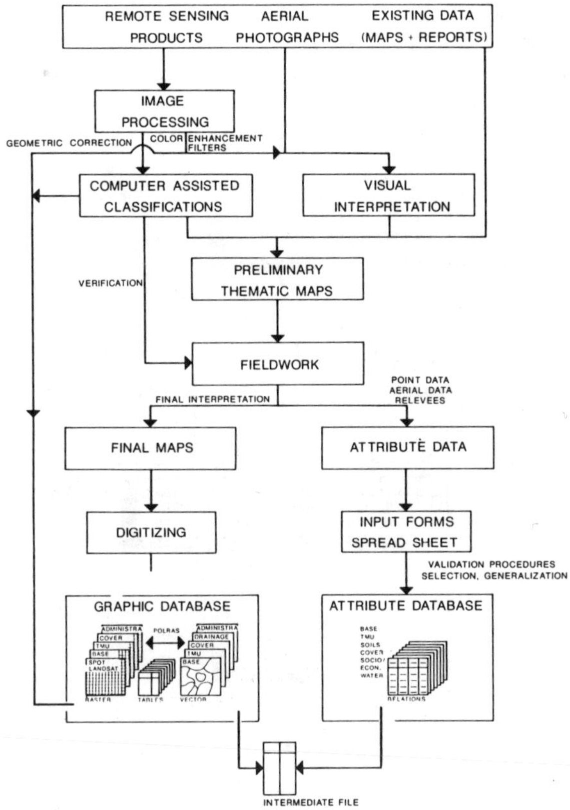

Figure 2. Remote sensing products in a GIS (from Valenzuela, 1988)

Basic enhancement techniques included band stretching and edge detection filtering. Different values for both stretch bounds and 3x3 filter kernels were tested. Figure 4 shows band 5 enhanced using an edge enhancement filter. The best colour combination was a pseudo-natural colour composite consisting of two stretched, filtered bands (band 5 displayed in red and band 4 in green) and a stretched, non-filtered band (band 3 displayed in blue).

2.2.MULTISPECTRAL CLASSIFICATION

Spectral classes were selected according to the interpreter's knowledge of the area (field verified aerial photos and maps), using the enhanced colour composite displayed on the screen.

The variability within and between spectral classes was analysed by evaluating the mean and standard deviation of each spectral class and the distances between classes during the entire sampling procedure.

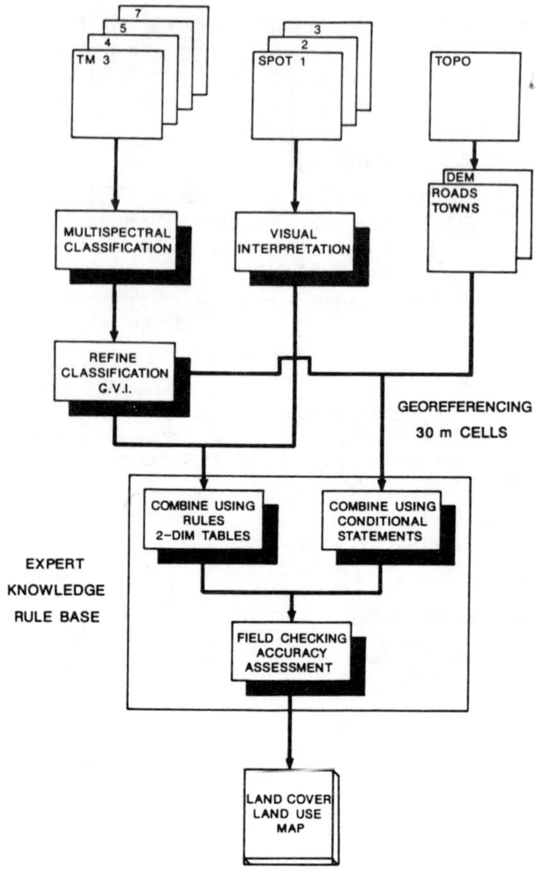

Figure 3. Schematical representation of the analysis procedure

These classes were refined by adding or deleting individual samples through a continuous analysis of the class statistics. The separability between classes was evaluated by displaying two-dimensional feature spaces of relevant band combinations on the screen.

The performance of these spectral classes and the possible need for additional classes were checked by running a classification algorithm requiring limited computer time. This proved useful in identifying the unclassified pixels and the major misclassification problem. The evaluation also helped in defining classes to be added and in identifying those classes that should be deleted.

This analysis generated 26 spectral classes that were expected to model the spectral variability of the area. A maximum likelihood classifier was run on four selected spectral bands (TM3, TM4, TM5 and TM7) chosen after evaluating their potential information content as indicated in their histograms.

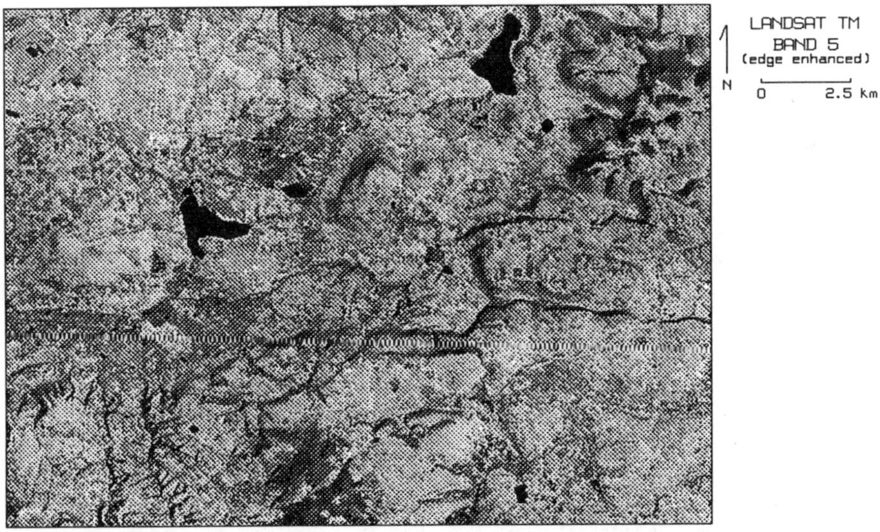

Figure 4. TM band 5 edge enhanced image

The 26 spectral classes were labelled to 9 information classes, so that the variability of each information class was represented by its constituent spectral classes forming the information class. The classification results are presented in Figure 5 and Table 1.

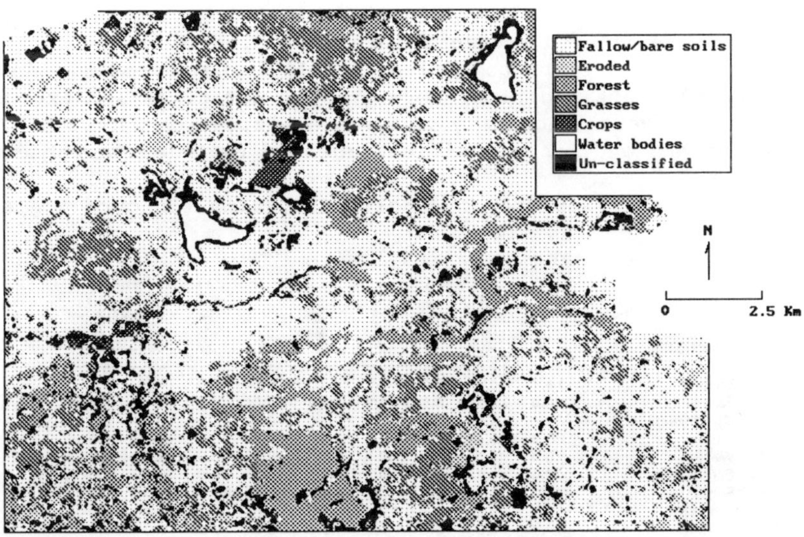

Figure 5. Multispectral classification of land cover

Table 1. Multispectral cover classification, TM data

Spectral Class	Information class	% Area	Misclassifications
0	0 Unclassified	9.0	Open forest, bare, fallow
1	1 Bare soil 1		Bare/fallow, fallow, eroded
2	1 Bare soil 2	12.2	-
3	2 Bare/fallow 1		Open forest, fallow
4	2 Bare/fallow 2		-
5	2 Bare/fallow 3	9.7	-
6	3 Open forest 1		Bare/fallow
7	3 Open forest 2		-
8	3 Open forest 3		-
9	3 Open forest 4	10.9	-
10	4 Fallow/bare 1		Grasses, open forest
11	4 Fallow/bare 2		-
12	4 Fallow/bare 3		-
13	4 Fallow/bare 4	30.9	-
14	5 Water 1		Polluted water, unclassified
15	5 Water 2	0.9	-
16	6 Crops 1		Open forest, fallow
17	6 Crops 2	1.2	-
18	7 Forest 1		Open forest
19	7 Forest 2		-
20	7 Forest 3	3.7	-
21	8 Grasses 1		Fallow/bare, open forest
22	8 Grasses 2		-
23	8 Grasses 3		-
24	8 Grasses 4	16.5	-
25	9 Eroded soils 1		Bare soil, towns, roads
26	9 Eroded soils 2	5.0	-

The results were assessed quantitatively by constructing an error matrix (Table 2). Sampling points were defined in a grid of 33 columns and 25 rows (every 20 pixels, i.e., every 600 m on the ground). The grid was overlaid on the classified image and the spectral class for each of the 825 points of the grid was automatically read from the classification. Each point was verified by visual interpretation of the field checked aerial photos and the colour composite. Finally, the scores for successful and unsuccessful classifications were grouped per class and expressed as percentages. The overall accuracy of the classification was 75 %. The main misclassifications and non-classifications are included in table 1.

Table 2. Error matrix for the cover classification (in percentages)

Spectral/Visual	Bare	Bare/Fallow	Op. forest	Fallow	Water	Crops	Forest	Grasses	Eroded
Bare	74.7	8.5	3.2	4.2	1.0	0.0	0.0	4.2	4.2
Bare/Fallow	0.0	79.2	9.7	6.9	1.4	0.0	0.0	1.4	1.4
Op. forest	1.0	6.9	81.2	7.9	0.0	0.0	2.0	1.0	0.0
Fallow	3.5	5.9	2.3	77.7	0.4	0.8	0.4	9.0	0.0
Water	0.0	12.5	0.0	12.5	62.5	0.0	0.0	12.5	0.0
Crops	0.0	0.0	14.3	14.3	0.0	71.4	0.0	0.0	0.0
Forest	0.0	0.0	12.1	0.0	0.0	0.0	87.9	0.0	0.0
Grasses	8.1	3.7	2.3	20.0	1.5	0.0	0.0	64.4	0.0
Eroded	21.4	0.0	0.0	0.0	0.0	0.0	0.0	0.0	78.6
Unclassified	20.0	20.0	43.3	6.7	0.0	6.7	0.0	0.0	3.3

2.3.RECLASSIFYING WATER WITH ENHANCED IMAGES

Water bodies (in this area irrigation dams) are difficult to classify because of different quantities and characteristics of suspended sediments, and several spectral classes are needed to represent the variability of a single information class. This increases sampling and classification efforts, including computer time. Moreover, some of these spectral classes will result in misclassification of other covers (such as towns, eroded terrain and bare soils).

Because we did not need to differentiate among "types" of water, simple image processing techniques (enhanced infrared bands, Green Vegetation Index (GVI) in semi-arid areas, principal components, density slicing) could be applied using GIS procedures to classify water. An object-oriented (water) database was created as follows:

- Perform the GVI (TM4-TM3)/(TM4+TM3).
- Using density slicing, assess the values of GVI for water (in this instance, 0 to 100).
- Define the rule: if the values of GVI are < 100, then all classified pixels within that range are classified as water, else the classification remains unchanged.

This procedure is quick and straightforward when the remote sensing module is a part of the GIS.

If the main target of mapping are agricultural soils, a similar approach can be used to define a soils database by masking green vegetation and water using a GVI. Classification efforts can then concentrate on only the remaining area.

2.4.IMPROVING THE SPECTRAL CLASSIFICATION USING GEOMORPHOLOGIC INFORMATION

There is a strong relationship between land use and landforms that can be captured by visual mapping of stereoscopic imagery. In this study, enhanced SPOT false colour stereopairs were used to perform a field verified TMU map. TMUs are homogeneous in landform, lithology and soils (Meijerink, 1988). The relationship between geomorphic unit and dominant land use is established at unit level (Figure 6, Table 3).

On the basis of the knowledge contained in the TMU map, the spectral map was reclassified using a two-dimensional table (Table 4) and a new map was generated. A two-dimensional table constitutes an interactive and powerful tool that allows the combination of ancillary data and the knowledge of the applications specialist to reclassify the multispectral classification. In this case, the TMU map was used as ancillary data. The new information classes are obtained by matching each combination of multispectral class with the TMU class (Table 4), as in the following example:

- If spectral class is crops and TMU is steep lavic edge, then the new information class is open forest.

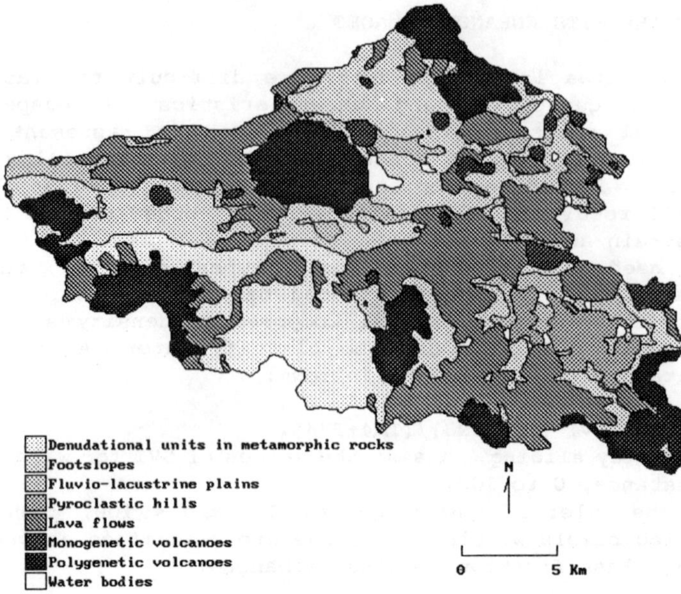

Figure 6. Terrain mapping units

Table 3. Some terrain mapping units and dominant land use

UNIT No.	UNIT NAME	DOMINANT LAND USE
1.	Denudational units (on metamorphic rocks)	Rain-fed agriculture, locally forest
2.	Stratovolcanoes (andesitic)	Forest, locally grasses
3.	Complex volcanoes (basaltic)	Rain-fed agriculture, grasses
4.	Monogenetic cones (basaltic)	Open forest, grasses
5.	Basaltic fields	Rain-fed agriculture, grasses
6.	Volcanic footslopes	Rain-fed agriculture, grasses
7.	Ignimbritic flows	Rain-fed agriculture, grasses
8.	Andesitic flows	Rain-fed agric., open forest
9.	Basaltic flows	Rain-fed agric., open forest
10.	Fluvio-lacustrine plains	Irrigated agriculture

2.5. RECLASSIFICATIONS USING SLOPE GRADIENT DATA

Slope steepness is a critical variable that governs important changes in land use and erosion. Using an approach similar to that described above, slope gradient data can be used to improve the identification of informational classes.

Table 4. Two-dimensional table. X-axis, TMUs; Y-axis, spectral classes

	X AXIS	Background	Stratovolcanoes	Complex volcanoes	Lavic cones	Scoria cones	Andesitic flows	Ignimbritic flows	Rhyolitic flows	Basaltic flows	Basaltic fields	Volcanic footslopes	Pyroclastic hills	Slopes on metamorphic ro	Non volcanic footslopes	Fluvio-lacustrine plains	Reservoirs
Y AXIS	Unclassified	0	3	4	3	3	4	8	3	3	2	1	1	1	1	2	5
	Bare soil 1	0	1	1	1	1	1	1	1	1	1	1	1	1	1	?	5
	Bare soil 2	0	1	1	1	1	1	1	1	1	1	1	1	1	1	2	5
	Bare/Fallow 1	0	2	2	2	2	2	2	2	2	2	2	2	2	2	2	5
	Bare/Fallow 2	0	2	2	2	2	2	2	2	2	2	2	2	2	2	2	5
	Bare/Fallow 3	0	2	2	2	2	2	2	2	2	2	2	2	2	2	2	5
	Open forest 1	0	3	3	3	3	3	3	3	3	3	3	3	3	3	4	5
	Open forest 2	0	3	3	3	3	3	3	3	3	3	3	3	3	3	4	5
	Open forest 3	0	3	3	4	4	3	3	3	3	3	3	3	3	3	4	5
	Open forest 4	0	3	3	3	3	3	3	3	3	3	3	3	3	3	4	5
	Fallow/Bare 1	0	8	4	4	4	4	4	4	4	4	4	4	4	4	4	5
	Fallow/Bare 2	0	8	4	4	4	4	4	4	4	4	4	4	4	4	4	5
	Fallow/Bare 3	0	8	4	4	4	4	4	4	4	4	4	4	4	4	4	5
	Fallow/Bare 4	0	8	4	4	4	4	4	4	4	4	4	4	4	4	4	5
	Water 1	0	5	5	5	5	5	5	5	5	5	5	5	5	5	5	5
	Water 2	0	5	5	5	5	5	5	5	5	5	5	5	5	5	5	5
	Crops 1	0	6	6	6	6	6	6	6	6	6	6	6	6	6	6	5
	Crops 2	0	3	6	6	6	6	6	3	6	6	6	6	6	6	6	5
	Forest 1	0	7	4	7	7	7	7	7	7	3	3	4	7	7	6	5
	Forest 2	0	7	3	7	7	7	7	7	7	7	7	7	7	7	4	5
	Forest 3	0	7	7	7	7	7	7	7	7	7	7	7	7	7	4	5
	Grasses 1	0	8	8	8	8	8	8	8	8	8	8	8	8	8	4	5
	Grasses 2	0	8	8	8	8	8	8	8	8	8	8	8	8	8	4	5
	Grasses 3	0	8	8	8	8	8	8	8	8	8	8	8	8	8	4	5
	Grasses 4	0	8	8	8	8	8	8	8	8	8	8	8	8	8	4	5
	Eroded soils 1	0	9	9	9	9	9	9	9	9	9	9	9	9	9	9	5
	Eroded soils 2	0	9	9	9	9	9	9	9	9	9	9	9	9	9	9	5

Legend new map :

1 Bare soils
2 Bare/Fallow
3 Open forest
4 Fallow/Bar
5 Water bodies
6 Crops
7 Forest
8 Grasses
9 Eroded soils

A map of slope gradient was generated using the following procedure:

- Contour lines (50 m interval) were digitized from the topographic map at 1:50,000 scale and an Digital Elevetion Model was generated. Figure 7, illustrates a hillshading representation of the DEM and Figure 8 shows the landsat TM scene, both in a two dimensional representation and draped onto the DEM.
- The raster map was interpolated to obtain an elevation map, from which the slope gradient map was calculated using standard filtering procedures.
- The slope map was classified into slope steepness ranges, where the areas for each slope range became the units of a new map.

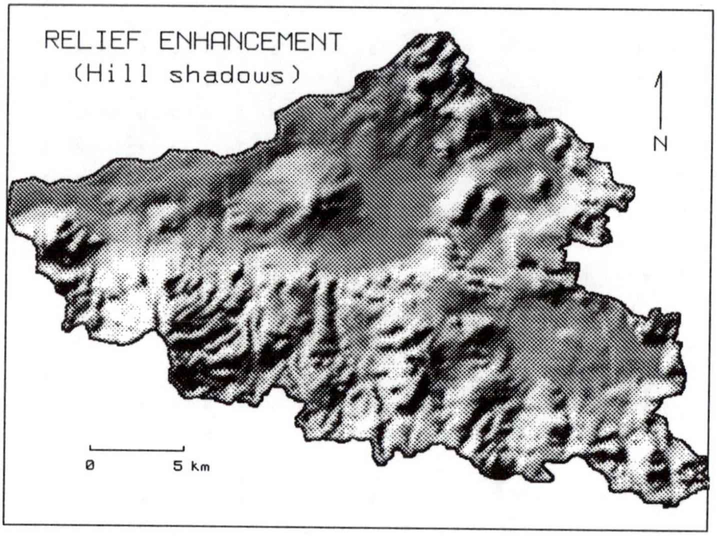

Figure 7. Hillshading representation of the watershed

Field knowledge indicated at least two important facts regarding the influence of slope gradient on the cover:

(1) Abrupt changes in cover occur at significant breaks of slope; for example, between the top of a lava flow (less than 5 % slope with agricultural residues (fallow) or bare soils) and its edges (steeper slopes under forest or open forest). If fallow is misclassified as open forest on top of volcanic flows (or conversely, open forest as fallow or grasses on the steep slopes), the cover on those areas can be reclassified according to the following decision rule:

- If the TMU is a volcanic flow, and the slope steepness is < 5 %, and the spectral class is open forest, then map it as fallow, else do not change the spectral map; or, conversely:
- If the TMU is a volcanic flow, and the steepness is > 5 %, and the spectral class is either fallow or grasses, then map it as open forest, else do not change the spectral map.

(2) Research in the study area, using aerial photogeomorphologic interpretation and intensive field checking, indicated that most accelerated erosion processes occurs on slopes of less than 15 % (Bocco, 1990). On the other hand, previous experience with a classification approach based strongly on clustering showed that with Landsat TM data there is a tendency to overestimate eroded terrain by misclassifying bare soils as eroded, probably because of the high reflectance of sealed clayey soils (Bocco and Valenzuela, 1988). This was also true in the present spectral map. One way of at least partially solving this problem is to use the following decision rule:

- If the slope is > 15 %, and the spectral class is eroded soils, then map it as bare, else unchanged.

Cultural features (such as towns, roads, canals, etc.) can also be misclassified because their spectral responses are close to those of bare and eroded soils. Using digitized/rasterized infrastructure data (towns, roads) and a similar approach as that described for slopes, this confusion may be overcome. The final results, including those obtained after combinations with the TMU and slope maps, are shown in Figure 9.

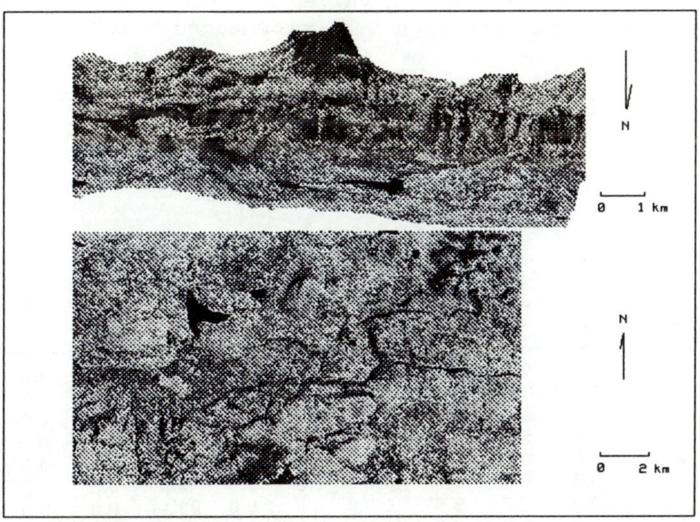

Figure 8. Landsat TM image in two-dimensional representation and draped onto a digital elevation model

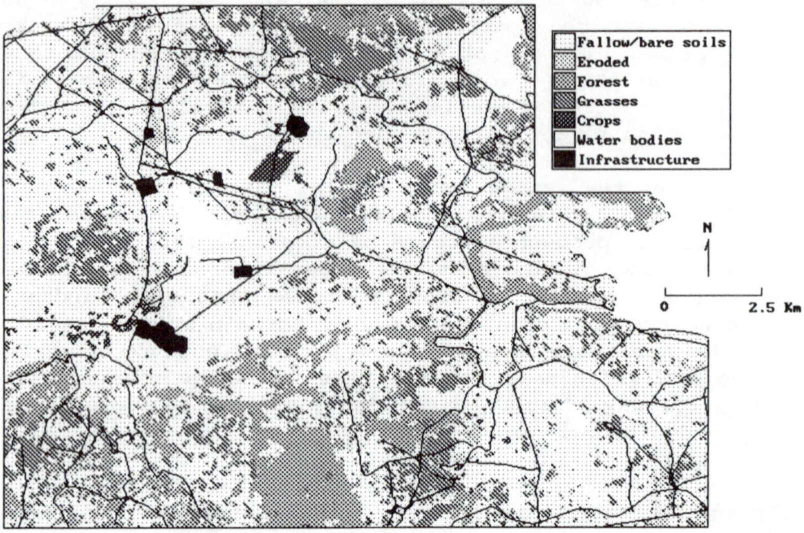

Figure 9. Final land use/erosion map based on spectral cover and field knowledge

3.Results and conclusions

The final map was assessed quantitatively with an error matrix (Table 5). The improvements provided by the GIS manipulations were evaluated by calculating the differences between values of both matrices (Tables 2 and 5). The changes between corresponding percentages from each matrix are indicated in Table 6.

Table 5. Error matrix for the final map

Final map Visual	Bare	Bare/Fallow	Op. forest	Fallow	Water	Crops	Forest	Grasses	Eroded
Bare	76.6	6.1	4.3	3.5	0.0	2.6	0.0	2.6	4.3
Bare/Fallow	0.0	81.4	9.3	6.9	0.0	0.0	0.0	1.2	1.2
Op. forest	1.0	6.3	85.5	5.2	0.0	0.0	1.0	1.0	0.0
Fallow	2.3	5.2	3.2	80.2	0.0	0.8	0.4	7.5	0.4
Water	0.0	0.0	0.0	0.0	100	0.0	0.0	0.0	0.0
Crops	0.0	0.0	0.0	16.7	0.0	83.3	0.0	0.0	0.0
Forest	0.0	0.0	6.9	0.0	0.0	0.0	93.1	0.0	0.0
Grasses	8.3	2.5	1.7	14.2	0.0	0.0	0.0	73.3	0.0
Eroded	8.8	0.0	0.0	0.0	0.0	0.0	0.0	0.0	91.2

Table 6. Changes in classification accuracies

	Bare	Bare/Fallow	Op. forest	Fallow	Water	Crops	Forest	Grasses	Eroded
Bare	+1.9	+2.4	−1.1	+0.7	+1.0	−2.6	0.0	+1.6	−0.1
Bare/Fallow	0.0	+2.2	+0.4	0.0	+1.4	0.0	0.0	+0.2	+0.2
Op. forest	0.0	+0.6	+4.5	+2.7	0.0	0.0	+1.0	0.0	0.0
Fallow	+1.2	+0.7	−0.9	+2.5	+0.4	0.0	0.0	+1.5	0.0
Water	0.0	+12.5	0.0	+12.5	+37.5	0.0	0.0	+12.5	0.0
Crops	0.0	0.0	+14.3	−2.4	0.0	+11.9	0.0	0.0	0.0
Forest	0.0	0.0	+5.2	0.0	0.0	0.0	+5.2	0.0	0.0
Grasses	−0.2	+1.2	+0.6	+5.8	+1.5	0.0	0.0	+8.9	0.0
Eroded	+12.6	0.0	0.0	0.0	0.0	0.0	0.0	0.0	+12.6

The overall accuracy of the final map increased from 75 to 85 % and spectrally unclassified pixels (9 % of the total area) were satisfactorily allocated to their appropriate information classes. In only 6 (negative values in table 6) of the 81 possible combinations was the final map less accurate than the spectral map. The decrease of accuracy was always small and did not affect correct classifications (the diagonal of the matrix). A GIS, especially when it is integrated with image processing capabilities, is a powerful tool for computer-assisted land use mapping. It allows the full incorporation of the interpreter's field knowledge by interactively guiding and analyzing the information extraction procedures. This method was tested in two areas in Argentina and Brazil with similar improvements in the identification of information classes (Bedendo, 1990).

Contextual procedures, such as those reported here, should be used with caution, however. All occurrences of anomalous land cover are re-labelled to the "expected" value (e.g., crops on edges of lava flows reclassified as forest). It is possible that it is such anomalies, rather than the predicted pattern of land cover, that is of most interest to the applications specialist.

Acknowledgements

The research described in this paper was funded by the University of Mexico (Institute of Geography) and ITC. Ann Stewart and Hans de Brouwer contributed to the presentation of the final version.

References

Anuta, P.E., L.A. Bartolucci, M.E. Dean, D.F. Lozano, E. Malaret, C.D. McGillem, J.A. Valdes and C.R. Valenzuela. 1984. Landsat-4 MSS and Thematic Mapper data quality and information content analysis. IEEE Trans. Geoscience and Remote Sensing. 22(3):222-236.

Bedendo, D. 1990. Use of Geographic Information Systems and Remotely-Sensed Data in the Assessment of Soil Degradation. Unpubl. MS thesis. ITC, Enschede.

Bocco, G. 1990. Gully erosion analysis using Geographic Information Systems and Remote Sensing. PhD dissertation. University of Amsterdam.

Bocco, G. and C.R. Valenzuela. 1988. Integration of GIS and image processing in soil erosion studies using ILWIS. ITC Journal 1988-4:309-319.

Dirección General de Geografía (DGG). 1977. Carta Topografica El Oro de Hidalgo, 1:50,000. INEGI, SPP. Mexico.

Gorte, B., R. Liem and J. Wind. 1988. The ILWIS software kernel.ITC Journal 1988-1:15-22.

Jadkowski, M.A. and M. Ehlers. 1989. GIS analysis of SPOT image data. Proc. ASPRS/ACSM Annual Convention. Baltimore. 4:65-74.

Marble, D.F. 1984. GIS: An overview. G.Pecora Proc.:18-24

Marble, D.F. and D.J. Peuquet (eds.). 1983. Geographic Information Systems and Remote Sensing. In: R.N Colwell (ed.). Manual of Remote Sensing:923-958. ASPRS. Falls Church, Va.

Meijerink, A.M. 1988. Data acquisition and data capture through terrain mapping units. ITC Journal 1988-1:23-44.

Thompson, D.R. and K.E. Henderson. 1984a. Evaluation of Thematic Mapper for detecting soil properties under grassland vegetation. IEEE Trans. Geoscience and Remote Sensing. 22(3):319-322.

Thompson, D.R. and K.E. Henderson. 1984b. Detecting soils under cultural vegetation using digital Landsat Thematic Mapper data. Soil Sci. Soc. Am. J. 48:1316-1319.

Valenzuela, C.R. 1988. ILWIS overview. ITC Journal 1988-1:3-14.

Walsh, S.J., D.R. Lightfood and D.R. Butler. 1987. Recognition and assessment of error in Geographic Information Systems. Photog. Eng. & R.S. 53 (10):1423-1430.

GIS EDUCATION AND TRAINING[1]

Michael J.C. Weir
I.T.C.
P.O. Box 6, Boulevard 1945
7500 AA Enschede
The Netherlands

1. Introduction

The widespread application of GIS technology in natural resource management in recent years can be attributed on the one hand to the increasingly favourable price/performance ration of computer systems, and on the other to remarkable improvements in the power of applications software. As these trends continue, the benefits offered by GIS technology lie within the reach of most resource management agencies, including those in the developing countries. Indeed, it is in the developing countries, where the need for reliable and timely information is most acute, that GIS technology has much to offer in assisting the task of resource management and development planning.

Although the availability of relatively inexpensive, but powerful, computing facilities can offer considerable benefits, there are also a number of serious constraints which cannot be ignored (Weir, 1989). Among the most important of these constraints is a shortage of well-trained personnel.

As long as applications were restricted to large, expensive mainframe- or mini- computers, there was a reasonable guarantee that such systems would be managed and operated by expert personnel. Today, however, the trend is increasingly towards the use of PC-based systems. Under these circumstances, GIS is no longer the preserve of a select group of experts but is, in principle, available to all staff at every level of management from ministries down to local offices. It is therefore necessary to prepare such staff for the transition from "traditional" systems of handling management information to GIS. Although features such as online help functions and menu-driven software may make a system more "user friendly" they do not remove the need for sound education and training in fundamental theoretical concepts and practical skills.

---

[1]This chapter is adapted from the author's paper "What should the forest manager know about geographical information systems", IUFRO Symposium on "New Technologies in Forest Management Planning: Remote Sensing - Integrated Data Bases - Geographic Information Systems" Luxemburg, 18-22 September 1989.

A. S. Belward and C. R. Valenzuela (eds.),
Remote Sensing and Geographical Information Systems for Resource Management in Developing Countries, 491–497.

Insofar as it is considered to involve some degree of automation, GIS is a new technology. Yet, as table 1 shows, many GIS operations have a parallel in "traditional" systems for handling spatial data.

Table 1: Comparison between GIS and traditional approaches for handling spatial data

GIS operations	Traditional tasks
data input	
- map digitizing, scanning	map drafting
- data entry from keyboard	writing/typing tables, etc.
map editing, e.g. with digitizer	erase/redraw map features
map retrieval from database	search in map archive
map aggregate	recompilation, usually involving scale change and generalization
area computation from coordinates	dot grid or planimeter measurement
distance computation	map measurement with ruler or opisometer (curvimeter)
spatial search from database	visual examination of maps and associated data
route selection using digital elevation model, etc.	measurement of terrain slopes on contours of topographic map
map overlay	physical overlay of maps (usually on transparent paper or film) and tracing of relevant combinations
data display	
- map display on monitor or production of hard copy on printing), plotter/printer	cartographic design and construction map reproduction (e.g. diazo and sometimes hard colouring of maps
- output of tables	data compilation, typing
- production of charts, graphs	data compilation, hand drawing

The function of any information system is to improve the user's ability to make decisions in research, planning and management. In the case if a GIS, Smith et. al. (1987) identify the following general requirements which should be satisfied:

i) an ability to handle large, multilayered, heterogeneous data bases of spatially-indexed data;
ii) an ability to query such data bases about the existence, location and properties of a wide range of spatial objects;
iii) an efficiency in handling such queries that permits the system to be interactive;
iv) an ability of the system to "learn" in a significant way about the spatial objects in its knowledge and data bases during use of the system.

Such a system is essentially a management tool which has an ability to create new information rather than simply retrieve previously encoded information. "This ability to both automatically synthesize existing layers of geographic data and to update a database for spatial entities is the key to a functional definition of a GIS" (Cowen, 1988, p 1553).

2.Benefits and pitfalls.

In recent years the published literature has abounded with reports of the successful application of GIS in resource management. Among the most promising benefits are:

i) the ability to greatly speed up certain tedious, but essential tasks, notably area measurement and map updating;
ii) the ability to forecast the result of different management options;
iii) the ability to optimize resource management operations.

In the case of a large agency, such benefits can yield considerable economic returns. In British Columbia, for example, the Ministry of Forestry's forest land information system is expected to enable wood delivery costs to be reduced by $2.00 per cubic meter, resulting in cost savings of $140 million per year (Heygi and Sallaway, 1986).

On the other hand, the introduction of GIS does not guarantee that such benefits will be achieved. In some cases, especially in developing countries, the successful implementation of GIS may be hampered by organizational and financial difficulties. These include problems of data supply (unavailable, unreliable, inaccessible data), financial constraints, inadequate technical support and a lack of trained personnel. Successful implementation of a GIS depends, therefore, on the existence of a proper support structure (Table 2).

Table 2: Support structures for information utilization (De Man 1988).

Professionals and users	Mobilization and preparation of the users of the envisaged information system Educational support provides for training of professionals as well as consumers of information
System maintenance	Technical support and maintenance deal with hardware, software and data files Financial support deals with available funds and reliable and stable funding Logistic support deals with the availability of all necessary facilities for the operation of the information system
Organizations	Control structures ensure that goals are specified and reformulated when required, and that the various activities are coordinated towards the fulfillments of these goals Protection structures deal with "gate-keeping" functions, such as licensing and patenting Change or growth structures prepare for new developments, expected or unexpected

In addition to organizational difficulties, there are a number of "operational" pitfalls which may be encountered, notably:

i) problems and difficulties associated with combining data from different sources and especially unquestioning acceptance of the accuracy of such data;
ii) disregarding the propagation of errors during the various stages of data input, manipulation and analysis (in particular, during map overlay);
iii) failure to understand the nature of certain functions and models through misuse of logic and inappropriate operations.

Failure to recognize and avoid such pitfalls will almost certainly results in erroneous and misleading information.Although features designed to make GIS software more user friendly are generally to be welcomed, it is a fallacy to think that such improvements will ensure better results. A GIS will only provide reliable information when it is operated and maintained by personnel with a sound education in fundamental concepts of spatial analysis.

3.Standards of GIS education and training

Resource managers may receive GIS education and training either as part of a formal education programme at university or college, or through in-house training. In the latter case, training may be provided by the vendor of the system. Since such training tends to concentrate on the capabilities of the system in question, it is usually limited in scope. The starting point for designing any course is a definition of the required standards of education and training. Such standards can best be defined by identifying those theoretical concepts and practical skills which the student, or course participant, is expected to master on completion of the course. An example of this approach to defining suggested standards of education and training in photogrammetry for the forestry profession can be found in Weir (1985). Since most resource management agencies seem likely to make use of GIS in the foreseeable future, there is an urgent need to define appropriate standards of education and training in this field. The following standards are suggested:

i) have a sound understanding of basic GIS concepts, including hardware requirements, data structures (both spatial and attribute) and database management systems;
ii) have a good working knowledge and practical experience of basic GIS operations, including aspects of digital mapping and digital image processing;
iii) understand the concepts of the main GIS analysis functions, such as spatial search and map overlay, and be able to apply such functions for modelling and planning resource management operations;
iv) understand the nature and propagation of errors in a GIS.

The last of these is an often neglected, yet important, requirement for the successful operation of a GIS. Firstly, since many GIS users may tend to assume the contents of the database to error-free, there is a need to ensure that basic input operations such as map digitizing and image processing are carried out with the utmost care. This is particularly difficult in the case of a map digitizing, a tedious operation which tends to be the task of more junior personnel, but which accounts for perhaps 80% of all GIS activity. Secondly, GIS users must be aware of errors arising from processing operations such as vector to raster conversion. Finally, GIS users (and especially decision makers using the results of spatial analysis) must be aware of the way in which errors are propagated during map overlay and other operations (Bailey, 1988; Walsh et.al., 1987).

4.Organizational implications

With the implementation of GIS in an there will be a need for personnel to acquire new skills and adapt to new roles. Suggested standards of education and training for such personnel have been outlined above. Although the availability well trained personnel is a prerequisite for the successful application of GIS in resource management, there are also a number of organizational aspects which cannot be ignored. This begins with the decision to implement a GIS. The implications of implementing GIS within an organization are not restricted to training and personnel management, but will be felt in all aspects of the organization. As De Man (1988) points out, it is necessary to view the whole environment of bureaucracy, culture and legislation in which the information system is to be established (figure 1).

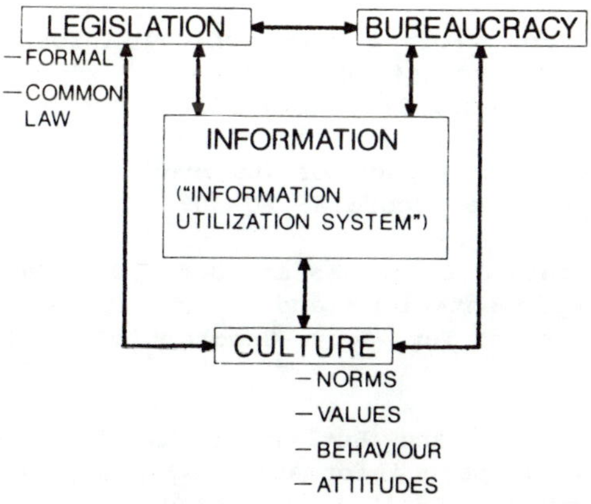

Figure 1: The environment of information utilization (De Man, 1988)

At the outset it is essential to clearly define the intended role of the proposed system, i.e. is it intended to support long term strategic planning or short term decision making at a local level? "The basic informational needs for each type of decision making vary appropriately, and failure to realize this leads to problems in effectively using GIS technology" (Newton, 1986, p21).

The intended role of the system will determine whether it is to be located in headquarters (perhaps-running on a large central computer) or dispersed throughout the organization. In the former case there may be a danger of inter-personnel conflict between a well paid "GIS elite" and other staff (Newton, op. cit.). In the latter case there will be a need for more widespread training, especially to maintain consistency in the quality of data and the way in which such data are analyzed. GIS must also bee seen within the context of the long term functioning of the organization. The useful lifespan of most computer hardware (and to a lesser extent software) in only a few years. There is a very real danger that a hastily designed system will become obsolete before it has provided real benefits.

References

Baily, R.G. 1988. Problems with using overlay mapping for plan-ning and their implications for geographic information systems,Environmental Management, 12(1):11-17.

Cowen, D.J. 1988. GIS versus CAD versus DBMS: what are the dif-ferences? Photogrammetric Engineering and Remote Sensing, 54(11):1551-1555.

Heygi, F. and Sallaway, P. 1986. The integrated three-dimensionalforest land information system of B.C., Canada, IUFRO Study Group6.05, XVIII World Congress, Ljubljana: 7 p.

De Man, W.H.E. 1988. Establishing a geographical informationsystem in relation to its use: a process of strategic choices,Int. Jour. Geographical Information Systems, 2(3): 245-261.

Newton, C.M. 1986. Management impact of geographic informationsystem, Proc. ASP GIS Workshop, Atlanta, Georgia: 20-30.

Smith, T.R., Menon, S., Star, J.L. and Estes, J.E. 1987. Require-ments an principles for the implementation and construction oflarge-scale geographic information systems, Int. Jour. GeographicInformation Systems, 1 (1): 13-31.

Walsh, S.J., Lightfoot, D.R. and Butler, D.R. 1987. Recognitionand assessment of error in geographic information systems, Photo-grammetric Engineering and Remote Sensing, 53(10): 1423-1430.

Weir, M.J.C. 1985. Photogrammetric education for foresters. Proc.Colloquium on Surveying and Mapping Education, UNB, Fredericton,Canada: 343-352.Weir, M.J.C. 1989. More advance technology education needed,Journal of Forestry, 87(7): 20-25.

Subject Index